NEW FRONTIERS IN NANOCHEMISTRY

Concepts, Theories, and Trends

Volume 1
Structural Nanochemistry

NEW FRONTIERS IN NANOCHEMISTRY

Concepts, Theories, and Trends

Volume 1
Structural Nanochemistry

Edited by

Mihai V. Putz, PhD, Dr.-Habil., MBA

Professor, Faculty of Chemistry, Biology, Geography,
Laboratory of Computational and Structural Physical Chemistry for
Nanosciences and QSAR, West University of Timişoara, Romania

PI-1, Laboratory of Renewable Energies - Photovoltaics,
National Research and Development Institute for Electrochemistry
and Condensed Matter (INCEM), Timişoara, Romania

Apple Academic Press Inc.
4164 Lakeshore Road
Burlington ON L7L 1A4
Canada

Apple Academic Press Inc.
1265 Goldenrod Circle NE
Palm Bay, Florida 32905
USA

© 2020 by Apple Academic Press, Inc.

First issued in paperback 2021

Exclusive worldwide distribution by CRC Press, a member of Taylor & Francis Group

No claim to original U.S. Government works

New Frontiers in Nanochemistry: Concepts, Theories, and Trends, Volume 1: Structural Nanochemistry

ISBN 13: 978-1-77463-174-4 (pbk)
ISBN 13: 978-1-77188-777-9 (hbk)

New Frontiers in Nanochemistry, 3-volume set
International Standard Book Number-13: 978-1-77188-780-9 (Hardcover)
International Standard Book Number-13: 978-0-36707-782-2 (eBook)

Library and Archives Canada Cataloguing in Publication

Title: New frontiers in nanochemistry : concepts, theories, and trends / edited by Mihai V. Putz, PhD, Dr.-Habil., MBA.

Names: Putz, Mihai V., editor.

Description: Includes bibliographical references and indexes. | Contents: Volume 1. Structural nanochemistry -- Volume 2. Topological nanochemistry -- Volume 3. Sustainable nanochemistry.

Identifiers: Canadiana (print) 20190106808 | Canadiana (ebook) 20190106883 | ISBN 9781771887779 (v. 1 ; hardcover) | ISBN 9781771887786 (v. 2 ; hardcover) | ISBN 9781771887793 (v. 3 ; hardcover) | ISBN 9780429022937 (v. 1 ; ebook) | ISBN 9780429022944 (v. 2 ; ebook) | ISBN 9780429022951 (v. 3; ebook)

Subjects: LCSH: Nanostructured materials. | LCSH: Nanochemistry.

Classification: LCC TA418.9.N35 N49 2019 | DDC 541/.2—dc23

Library of Congress Cataloging-in-Publication Data

Names: Putz, Mihai V., editor.

Title: New frontiers in nanochemistry : concepts, theories, and trends / editor, Mihai V. Putz, PhD, Dr.-Habil., MBA (professor, Faculty of Chemistry, Biology, Geography, Laboratory of Computational and Structural Physical Chemistry for Nanosciences and QSAR, West University of Timisoara, Romania, PI-1, Laboratory of Renewable Energies--Photovoltaics, National Research and Development Institute for Electrochemistry and Condensed Matter (INCEM), Timisoara, Romania).

Description: Toronto : Apple Academic Press, 2020. | Includes bibliographical references and index. | Contents: volume 1. Structural nanochemistry -- volume 2. Topological nanochemistry -- volume 3. Sustainable nanochemistry.

Identifiers: LCCN 2019021643 (print) | LCCN 2019022246 (ebook) | ISBN 9781771887779 (hardcover : alk. paper : v. 1) | ISBN 9781771887786 (hardcover : v. 2) | ISBN 9781771887793 (hardcover : v. 3)

Subjects: LCSH: Nanochemistry. | Surface chemistry. | Nanoscience. | Chemistry, Analytic.

Classification: LCC QD506 .N475 2020 (print) | LCC QD506 (ebook) | DDC 541/.2--dc23

LC record available at https://lccn.loc.gov/2019021643

LC ebook record available at https://lccn.loc.gov/2019022246

Apple Academic Press also publishes its books in a variety of electronic formats. Some content that appears in print may not be available in electronic format. For information about Apple Academic Press products, visit our website at **www.appleacademicpress.com** and the CRC Press website at **www.crcpress.com**

About the Editor

Mihai V. Putz, PhD, Dr.-Habil., MBA

Professor, Laboratory of Computational & Structural Physical Chemistry for Nanosciences & QSAR, West University of Timișoara; Laboratory of Renewable Energies-Photovoltaics, National Research & Development Institute for Electrochemistry & Condensed Matter (INCEM), Romania

Mihai V. Putz, PhD, MBA, Dr.-Habil, is a laureate in physics (1997), with a post-graduation degree in spectroscopy (1999), and a PhD degree in chemistry (2002). He bears with many postdoctoral stages: in chemistry (2002–2003) and in physics (2004, 2010, 2011) at the University of Calabria, Italy, and Free University of Berlin, Germany, respectively. He is currently a Full Professor of theoretical and computational physical chemistry at his alma mater, West University of Timisoara, Romania.

He has made valuable contributions in computational, quantum, and physical chemistry through seminal works that appeared in many international journals. He is an Editor-in-Chief of the *International Journal of Chemical Modeling* and *New Frontiers in Chemistry*. In addition, he has authored and edited many books. He is a member of many professional societies and has received several national and international awards from the Romanian National Authority of Scientific Research (2008), the German Academic Exchange Service DAAD (2000, 2004, 2011), and the Center of International Cooperation of Free University Berlin (2010). He is the leader of the Laboratory of Computational and Structural Physical Chemistry for Nanosciences and QSAR at the Biology-Chemistry Department of West University of Timisoara, Romania, where he conducts research into the fundamental and applicative fields of quantum physical chemistry and QSAR. He is also a PI-1 at the Laboratory of Renewable Energies – Photovoltaics, National Research and Development Institute for Electrochemistry and Condensed Matter (INCEM), Timișoara, Romania. Among his numerous awards, in 2010 Mihai V. Putz was declared, through a national competition, the Best Researcher of Romania, while in 2013 he was recognized among the first Dr.-Habil. in Chemistry in Romania. In 2013, he was appointed

Scientific Director of newly founded Laboratory of Structural and Computational Physical Chemistry for Nanosciences and QSAR at his alma mater university. In 2014, he was recognized by the Romanian Ministry of Research as Principal Investigator of the first degree at the National Institute for Electrochemistry and Condensed Matter (INCEMC), Timisoara, and was also granted full membership in the International Academy of Mathematical Chemistry.

Recently, Mihai V. Putz expanded his interest to strategic management in general and to nanosciences and nanotechnology strategic management in particular. In this context, between 2015–2017, he attended and finished as the promotion leader of the MBA in Strategic Management of Organizations – The Development of the Business Space specialization program at the West University of Timişoara, the Faculty of Economics and Business Administration. While from 2016, he was engaged in the doctoral school of the same faculty, advancing new models of strategic management in the new economy based on frontier scientific inclusive ecological knowledge.

Contents

Contributors

Caglar Celik Bayar
Bulent Ecevit University, Department of Metallurgical and Materials Engineering, 67100 Zonguldak, Turkey

Attila Bende
Molecular and Biomolecular Physics Department, National Institute for Research and Development of Isotopic and Molecular Technologies, Donat Street, No. 67-103, RO-400293 Cluj-Napoca, Romania, Tel.: 0040-264-584037, Fax: 0040-264-420042, E-mail: bende@itim-cj.ro

Aranya B. Bhattacherjee
Department of Physics, Birla Institute of Technology and Science, Pilani, Hyderabad Campus, India

Pablo M. Blanco
Department of Material Science and Physical Chemistry and Research Institute of Theoretical and Computational Chemistry (IQTCUB) of Barcelona University, C/Martí i Franquès, 1, 08028-Barcelona, Spain

Sorana D. Bolboacă
Iuliu Haţieganu University of Medicine and Pharmacy Cluj-Napoca, Romania

Bogdan Bumbăcilă
Laboratory of Computational and Structural Physical Chemistry for Nanosciences and QSAR, Biology-Chemistry Department, Faculty of Chemistry, Biology, Geography at West University of Timişoara, Pestalozzi Street No. 16, Timişoara, RO-300115, Romania

Gloria Castellano
Department of Experimental Sciences and Mathematics, Faculty of Veterinary and Experimental Sciences, Valencia Catholic University Saint Vincent Martyr, Guillem de Castro-94, E-46001 Valencia, Spain

Edgar G. DuCasse
Mathematics Department, Pace University, New York, NY 10038, USA, E-mail: egducasse@pace.edu

Nicoleta A. Dudaş
Laboratory of Computational and Structural Physical Chemistry for Nanosciences and QSAR, Biology-Chemistry Department, West University of Timişoara, Pestalozzi Str. No. 16, Timişoara 300115, Romania, E-mail: nicole_suceveanu@yahoo.com

Corina Duda-Seiman
Laboratory of Computational and Structural Physical Chemistry for Nanosciences and QSAR, Biology-Chemistry Department, Faculty of Chemistry, Biology, Geography at West University of Timişoara, Pestalozzi Street No. 16, Timişoara, RO-300115, Romania

Daniel Duda-Seiman
Department of Medical Ambulatory, and Medical Emergencies, University of Medicine and Pharmacy 'Victor Babes,' Avenue C.D. Loga No. 49, RO-300020 Timisoara, Romania

Suman Dudeja
Department of Chemistry, ARSD College, University of Delhi, New Delhi–110021, India

Josep Lluís Garcés
Department of Chemistry, University of Lleida (UdL), Av. Alcalde Rovira Roure 191,
25198-Lleida, Spain

Mirela I. Iorga
National Institute for Research and Development in Electrochemistry and Condensed Matter,
Department of Applied Electrochemistry, Laboratory of Renewable Energies-Photovoltaics, 300569
Timisoara, Romania, E-mail: mirela_iorga@yahoo.com; mirela.iorga@incemc.ro

Nazmul Islam
Department of Chemistry and Theoretical and Computational Chemistry Research Laboratory,
Techno Global-Balurghat, Balurghat, D. Dinajpur, 733103, India,
E-mail: nazmul.islam786@gmail.com; nazmulislam@tgbtechno.org

Adriana Isvoran
Department of Biology-Chemistry, West University of Timisoara, 16 Pestalozzi, Timisoara,
Romania/Advanced Environmental Research Laboratory, 4 Oituz, Timisoara, Romania

Lorentz Jäntschi
Technical University of Cluj-Napoca, Romania, Tel.: +40-264-401775,
E-mail: lorentz.jantschi@gmail.com

István László
Department of Theoretical Physics, Budapest University of Technology and Economics,
H-1521 Budapest, Hungary, E-mail: laszlo@eik.bme.hu

Sergio Madurga
Department of Material Science and Physical Chemistry and Research Institute of Theoretical and
Computational Chemistry (IQTCUB), University of Barcelona, C/Martí i Franquès, 1,
08028-Barcelona, Spain

Francesc Mas
Department of Material Science and Physical Chemistry and Research Institute of Theoretical and
Computational Chemistry (IQTCUB), University of Barcelona, C/Martí i Franquès, 1,
08028-Barcelona, Spain, E-mail: fmas@ub.edu

Laura Pitulice
Department of Biology-Chemistry, West University of Timisoara, 16 Pestalozzi, Timisoara,
Romania/Advanced Environmental Research Laboratory, 4 Oituz, Timisoara, Romania

Nicolina Pop
Politehnica University of Timisoara, Department of Fundamental of Physics for Engineers,
Vasile Pârvan Blv., 300223 Timisoara, Romania, E-mail: nicolina.pop@upt.ro

Mihai V. Putz
Laboratory of Computational and Structural Physical Chemistry for Nanosciences and QSAR,
Biology-Chemistry Department, Faculty of Chemistry, Biology, Geography at West University of
Timişoara, Pestalozzi Street No. 16, 44, Timişoara, RO-300115, Romania / Laboratory of Renewable
Energies-Photovoltaics, R&D National Institute for Research and Development in Electrochemistry
and Condensed Matter, Department of Applied Electrochemistry, Dr. A. Paunescu Podeanu
Str. No. 144, RO-300569 Timişoara, Romania, Tel.: +40-256-592-638; Fax: +40-256-592-620;
E-mail: mv_putz@yahoo.com; mihai.putz@e-uvt.ro

Louis V. Quintas
Mathematics Department, Pace University, New York, NY 10038, USA, E-mail: lvquintas@gmail.com

James Garnett Sawyer II
241, Lexington Avenue, Buffalo, NY 14222, USA

Francisco Torrens
Institute for Molecular Science, University of Valencia, PO Box 22085, E-46071 Valencia, Spain,
Tel.: +34-963-544-431, Fax: +34-963-543-274, E-mail: torrens@uv.es

Marina A. Tudoran
Laboratory of Renewable Energies–Photovoltaics, R&D National Institute for Electrochemistry
and Condensed Matter, Dr. A. Paunescu Podeanu Str. No. 144, Timisoara, RO-300569, Romania

Lemi Turker
Middle East Technical University, Department of Chemistry, 06800 Ankara, Turkey,
Tel.: +90 312 210 3244, Fax: +90 312 210 3200, E-mail: lturker@metu.edu.tr; lturker@gmail.com

Margherita Venturi
Department of Chemistry "Giacomo Ciamician," University of Bologna, Italy

Eudald Vilaseca
Department of Material Science and Physical Chemistry and Research Institute of Theoretical and
Computational Chemistry (IQTCUB), University of Barcelona, C/Martí i Franquès,
1, 08028-Barcelona, Spain

Marla Jeanne Wagner
407, Norwood Avenue, Buffalo, NY 14222, USA

Cynthia Whitney
Galilean Electrodynamics, 11660 239th Ave. NE, Redmond, WA 98053-5613, USA,
E-mail: Galilean_Electrodynamics@Comcast.net

Abbreviations

^{1}H-NMR	proton nuclear magnetic resonance
2D	two-dimensional
2DL	two-dimensional layered
3D	three-dimensional
3DT	three dimensions triangular plane
6D	six dimensions
^{13}C-NMR	carbon-13 nuclear magnetic resonance
AC	algebraic chemistry
ACs	activated carbons
ADM	adomian decomposition method
AIB	adjacency in bonding
AIM	atoms in molecule(s)
ASE	aromatic stabilization energy
ATP	adenosine triphosphate
AuNPs	gold nanoparticles
B3LYP	Becke, 3-parameter, Lee-Yang-Parr exchange-correlation functional
BEC	Bose-Einstein condensates
BG-CSs	Barut-Girardello coherent states
CC	coupled-cluster theory
CPs	critical points
CQDs	carbon quantum dots
CS	contact surface
CV	cyclic voltammetry
DCS	dual coordinate system
DFT	density functional theory
DFT	discrete Fourier transform
DNA	deoxyribonucleic acid
DRT	distribution of relaxation times
E&EM	entropy & EM theory
EA	electron affinity
EBSs	excited binomial states
EC	equipartition conjecture
EE	electronegativity equalization principle
EEM	electronegativity equalization method

EFG	electric field gradient
EIS	electrochemical impedance spectroscopy
EL	equipartition line
ELF	electron localization function
EMR	electromagnetic radiation
EMS	electromagnetic signals
EO	electron orbital
EP	entropy production
FD	fractal dimension
FK	fractal kinetics
GEC	global equipartition conjecture
GMR	giant magnetoresistance
GQDs	graphene quantum dots
GR	grapheme
GSB	golden stone symmetry breaking
HB	Hausdorff-Besicovich
H-FA	hydrogen: the first atom
HFCO	Hartree-Fock crystal orbital
HO	harmonic oscillator
HOMO	high occupied molecular orbital
HPM	homotopy perturbation method
HSAB	hard and soft acids and bases
HSRE	Hess–Schaad resonance energy
HTT	heat treatment temperature
HUPD	hydrogen under potential deposition
ICs	ionic configurations
IE	information entropy
IPs	ionization potentials
IR	infrared
ISQW	infinite square quantum well
IT-ELF	information-theoretic electron localization function
KC	Karlsruhe congress
KP-CSs	Klauder–Perelomov coherent states
LAD	local density approximation
LCAO	linear combination of atomic orbitals
LST	linear systems theory
LUMO	lowest unoccupied molecular orbital
MAD	mean absolute deviation
MC	Monte Carlo method
MD	mass fractal dimension

MD	molecular dynamics
MH	maximum hardness
MI	miller indices system
MLR	multiple linear regressions
MM	molecular mechanics
MO	molecular orbital
MO	Morse oscillator
MS	mirror-symmetric
NA	non-Archimedean
NET	new energy technology
NICS	the nucleus-independent chemical shift
NP	nanoparticle
NQR	nuclear quadrupole resonance
NS	nanosystems
NT	nanotube
OS	oxidation state
PA	periodic arch
PAH	polycyclic aromatic hydrocarbons
PBC	periodic boundary condition
PEDOT	polymer poly(3,4-ethylenedioxythiophene)
PES	potential energy surface
PHO	pseudoharmonic oscillator
PL	periodic law
PNP	Poisson–Nernst–Planck equation
POL	polarizability
PP	periodic property
PR	pattern recognition
PTE	periodic table of the elements
PVA	poly(vinyl alcohol)
PVC	photovoltaic cell
QDs	quantum dots
QL	quantum loop
QSAR	quantitative structure-activity relationship
QSPR	quantitative structure-property relationship
QT	quantum theory
RNA	ribonucleic acid
RT	room temperature
SBI	signals between ions
SD	surface fractal dimension
SLR	simple linear regressions

SO	self-organization
Spintronics	spin transport electronics
SSB	spontaneous symmetry breaking
SW	stone-Wales
TDCS	tetrahedral coordinate system
TISE	time-independent Schrödinger equation
TMD	transition metal dichalcogenide
TRE	topological resonance energy
UV-vis	UV-visible
VEDCS	vector equilibrium dual coordinate system
VESPR	valence shell electron pair repulsion
VIS	visible
VIT	variational iteration method
VMP2	vibrational Møller-Plesset second-order perturbation theory
VSCF	vibrational self-consistent field theory
VUV	vacuum ultra-violet
XC	exchange-correlation functional

General Preface to Volumes 1–3

The nanosciences, just born on the dawn of the 21st century, widely require dictionaries, encyclopedias, handbooks to fulfill its consecration, and development in academia, research, and industry.

Therefore, this present editorial endeavor springs from the continuous demand on the international market of scientific publications for having a condensed yet explicative dictionary of the basic and advanced up-to-date concepts in nanochemistry. It is viewed as a combination (complementing and overlapping) on various notions, concepts, tools, and algorithms from physical-, quantum-, theoretical-, mathematical-, and even biological-chemistry. The definitions given in the integrated volumes are accompanied by essential examples, applications, open issues, and even historical notes on the persons/subjects, with relevant literature and scholarly contributions.

The current mission is to prepare a premiere referential work for graduate students, PhDs and post-docs, researchers in academia, and industry dealing with nanosciences and nanotechnology. From the book format perspective, the volumes are imagined as practical and attractive as possible with about 130 essential terms as coming from about 60 active scholars from all parts of the globe, explaining each entry with a minimum of five pages (viewed as scientific letters/essays or short course about it), containing definitions, short characterizations, uses, usefulness, limitations, etc., and references – while spanning more than 1,600 pages in the present edition. This effort resulted in this unique and up-to-date *New Frontiers in Nanochemistry: Concepts, Principles, and Trends*, a must for any respected university and the individual updated library! It will also support the future more than necessary decade-editions with new additions in both entries and contributors worldwide!

On the other side, the broad expertise of the editor – in the non-limitative fields of quantum physical-chemical theory, nanosciences and quantum roots of chemistry, computational and theoretical chemistry, quantum modeling of chemical bonding, atomic periodicity and scales of chemical indices, molecular reactivity by chemical indices, electronegativity theory of atoms in molecule, density functionals of electronegativity, conceptual density functional theory, graphene and topological defects of graphenic ribbons, quantitative structure-activity/property relationships (QSAR/QSPR), effector-receptor complex interaction and quantum/logistic enzyme kinetics, and many other related scientific branches – assures that both the educated and generally interested reader in science and technology will be benefited from the book in many ways. They may be listed as:

- an introduction to general nanochemistry;
- an introduction to nanosciences;
- a resource for fast clarification of the basic and modern concepts in multidisciplinary chemistry;
- inspiration for a new application and transdisciplinary connection;
- an advocate for unity in natural phenomena at nano-scales, merging between mathematical, physics and biology towards nanochemistry;
- a reference for both academia (for lessons, essays, exams) as well in R&D industrial sectors in planning and projecting new materials, with aimed properties in both structure and reactivity;
- a compact yet explicated collection of updated scientific knowledge for general, university, and personal libraries; and
- a self-contained book for personal, academic, and technological instruction and research.

Accordingly, the specific aims of the present book are:

- to be a concise and updated source of information on the nanochemistry fields;
- to comprehensively cover nanosystems: from quantum theory to atoms, to molecules, to complexes and chemical materials;
- to be a necessary resource for every-day use by students, academics, and researchers;
- to present informative and innovative contents alike, presented in a systematical and alphabetically manner;
- to present not only definitions but consistent explicative essays on nanochemistry in a short essay or paper/chapter format; and
- to be written by leading and active experts and contributors in nanochemistry worldwide.

All in all, the book is aimed to give supporting material for relevant multi- and trans-disciplinary courses and disciplines specific to nanosciences and nanotechnology in all the major universities in Europe, the Americas, and Asia-Pacific, among which nanochemistry is the core for the privileged position in between physics (elementary properties of quantum particles) and biology (manifested properties of bodies by the environment/forces/substances influences). They may be non-exclusively listed as:

- *Nanomaterials: Chemistry and Physics*
- *Introduction to Nanosciences*
- *Bottom-Up Technology: Nanochemistry*
- *Nanoengineering: Chemistry and Physics*
- *Sustainable Nanosystems*
- *Ecotoxicology*
- *Environmental Chemistry*
- *Quantum Chemistry*
- *Stuctural Chemistry*
- *Physical Chemistry–Chemical Physics*

With the belief that, at least partly, this present editorial endeavor succeeds in the above mission and purposes. The editor of this publishing event heartily thank all the contributors for their dedication, inspiration, insight, and generosity, along with the truly friendly and constructive Publisher, Apple Academic Press (jointly with CRC Press at Taylor & Francis), and especially to the President, Mr. Ashish Kumar, and the rest of the AAP team.

Special Preface to Volume 1: Structural Nanochemistry

The nanosciences in general and nanochemistry discipline in particular should cover the fundamental domains of mathematics and natural science, with fundamental research objectives such as the multidisciplinary chemistry (integrating in a nonlimitative the subdisciplines of physical chemistry, informatics-chemistry, mathematical-chemistry, physical-organic chemistry, nano-inorganic chemistry, biology-chemistry, biochemistry, informatics-biology, pharmaceutical chemistry, medical chemistry, ecotoxicology, geochemistry, QSAR, etc.) as applied to phenomena and nano-structural properties of matter, in isolated state and in reciprocal and with medium interaction (QS[A-activity/P-property/T-toxicity]R), explained and controlled by physics laws, along with mathematics and computational aid.

In this context, one gives the eminent example of the "bosonic" approach of the matter and the chemical bonding in particular (*see* Chapter 6), as it is known that in this quantum state, the matter can become condensed in limited spaces, but with accumulating energy, on the nano-atomic level. On the theoretical level, the available models are currently applied to extended nanosystems of graphenes, silicenes, germaneness types, with description and prediction of the phase transfer for topological types defects (Stone-Wales). On the other side, at the experimental level, one is looking for observing the predicted phenomena by forming the equivalent bondons through a unique experimental assembly (now in the project phase) with the aid of quantum optics and photonics integrated at the micro-electronic level, but with possible extensions at the micro-nano-biosystems (MNBS), considered as being the key for future technologies (KET: key enabling technologies), which include the quantum computer molecular-based, molecular electronics (moletronics), etc. This way, considerable economies the non-renewable or hard-renewable natural resources (minerals, Cu, Al, Au, Ag, etc.) of Terra in general, including the respective exploitation costs.

In the same spirit, the other included voices of this first volume of the multi-volume package *New Frontiers in Nanochemistry: Concepts, Theories, and Trends* unfold the general (and universal) themes relevant to it, accordingly, with the above general definition (the first paragraph). The entries all came from eminent international scientists and scholars and span from A-to-Z nanochemistry fundamentals and vanguardist entries such as: algebraic chemistry, Bose-Einstein condensation (multiple Noble Prizes in Physics) and bondonic theory of chemistry, localization and delocalization of electrons, fractal dimension and kinetics, electromagnetism and electrochemistry, quantum physics including coherent states, excitons, quantum dots and tight binding, molecular descriptors, atomic theory, chemical compounds descriptions, molecular classifications, molecular machines (Nobel Prize in Chemistry for 2016), self-organization and its processes, supramolecular chemistry, electronic orbitals in more than three dimensions, quantum electrochemistry, spintronics, periodic law and periodic table, nanotubes and new energy technology, and many more.

The present volume contains about 50 entries, in more than 550 pages, with contributors coming from three continents (Europe, America, Asia), from six countries (USA, India, Spain, Italy, Hungary, and Romania) in multiple first-rate explicative dictionary voices. Let's hope they will be heard worldwide and will positively (in an ecology-progressist manner) influence the 21st-century macro-destiny of the Earth from the nanostructure and dynamics!

Fugit irreparabile tempus!

Heartily Yours,
—**Mihai V. PUTZ**
(Timisoara, Romania)

CHAPTER 1

Algebraic Chemistry

CYNTHIA WHITNEY

Galilean Electrodynamics, 11660 239th Ave. NE, Redmond,
WA 98053-5613, USA, E-mail: Galilean_Electrodynamics@Comcast.net

1.1 DEFINITION

Algebraic chemistry (AC) is a short name appropriate for any approach to chemistry that relies more on the exploitation of numerical patterns in experimental data than on deep quantum mechanical concepts and the complex computer calculations they often require. Such approaches are needed for many applications in chemical engineering. The particular realization of AC described here takes advantage of measured ionization potentials (IPs) of all elements (*see* Chapter 26). Clear patterns exist in the raw data, and the patterns are very much sharpened and simplified when the data are appropriately scaled. For the element with nuclear charge Z and nuclear mass M, the appropriate scale factor is M/Z. This scaling produces a new numerical variable for IP, where the raw data were about specific elements, in their natural range of isotopes, the scaled data are "population generic." The revealed patterns suggest a simple algorithm for estimating the energy requirements for removing any number of electrons from, or adding any number of electrons to, any neutral atom or isotope thereof. From there, one can estimate the energy requirements for making the ions that can serve as constituents of any molecule of interest. Indeed, for any molecule, one can compare different candidate ionic configurations, and identify the one particular ionic configuration that has the lowest energy, and hence the highest rate of occurrence. Then, by analyzing input molecules and output molecules, one can understand many problems, such as why reactions proceed or not, or how catalysis can enable an otherwise blocked reaction to proceed, etc. All of that comprises the domain of the AC discussed here.

1.2 HISTORICAL ORIGIN(S)

Chemistry presents many problems that seem very difficult to solve, even with the most powerful of computational resources. That situation motivates the present effort to develop algebraic tools for analysis of chemical situations.

The origin for AC as described here lies in mathematical physics, and in engineering science (*see* Chapter 11).

AC here uses the measured IPs of all elements for which measurements exist, and the patterns that emerge when those raw data are scaled by M/Z, where M is atomic mass and Z is a nuclear charge. Such scaling basically removes from the data and the effects of the various specific neutron counts of different elements, and even the isotopes thereof. The M/Z scaling can be understood as converting 'element-specific' information into 'population-generic' information. The latter is useful for anticipating behaviors of different isotopes or elements.

1.3 NANO-SCIENTIFIC DEVELOPMENT(S)

The population-generic IPs exhibit a great deal of regularity and pattern (*see* Chapter 26). It is an intriguing and on-going task to understand what the patterns are exactly and why they are. But even in the advance of completing that task, one can see how to use information known about one element to infer information about ions of another element, or even about different isotopes of the same element, or another element.

The population generic IPs generally need subscripts; IO for ionization order and Z for element nuclear charge. So, in general, we speak of $IP_{IO,Z}$. But all the information about $IO > 1$ can be determined from information about $IO = 1$ (*see* Chapter 26). So, we only need input data about $IP_{1,Z}$.

One fact about population-generic IPs is that they can all be written in the form:

$$IP_{1,Z} = IP_{1,1} + \Delta IP_{1,Z} \tag{1}$$

Most of them $\Delta IP_{1,Z}$ are positive, but at the beginnings of periods, a few are negative.

Another fact is that all IPs are positive numbers. So for all Z,

$$\left|\Delta IP_{1,Z}\right| < IP_{1,1} \tag{2}$$

Now, how can we detail a best scenario in which electrons are removed, or added, one-at-a-time? There is a pattern to follow, and here is the most efficient statement of the pattern that has yet been developed:

1) Given an atom of the element with nuclear charge Z, removing one electron from the nuclear field requires some positive energy. Call that required energy W, for 'work' required. It is:

$$W_{\text{removing } e_1 \text{ from the neutral atom}} = IP_{1,1}(Z/M_Z) \qquad (3)$$

   Observe that the $IP_{1,1}$ comes from Hydrogen, with $M_1/1$ scaling to population generic information, and that the re-scaling by Z/M_z converts that population generic information into information about element Z.

2) After the removal of one electron, the remaining electrons have to readjust. That involves some energy, possibly positive or possibly negative. This readjustment is not under external control, so all the energy involved H for 'heat.'

$$H_{\text{removing } e_1 \text{ from the neutral atom}} = (\Delta IP_{1,Z} - \Delta IP_{1,Z-1})(Z/M_Z) \qquad (4)$$

Note that, for the special case $Z = 1$, $\Delta IP_{1,Z-1} = \Delta IP_{1,0}$ does not exist, and $\Delta IP_{1,Z} = \Delta IP_{1,1} = 0$. And for $Z = 2$, $\Delta IP_{1,Z-1} = \Delta IP_{1,1} = 0$.

Thus, the total energy involved in removing one electron from an atom with nuclear charge Z is:

$$(W+H)_{\text{removing } e_1 \text{ from the neutral atom}} = \left[IP_{1,1}Z + \Delta IP_{1,Z}Z - \Delta IP_{1,Z-1}(Z-1) \right]/M_Z \qquad (5)$$

Now suppose we remove a second electron, then the work that has to be supplied because of the nuclear field is:

$$W_{\text{removing } e_2 \text{ after } e_1} = IP_{1,1}\sqrt{Z(Z-1)}/M_Z \qquad (6)$$

Observe the factor of $\sqrt{Z(Z-1)}$, in place, the factor of Z that occurs for the removal of the first electron. For the removal of the first electron, we had nuclear charge Z and electron count Z, so we had Coulomb attraction proportional to $Z \times Z$. So, the Z in Z/M scaling should be understood as $Z = \sqrt{Z \times Z}$. For the removal of the second electron, we have Coulomb attraction proportional to $Z \times (Z-1)$. So the occurrence of Z scaling in the removal of the first electron suggests the use of $\sqrt{Z \times (Z-1)}$ scaling in the removal of the second electron. Then for removal of the third electron, we have $\sqrt{Z \times (Z-2)}$, and so on.

The heat then involved in electron readjustment is:

$$H_{\text{removing } e_2 \text{ after } e_1} = \Delta IP_{1,Z-1} \times (Z-1)/M_Z - \Delta IP_{1,Z-2} \times (Z-2)/M_Z \tag{7}$$

The total energy involved in this step is then:

$$(W+H)_{\text{removing } e_2 \text{ after } e_1} = IP_{1,1}\sqrt{Z \times (Z-1)}/M_Z \\ + \Delta IP_{1,Z-1} \times (Z-1)/M_Z - \Delta IP_{1,Z-2} \times (Z-2)/M_Z \tag{8}$$

The total energy involved in removing two electrons is then:

$$(W+H)_{\text{removing } e_1 \,\&\, e_2} = IP_{1,1}\left[Z + \sqrt{Z \times (Z-1)}\right]/M_Z \\ + \left[\Delta IP_{1,Z} \times Z - \Delta IP_{1,Z-2} \times (Z-2)\right]/M_Z \tag{9}$$

Note the cancellation of $\Delta IP_{1,Z-1} \times (Z-1)/M_Z$ terms; similar cancellations simplify the formulae for all higher orders of ionization. The grand total energy involved in removing three electrons will be just:

$$(W+H)_{\text{removing } e_1, e_2, \,\&\, e_3} = IP_{1,1}\left[Z + \sqrt{Z \times (Z-1)} + \sqrt{Z \times (Z-2)}\right]/M_Z \\ + \left[\Delta IP_{1,Z} \times Z - \Delta IP_{1,Z-3} \times (Z-3)\right]/M_Z \tag{10}$$

and so on.

Now consider adding electrons. The work involved in adding one electron into the nuclear field of the atom with nuclear charge Z is:

$$W_{\text{adding } e_1 \text{ to neutral atom}} = -IP_{1,1}\sqrt{Z \times (Z+1)}/M_Z \tag{11}$$

The heat involved is then re-adjusting the electron population:

$$H_{\text{adding } e_1 \text{ to neutral atom}} = -\Delta IP_{1,Z+1} \times (Z+1)/M_Z + \Delta IP_{1,Z} \times Z/M_Z \tag{12}$$

Then altogether:

$$(W+H)_{\text{adding } e_1 \text{ to neutral atom}} = -IP_{1,1}\sqrt{Z \times (Z+2)}/M_Z \\ + \left[\Delta IP_{1,Z} \times Z - \Delta IP_{1,Z-1} \times (Z+1)\right]/M_Z \tag{13}$$

If we add a second electron to the singly charged ion, it will require additional work:

$$W_{\text{adding } e_2 \text{ after } e_1} = -IP_{1,1}\sqrt{Z \times (Z+2)}/M_Z \tag{14}$$

And it will cause a further heat adjustment:

$$H_{\text{adding } e_2 \text{ after } e_1} = -\Delta IP_{1,Z+2} \times (Z+2)/M_Z + \Delta IP_{1,Z+1} \times (Z+1)/M_Z \tag{15}$$

Putting both steps together, the grand total energy involved in adding two electrons is:

$$(W + H)_{\text{adding } e_1 \& e_2} = -IP_{1,1}\left[\sqrt{Z \times (Z+1)} + \sqrt{Z \times (Z+2)}\right]/M_Z$$
$$+ \left[\Delta IP_{1,Z} \times Z - \Delta IP_{1,Z+2} \times (Z+2)\right]/M_Z \tag{16}$$

Similarly, the total energy involved in adding three electrons is:

$$(W + H)_{\text{adding } e_1, e_2 \& e_3} = -IP_{1,1}\left[\sqrt{Z \times (Z+1)} + \sqrt{Z \times (Z+2)} + \sqrt{Z \times (Z+3)}\right]/M_Z$$
$$+ \left[\Delta IP_{1,Z} \times Z - \Delta IP_{1,Z+3} \times (Z+3)\right]/M_Z \tag{17}$$

1.4   NANO-CHEMICAL APPLICATION(S)

The algorithm described above permits efficient exploration of a problem domain, such as determining what the likely ionic configuration of a molecule is, showing why the shape of a molecule is what it is, evaluating whether a particular chemical reaction will transpire or not, showing how different reaction paths may compete, and how catalysis may help a reaction transpire.

1.5   MULTI-/TRANS-DISCIPLINARY CONNECTION(S)

Practical success with AC may encourage new work in theoretical physics, especially in quantum mechanics and in relativity theory.

KEYWORDS

- **computational chemistry**
- **ionization potentials**
- **scaling laws in chemistry**

REFERENCE AND FURTHER READING

Whitney, C., (2013). *Algebraic Chemistry Applications and Origins*, Chapter 3; Nova Publishers: New York, USA.

CHAPTER 2

Alkanes

LOUIS V. QUINTAS and EDGAR G. DUCASSE

Mathematics Department, Pace University, New York, NY 10038, USA,
E-mail: lvquintas@gmail.com, egducasse@pace.edu

2.1 DEFINITION

A molecule that contains only hydrogen and carbon atoms is called a *hydrocarbon* and is *saturated* if it has no double or triple covalent bonds. An *alkane* is a saturated hydrocarbon.

2.2 ORIGINS AND DEVELOPMENT

Alkanes are molecules that consist only of hydrogen and carbon atoms and whose bonds are all single covalent bonds. Formally, such molecules are called saturated hydrocarbons. Depending on their number of carbons, they exist in all states: gas, liquid, and solid. The molecular formula for an alkane is C_nH_{2n+2}.

Alkanes are of importance because of their utility in all of their states. For small values of n, that is, for alkanes with a small number of carbons such as CH_4 (methane), C_2H_6 (ethane), C_3H_8 (propane), and C_4H_{10} (butane), which exist in the gas state are used as propellants for aerosol sprays, some in particular, butane and propane, are liquefied for use as fuel for cooking and disposable cigarette lighters. A pentane/hexane/heptane mixture (n = 5, 6, and 7) is used as a solvent in dry cleaning. The liquid alkanes C_7H_{16} through C_9H_{20} (heptane, octane, through nonane) are used as gasoline, for various internal combustion engines, diesel, and jet engine fuels. The substance called kerosene is made up of mixtures of the alkanes decane to hexadecane ($C_{10}H_{22}$ through $C_{16}H_{34}$). Mixtures of alkanes containing 18 or more carbons, although classified as solids, are considered and used as heavy fuel oil. From alkanes with still higher carbon content, lubricating oil, vaseline, and solid waxes are obtained. Alkanes with more than 35 carbons currently have little commercial value and if dealt with are generally broken down to smaller alkanes. On the other hand, polymerization of alkanes produces synthetic materials made up of alkanes with thousands of carbons. These materials are manufactured in large quantities and have many uses; for example, as fabrics and plastics.

Methane and ethane are detected in the atmospheres of the Earth, Jupiter, Saturn, Uranus, Neptune, and in the tails of comets. Liquid alkanes were formed deep inside the earth's surface and took many millions of years to develop. These are obtained via commercial drilling. However, the use of the products made from these sources far exceeds the amount of alkanes that need to be generated to replenish the amount used. This and the following are matters of serious concern. Unfortunately, methane is a potent greenhouse gas; thus its leakage from industrial sources into the atmosphere has to be minimized (Schwartz, 2015). In addition, the heavier petroleum products play a significant role in pollution generating accidents, both in the obtaining of the associated raw materials and in their transportation.

The study of alkanes and their variations is extensive with many subdivisions, including a systematic way of naming these molecules. The latter has been formalized by the International Union of Pure and Applied Chemistry (IUPAC) with the objective of having a standardized and universal nomenclature system for all chemical compounds (Gutsche & Pasto, 1975, pp. 63, 465–473).

The study of alkanes commences with the molecular formula extended to the *structural formula*, a representation that shows the covalent bonding among the atoms of the alkane. The first three alkanes, i.e., methane, ethane, and propane corresponding to $n = 1, 2$, and 3, respectively, have the expected simple structure as shown in Figure 2.1.

Another notational descriptor is the *condensed structural formula*, where the covalent bond structure is shown via an extension of the molecular formula. For example, the molecular formula CH_4 for methane is the same as its condensed structural formula, but the condensed structural formula for ethane is CH_3CH_3. From the condensed structural formula the diagrammatic structural formula can be obtained. More details and examples will be given in what follows.

It is not surprising that two alkanes with different molecular formulas have different physical properties. However, it is also a fact that two alkanes may have the same molecular formula but have a different structural formula, and consequently, a different condensed structural formula indicating different covalence bonding. This difference assures the alkanes have different physical properties. Two alkanes with the same molecular formula but different structural formulas are said to be *structural isomers*. The first example of structural isomerism for alkanes occurs with butane and isobutane. Both have the molecular formula C_4H_{10} but have the distinct structural formulas shown in Figure 2.2.

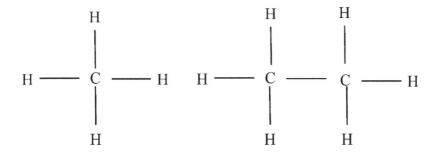

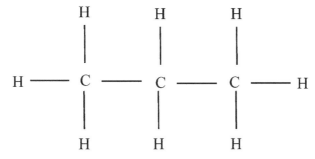

FIGURE 2.1   The structural formulas for methane, ethane, and propane.

The condensed structural formula for butane is $CH_3(CH_2)_2CH_3$ whereas for isobutane one has $CH_3C(CH_3)_2H$ as its condensed structural formula.

It is clear that the description of alkanes via the structural and condensed structural formulas gets cumbersome very quickly. An important simplification is obtained when one notices that the substructure of the carbons in an alkane completely determines the entire alkane. To be more specific, if one takes the structural formula for an alkane and deletes all hydrogen atoms and the bonds to their unique adjacent carbon atom, the result is a connected configuration made up of only carbon atoms and the covalent bonds between them. In the language of graph theory, this is a connected graph without cycles and is called a *tree*. In chemistry language, this is called the *carbon skeleton* of the alkane. If hydrogen atoms are added to fill out an alkane carbon skeleton, the structural formula obtained is precisely that of the original alkane. That is, the trees representing carbon skeletons of alkanes are in a one-to-one correspondence with their associated alkanes. For the carbon skeletons of the alkanes with up to six carbons, drawn as graph-theoretic trees, *see* Figure 2.3 and Cha et al. (2015).

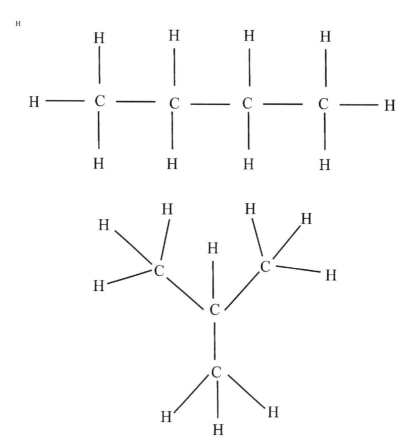

FIGURE 2.2   The structural formulas for butane and isobutane.

2.3   HISTORICAL COMMENTS

An excellent overview of the origin and development of organic chemistry, the domain of the alkanes, is given in Gutsche & Pasto (1975, pp. 8–16).

This development took place in the 1800s and its pioneers, Friedrich Wöhler (1800–1882), Justus Liebig (1803–1873), Archibald Scott Couper (1831–1892), Friedrich August Kekulé (1829–1896), Hermann Kolbe (1818–1887), William Henry Perkin (1838–1907), and Emil Fischer (1852–1919), and their contributions are duly credited.

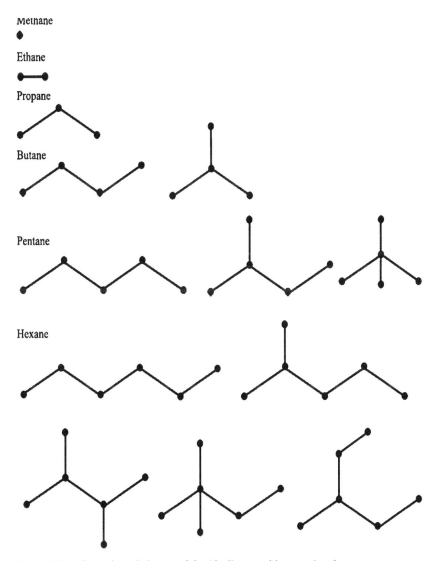

FIGURE 2.3   The carbon skeletons of the 13 alkanes with up to 6 carbons.

It is noted that many products obtained from alkanes and their derivatives are associated with strong smells (e.g., gasoline) and distinctive colors (e.g., petroleum and asphalt), however pure alkanes themselves are relatively odorless and essentially colorless. The usefulness of these products has motivated chemists to study the structure and physical properties of alkanes in great depth. A. S. Couper and F. A. Kekulé are given credit for representing a covalent bond with a line between a pair of atoms, each of which is represented by a point (Balaban, 1976). This resulted in a chemical graph (in the graph theory sense) and in particular to the structural formula for a molecule, which in turn led to the concept of structural isomers.

The beginning of a systematic enumeration of alkane isomers can be attributed to Arthur Cayley (1821–1895) who, in 1857, approached the problem via the concept of trees (connected graphs with no cycles) with vertices either having exactly one adjacency (degree 1) or with exactly four adjacencies (degree 4) (Cayley, 1857, 1874). These trees are the structural formulas for alkanes and are called (1, 4)-trees. Cayley further simplified his enumeration search by showing that counting (1, 4)-trees were equivalent to counting trees with no vertex of degree greater than four (4-trees) (DuCasse & Quintas, 2016). The latter trees are the representations of the carbon skeletons of alkanes. Cayley did not solve the problem completely. He was followed by Camille Jordan (1838–1922) who, in 1869, obtained a more rigorous enumeration of trees, although not in a chemical context (Jordan, 1869). Progress on this enumeration problem continued through work, in the early 1930s (Blair & Henze, 1931, 1933). The definitive breakthrough was made by George Pólya (1838–1922) with what is considered the flagship theorem on enumeration now known as Pólya's Enumeration Theorem (Pólya, 1937). This result was used by Frank Harary (1921–2005) and his disciples to enumerate a vast spectrum of structures (Harary, 1969; Harary & Palmer, 1973). Most notable for interests here, Ronald C. Read enumerated all the acyclic chemical compounds, which in addition to the alkanes, included the number of acyclic hydrocarbons having a given number of carbons and a given number of double and triple covalent bonds (Read, 1976).

Noted is the fact that in the above, the focus is on the enumeration of structural isomers, but the work involved enumeration in general and in particular, the enumeration of graphs. Thus, it is important to emphasize the significance of graphs in chemistry (Harary, 1969; Harary & Palmer, 1973; Balaban, 1976), wherein the preceding discussion (Harary & Palmer, 1973) is fleshed out with credit given to other major researchers in the enumeration.

2.4   ISOMERISM

The enumeration of structural isomers of alkanes is the enumerations of trees with no vertex of degree greater than 4, that is, counting 4-trees, *see* Balaban (1976) and *f*-trees in DuCasse & Quintas (2016). The sophisticated enumeration methods and computational facilities currently available have simplified this task in contrast to the days of the pioneer (Cayley, 1857). Such enumerations are now readily available; for example, in R. C. Read's work (Read, 1976), where this mathematical work is done in a chemistry context. The number of structural isomers of the alkanes for up to $n = 25$ are shown in Table 2.1.

TABLE 2.1   The Number N of Structural Isomers of the Alkanes Having n Carbons (n, N) for up to $n = 25$

(1,1)	(2,1)	(3,1)	(4,2)	(5,3)	(6,5)
(7,9)	(8,18)	(9,35)	(10,75)	(11,159)	(12,355)
(13,802)	(14,1858)	(15,4347)	(16,10359)	(17,24894)	(18,60523)
(19,48284)	(20,366319)	(21,910726)	(22,2278658)		
(23,731580)	(24,4490245)	(25,36797588)			

A structural isomer is said to have a *stereoisomer* if it has at least two orientations in 3-space which are not superimposable on each other. A molecule that has a non-superimposable mirror image is classified as a *chiral molecule*. If the mirror image is superimposable, the molecule is called *achiral* and is considered the same as the original molecule. Chirality, unless otherwise stated, is what is generally understood to be the concept involved in stereoisomerism (Gutsche & Pasto, 1975).

For a specific example of stereoisomerism, one has 3-methylhexane C_7H_{16}, a molecule that has a non-superimposable pair of mirror images (*see* Figure 2.4).

R. C. Read has rigorously discussed stereoisomerism and has provided an enumeration of the stereoisomers of the alkanes, together with the acyclic hydrocarbons with double and triple carbon bonds [*see* Tables 2–6 in Read (1976); For the number of stereoisomers of the alkanes, *see* Table 2.2].

The third type of isomer is obtained when the alkanes having n carbons are partitioned into classes having a specified number of *primary, secondary, tertiary,* and *quarternary carbons* (Gutsche & Pasto, 1975). This partitioning is the same as partitioning the 4-trees of order n (carbon skeletons) into vertex degree classes, where the degrees 1, 2, 3, and 4 in the 4-trees correspond to primary, secondary, tertiary, and quarternary carbons, respectively. The 4-trees with the same degree sequence are equivalent. This enumeration was carried out by William J. Martino, Louis V. Quintas, and Joshua Yarmish (Martino et al., 1980), using a 3-variable form of Pólya's Enumeration Theorem applied to 4-trees. This enumeration is summarized in Table 2.3. Namely, the number of equivalence classes of valence isomers for a given n is given. In Martino et al. (1980), the number of structural isomers in a given valence class is also provided.

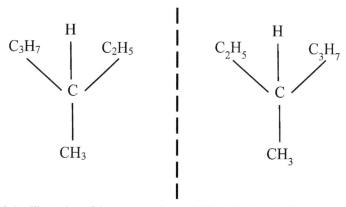

FIGURE 2.4   Illustration of the non-superimposability of the 3-methylhexane molecule.

TABLE 2.2   The Number S of Stereoisomers of Alkanes Having n Carbons (n, S) for up to $n = 25$

(1,1)	(2,1)	(3,1)	(4,2)	(5,3)	(6,5)
(7,11)	(8,24)	(9,55)	(10,136)	(11,345)	(12,900)
(13,2412)	(14,6563)	(15,18127)	(16,50699)	(17,143255)	
(18,408429)	(19,1173770)	(20,3396844)	(21,9892302)	(22,28972080)	
(23,85289390)	(24,252260276)	(25,749329719)			

Note that for a given n, the number of stereoisomers of alkanes is greater than or equal to the number of structural isomers and that the number of valence isomers is less than or equal to the number of structural isomers. In particular, for $n = 5$, 6, and 7 the number of stereoisomers is 3, 5, and 11, the number of structural isomers is 3, 5, and 9, and the number of valence isomers is 3, 4, and 5. For $n \geq 7$, the number of stereoisomers is always strictly greater than the number of structural isomers, and the number of structural isomers is strictly greater than the number of valence isomers.

Additional discussion concerning valence isomers of alkanes can be found in Quintas et al., (1979, 1980) and Quintas & Yarmish (1979, 1981).

TABLE 2.3 The Number V of Valence Isomers of Alkanes Having n Carbons (n, V) for up to $n = 25$

(1,1)	(2,1)	(3,1)	(4,2)	(5,3)	(6,4)
(7,5)	(8,7)	(9,8)	(10,10)	(11,12)	(12,14)
(13,16)	(14,19)	(15,21)	(16,24)	(17,27)	(18,30)
(19,33)	(20,37)	(21,40)	(22,44)	(23,48)	
(24,52)	(25,56)				

2.5 OPEN ISSUES

The important role of alkanes in so many areas has naturally stimulated research in generating alkanes in the laboratory. In particular, it is of interest to be able to convert methane directly to higher order alkanes. Presently, methane conversion is successfully achieved in the industry by the well-established indirect Fisher-Tropsch (FT) process (Yuliati & Yoshida, 2008; Keil, 2013). This method initially achieved in 1936 has been developed to produce large quantities of alkanes in the C_3 to C_{30} range. However, the relatively high cost of this conversion has continued to spur chemists to develop other methods (Crabtree, 1995). Viable conversion of methane directly to higher alkanes was not discovered until 1991 (Belgued et al., 1992; Koerts et al., 1992; Amariglio et al., 1995; Soulivong et al., 2008). For an overview, *see* Cha et al. (2015). These recent successes in the direct coupling of methane molecules to produce ethane and higher order alkanes suggest that research is directed to finding processes that can reliably produce higher order alkanes in large quantities and at low cost.

KEYWORDS

- **alkane**
- **hydrocarbon**
- **isomers**
- **saturated**

REFERENCES

Amariglio, H., Saint-Just, J., & Amariglio, A., (1995). Homologation of methane under non-oxidative conditions, *Fuel Processing Technology, 42*, 291–323.

Balaban, A. T., (1976). *Chemical Applications of Graph Theory*, Academic Press, New York.

Belgued, M., Amariglio, H., Pareja, P., Amariglio, A., & Saint-Just, J., (1992). Low-temperature catalytic homologation of methane on platinum, ruthenium, and cobalt. *Catalysis Today, 13*, 437–445.

Blair, C. M., & Henze, H. R., (1931). The number of structurally isomeric alcohols of the methanol series. *J. Amer. Chem. Soc., 53*, 3042–3046.

Blair, C. M., & Henze, H. R., (1933). The number of structurally isomeric hydrocarbons of the ethylene series. *J. Amer. Chem. Soc., 55*, 680–686.

Cayley, A., (1857). On the theory of the analytical forms called trees. *Philos. Mag., 13*, 172–176.

Cayley, A., (1874). On the mathematical theory of isomers, *Philos. Mag., 47*, 444–447.

Cha, S. H., DuCasse, E. G., Quintas, L. V., Kravette, K., & Mendoza, D. M., (2015). A graph model related to chemistry: Conversion of methane to higher alkanes. *International J. Chemical Modeling, 7*(1), 17–29.

Crabtree, R. H., (1995). Aspects of methane chemistry. *Chemical Reviews, 95*, 987–1007.

DuCasse, E. G., & Quintas, L. V., (2016). F-graphs. In: *The Explicative Dictionary of Nanochemistry*.

Gutsche, C. D., & Pasto, D. J., (1975). *Fundamentals of Organic Chemistry*, Prentice-Hall, Englewood Cliffs, New Jersey.

Harary, F., & Palmer, E., (1973). *Graphical Enumeration*, Academic Press, New York.

Harary, F., (1969). *Graph Theory*. Addison-Wesley Publishing Company, Reading, Massachusetts.

IUPAC, http://www.chem.qmul.ac.uk/iupac/ (accessed on 2 February 2019).

Jordan, C., (1869). Sur les assemblages de lignes, *J. Reine Angew. Math., 70*, 185–190.

Keil, F. J., (2013). Methane activation: Oxidation goes soft. *Nature Chemistry, 5*, 91–92.

Koerts, T., Deelen, M. J., & Van Santen, R. A., (1992). Hydrocarbon formation from methane by a low-temperature two-step reaction sequence. *Journal of Catalysis 138*, 101–114.

Martino, W. J., Quintas, L. V., & Yarmish, J., (1980). *Degree Partition Matrices for Rooted and Unrooted 4-Trees*. Mathematics Department, Pace University, New York, NY.

Pólya, G., (1937). Kombinatorische Anzahlbestimmungen für Gruppen, Graphen und chemische Verbindungen, *Acta Math., 68*, 145–254.

Quintas, L. V., & Yarmish, J., (1981). Valence isomers for chemical trees. *MATCH, 12*, 65–73.

Quintas, L. V., Schiano, M. D., & Yarmish, J., (1980). *Degree Partition Matrices for Identity 4-Trees*. Mathematics Department, Pace University, New York, NY.

Quintas, L. V., Stehlik, M., & Yarmish, J., (1979). *Congressus Numerantium* XXIV, 795–821.

Quintas, L. V., & Yarmish, J., (1979). The number of chiral alkanes having given diameter and carbon automorphism group a symmetric group. *Annals New York Academy of Sciences, 319*, 436–443.

Read, R. C., (1976). The enumeration of acyclic chemical compounds, Chapter 4. In: Balaban, A. T., (ed.), *Chemical Applications of Graph Theory* (pp. 25–61). Academic Press, New York.

Schwartz, J., (2015). New study challenges methane leak estimates. *New York Times,* A10.

Soulivong, D., Norsic, S., Taoufik, M., Coperet, C., Thivole-Caza, J., Chakka, S., & Basset, J. M., (2008). Non-oxidative coupling reaction of methane to ethane and hydrogen catalyzed by the silica-supported tantalum hydride: $(\equiv SiO)_2 Ta{-}H$. *Journal of the American Chemical Society, 130,* 5044–5045.

Yuliati, L., & Yoshida, H., (2008). Photocatalytic conversion of methane. *Chemical Society Reviews, 37,* 1592–1602.

CHAPTER 3

The Anharmonic Correction of Normal Mode Vibrations

ATTILA BENDE

Molecular and Biomolecular Physics Department, National Institute for Research and Development of Isotopic and Molecular Technologies, Donat Street, No. 67–103, RO-400293 Cluj-Napoca, Romania

3.1 DEFINITION

The molecular vibrations can be described in the framework of the quantum theory of the harmonic oscillator. This approximation, however, ignores the coupling of different normal mode vibrations as well as the non-parabolic profile of the potential energy in which the collective oscillation of the atoms happens. To accurately reproduce the experimental IR or Raman vibrational spectra, one needs to go beyond the model of the harmonic oscillator and include effects of the anharmonicity induced by the small perturbations (deviations) from the ideal case of the perfectly harmonic potential. Obviously, this means changes (shifts) in vibrational frequency values with respect to the magnitude of harmonic frequencies. Since anharmonicities are considered as small perturbational corrections, this approximation will not be valid if the shape of the potential energy valley significantly differs from the parabolic form.

3.2 HISTORICAL ORIGIN(S)

The complex structure of the IR (or Raman) absorption spectra requires methods able to correctly interpret the relationship between the normal mode vibrations and the experimental frequency values given by different spectral peaks. It was well-known even at the early period of the development of the molecular vibration theory that the model of the coupled harmonic oscillator is not suitable to describe the coordinated motion of the atoms around their equilibrium geometry in a molecular system. Therefore, there is a need for a more sophisticated model which includes all effects derived from the distorted harmonic (or anharmonic) potential energy profiles and is able to properly reproduce the most part of the spectral characteristics given by the experimental measurements. Accordingly, a huge effort has been made in the past 20–30 years in order to fully develop powerful theoretical models and to make them available for a large number of a researcher who then successfully used in their different studies.

3.3 NANO-SCIENTIFIC DEVELOPMENT(S)

Both infrared (IR) and Raman spectroscopy methods provide a unique spectral fingerprint of a certain molecule. This uniqueness is given by the correlated motion of coupled atomic oscillators (pattern of motion) called normal mode vibrations; each of them with a well-defined frequency value (Herzberg, 1945). Accordingly, the spectral positions of the peaks from the IR (or Raman) absorption spectra are distinctively defined by the frequency values of the normal mode vibrations (Hollas, 2004) and the magnitude of the peaks, called intensities, are given by derivatives of the molecular dipole moments with respect to the normal coordinates (Steel, 1971).

The Schrödinger equation for a particle in motion in a harmonic potential, called one-dimensional harmonic oscillator is

$$\left(\frac{-\hbar}{2m} \frac{\partial^2}{\partial x^2} + \frac{1}{2} m\omega^2 x^2 \right) |\psi(x)\rangle = E |\psi(x)\rangle \tag{1}$$

where ω is the frequency of the classical oscillator, and m the particle mass. Solving the differential equation given by Eq. (1), one obtains a set of discrete values for the energy

$$E_n = \left(n + \frac{1}{2} \right) \hbar\omega \tag{2}$$

Any molecule is built by a number of N atoms defined by $3N$ atomic coordinates, from which a set of three coordinates define the translational motion of the center of mass, three additional coordinates characterize the molecular rotation around its center of mass, while the remained $3N$-6 coordinates describe the relative motion of atoms inside the molecule. In this way, instead of a simple one-particle oscillator case, one has a system of differential equations similar to Eq. (1). Using the matrix formalism of the Wilson's GF method (Wilson et al., 1955) the $3N - 6$ frequency values of the normal mode vibrations can be obtained almost in a similar manner than for the one-particle oscillator case, with the condition that force constants (derivatives of the potential energy with respect of the atomic coordinates) defined by the atomic bonds, angles or torsions are already known. With the rapidly developed quantum chemical softwares and computational hardwares, the calculation of force constants (a number of [(3N–6)x(3N)] energy second derivatives) becomes available with moderate computational efforts, even for medium-sized molecular systems ($\approx$ 100–150 atoms). In this way, the harmonic frequency calculation and implicitly the drawing of theoretical IR and Raman absorption spectra are generally feasible for most of the case of studies.

The harmonic oscillator is a highly idealized model, where the oscillator's fundamental frequency of vibration is independent of the amplitude of the vibrations. The atoms in a molecule or a solid oscillate around their equilibrium positions defined by their potential energy profile. It is well-known that this energy curve has a harmonic profile only if the atoms oscillate with small vibrational amplitudes. So, for the general case, one needs to consider distorted harmonic (or anharmonic) potential energy profiles. In order to include corrections originated from the deviation from the harmonic oscillator model, one can follow two different strategies. The first is, when knowing a huge number of empirical frequency data for several previous molecular cases, for the given level of theory (theoretical method + basis set) we are trying to find out the so-called scaling factor, which used as a multiplier in the case of new absorption spectra calculations will reproduce the experimental frequency values with good accuracy (Scott et al., 1996; Temelso et al., 2011). The second possibility is to develop further the theoretical model and includes the anharmonicities as small perturbations to the harmonic approximation. This perturbative expansion can be done in first- (or VSCF) (Clabo Jr. et al., 1988; Jung et al., 1996; Roy et al., 2013) or second-order (or VMP2) (Norris et al., 1996; Christiansen, 2003; Barone, 2005) approximation. The anharmonic corrections can be explicitly calculated considering different third and semi-diagonal fourth order energy derivatives. To evaluate third-order derivatives of the energy with respect to normal coordinates, one should perform numerical differentiation of analytical Hessian matrices at geometries displaced by small increments δq from the reference geometry (Barone, 2005):

$$f_{ijk} = \frac{1}{3}\left(\frac{f_{ij}(q+\delta q_k)-f_{ij}(q-\delta q_k)}{2\delta q_k} + \frac{f_{jk}(q+\delta q_i)-f_{jk}(q-\delta q_i)}{2\delta q_i} + \frac{f_{ki}(q+\delta q_j)-f_{ki}(q-\delta q_j)}{2\delta q_j} \right)$$

(3)

The semi-diagonal fourth-order derivatives of energy can be computed as:

$$f_{ijkk} = \frac{f_{ij}(q+\delta q_k)-2f_{ij}(q)+f_{ij}(q-\delta q_k)}{\delta q_k^2}$$

$$f_{iikk} = \frac{1}{2}\left(\frac{f_{ii}(q+\delta q_k)-2f_{ii}(q)+f_{ii}(q-\delta q_k)}{\delta q_k^2} + \frac{f_{kk}(q+\delta q_i)-2f_{kk}(q)+f_{kk}(q-\delta q_i)}{\delta q_i^2} \right)$$

(4)

Using the previously defined cubic (f_{ijk}) by Eq. (3) and quartic (f_{ijkk} and f_{iikk}) force constants by Eq. (4), the diagonal (x_{ii}) and off-diagonal (x_{ij}) vibrational anharmonic constants for an asymmetric top molecule can be obtained as:

$$x_{ii} = \frac{1}{16}f_{iiii} - \frac{1}{32}\sum_j f_{iij}^2 \left[\frac{1}{2\omega_i + \omega_j} - \frac{1}{2\omega_i - \omega_j} + \frac{4}{\omega_j} \right]$$

(5)

and

$$x_{ij} = \frac{1}{4}f_{iijj} - \frac{1}{4}\sum_k \frac{f_{iik}f_{kkj}}{\omega_k} - \frac{1}{8}\sum_k f_{ijk}^2 \left[\frac{1}{\omega_i+\omega_j+\omega_k} - \frac{1}{\omega_i+\omega_j-\omega_k} \right.$$
$$\left. + \frac{1}{\omega_i-\omega_j+\omega_k} - \frac{1}{\omega_i-\omega_j-\omega_k} \right] +$$
$$+ \left(A_e(\zeta_{ij}^a)^2 + B_e(\zeta_{ij}^b)^2 + C_e(\zeta_{ij}^c)^2 \right)\left(\frac{\omega_i}{\omega_j} + \frac{\omega_j}{\omega_i} \right)$$

(6)

where ω_i are the harmonic frequencies; A_e, B_e, and C_e are the equilibrium rotational constants; and ζ_{ij} are the Coriolis zeta constants.

Once the diagonal and off-diagonal anharmonic constants were obtained, the full anharmonic correction (Δ_i) and the fundamental vibrational frequency (v_i) can be easily calculated [for details, *see* Neugebauer et al., (2003) and references therein],

$$v_i = \omega_i + \ddot{A}_i = \omega_i + 2x_{ii} + \frac{1}{2}\sum_{j,i\neq j} x_{ij} \tag{7}$$

Studies based on DFT calculations including different XC functional (Barone, 2004; Biczysko, 2010) have shown that the MAD from the experimental values of the anharmonically corrected normal mode frequencies typically scatters around 10–20 cm^{-1}, depending on the XC functional, basis set and the molecule of interest. The method based on the second-order perturbative treatment including quadratic, cubic, and semi-diagonal quartic force constants (Barone, 2005) can be easily applied to systems up to 30–40 atoms (Bende, 2010; Bende et al., 2011, 2014; Fornaro et al., 2014). If one needs higher accuracies, one should switch to wavefunction-based electronic structure calculations, e.g., CC (Christiansen, 2004; Hrenar et al., 2007; Rauhut et al., 2009) or vibrational multireference methods (Pfeiffer et al., 2014), which leads to a further reduction of the final deviation, but also results in a significant increase of the computational cost.

Furthermore, size-extensive generalizations at different levels of theories (SCF, MP2, and CC) were also made to anharmonic lattice vibrations of extended periodic systems (solids) (Hirata et al., 2010).

3.4 NANO-CHEMICAL APPLICATION(S)

In order to unequivocally identify the molecular fingerprint given by the vibrational absorption spectrum one needs to correctly assign the normal mode frequency vibrations to different absorption peaks from the spectrum. Mainly, this can be done by reproducing with good accuracy the absorption spectrum using high-level theoretical methods. It is generally accepted that vibrational frequencies calculated in the harmonic approximation do not agree well with the experimental values. Accordingly, one needs to go beyond the harmonic potential model and use different correction schemes to achieve the desired accuracy. In the scientific literature, there are tons of articles where different levels of anharmonic corrections were applied in order to accurately calculate the theoretical IR (or Raman) absorption spectrum for a certain molecule. Usually, the size of the systems is ranked between small and medium-large (up to 50–60 atoms) molecular cases, but among the cases the most intensively studied systems are the guanine-cytosine and adenine-thymine DNA base pairs (Špirko et al., 1997; Bauer et al., 2005; Wang et al., 2009; Bende, 2010; Bende et al., 2011; Fornaro et al., 2014). Based on these studies, we can draw a general conclusion that the large anharmonic shifts for the normal mode frequencies as well as strong anharmonic couplings between the normal modes. These conclusions were also confirmed experimentally (Peng et al., 2011; Greve et al., 2013; Nosenko et al., 2013).

The spectroscopic fingerprint can also be used to reveal different cluster configurations of the supramolecular assemblies. Different cluster configurations can be distinguished between them by the relative intermolecular orientation of the monomers. Each intermolecular interaction between the molecular constituents, whether they are H⁻, halogen-bond or van der Waals-types, carries specific structural features of the cluster and they are also "visible" in their IR (or Raman) absorption spectrum (Bende et al., 2005; Dunn et al., 2006; Bende, 2010; Temelso et al., 2011; Howard et al., 2014; Oschetzki et al., 2014).

Software modules dedicated to anharmonic frequency corrections are included in several quantum chemistry program packages: GAUSSIAN09 (www.gaussian.com), Molpro 2015.1 (https://www.molpro.net/), GAMESS (http://www.msg.ameslab.gov/ga-mess/), CFour (http://www.cfour.de/), or NWChem (http://www.nwchem-sw.org).

3.5   MULTI-/TRANS-DISCIPLINARY CONNECTION(S)

Anharmonic frequency calculations are widely used in several fields of science like physics, chemistry, or biology, where the computation of the theoretical IR (or Raman) absorption spectra is utilized to identify different molecular or supramolecular structures. Its mathematical development is based on different advanced numerical analysis techniques.

3.6   OPEN ISSUES

Although the inclusion of the anharmonic corrections in computing the theoretical IR (or Raman) absorption spectra is very important, there are some limitations for the applicability of this correction. (i) It requires huge computational resources, due to the computation of the (analytical) second- or the (numerical) third- and fourth-order derivatives of the energy. It is considered that for DFT methods and medium-size basis set (a simple triple ζ–quality basis set without polarization or diffuse functions), its applicability domain is up to 50–60 atoms or up to 20–30 atoms if we use higher level theoretical methods and larger basis sets. There are some initiatives for easier and faster computation of these frequency corrections, but it is not yet ready for widespread use (Hanson-Heine et al., 2012; Thomas et al., 2013; Hermes et al., 2015). Due to this limitation problem, some authors have considered that, for large molecular systems using smaller basis sets, "anharmonic calculations are no more accurate than scaled harmonic calculations" (Jacobsen et al., 2013). (ii) Being a perturbation theory, it cannot be used if the potential energy surface around the equilibrium molecular geometry strongly differs from the harmonic profile. This could be the case when we study proton (H^+) bridge vibrations in some ionized molecular clusters (Bende et al., 2011, 2014). Here, vibrations where the H^+ is involved, under some circumstances, could have even chaotic behavior.

KEYWORDS

- **anharmonic frequency shifts**
- **anharmonicity**
- **cubic and quartic force constants**
- **vibrational couplings**

REFERENCES AND FURTHER READING

Barone, V., (2004). Accurate vibrational spectra of large molecules by density functional computations beyond the harmonic approximation: The case of azobenzenes. *J. Phys. Chem. A108*, 4146–4150.

Barone, V., (2005). Anharmonic vibrational properties by a fully automated second-order perturbative approach. *J. Chem. Phys., 122*, 014108.

Barone, V., Biczysko, M., & Bloino, J., (2014). Fully anharmonic IR and Raman Spectra of medium-size molecular systems: Accuracy and interpretation. *Phys. Chem. Chem. Phys., 16*, 1759–1787.

Bende, A., (2010). Hydrogen bonding in the urea dimers and adenine–thymine DNA base pair: Anharmonic effects in the intermolecular H-bond and intramolecular H-stretching vibrations. *Theor. Chem. Acc., 125*(3–6), 253–268.

Bende, A., & Muntean, C. M., (2014). The influence of anharmonic and solvent effects on the theoretical vibrational spectra of the guanine-cytosine base pairs in Watson–Crick and Hoogsteen Configurations. *J. Mol. Model., 20*, 2113.

Bende, A., & Suhai, S., (2005). BSSE-corrected geometry and harmonic and anharmonic vibrational frequencies of formamide–water and formamide–formamide dimers. *Int. J. Quantum Chem., 103*, 841–853.

Bende, A., Bogdan, D., Muntean, C. M., & Morari, C., (2011). Localization and anharmonicity of the vibrational modes for the GC Watson-Crick and Hoogsteen Base Pairs. *J. Mol. Model., 17*(12), 3265–3274.

Biczysko, M., Panek, P., Scalmani, G., Bloino, J., & Barone, V., (2010). Harmonic and anharmonic vibrational frequency calculations with the double-hybrid B2PLYP method: Analytic second derivatives and benchmark studies. *J. Chem. Theory Comput., 6*, 2115–2125.

Brauer, B., Gerber, R. B., Kabeláč, M., Hobza, P., Bakker, J. M., Riziq, A. G. A., & de Vries, M. S., (2005). Vibrational spectroscopy of the G-C base pair: Experiment, harmonic and anharmonic calculations, and the nature of the anharmonic couplings. *J. Phys. Chem., A109*, 6974–6984.

Christiansen, O., (2003). Møller–Plesset perturbation theory for vibrational wavefunctions. *J. Chem. Phys., 119*(12), 5773–5781.

Christiansen, O., (2004). Vibrational coupled cluster theory. *J. Chem. Phys., 120*, 2149–2159.

Clabo, Jr., D. A., Allen, W. D., Remington, R. B., Yamaguchi, Y., & Schaefer, III, H. F., (1988). A systematic study of molecular vibrational anharmonicity and vibration-rotation interaction by self-consistent-field higher derivative methods: Asymmetric top molecules. *Chem. Phys., 123*, 187–239.

Dunn, M. E., Evans, T. M., Kirschner, K. N., & Shields, G. C., (2006). Prediction of accurate anharmonic experimental vibrational frequencies for water clusters, $(H2O)_n$, $n = 2$–5. *J. Phys. Chem., A110*, 303–309.

Fornaro, T., Biczysko, M., Monti, S., & Barone, V., (2014). Dispersion corrected DFT approaches for anharmonic vibrational frequency calculations: Nucleobases and their dimers. *Phys. Chem. Chem. Phys., 16*, 10112–10128.

Greve, C., Preketes, N. K., Fidder, H., Costard, R., Koeppe, B., Heisler, I. A., et al., (2013). N–H stretching excitations in adenosine-thymidine base pairs in solution: Pair geometries, infrared line shapes, and ultrafast vibrational dynamics. *J. Phys. Chem. A., 117*(3), 594–606.

Hanson-Heine, M. W. D., George, M. W., & Besley, N. A., (2012). Rapid anharmonic vibrational corrections derived from partial hessian analysis. *J. Chem. Phys., 136*, 224102.

Hermes, M. R., & Hirata, S., (2015). Diagrammatic theories of anharmonic molecular vibrations. *Int. Rev. Phys. Chem., 34*(1), 71–97.

Herzberg, G., (1945). *Molecular Spectra and Molecular Structure: Infrared and Raman Spectra of Polyatomic Molecules* (3rd edn.). Van Nostrand Reinhold Inc.: New York City, USA.

Hirata, S., Keçeli, M., & Yagi, K., (2010). First-principles theories for anharmonic lattice vibrations. *J. Chem. Phys., 133*, 034109.

Hollas, J. M., (2004). *Modern Spectroscopy* (4th edn., pp. 137–198). John Wiley & Sons Ltd: West Sussex, England.

Howard, J. C., Gray, J. L., Hardwick, A. J., Nguyen, L. T., & Tschumper, G. S., (2014). Getting down to the fundamentals of hydrogen bonding: Anharmonic vibrational frequencies of $(HF)_2$ and $(H_2O)_2$ from Ab initio electronic structure computations. *J. Chem. Theory Comput., 10*, 5426–5435.

Hrenar, T., Werner, H. J., & Rauhut, G., (2007). Accurate calculation of anharmonic vibrational frequencies of medium-sized molecules using local coupled cluster methods. *J. Chem. Phys., 126*, 134108.

Jacobsen, R. L., Johnson, III, R. D., Irikura, K. K., & Kacker, R. N., (2013). Anharmonic vibrational frequency calculations are not worthwhile for small basis sets. *J. Chem. Theory Comput., 9*, 951–954.

Jung, J. O., & Gerber, R. B., (1996). Vibrational wave functions and spectroscopy of $(H_2O)_n$, n = 2,3,4,5: Vibrational self-consistent field with correlation corrections. *J. Chem. Phys., 105,* 10332–10348.

Neugebauer, J., & Hess, B. A., (2003). Fundamental vibrational frequencies of small polyatomic molecules from density-functional calculations and vibrational perturbation theory. *J. Chem. Phys., 118,* 7215–7225.

Norris, L. S., Ratner, M. A., Roitberg, A. E., & Gerber, R. B., (1996). Møller–Plesset perturbation theory applied to vibrational problems. *J. Chem. Phys., 105,* 11261–11267.

Nosenko, Y., Kunitski, M., Stark, T., Göbel, M., Tarakeshwarc, P., & Brutschy, B., (2013). Vibrational signatures of Watson Crick base pairing in adenine–thymine mimics. *Phys. Chem. Chem. Phys., 15,* 11520–11530.

Oschetzki, D., & Rauhut, G., (2014). Pushing the limits in accurate vibrational structure calculations: Anharmonic frequencies of lithium fluoride clusters $(LiF)_n$, n = 2–10. *Phys. Chem. Chem. Phys., 16,* 16426–16435.

Peng, C. S., Jones, K. C., & Tokmakoff, A., (2011). Anharmonic vibrational modes of nucleic acid bases revealed by 2D IR spectroscopy. *J. Am. Chem. Soc., 133,* 15650–15660.

Pfeiffer, F., & Rauhut, G., (2014). Multi-reference vibration correlation methods. *J. Chem. Phys., 140,* 064110.

Rauhut, G., Knizia, G., & Werner, H. J., (2009). Accurate calculation of vibrational frequencies using explicitly correlated coupled-cluster theory. *J. Chem. Phys., 130,* 054105.

Roy, T. K., & Gerber, R. B., (2013). Vibrational self-consistent field calculations for spectroscopy of biological molecules: New algorithmic developments and applications. *Phys. Chem. Chem. Phys., 15,* 9468–9492.

Scott, A. P., & Radom, L., (1996). Harmonic vibrational frequencies: An evaluation of Hartree-Fock, Møller-Plesset, quadratic configuration interaction, density functional theory, and semi-empirical scale factors. *J. Phys. Chem., 100,* 16502–16513.

Špirko, V., Šponer, J., & Hobza, P., (1997). Anharmonic and harmonic intermolecular vibrational modes of the DNA base pairs. *J. Chem. Phys., 106,* 1472–1479.

Steele, D., (1971). *Theory of Vibrational Spectroscopy,* W. B. Saunders Inc.: Philadelphia, USA.

Temelso, B., & Shields, G. C., (2011). The role of anharmonicity in hydrogen-bonded systems: The case of water clusters. *J. Chem. Theory Comput., 7,* 2804–2817.

Thomas, M., Brehm, M., Fligg, R., Vöhringer, P., & Kirchner, B., (2013). Computing vibrational spectra from Ab initio molecular dynamics. *Phys. Chem. Chem. Phys., 15,* 6608–6622.

Wang, G. X., Ma, X. Y., & Wang, J. P., (2009). Anharmonic vibrational signatures of DNA bases and Watson Crick base pairs. *Chin. J. Chem. Phys., 22*(6), 563–570.

Wilson, Jr., E. B., Decius, J. C., & Cross, P. C., (1955). *Molecular Vibrations: The Theory of Infrared and Raman Vibrational Spectroscopy.* McGraw-Hill Book Co.: New York, USA.

CHAPTER 4

Aromaticity

MIHAI V. PUTZ[1,2] and MARINA A. TUDORAN[2]

[1]*Laboratory of Structural and Computational Physical Chemistry for Nanosciences and QSAR, Biology-Chemistry Department, West University of Timisoara, Pestalozzi Street No. 44, Timisoara, RO-300115, Romania, Tel.: +40-256-592638, Fax: +40-256-592620, E-mail: mv_putz@yahoo.com, mihai.putz@e-uvt.ro*

[2]*Laboratory of Renewable Energies-Photovoltaics, R&D National Institute for Electrochemistry and Condensed Matter, Dr. A. Paunescu Podeanu Str. No. 144, Timisoara, RO-300569, Romania*

4.1 DEFINITION

According to the IUPAC definitions, the aromaticity concept is referring to a spatial and electronic structure of cyclic molecular systems, which displays the cyclic electron delocalization effects. An aromatic compound present enhanced thermodynamic stability, which is relative to acyclic structural analogs, and tends to retain the structural type during the chemical transformations.

4.2 HISTORICAL ORIGIN(S)

The word "aromatic" as a chemical term was used for the first time by August Wilhelm Hofmann in 1855 and was referring to compounds containing the phenyl radical (Hofmann, 1855). In 1865, August Kekulé proposed the cyclo-hexatriene structure for benzene, structure accepted by most chemists, due to the fact that it respects the isomeric relationships of aromatic chemistry known at that time, even if the unsaturated molecule of benzene was unreactive toward addition reactions. After J.J. Thomson, who discover the electron, placed three equivalent electrons between each carbon atom in benzene in

1921, and along with the introduction of the term aromatic sextet as a group of six electrons resisting disruption by Sir Robert Robinson in 1925 (Armit and Robinson, 1925), the stability of benzene was explained. Still, the quantum mechanical origins of aromaticity were modeled for the first time in 1931 by Hückel who separate the bonding electrons in sigma and pi electrons.

Aromaticity can be described using a collection of physicochemical properties which determine specific features of cyclic and polycyclic molecules containing π-electrons (Elvidge and Jackman, 1961; Sondheimer, 1964; Krygowski et al., 2000).

In order to be considered aromatic, a molecule will display the following features: its conformation will be cyclic or polycyclic (Figure 4.1); it will present low bond lengths alternation; it will be more stable comparing with its acyclic analog; due to the ring current induced by the magnetic field, it will display increased diamagnetic susceptibility and diatropic chemical shift in ¹H NMR spectra and in an external magnetic field; it will have a high affinity for participating in substitution reactions than in addition reactions.

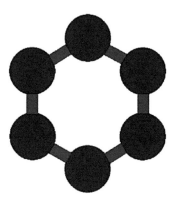

FIGURE 4.1   The cyclic conformation of benzene.

4.3   NANO-SCIENTIFIC DEVELOPMENT(S)

In literature, aromatic molecules are defined as cyclic, planar molecules which present a complete delocalization of $4n + 2\pi$ electrons and undergo electrophilic substitution. Experimentally, the first attempt in describing the resonance energy was made by Krisiakowsky et al. (1936), starting from the experimental enthalpies Eq. (1), this type of study showing that there is a link between the aromaticity and some physicochemical properties.

$$\text{Benzene} + 2 \text{ cyclohexene} = 3 \text{ cyclohexene} + 36 \text{ kcal mol}^{-1} \qquad (1)$$

A new approach for expressing the aromaticity concept was introduced by Putz (2010a, 2010b) and uses the approximation of electronegativity and chemical hardness as the compact finite difference. This method is based on the molecular frontier orbitals and expresses both the electron accepting (electrophilic) and the electron donating (nucleophilic) states using the differential expansion of the energy considered around its isolated value. The bond-forming stage is described by the superposition of atoms in the molecule and is based on the electronegativity equalization principle, while the post-bonding molecular stability is based on the maximum hardness principle. For a molecular system, its stability is determined from the excess chemical information obtained from the difference between the two stages of a molecular bonding applied to a given reactivity index. In this context, the absolute aromaticity (A_a) is defined as the difference between the superposition of atomic information in bonding (Π_{AIM}) and the information of the molecular orbital (Π_{MO}), the obtained result being the chemical information which characterizes the chemical bond stability (Putz, 2010a). As a consequence of the fact that the aromaticity character appears to be directly related with the excited and valence orbitals of a molecule and not with its ground state energy, a new form of aromaticity, the compactness aromaticity (A_c) is proposed and defined as the report between the information from the pre-bonding stage (represented by the atoms-in-molecules, Π_{AIM}) and the information from the post-bonding stage (represented by the molecular orbitals, Π_{MO}). This approach describes the aromaticity in terms of molecular stability in the same way as the crystal stability orderings is given by the compactness of rigid spheres in unit cells (Putz, 2010b).

Another way of describing absolute aromaticity starting from its formula proposed by Putz (2010a) is defined by replacing the prebonding stage *AIM* with the Specific-Adjacency-in-Bonding (*SAIB*) (Tudoran and Putz, 2015) and the molecular (*MOL*) manifestation of a given property with the LUMO-HOMO value (*LH*), being obtained the following formula:

$$A_{abs} = \left[\Pi_{AIM} \to \Pi_{AIB} \right] - \left[\Pi_{Mol} \to \Pi_{Homo-Lumo(H-L)} \right] = \Pi_{AIB} - \Pi_{HL} \qquad (2)$$

Then, Eq. (2) can be further specialized for topo-reactivity studies (Putz and Tudoran, 2016), i.e., for electronegativity, and will result in the formula:

$$\left(A_\chi \right)_{abs} = \chi_{AIB} - \chi_{HL} \qquad (3)$$

In the same manner, the compactness aromaticity (Putz and Tudoran, 2016) can be rewritten as:

$$A_{comp} = \frac{\Pi_{AIM} \rightarrow \Pi_{(AIB)}}{\Pi_{Mol} \rightarrow \Pi_{Homo-Lumo(H-L)}} = \frac{\Pi_{AIB}}{\Pi_{HL}} \tag{4}$$

with the following form for electronegativity:

$$\left(A_{\chi}\right)_{comp} = \frac{\chi_{AIB}}{\chi_{HL}} \tag{5}$$

As for organic molecules, different studies in this area show that when the benzene is interacting with cyclobutadiene, its ring will adopt a shifted, parallel conformation, but when is interacting with another benzene ring will have a tilted T-shaped orientation, meaning that the aromaticity may be correlated with the tilt angle. A further research (Shee and Datta, 2015) came with the idea that the angular conformation is possible only for the interaction with an aromatic system, but is not possible for the interaction with an anti-aromatic system. To verify the validity of this idea, the interaction of benzene with ten systems of the form C_4H_4X (with $X = BH$, AlH, SiH_2, CH_2, O, S, PH, NH, SiH, CH^-) was studied, by analyzing the obtained values for different index of aromaticity (Table 4.1).

In 1996, Schleyer and co-workers (Schleyer et al., 1996; Shee and Datta, 2014) introduce the nucleus independent chemical shift (NICS), one of the most effective indices of aromaticity, and its value at the center of the planar ring, NICS(0), appears to be linearly related to θ^{-1}, more exactly, a larger value of the tilt angle θ implies a higher aromatic character. Using the information obtained by analyzing the tilt angle in the work of Shee and Datta (2015) made the following observations: pyridine is less aromatic than benzene; imidazole is less aromatic that pyridine and benzene; cyclopropane is aromatic, but cyclobutane is not aromatic.

Another method for evaluating aromaticity is by using the electric field gradient (EFG), which was first introduced to analyze the nuclear quadrupole resonance (NQR) in molecules (De Luca et al., 1999; Bailey, 2000, Amini et al., 2004; Elmi and Hadipour, 2005; Behzadi et al., 2008). This method is based on the electric field gradient tensor, which represents a symmetric tensor of second rank zero off-diagonal elements (Pakiari and Bagheri, 2011). The tensor can be express in an axis system, named the principal axis of the field gradient tensor, with non-zero compounds on its diagonal ($\partial^2 V/\partial z^2$, $\partial^2 V/\partial y^2$, $\partial^2 V/\partial x^2$) with the property (Lucken, 1969) that:

$$\partial^2 V / \partial z^2 + \partial^2 V / \partial y^2 + \partial^2 V / \partial x^2 = 0 \tag{6}$$

TABLE 4.1 The Values Obtained for Several Indexes of Aromaticity for Benzene in Interaction with Molecules of Type C_4H_4X (Shee and Datta, 2015)

Molecule	Confirmation	SA × 10²	Tilt angle	Mo-Schleyer resonance energy	NCIS(0)
BH	Angular	0.8842	95.4	-8.2	17.2
	Planar displaced		–		
AlH	Angular	0.6493	103.2	-3.5	6.9
	Planar displaced				
SiH₂	Angular	Not available	103.2	-0.8	0.05
	Planar displaced		–		
CH₂	Angular	0.4707	118.5	3.4	-4.2
	Staked				-2.34
O	Angular	0.2516	144.5	12.8	-13.9
	Staked		–		-2.14
S	Angular	0.1099	149.6	12.7	-14.7
	Staked		–		
SiH⁻	Angular on the same lone pair	Not available	163.4	17.2	Not relevant
	Angular on the opposite of the lone pair		158.2		
PH	Angular on the same lone pair	0.3005	158.4	17.9	Not relevant
	Angular on the opposite of the lone pair		90.5		
NH	Angular	0.0920	148.5	17.9	-17.3
CH⁻	Angular	3×10⁻⁸	158.8	19.1	-19.4

Starting from this, one can assume that the EFG value is given by the quantity:

$$q = \partial^2 V / \partial z^2 \tag{7}$$

with V as the electrostatic potential.

Applied to aromaticity, the existence of "delocalization" of π bonds in benzene was demonstrated by calculating the $EFG^{(0.5)}$ values of all six C-C bonds. Results showed the same value for all of them, 1.1986, leading to the following equation for determining the EFG π-delocalization contribution:

$$\sum EFG^{(0.5)} = \left(6 \times EFG^{(0.5)}\right)_{\text{of benzene}} - \left(6 \times EFG^{(0.5)}\right)_{\text{of ethane}}$$

Moving forward, $\Delta EFG^{(0.5)}$ for benzene will have the form:

$$\Delta EFG^{(0.5)} = 1.7076 - \left(3 \times 0.4743\right)_{\text{of ethylenic}} = 0.2874 \tag{8}$$

In their study, Pakiari and Bagheri (2011) determined that in the case of aromatic compounds, has positive values. On the other hand, by comparing the results from both $\Delta EFG^{(0)}$ and $\Delta EFG^{(0.5)}$, one can say that the cyclic π electron delocalization presents a small change when a substitution is made on benzene, this leading to a small change in aromaticity.

The authors also prove from the EFG studies that the aromaticity can be increased by the molecular symmetry in the case of benzene derivatives. Applied to PAHs, $\Delta EFG^{(0)}$ and $\Delta EFG^{(0.5)}$ values are presented in Table 4.2. In the case of naphthalene, the results show that none of its rings do not present values higher than in the case of benzene, for pyrene and anthracene the A ring have a higher aromaticity than the B ring, and for the coronene case, only the A ring is aromatic while the B ring is not. Nevertheless, *EFG* method has a range of error, due to the fact that σ-bond *EFG* contributions of benzene and ethane are allegedly equal, this error giving untrusting results for small ΔEFG of compounds from the borderline of the antiaromatic and aromatic domain (Pakiari and Bagheri, 2011).

4.4 NANO-CHEMICAL APPLICATION(S)

Graphene is one of the recent discovered materials, known as a honeycomb crystal lattice constituted of a planar sheet of one-atom-thick of carbon atoms (Novoselov et al., 2004, 2005), with specific properties, namely exceptional Young modulus values, high carrier mobilities and large force constants (Unarunotai et al., 2010; Morozov et al., 2008; Frank et al., 2007; Faccio et al., 2009) its chemical bonding having an important role in developing new materials based on graphene and graphene derivatives. There are two types of bonds in graphene: σ *type bond*, represented by a rigid honeycomb framework consisting in two-center two-electron (2c-2e) carbon-carbon σ-bonds, and the π-*type bond*, which is assumed to be delocalized. In his works, Kekule (1865, 1866a, 1866b) associate delocalized bonding in polycyclic hydrocarbons (especially benzene) with aromaticity, precisely with the existence of C_6 units. In his theory, for a compound to be aromatic, it was essential to be similar with benzene. Starting from this considerate, Moran et al., (2003) states that graphite presents higher resonance energy per π-electron and can be considered an ultimate expression of benzene, as a prototypical aromatic molecule.

TABLE 4.2 $\Delta EFG^{(0)}$ and $\Delta EFG^{(0.5)}$ Values for Several PAHs (Pakiari and Bagheri, 2011)

Molecule	$\Delta EFG^{(0)}$		$\Delta EFG^{(0.5)}$	
	Ring A	**Ring B**	**Ring A**	**Ring B**
Naphthalene	0.4666	–	0.1461	–
Anthracene	0.2803	0.1061	0.0384	0.0202
Phenanthrene	0.4712	0.1669	0.3303	0.2474
Pyrene	0.3930	0.0661	0.1996	0.0994
Coronene	0.1016	−0.5375	0.1010	−0.1081

Popov et al. (2012) present an analysis of graphene chemical bonding with the aid of a fragmental-based method. For this method three fragments were selected, $C_6H_6^{4+}$, $C_{24}H_{12}^{10+}$ and $C_{54}H_{18}^{16+}$, is based on the fact that for the graphene lattice each atom is presented only in three adjacent hexagons, so that a specific hexagon will have 1/3 of its π-electron or, in other words, two π-electron per hexagon. For calculation were used the distances 1.42 Å for fixed C-C [determined for graphite experimentally (Donohue, 1974)] and 1.10 Å for C-H, along with the adaptive natural density partitioning method (AdNDP) proposed by Zubarev and Boldyrev (2008a).

Results show that the $C_6H_6^{4+}$ fragment presents one complete delocalized 6c-2e-bond, six C-C classical σ bonds, and six C-H classical σ bonds; the $C_{24}H_{12}^{10+}$ fragment presents seven 6c-2e π-bonds, thirty 2c-2e C-C bonds, and twelve 2c-2e classical C-H σ bonds—a result different from the one obtained for its neutral correspondent (Zubarev and Boldyrev, 2008b); and the $C_{54}H_{18}^{16+}$ fragment nineteen 6c-2e π-bonds, seventy-two 2c-2e C-C bonds, and eighteen 2c-2e classical C-H σ bonds—in this type of fragment the π-bonding also being different from the neutral circumcoronene. All these calculations lead to the conclusion that graphene is indeed aromatic, but its aromaticity character is not the same as the one of the neutral species (e.g., benzene, coronene, and circumcoronene). In other words, even if there is no global π-delocalization, every hexagon ring has two π-electrons which give local aromaticity. Results also reveal that over the twelve carbon atoms, a π-electron pair is a completely delocalized eve if its density is placed over the hexagon in the proportion of 92–94% (Popov et al., 2012).

In the case of fullerenes, the aromaticity character cannot be determined by following the same rules as the other molecules. An example is a study conducted by J. Aihara and H. Kanno (2005) on C_{32} fullerene. Their starting point was the new magnetic criterion for aromaticity introduced by Schleyer et al., in 1996, namely the nucleus-independent chemical shift (*NICS*). According to his theory, aromatic rings will present negative *NICS* values (shielded rings), while antiaromatic rings will present positive *NICS* values (unshielded rings). Later on, in the same direction, Hirch et al. (Hirsch et al., 2000, 2001; Chen et al., 2002; Reiher and Hirsch, 2003) determined a large NICS value for neutral and charged fullerenes with $2(N+1)^2\pi$ electrons, on the cage center, and consider those fullerenes as highly aromatic, naming this magnetic behavior as the $2(N+1)^2$ rule of spherical aromaticity (Aihara and Kanno, 2005). On the other hand, topological resonance energy (TRE) is defined as a type of stabilization energy (Aihara, 1976), which applied to small fullerene cages with $2(N+1)^2\pi$ electrons contradict the spherical aromaticity rule. According to Chen et al., (2001, 2002), even if a fullerene has low symmetry, the spherical aromaticity is still present.

Applied to C_{32} fullerene isomers, which respect the $2(N+1)^2$ rule of spherical aromaticity for $N = 3$, if topological resonance energy (*TRE*) (Table 4.3) and the Hess-Schaad resonance energy (*HSRE*), (Table 4.2) are considered, one can conclude from their negative values that the fullerene isomers are mostly antiaromatic, meaning that the aromaticity cannot always be attributed when a large negative *NICS* value is determined at the center of a fullerene cage (Aihara and Kanno, 2005).

TABLE 4.3 Calculated Values for Resonance Energies and NICS for the Isomers of C_{32} Fullerene (Aihara and Kanno, 2005)

Isomer	TRE|β|	HSRE|β|	NCIS[a]
$^1(C_{32})$	− 0.559	− 0.461	− 22.9
$^2(C_{32})$	− 0.567	− 0. 461	− 24.2
$^3(C_{32})$	− 0.621	− 0.517	− 39.3
$^4(C_{32})$	− 0.254	− 0.165	− 48.8
$^5(C_{32})$	− 0.321	− 0.231	− 37.9
$^6(C_{32})$	− 0.133	− 0.049	− 53.2
$^7(C_{32})$	− 0.089	+ 0.025	–
$^8(C_{32})$	− 0.107	+ 0.007	–
$^9(C_{32})$	+ 0.035	+ 0.177	–
$^{10}(C_{32})$	− 2.263	− 0.953	–

[a]The NCIS values were calculated by Chen et al., (Chen and Thiel, 2003) at the GIAO–SCF/6–31G*//B3LYP–31G*.

4.5 MULTI-/TRANS-DISCIPLINARY CONNECTION(S)

Recent studies in the astrochemistry field have determined that not only PAHs are populating the interstellar medium, but also the fullerenes, and there are studies which suggest that the second one may be the most abundant carbon form from the interstellar space (Ehrenfreund and Foing, 1995). In their study, Salcedo and Fomina (2011) analyze the stability of fullerene and PAHs in order to establish which one is the most abundant in space.

As a working method, they evaluated the aromatic stabilization energy (*ASE*) in fullerene and adopted the alternative equations to become similar with homodesmotic reactions required for PAHs. This principle is based on two considerations:

- PAHs molecules have quaternary sp^2 carbons, while fullerene have carbon atoms with a certain degree of pyramidalization; and
- the intrinsic aromaticity of the sphere in fullerene needs to be taking into consideration even if it is not a characteristic for PAHs molecules.

An indirect estimation for estimating the spherical aromaticity of fullerene can be made using the following homodesmotic reaction:

$$20\ CH_2 = CH - CH = CH_2 + 10\ (CH_2 = CH)_2\ C = C(CH = CH_2)_2$$
$$\downarrow$$
$$C_{60} + 60\ CH_2 = CH_2 \tag{9}$$

The correspondent homodesmotic reaction used in a study the PAHs molecules has the following form:

$$\frac{2a-b}{2}CH_2 = CH - CH = CH_2 + (a-b)C_4H_8$$
$$\downarrow$$
$$C_aH_b + \frac{a}{2}CH_2 = CH_2 + 2(a-b)C_3H_6 \tag{10}$$

The obtained results prove that the fullerene is indeed more abundant in the interstellar space than PAHs, due to the fact that presents a high ionization potential, its ions are more stable, and its tendency to react and transform in large molecules is limited.

The aromaticity concept can also be extended on chars (Wiedemeier et al., 2015). Known for their ability to improve fertility (Biederman and Harpole, 2013) and to immobilize hazardous compounds (Beesley et al., 2011), chars can also be used in producing green energy (biochars), offering an ecological perspective to the concept of organic waste management (Meyer et al., 2011). One of their defining property is represented by the aromatic C structure (Lehmann and Joseph, 2009), divided into two distinct aromatic C phases: an amorphous phase (randomly organized aromatic rings) and a crystalline phase (condensed polyaromatic sheets turbostratically aligned) (Franklin, 1951; Cohen-Ofri et al., 2006; Keiluweit et al., 2010). There are several methods used in measuring the aromaticity in chars, such as: spectroscopic assessment of functional groups (MIR_{index}), measurement of aromatic molecular markers ($BPCA_{index}$, PAH_{index}) or determination of elemental composition ($O-C_{index}$, $H-C_{index}$) (Wiedemeier et al., 2015). The MIR_{index} can be used as a screening method for unknown char samples categorization, because of its ability to distinguishing the high-temperature chars from the low-temperature ones (Harvey et al., (2012a). Also for distinguishing the high-temperature grass chars from the wood chairs, one can use the $BPCA_{index}$, but the $BPCA$ is a method which quantifies and separate aromatic moieties derived from condensed aromatic structures, so one cannot use it on lignin. On the other hand, the $O–C_{index}$ and $H–C_{index}$ are used in indirect measurements of aromaticity, based on the fact that in a sample, the C proportion is related to HTT (heat treatment temperature), allowing H and O structures to be formed from dehydration, depolymeriza- tion volatilization phenomena (Keiluweit et al., 2010; Wang et al., 2013). $H–C_{index}$ index can also be used in distinguishing between wood and grass char when the temperature interval is low (Rutherford et al., 2012). Overall, studies on char aromaticity (Wiedemeier et al., 2015) have proven to be of a high interest for a better understanding of their sequestration potential or for determining the optimal period in which they can be beneficial to the soil (Nguyen et al., 2010).

Another class of compounds, the methylated derivatives of xanthine (caffeine, theobromine, paraxanthine, and theophylline) are well known for their mild stimulant properties and are found in chocolate, tea, or coffee. In order to analyze their aromaticity, there are considered two characteristics: they participate in π-stacking interactions, and the heavy atom presents a planar conformation. Based on the magnetic criterion, the aromatic character can be confirmed if the ring currents can be visualized and calculated (Gibson and Fowler, 2014). A new method proposed for this type of calculations is represented by the isocentric approach, proposed by Keith and Bader (Steiner and Fowler, 2001a; Keith and Bader, 1993) which can be applied to π systems to obtain a physically realistic current pattern. The specific characteristic of this method is given by the fact that one can obtain a frontier-orbital theory from the aromaticity perspective due to the fact that the calculated current density separated in non-redundant orbital contributions (Steiner and Fowler, 2001a, 2004). The symmetry of the HOMO-LUMO transition can be used to determine the type of current as follow: if the transition from an occupied orbital to an empty orbital preserve the angular nodes number than the system present two-electron paratropic current and is considered to be antiaromatic; on the other hand, if the transition increase by one the angular nodes number than the system present 4-electron diatropic current is considered to be aromatic (Steiner and Fowler, 2001b). Applied to their set of molecules, Gibson and Fowler (Gibson and Fowler, 2014) determine that around their imidazole subunits there are strong delocalized π ring currents, and by considering separately the localized group and delocalized ring contributions, one can construct the total π response for each molecule, with applications in other fields (Gibson and Fowler, 2014).

4.6   OPEN ISSUES

Although the aromaticity concept is used in many studies for rationalizing chemical bonding, classifying empirical data or predicting reactivity and stability for the organic compounds, there is a new trend among scientist to use it also on inorganic molecules, knowing that for inorganics, the planarity of molecule, which is a condition of aromatic character, is no longer applied. As a consequence, for organic aromatic substitution reactions were developed inorganic analogs, leading to designing of new materials (Zhen, 2006).

KEYWORDS

- **aromatic C structure**
- **benzene**
- **electron localization function**
- **graphene**
- **nucleus-independent chemical shift**

REFERENCES AND FURTHER READING

Aihara, J., (1976). A new definition of Dewar-type resonance energies. *Journal of the American Chemical Society, 98*, 2750–2758.

Aihara, J., & Kanno, H., (2005). Aromaticity of C_{32} fullerene isomers and the 2(NC1)2 rule. *Journal of Molecular Structure: THEOCHEM, 722*, 111–115.

Amini, S. K., Hadipour, N. L., & Elmi, F., (2004). A study of hydrogen bond of imidazole and its 4–nitro derivative by ab initio and DFT calculated NQR parameters. *Chemical Physics Letters, 391*, 95–100.

Armit, J. W., & Robinson, R., (1925). CCXI—Polynuclear heterocyclic aromatic types, Part II, Some anhydronium bases. *Journal of the Chemical Society, 127*, 1604–1618.

Bailey, W., (2000). DFT and HF-DFT calculations of 14N quadrupole coupling constants in molecules. *Chemical Physics, 252*(1/2), 57–66.

Behzadi, H., Spoel, D., Esrafili, M. D., Parsafar, G. A., & Hadipour, N. L., (2008). Role of spin state on the geometry and nuclear quadrupole resonance parameters in Hemin complex. *Biophysical Chemistry, 134*, 200–206.

Chen, Z., Jiao, H., & Hirsch, A., (2002). Spherical aromaticity–overview development. *Fullerene Science, 4*, 121–135.

Chen, Z., Jiao, H., Hirsch, A., & Thiel, W., (2001). The $2(N+1)^2$ rule for spherical aromaticity: A further validation. *Journal of Molecular Modeling, 7*, 161–163.

De Luca, G., Russo, N., Koster, A. M., & Calaminici, P., Ju, K., (1999). Density functional theory calculations of nuclear quadrupole coupling constants with calibrated O-17 quadrupole moments. *Molecular Physics, 97*, 347–354.

Donohue, J., (1974). *The Structures of the Elements*. Wiley-Interscience: New York.

Ehrenfreund, P., & Foing, B. H., (1995). Search for fullerenes and PAHs in the diffuse interstellar medium. *Planetary and Space Science, 43*, 1183–1187.

Elmi, F., & Hadipour, N. L., (2005). A study on the intermolecular hydrogen bonds of α–glycylglycine in its actual crystalline phase using ab initio calculated 14N and 2H nuclear quadrupole coupling constants. *The Journal of Physical Chemistry A., 109*, 1729–1733.

Elvidge, J. A., & Jackman, L. M., (1961). Studies of aromaticity by nuclear magnetic resonance spectroscopy. Part I. 2–Pyridones and related systems. *Journal of the Chemical Society*, 859–866.

Faccio, R., Denis, P. A., Pardo, H., Goyenola, C., & Mombru, A. W., (2009). Mechanical properties of graphene nanoribbons. *Journal of Physics: Condensed Matter, 21*, 285304.

Frank, I. W., Tanenbaum, D. M., Van der Zanda, A. M., & McEuen, P. L., (2007). Mechanical properties of suspended grapheme sheets. *Journal of Vacuum Science & Technology, B25*, 2558–2561.

Gibson, C. M., & Fowler, P. W., (2014). Aromaticity of caffeine, xanthine, and the dimethyl xanthines. *Tetrahedron Letters, 55*, 2078–2081.

Hirsch, A., Chen, Z., & Jiao, H., (2000). Spherical aromaticity in I_h symmetrical fullerenes: The $2(N+1)^2$ rule. *Angewandte Chemie International Edition, 39*, 3915–3917.

Hirsch, A., Chen, Z., & Jiao, H., (2001). Spherical aromaticity of inorganic cage molecules. *Angewandte Chemie International Edition, 40*, 2834–2838.

Keith, T. A., & Bader, R. F. W., (1993). Calculation of magnetic response properties using a continuous set of gauge transformations. *Chemical Physics Letters, 210*, 223–231.

Kekulé, A., (1865). Sur la constitution des substances aromatiques. *Bulletin de la Société Chimique de France (Paris), 3*, 98–110.

Kekulé, A., (1866a). Note sur quelques produits de substitution de labenzene. *Bulletins de l'Academie Royale des Sciences, des Lettres et des Beaux Arts de Belgique, 119*, 551–563.

Kekulé, A., (1866b). Untersuchungen über aromatische Verbindungen. *Justus Liebigs Annalen der Chemie, 137*, 129–136.

Kistiakowsky, G. B., Ruhoff, J. R., Smith, H. A., & Vaughan, W. E., (1936). Heats of organic reactions. IV. Hydrogenation of some dienes and of benzene. *Journal of the American Chemical Society, 58*, 146–153.

Krygowski, T. M., Cyranski, M. K., Czarnocki, Z., Haefelinger, G., & Katritzky, A. R., (2000). Aromaticity: A theoretical concept of immense practical importance. *Tetrahedron, 56*, 1783–1796. Tetrahedron Report no. 520.

Lucken, E A. C., (1969). *Nuclear Quadrupole Coupling Constants*. Academic, London.

Moran, D., Stahl, F., Bettinger, H. F., Schaefer, H. F. III., & Schleyer, P. V. R., (2003). Towards graphite: Magnetic properties of large polybenzenoid hydrocarbons. *Journal of the American Chemical Society, 125*, 6746–6752.

Morozov, S. V., Novoselov, K. S., Katsnelson, M. I., Schedin, F., Elias, D. C., Jaszczak, J. A., & Geim, A. K., (2008). Giant intrinsic carrier mobilities in graphene and its bilayer. *Physical Review Letters, 100*, 016602.

Novoselov, K. S., Geim, A. K., Morozov, S. V., Jiang, D., Katsnelson, M. I., Grigorieva, I. V., Dubonos, S. V., & Firsov, A. A., (2005). Two-dimensional gas of massless Dirac fermions in graphene. *Nature, 438*, 197–200.

Novoselov, K. S., Geim, A. K., Morozov, S. V., Jiang, D., Zhang, Y., Dubonos, S. V., Grigorieva, I. V., & Firsov, A. A., (2004). Electric field effect in atomically thin carbon films. *Science, 306*, 666–669.

Pakiari, A. H., & Bagheri, N., (2011) A new approach for aromaticity criterion based on electrostatic field gradient. *Journal of Molecular Modeling, 17*, 2017–2027.

Popov, I. A., Bozhenko, K. V., & Boldyrev, A. I., (2012). Is Graphene Aromatic? *Nanoresearch, 5*(2), 117–123.

Putz, M. V., (2010a). On absolute aromaticity within electronegativity and chemical hardness reactivity pictures. *MATCH Communications in Mathematical and in Computer Chemistry, 64*, 391–418.

Putz, M. V., (2010b). Compactness aromaticity of atoms in molecules. *International Journal of Molecular Science, 11*, 1269–1310.

Putz, M. V., & Tudoran, M. A., (2016). Carbon-based specific adjacency-in-bonding (SAIB) isomerism driving aromaticity. *Fullerenes, Nanotubes and Carbon Nanostructures*, accepted the paper.

Reiher, M., & Hirsch, A., (2003). From rare gas atoms to fullerenes: Spherical aromaticity studied from the point of view of atomic structure theory. *Chemistry–A European Journal, 9*, 5442–5452.

Salcedo, R., & Fomina, L., (2011). Aromaticity under star dust conditions. *Structural Chemistry, 22*, 971–975.

Schleyer, P. V. R., Maerker, C., Dransfeld, A., Jiao, H., & Hommes, N. J. R. V. E., (1996). Nucleus independent chemical shifts: A simple and efficient aromaticity probe. *Journal of the American Chemical Society, 118*, 6317–6318.

Shee, N. K., & Datta, D., (2014). Nucleus independent chemical shifts and side dependent aromaticity of phospholes and their anionic silicon counterparts. *Computational and Theoretical Chemistry, 1046*, 70–72.

Sondheimer, F., (1964). Recent advances in the chemistry of large-ring conjugated systems. *Pure and Applied Chemistry, 7*, 363–388.

Steiner, E., & Fowler, P. W. J., (2001a). Patterns of ring currents in conjugated molecules: A few-electron model based on orbital contributions. *The Journal of Physical Chemistry, A105*, 9553–9562.

Steiner, E., & Fowler, P. W., (2001b). Four- and two-electron rules for diatropic and paratropic ring currents in monocyclic π systems. *Chemical Communications, 21*, 2220–2221

Steiner, E., & Fowler, P. W., (2004). On the orbital analysis of magnetic properties. *Physical Chemistry Chemical Physics, 6*, 261–272

Tudoran, M. A., & Putz, M. V., (2015). Molecular graph theory: From adjacency information to colored topology by chemical reactivity. *Current Organic Chemistry, 19*, 359–386.

Unarunotai, S., Murata, Y., Chialvo, C. E., Mason, N., Petrov, I., Nuzzo, R. G., Moore, J. S., & Rogers, J. A., (2010). Conjugated carbon monolayer membranes: Methods for synthesis and integration. *Advanced Materials, 22*, 1072–1077.

Wiedemeier, D. B., Abiven, S., Hockaday, W. C., Keiluweit, M., Kleber, M., Masiello, C. A., et al., (2015). Aromaticity and degree of aromatic condensation of char. *Organic Geochemistry, 78*, 135–143.

Zhen, H., (2006). Aromaticity in inorganic chemistry: The Hückel and Hirsch rules. *Literature Seminar*.

Zubarev, D. Y., & Boldyrev, A. I., (2008a). Developing paradigms of chemical bonding: Adaptive natural density partitioning. *Physical Chemistry Chemical Physics, 10*, 5207–5217.

Zubarev, D. Y., & Boldyrev, A. I., (2008b). Revealing intuitively assessable chemical bonding patterns in organic aromatic molecules via adaptive natural density partitioning. *The Journal of Organic Chemistry, 73*, 9251–9258.

CHAPTER 5

Bose-Einstein Condensate

ARANYA B. BHATTACHERJEE

Department of Physics, Birla Institute of Technology and Science, Pilani, Hyderabad Campus, India

5.1 DEFINITION

Bose-Einstein condensate (BEC) is a state of matter in which all the neutral atoms or subatomic particles are cooled down to a temperature very near to absolute zero (i.e., 0 K). At such a low temperature, the thermal de-Broglie wavelength of the individual atoms starts overlapping with each other which results into the phase transition of the system such that a large number of bosons occupies the lowest energy level of the system or the lowest quantum state. In such a case, atoms can be illustrated by a single wave function on a near macroscopic scale. Thus, quantum effects become apparent on a macroscopic scale at the critical temperature which leads to macroscopic quantum phenomena.

5.2 HISTORICAL ORIGIN(S)

This state of matter was first predicted by Satyendra Nath Bose and Albert Einstein in 1924–25. The Indian physicist S. N. Bose was the first to describe the phenomena of Black Body Radiation. In the same year, Albert Einstein extended the work done by S. N. Bose to non-interacting particles. The efforts made by Bose and Einstein introduced the novel idea of a Bose gas obeying the quantum statistics known as Bose-Einstein statistics. These statistics depict the distribution of identical particles having integral spin which are now well-recognized as bosons. Bosons comprise the photon as well as atoms such as helium-4 (^{4}He) that are allowed to share quantum states with each other. In this field, Eric A. Cornell, Wolfgang Ketterle, and Carl E. Wieman got the Noble Prize in Physics in 2001 for their outstanding work. In 2010, the first photon Bose-Einstein condensate was formed.

5.3 NANO-SCIENTIFIC DEVELOPMENT(S)

Cavity quantum optomechanics is an emerging field which controls the quantum mechanical interaction between electromagnetic radiation and bulk mechanical resonators. This field acts a boundary between quantum optics and nanoscience. The mechanical oscillator at the scale of micro- and nanometer are commonly used for a wide variety of applications. They are being normally employed as sensors or actuators in integrated optical, electrical and optomechanical systems. The optomechanical cavity is basically an optical cavity in which one of the mirrors is mechanically compliant. The simplest optomechanical system in which radiation pressure provides the dominant optomechanical coupling is a Fabry-Perot cavity with one mirror fixed and another mirror movable (Figure 5.1).

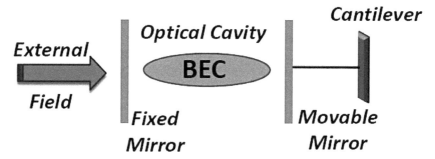

FIGURE 5.1   Schematic representation of the optomechanical. It involves the Bose-Einstein condensate inside a lossless optical cavity driven by an external field. One of the cavity mirrors is movable.

Systems composed of ultracold atoms or BEC inside a high-finesse cavity coupled to the radiation field are the new types of hybrid nano-optomechanical systems. One of the significant applications of an optomechanical cavity system is recognized as weak force detection.

Weak force measurement is a significant task for nanomechanical systems, both, in experiments of quantum mechanics and when using devices as sensitive detectors. The weak force may arise due to several things, including a nuclear or electronic spin, or a single charge of an electron. The fundamental problem is to find how big the force must be in order to see its consequence on the displacement of the mechanical resonator from its equilibrium position in the presence of noise. For example, the external weak force can be due to quantum fluctuations in the radiation pressure in a quantum experiment which are usually overlapped by background thermal noise.

Experimentally, several experiments have set hallmark in achieving fine force sensitivity. The primary interest of a wide range of applications like magnetometry of nanoscale magnetic particles, femtojoule calorimetry, and other numerous types of force microscopy is to detect weak forces using a cantilever.

5.4 NANO-CHEMICAL APPLICATION(S)

A possible nanochemical application is a new form of calorimeter coupled with an opto-nano-mechanical hybrid system for use in gaseous and vacuum environments which can sense chemical reactions. Heat fluxes generated from the chemical reaction can be detected by measuring the cantilever deflection induced by the differential thermal expansion of the lever (bimetallic effect) using the opto-nano-mechanical hybrid system.

5.5 MULTI-/TRANS-DISCIPLINARY CONNECTION(S)

Cavity optomechanical system consisting of BEC has extremely high potential applications in the areas of quantum communications and quantum information processing. A strong quantum entanglement can be achieved between the mechanical mode of the BEC and the optical mode. This strongly correlated two-mode optomechanical system is very useful in the next generation of quantum communications and quantum information processing units since the mechanical modes can store energy for a longer duration and the optical modes can exchange energy over long distances.

5.6 OPEN ISSUES

One of the major limitations is that the apparatus involving BEC are extremely expensive and bulky. Moreover, the experiments are very sensitive towards external environments like temperature and pressure. The real challenge is to make compact hybrid nano-opto-mechanical devices.

KEYWORDS

- **Bose-Einstein condensate**
- **cavity opto-mechanics**
- **nano-opto-mechanical system**
- **quantum degenerate gas**

REFERENCES AND FURTHER READING

Aggarwal, N., Mahajan, S., & Bhattacherjee, A. B., (2013). Optomechanical effect on the Dicke quantum phase transition and quasi-particle damping in a Bose-Einstein condensate: A new tool to measure weak force. *Journal of Modern Optics, 60,* 1263.

Aggarwal, N., Mahajan, S., Bhattacherjee, A. B., & Mohan, M., (2014). Excitations of optomechanically driven Bose-Einstein condensate in a cavity: Photo-detection measurements. *Chinese Physics B., 23,* 100305.

Anderson, M. H., Ensher, J. R., Matthewes, M. R., Wieman, C. E., & Cornell, E. A., (1995). Observation of Bose-Einstein condensation in a Dilute Atomic Vapor. *Science, 269,* 198.

Arlett, J. L., Maloney, J. R., Gudlewski, B., Muluneh, M., & Roukes, M. L., (2006). Self-sensing micro- and nanocantilevers with Attonewton-scale force resolution, *Nano Letters, 6,* 1000.

Bhattacherjee, A. B., Aggarwal, N., & Mahajan, S., (2014). Recent trends in nano-optomechanical systems. In: *Quantum Nanosystems, Structure, Properties and Interactions.* Apple Academic Press (New Jersey, USA).

Davis, K. B., Mewes, M. O., Andrews, M. R., Van Druten, N. J., Durfee, D. S., Kurn, D. M., & Ketterle, W., (1995). Bose-Einstein condensation in a gas of sodium atoms. *Phys. Rev. Lett., 75,* 3969.

Gimzewski, J. K., Gerber, C., Meyer, E., & Schlittler, R. R., (1994). Observation of a chemical reaction using a micromechanical sensor. *Chem. Phys. Lett., 217,* 589.

Mahajan, S., Aggarwal, N., Bhattacherjee, A. B., & Mohan, M., (2013). Achieving the quantum ground state of a mechanical oscillator using a Bose-Einstein condensate with back-action and cold damping feedback schemes. *J. Phys. B: At. Mol. Opt. Phys., 46,* 085301.

Mahajan, S., Kumar, T., Bhattacherjee, A. B., & Mohan, M., (2013). Ground-state cooling of a mechanical oscillator and detection of a weak force using a Bose-Einstein condensate. *Phys. Rev. A., 87,* 013621.

Teufel, J. D., Donner, T., Castellanos-Beltran, A., Harlow, J. W., & Lehnert, K. W., (2009). Nanomechanical motion measured with an imprecision below that at the standard quantum limit. *Nature Nanotechnology, 4,* 820.

CHAPTER 6

Bondonic Theory

MIHAI V. PUTZ[1,2]

[1]*Laboratory of Structural and Computational Physical Chemistry for Nanosciences and QSAR, Biology-Chemistry Department, West University of Timisoara, Pestalozzi Street No. 44, Timisoara, RO-300115, Romania, Correspondent author: Tel.: +40-256-592638, Fax: +40-256-592620, E-mail: mv_putz@yahoo.com, mihai.putz@e-uvt.ro*

[2]*Laboratory of Renewable Energies-Photovoltaics, R&D National Institute for Electrochemistry and Condensed Matter, Dr. A. Paunescu Podeanu Str. No. 144, Timisoara, RO-300569, Romania*

6.1   DEFINITION

The bondons describe chemical particles, which are associated with electrons implicated in the chemical bond, as single, delocalized, or lone pair; they have orientated direction of movement and depend on a chemical field. Another definition states that the bondon is a quantum particle, which is correspondent to the effect or distribution of superimposed electronic pairing in chemical bonds, having a behavior similar with a neutrino in mass, with a photon in velocity and whit an electron in charge. For a nanosystem, it presents an observable lifetime at the fem to second level and a chemical bonding at an order of hundred Å and thousands of kcal/mol. The bondons can be found on the wave functions, on the covalent bonds, in dispersive–weak interactions (as in the ADN), on the mechanism of action between ligand and receptor, and even on the ionic interactions.

6.2   HISTORICAL ORIGIN(S)

The first attempt (Putz and Ori, 2015) in describing the chemical bonding belongs to Lewis (1916) who proposes a cubic model as representing the atoms in the molecule and develops the octet rule. A couple of years later, Langmuir (1919) developed the hypovalent (such as BF_3) and hypervalent (such as PF_5) molecules which represent the basis of the double quartet theory of Linnett (1961). Later, starting from octet and the electron pairing rules specific in bonding was first developed the intuitive valence shell electron pair repulsion (VESPR) model (Gillespie and Popelier, 2001) and later the electron localization function (ELF) model (Becke and Edgecombe, 1990; Putz, 2005). In order to understand the chemical bonding from the quantum perspective, Putz (2010, 2012a) propose the so-called *bondon* as a quasi-particle which correspond to the chemical field of bonding, as a new boson to represent the chemical bond (Putz and Ori, 2015).

6.3   NANO-SCIENTIFIC DEVELOPMENT(S)

In his work, Putz (2012a) developed an algorithm in 15 steps (Putz, 2010, 2012a, 2012b). In the beginning, in the *first step*, one can start with the quantum Schrodinger equation of motion for which the Broglie/Bohm electronic wave-function can be defined as (de Broglie, 1923, 1925, 1953, 1954):

$$\Psi_0(t,x) = R(t,x)\exp\left(i\frac{S(t,x)}{\hbar}\right) \tag{1}$$

where S represents the phase action factor and R represents the amplitude, and in terms of the total energy E, the electronic density ρ and the momentum p have the formulas (Putz, 2012a):

$$R(t,x) = \sqrt{\Psi_0(t,x)^2} = \rho^{1/2}(x) \tag{2}$$

$$S(t,x) = px - Et \tag{3}$$

The *second step* (Putz, 2012a) assumes to verify the charge current conservation law recovery, which is responsible with electronic fields' circulatory nature, as in equality:

$$\frac{\partial \rho}{\partial t} + \nabla \vec{j} = 0 \tag{4}$$

In the *third step* (Putz, 2012a), the quantum potential V_{qua} can be recognized, if appears, along with its correspondent equation:

$$V_{qua} = \frac{\hbar^2}{2m}\frac{\nabla^2 R}{R} \tag{5}$$

The electronic wave function is depicted in *step four* (Putz, 2012a), in term of the chemical field $\aleph$, as follows:

$$\Psi_G(t,x) = \Psi_0(t,x)\exp\left(\frac{i}{\hbar}\frac{e}{c}\aleph(t,x)\right) \tag{6}$$

with e the standard abbreviation with the formula:

$$e = e_0^2 \,/\, 4\pi\varepsilon_0 \tag{7}$$

Starting from the following chain equivalence:

$$\aleph_{\not B} \sim \frac{energy \times distance}{charge} \sim \frac{\left(charge \times \dfrac{potential}{difference}\right) \times distance}{charge} \sim \left(\frac{potential}{difference}\right) \times distance \tag{8}$$

one can express the bondons, carried by the chemical bonding field along the distance of bonding, having $\hbar c/e$ as the unit quanta. In *step five,* one can separate the terms with imaginary and real $\aleph$ chemical field in order to rewrite the equation of wave-function with the group object (Ψ_G). The chemical field charge current is identified in *step six*, which is the term in the context of the transformation group. For the Broglie-Bohm wave function, Ψ_0 from the Steps (1)–(3) one need to establish, in *step seven,* the global/local gauge transformations which is similar to it and in *step eight*, on the pattern quantum equation is imposed an invariant condition for the wave function Ψ_G. Next, the $\aleph$ chemical field equations are established (*step nine*), and the chemical field system is solved (*step ten*). The chemical bond at equilibrium is depicted in *step eleven* by assessing the stationary chemical field as follow (Putz, 2012a):

$$\frac{\partial\aleph}{\partial t} \equiv \partial_t\aleph = 0 \tag{9}$$

Along the bonding distance one can identify, in *step twelve*, the manifested Bondonic chemical field, noted with $\aleph_{bondon}$. After that, *step thirteen,* the possible flux condition of Bader is verified within the vanishing chemical bonding field (Putz, 2012; Bader, 1990), as follows:

$$\aleph_{\not B} = 0 \Leftrightarrow \nabla\rho = 0 \tag{10}$$

In *step fourteen* (Putz, 2012a), the velocity of electrons in bonding is obtained by the Heisenberg relation of time-energy relaxation-saturation through their kinetic energy:

$$v = \sqrt{\frac{2T}{m}} \sim \sqrt{\frac{2}{m}\frac{\hbar}{t}} \tag{11}$$

The bondons' mass is determined in step fifteen by equating the Bondonic chemical field along with the chemical field quanta $\aleph_0$, and the following equality is obtained (Putz, 2012):

$$\aleph_{\not{B}}(m_{\not{B}}) = \aleph_0 \tag{12}$$

By knowing the mass and velocity values for bondon, one can indicate, in general, which is the electronic behavior in bonding with the possibility of determining, in particular, the exact type of bonding (Putz et al., 2015). In order to predict, in the ground state, which are the values of energy and length necessary for a bondon to reach the unity of mass or its electronic charge (Putz, 2010, 2012a, 2012c, 2012d, 2012e), one needs to employ the following equations:

$$\varsigma_m = \frac{m_{\not{B}}}{m_0} = \frac{87.8603}{\left(E_{bond}[kcal/mol]\right)\left(X_{bond}\left[\mathring{A}\right]\right)^2} \tag{13}$$

$$\varsigma_e = \frac{e_{\not{B}}}{e} \sim \frac{1}{32\pi}\frac{\left(E_{bond}[kcal/mol]\right)\left(X_{bond}\left[\mathring{A}\right]\right)}{\sqrt{3.27817 \times 10^3}} \tag{14}$$

Results shows that, in the case of common chemical bonds, there is a small probability that the bondon will have the same charge as the electron, due to the fact that lower binding energy can span a longer distance having the same unit charge of electron and in the same time, being transmitted with the same relativistic velocity, phenomena also known as zitterbewegung.

6.4   NANO-CHEMICAL APPLICATION(S)

The link between the behavior of elements in the extended nanosystems such graphite layers and the macroscopic observable quantities can be realized with the aid of bondon. In this context, the bondon can be defined as the bosonic counterpart of the wave function of the bonding electrons (Putz, 2010, 2012a). In the ground state, the bondonic mass related with bond energy and length has the following form (Putz and Ori, 2012):

$$M_B = \frac{\hbar^2}{2E_{bond}X_{bond}^2} \tag{15}$$

For extended bonding systems, there is available an interdependency constraint of the Heisenberg type:

$$E_{bond}\left[kcal\,/\,mol\right]X_{bond}\left[\mathring{A}\right]=\alpha,\alpha=182,019 \tag{16}$$

By taking into consideration the previous formula (Eq. 16) one can obtain, for a nanosystem with hundred atoms and the energy of about 1000 kcal/mol, the value of 182.019 Å will be the spanned maximum bonding. This value can be also detected for a nanotube with the length equal with 18 nm across its end. In the same time, the bosonic condensation of bondons determines a bonding system to support along 100 Å, for example, along with a nanotube of 10 nm length, which is maximum of 1820 kcal/mol (Putz and Ori, 2012).

Due to the fact that, at the quantitative and qualitative description, the bondons support nonlocality spreading, one can be able to study the bondonic properties at extended systems, with further applications in nanomaterials engineering, e.g., silicone as nanostructure used in hydrogen storage (Putz and Ori, 2012; Jose and Datta, 2011).

In this context, the total bonding energy can be expressed as a sum of the thermal energy considered as bondon kinetic motion along with the topological contribution which describes the nanostructure physical information (Putz and Ori, 2012). The working equation is of the form:

$$\left\langle E_{Bond}\right\rangle\left[kcal\,/\,mol\right]=-\frac{\partial\ln z_{B}^{[0]}\left(\beta,E_{Bond}\right)}{\partial\beta}$$
$$=\frac{1}{2\beta}+\Xi=\begin{cases}\Xi,\dots\beta\to\infty\left(T\to0K\right)\\\infty,\dots\beta\to0\left(T\to\infty K\right)\end{cases} \tag{17}$$

With the specification that, i.e., there is a correspondence $V(x)\to\Xi$, i.e., for the periodic cell, there is a correspondence between driving potential and the topological descriptors.

In their work, Putz and Ori (2012) describe which is the role played by bondons in formation of Stone-Wales (SW) defects on the graphene fragments (e.g., defect-defect interactions and boundary conditions) using in this study only connectivity topological data to express specifically physical information instead of considering the density of the electronic distribution of the system. Defective hybrid nanoribbons systems were confirmed by first principle calculations (Dutta, 2009), with a single central SW defect in each super-cell, obtained after one diagonal bond is rotated in the net (Putz and Ori, 2012).

6.5 MULTI-/TRANS-DISCIPLINARY CONNECTION(S)

For a given system, the spontaneous symmetry breaking (SSB) determines its stable state to tend towards a different configuration which is characterized by the fact that possesses a lower symmetry than its Hamiltonian or Lagrangian symmetry correspondent, this characteristic being applied for both quantum and classical structures. The SSB method can be used in different types of theoretical Physics subjects, especially in modeling physicochemical phenomena. In honeycomb systems, the SSB effect determines the gap formation in the fermionic spectra (Araki, 2011), which is a similar phenomenon with the spontaneous break of the chiral symmetry from the theory of the strong coupling relativistic field. The SSB mechanism, same as the Higgs-like model states, is able to produce massive bosonic particle which can be used, in quantum systems, to mold the physical behavior. In their work, Putz and Ori (2015a) propose an original SSB algorithm designated to produce a certain type of Goldstone particles, i.e., with a mass related to the mass obtained by considering together all the electrons participating in the chemical bond. Apart from their typical occurrence, by symmetry breaking of the Lagrangian in terms of components of the chemical field, these Goldstone bosons can also be identified using the bondons, known as the quanta of the chemical bonding field (Putz, 2010, 2012a). This approach proposes the Goldstone symmetry breaking (GSB) mechanism as a method to obtain the general law of electronic and bondonic masses, applied to a set of several benzenoids for which their bondonic mass values were determined. For the bondonic mass, there are two topological descriptors used to construct its analytical expression: the sphericity (topological efficiency) index, noted with ρ, and the E-sphericity (extreme topological efficiency) index, noted with ρ^E. Starting from these considerate, one can define, first, the Lagrangian of the Schrodinger field (φ) which is produced by the potential V as the main index of the GSB algorithm, following the formula:

$$L = i\hbar \phi^* \dot{\phi} - \frac{\hbar^2}{2m_0}\left(\nabla\phi^*\right)\left(\nabla\dot{\phi}\right) - V\phi^*\dot{\phi} \qquad (18)$$

Using the ρ and ρ^E indices, one can also express the relationship between the potential φ^4 and the chemical field which is rooted in the chemical reactive/valence energy parabolic expression (Parr and Yang, 1989; Putz, 2011) as follow:

$$V\left(\phi\right) = -\rho\phi^2 + \frac{1}{2}\rho^E\phi^4 \qquad (19)$$

After several mathematical artifices, the chemical bonding field (φ) is given by the solution of the following differential equation:

$$i\hbar\dot{\phi} = 3\rho^E\phi^5 + \frac{15}{2}\rho\sqrt{\frac{\rho^E}{\rho}}\phi^4 + 6\rho\phi^3 + \frac{3\rho^2}{2\rho^E}\sqrt{\frac{\rho^E}{\rho}}\phi^2 - \frac{\rho^2}{\rho^E}\phi - \frac{\rho^3}{2(\rho^E)^2}\sqrt{\frac{\rho^E}{\rho}} \quad (20)$$

Considering the Hartree's atomic unit of energy of the form $Ha = 27.11$ eV, the spontaneous breaking symmetry bonding field can be rewritten in the expanded form of the first order:

$$\phi_B = \phi_\infty\left[1 + 6Ha\beta\left(\frac{1}{2}\rho^E\phi_\infty^4\right)\right] \quad (21)$$

The bondonic field can also be expressed in a simpler manner, as depending only on the length of the chemical field action, the E-sphericity and with the temperature considered an intensive physical parameter, with the expression:

$$\phi_B = e^{-\frac{r}{a_0}\sqrt{2\rho^E}}\left[1 + 3\beta\rho^E e^{-4\frac{r}{a_0}\sqrt{2\rho^E}}\right] \quad (22)$$

In the same manner, the bondonic mass can be determined to start from the expression:

$$m_B = \frac{m_0}{a_0}\int_0^\xi |\phi_B|^2\,dr \quad (23)$$

with obtaining its integrated form:

$$m_B = \frac{m_0}{a_0}\int_0^\xi e^{-2\frac{r}{a_0}\sqrt{2\rho^E}}\left[1 + 3\beta\rho^E e^{-4\frac{r}{a_0}\sqrt{2\rho^E}}\right]^2\,dr \quad (24)$$

$$= \frac{m_0}{10\sqrt{2\rho^E}}\left[5\left(1 - e^{-2\frac{\xi}{a_0}\sqrt{2\rho^E}}\right) + 10Ha\beta\rho^E\left(1 - e^{-6\frac{\xi}{a_0}\sqrt{2\rho^E}}\right) + 9(Ha)^2\beta^2(\rho^E)^2\left(1 - e^{-10\frac{\xi}{a_0}\sqrt{2\rho^E}}\right)\right]$$

Applied to a set of polycyclic aromatic hydrocarbons (PAH) molecules, the results show that the bondons are formed in molecules with $\rho \neq 0$, the first carbon bond creation determining a mass value closed to maximum. After that (the presence of first C-C bond) the sphericity become the one that is influencing the bondonic mass. There was also determined a relation between the aromaticity character and the bondonic mass, i.e., a highly aromatic molecule will present a higher value for the bondonic mass (Putz and Ori, 2015a).

6.6   OPEN ISSUES

There were many questions raised by the bondon, as a new concept in physical science, such as which is the maximum distance which can be considered a bond between two atoms, or what is exactly the bonding manifestation between to atoms (Putz and Tudoran, 2014). On the other hand, the next frontier in nanoscience appears to be the entangled chemistry on graphene, already predicted by Putz and Ori in their work (2015b), which will make possible the synthesis, control and molecular design on the graphene pattern and even the chemical bonding teleportation. In order to achieve these tasks, one will need to control first the transformation boson-fermion, at a large scale, and then, at the chemical bonding level, the wave-particle quantum symmetry breaking should also be controlled. Due to the fact that the quantum mechanics' principles can be used to describe unitary the condensation, delocalization and long-range phenomena, the bondonic chemistry becomes, in this context, an appropriate conceptual method for this type of research.

ACKNOWLEDGMENT

This contribution is part of the research project "Fullerene-Based Double-Graphene Photovoltaics" (FG2PV), PED/123/2017, within the Romanian National program "Exploratory Research Projects" by UEFISCDI Agency of Romania.

KEYWORDS

- bondon
- boson
- fermion
- goldstone symmetry breaking
- spontaneous symmetry breaking
- wave function

REFERENCES AND FURTHER READING

Araki, Y., (2011). Chiral symmetry breaking in monolayer graphene by strong coupling expansion of compact and non-compact U(1) lattice gauge theories. *Annals of Physics, 326*, 1408.

Bader, R. F. W., (1990). *Atoms in Molecules: A Quantum Theory* (1st edn., Vol. 22). The international series of monographs on chemistry. Clarendon, Oxford.

Becke, A. D., & Edgecombe, K. E., (1990). A simple measure of electron localization in atomic and molecular systems. *The Journal of Chemical Physics, 92*, 5397–5403.

Bohm, D., & Vigier, J. P., (1954). Model of the causal interpretation of quantum theory in terms of a fluid with irregular fluctuations. *Physical Review, 96*, 208–215.

De Broglie, L., (1923). Ondes et quanta. *Comptes Rendus de l'Académie des Sciences* (Paris), *177*, 507–510.

De Broglie, L., (1925). Sur la fréquence propre de l'électron. *Comptes Rendus de l'Académie des Sciences* (Paris), *180*, 498–500.

De Broglie, L., & Vigier, M. J. P., (1953). *La Physique Quantique Restera-t-elle Indeterministe?* Gauthier-Villars, Paris.

Dutta, S., Manna, A. K., & Pati, S. K., (2009). Intrinsic half-metallicity in modified graphene nanoribbons. *Physical Review Letters, 102*, 096601(4).

Gillespie, R. J., & Popelier, P. L. A., (2001). *Chemical Bonding, and Molecular Geometry.* Oxford University Press, New York.

Jose, D., & Datta, A., (2011). Structures and electronic properties of silicene clusters: A promising material for FET and hydrogen storage. *Physical Chemistry Chemical Physics, 13*, 7304–7311.

Langmuir, I., (1919). The arrangement of electrons in atoms and molecules. *Journal of the American Chemical Society, 41*, 868–934; ibid. Isomorphism, isosterism, and covalence. *Journal of the American Chemical Society, 41*, 1543–1559.

Lewis, G. N., (1916). The atom and the molecule. *Journal of the American Chemical Society, 38*, 762–785.

Linnett, J. W., (1961). A modification of the Lewis-Langmuir octet rule. *Journal of the American Chemical Society, 83*, 2643–2653.

Parr, R. G., & Yang, W., (1989). *Density Functional Theory of Atoms and Molecules.* Oxford University Press: New York.

Putz, M. V., (2005). Markovian approach of the electron localization functions. *International Journal of Quantum Chemistry, 105*, 1–11.

Putz, M. V., (2010). The bondons: The quantum particles of the chemical bond. *International Journal of Molecular Sciences, 11*(11), 4227–4256.

Putz, M. V., (2011). Electronegativity and chemical hardness: Different patterns in quantum chemistry. *Current Physical Chemistry, 1*(2), 111–139.

Putz, M. V., (2012a). *Quantum Theory: Density, Condensation, and Bonding.* Apple Academics & CRC Press–Taylor & Francis Group, Toronto & New Jersey.

Putz, M. V., (2012b). Nanoroots of quantum chemistry: Atomic radii, periodic behavior, and bondons. In: Castro, E. A., & Haghi, A. K., (eds.), *Nanoscience and Advancing Computational Methods in Chemistry.* IGI Global, Pasadena.

Putz, M. V., (2012c). From Kohn-Sham to Gross-Pitaevsky equation within Bose-Einstein condensation ψ-theory. *International Journal of Chemical Modeling, 4*(1), 1–11.

Putz, M. V., (2012d). Density functional theory of Bose-Einstein condensation: Road to chemical bonding quantum condensate. In: Putz, M. V., & Mingos D. M. P., (eds.), *Application of Density Functional Theory to Chemical Reactivity, Structure and Bonding* (Vol. 149, pp. 1–50). Springer.

Putz, M. V., (2012e). Chemical orthogonal spaces. In: *Mathematical Chemistry Monographs* (Vol. 14). Faculty of Science, University of Kragujevac.

Putz, M. V., & Ori, O., (2012). Bondonic characterization of extended nanosystems: Application to graphene's nanoribbons. *Chemical Physics Letters, 548*, 95–100.

Putz, M. V., & Ori, O., (2015a). Predicting bondons by goldstone mechanism with chemical topological indices. *International Journal of Quantum Chemistry, 115*, 137–143.

Putz, M. V., & Ori, O., (2015b). Bondonic chemistry: Physical origins and entanglement prospects. In: Putz, M. V., & Ori, O., (eds.), *Exotic Properties of Carbon Nanomatter, Advances in Physics and Chemistry, Carbon Materials: Chemistry and Physics* (Vol. 8, pp. 229–260). Springer, Chapter 10.

Putz, M. V., & Tudoran, M. A., (2014). Bondonic effects on the topo-reactivity of PAHs. *International Journal of Chemical Modeling, 6*(2/3), 311–346.

Putz, M. V., Duda-Seiman, C., Duda-Seiman, C., & Bolcu, C., (2015). Bondonic Chemistry: Consecrating Silanes as Metallic Precursors for Silicenes Materials. In: Putz, M. V., & Ori, O., (eds.), *Exotic Properties of Carbon Nanomatter, Advances in Physics and Chemistry, Carbon Materials: Chemistry and Physics* (Vol. 8, pp. 323–346). Springer, Chapter 12.

Putz, M. V., Ori, O., Diudea, M. V., Szefler, B., & Pop, R., (2016). Bondonic chemistry: Spontaneous Symmetry breaking of the topo-reactivity of graphene. In: Ashrafi, A. R., & Diudea, M. V., (eds.), *Distance, Symmetry, and Topology in Carbon Nanomaterials* (Vol. 9, pp. 345–389). Chapter 20.

CHAPTER 7

Chemical Modeling

CORINA DUDA-SEIMAN[1], DANIEL DUDA-SEIMAN[2], and
MIHAI V. PUTZ[1,3]

[1]*Laboratory of Computational and Structural Physical Chemistry for
Nanosciences and QSAR, Biology-Chemistry Department,
Faculty of Chemistry, Biology, Geography at West University of
Timişoara, Pestalozzi Street No. 16, Timişoara, RO-300115, Romania*

[2]*Department of Medical Ambulatory, and Medical Emergencies,
University of Medicine and Pharmacy 'Victor Babes,'
Avenue C. D. Loga No. 49, RO-300020 Timisoara, Romania*

[3]*Laboratory of Renewable Energies-Photovoltaics, R&D National
Institute for Electrochemistry and Condensed Matter,
Dr. A. Paunescu Podeanu Str. No. 144, RO-300569 Timişoara, Romania,
Tel.: +40-256-592-638; Fax: +40-256-592-620,
E-mail: mv_putz@yahoo.com, mihai.putz@e-uvt.ro*

7.1 DEFINITION

Knowledge in chemistry is based on experiment and observation. Beyond
this, after decades, theoretical chemistry evolved with the ability to make
enough accurate predictions that can be superimposed upon experimental
data (Gasteiger, 2016). Modeling, according to dictionary explanation, refers
to a plan or design. Concerning a chemical compound, it can be represented
using plastic models, unfortunately not always in perfect agreement with real
space conformations of the given structure. Computational modeling methods
overtook mechanical models (Hinchliffe, 2003), allowing to point out physical
and chemical properties in order to predict most stable conformations, and in
medicine and pharmacy the most bioactive compounds.

7.2 HISTORICAL ORIGINS

In the 60s, computers were introduced in chemistry in order to explain, model, and elucidate chemical reactions and phenomena. Afterwards, computational chemistry was developed with studies of quantitative structure-activity/property relationships (QSAR/QSPR) to predict physical, chemical, biological and environmental data of chemicals. Spectroscopic data provided structure information (computer-assisted structure elucidation, CASE) and then there were developed systems to design organic syntheses (computer-assisted synthesis design, CASD) (Gasteiger, 2016).

7.3 NANO-CHEMICAL IMPLICATIONS

In the early 2000s, a complex chemical modeling process identified three main types of activities with certain subactivities (Eggersmann et al., 2003):

a. *Synthesis:* has the main goal to create a specific design object (e.g., chemical structure, mathematical model, metabolic pathways, etc.); subordinate activities are: propose, add, remove, modify, merge, select, request.

b. *Analysis:* usually simulates a behavior and predicts the performance of a prospective design; subordinated activities are: calculate, estimate, determine, experiment, request.

c. *Decision:* points out if the design artifact is selected, rejected, or kept as a reliable alternative; subordinate activities are: select, evaluate, justify, request.

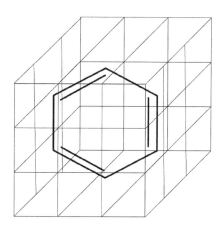

First ideas underlying QSAR studies (quantitative relationships chemical structure-biological activity) occurred after 1890 with the observations of Richet, Meyer, and Overton on the relationship concerning the relation between solubility in water/lipids and toxicity or narcosis, with the theory of Emil Fischer on the importance of stereo-configuration of the substrate in enzyme reactions, the antibacterial chemotherapy studies made by Ehrlich, which led him to formulate the concept of the receptor (Kubinyi, 1993). Research in this area was inspired by the study of reactions of structure-reactivity in organic chemistry and the studies of Hammett on the influence of donor or acceptor capacity of electrons, σ_x, of some substituents on the relative velocity of the reaction of the substituted compound, k_x, compared to the of the unsubstituted compound, k_H:

$$\log k_{R-X} = \log k_{R-H} + \rho\sigma_x \tag{1}$$

The work of Hansch and Fujita in 1964 can be considered the birth certificate of the QSAR, a work that describes the relations between physicochemical properties and biological activity; also the paper of Free and Wilson (1964) has to be taken into account, which refers to the additive contributions to determine biological activity.

The aspects of the relationship between chemical structure and biological activity of bioactive compounds had a real research basis and development in the past 20–30 years, various factors intervening in the determinism of time on the axis of performance in this field of scientific research with enormous theoretical and practical importance: the complexity of interactions between bioactive substances (so-called effectors, E) and biological macromolecules (receptor, R) is an element that has created enormous difficulties of mathematical and physical modeling of R–E interactions.

Recent years have revealed amazing aspects of science; the recorded progress is the natural consequence of the development of organic and physical chemistry (deciphering and more detailed knowing of the mechanisms of organic chemistry processes, refining of theoretical and practical methods of structural analysis).

The research domain of chemical structure-reactivity relations has developed faster and efficient. Hammett (1937) was the first who managed to model in a quantitative manner how the structural effects influence the reactivity and physical properties of organic compounds, "providing" the equation named after him. Hansen (1962) tried to apply the Hammett equation in the study of biological activities, while Zahradnik (1962) suggested using a similar equation.

The Hansch equation usually considers the modification of a reference structure through X characteristic elements, leading to a series of bioactive compounds, known as ligands, L. Their bioactivity is determined experimentally on an animal model and then correlated with structure. This is usually done by means of the correlation analysis (Shorter, 1973), using the multilinear regression, of the following form:

$$A_x = a + b_1 t_x + b_2 t_x^2 + b_3 \sigma_{I,x} + b_4 \sigma_{D,x} + b_5 S \qquad (2)$$

where A is the bioactivity, t is a transport parameter, σ_I and σ_D expresses localized electronic effects (field and/or inductive), and respective delocalized electronic effects (resonance), S measures steric effects. Bioactivity, A, is usually provided by $\log(1/C)$, where C is the molar concentration which determines a constant biological response.

Correlational equation (2) and various other relations derived from it, statistically validated, is called quantitative relationship structure-biological activity (or QSAR). If on the molecular structure of the effector L are grafted more substituents X, their effects may be or, on the contrary, each substituent may be treated separately. The difficulties of quantitative of a structure, using the t, S, σ parameters explain the great diversity of developed parametric scales.

The transport parameter t is traditionally represented by the logarithm of the partition coefficient of the bioactive substance (L), almost always determined between water and 1-octanol (simulating the hydrophobic lipid membrane), or a series of magnitudes derived from it. The chromatographic methods are more frequently used, as well as the obtained parameters. The t^2 term was introduced to take into account the parabolic dependence of bioactivity on the transport parameter.

There have been proposed and used other models, especially the bilinear model of Kubinyi (1979). In QSAR studies, some nonlinear behaviors are explained by the effector model that crosses a bio-membrane, being initially transferred from the aqueous phase in the lipidic phase of a biological membrane, and then transferred back to the other side again in the aqueous phase, etc.

Extra-thermodynamic models are the basis to quantify the electronic structure of bioactive substances, using the well-known σ-Hammett substituent constant (Hammett, 1935, 1937), with further refinements. This was and remains in most cases the easiest option of parameterization of electronic effects, existing vast libraries, now computerized, for such constants for a wide variety of substituents (Hansch et al. 1991). There are also additional/

complementary approach opportunities, an example being the use of quantum mechanics methods and of quantum indices of reactivity (e.g., the charge of atoms, etc.).

It was inferred since the late nineteenth century that chemical and biological interactions are produced according to the concept of fitting "key in lock," with the possibility of occurrence of various steric effects (Kubinyi, 1979). Despite all arguments, the concept was questioned for some time, being a decisive factor in delaying research on new drug-design. Taft (1956) was the first who realized in the 50s the separation of polar, resonance and steric effects within the methodology of linear relations of free energy and of the extra-thermodynamic model.

Currently, there are several available distinct methods to quantify in parametric terms steric effects via relations like (2). Thus, the first group of parameters is defined on the basis of the chemical reactivity (E_S – Taft parameters or I[st] generation parameters). IInd generation steric parameters were developed based on theoretical models, considering molecular geometries and van der Waals characteristics of atoms (Verloop et al., 1976). Scale extension of the first type of steric substituent constants was done using models based on IInd generation parameters (Charton, 1975), which led to the deep alteration of the physical significance of the introduced values for a wide range of substituents. Typological models and topological indices are valid and interesting alternatives to the above-mentioned parameters.

QSAR methods can be classified according to dimensionality and, respectively, to the type of used chemometric methods (Verma et al., 2010).

According to dimensionality, QSAR methods are (Verma et al., 2010):

- 1D-QSAR: biological activity is correlated with global molecular properties;
- 2D-QSAR: biological activity is correlated with structural patterns: connectivity indices, 2D-pharmacophores, etc.;
- 3D-QSAR: biological activity is correlated with non-covalent interaction fields surrounding the molecules;
- 4D-QSAR: includes the ensemble of ligand configurations in 3D-QSAR;
- 5D-QSAR: provides in an explicit way the representations of induced-fit models in 4D-QSAR;
- 6D-QSAR: contains different solvation models in 5D-QSAR.

According to the chemometric methods, QSAR recognizes (Verma et al., 2010):

- Linear methods: linear regression, multiple linear regression, partial least squares, and principal component analysis/regression;
- Non-linear methods: artificial neural networks, *k*-nearest neighbors, and Bayesian neural nets.

7.4   TRANS-DISCIPLINARY CONNECTIONS

Chemical and molecular modeling techniques are based on new evolving mathematical and computational methods, starting from chemical and physical principles involving a quantum level of confidence is given by using the newest and up-to-date notions of applied mathematics (Mezey, 2013).

7.5   OPEN ISSUES

Traditional chemical modeling methods (knowledge-based, neural network) tend to lose effectiveness due to the increasing complexity of chemical processes. A promising technique would be the *Agent* technique. The *agent* is an autonomous system in a specific environment; it has the ability to acquire information from its environment, to elaborate decisions and answers within the range of time when certain requirements appear in the system. Models based on *agents* are dynamic, with interactions between agents in terms of sharing information, knowledge, and tasks. This system to be working, the following steps should be undertaken: to classify the characteristics of a chemical process; to define the modeling elements (equipment, material, control system, environment, relationships between agents); to point out the semantics and to represent graphical each modeling element; to show the practical relevance over a real chemical process (Dong et al., 2013).

KEYWORDS

- **chemical modeling**
- **computer-assisted modeling**
- **structure**
- **synthesis**

REFERENCES AND FURTHER READING

Charton, M., (1975). Steric effects. I. Esterification and acid-catalyzed hydrolysis of esters. *J. Am. Chem. Soc., 97*(6), 1552–1556.

Dong, G., Xin, X., Beike, Z., Xin, M., & Chongguang, W., (2013). A framework for agent-based chemical process modeling. *J. Applied Sci., 13*(17), 3490–3496.

Eggersmann, M., Gonnet, S., Henning, G. P., Krobb, C., Leone, H. P., & Marquardt, W., (2003). Modeling and understanding different types of process design activities. *Latin American Applied Research, 33*(2), 167–175.

Free, S. M., & Wilson, J.W., (1964). A mathematical contribution to structure-activity studies. *J. Med. Chem., 7*(4), 395–399.

Gasteiger, J., (2016). Chemoinformatics: Achievements and challenges, a personal view. *Molecules, 21*, 151. doi: 10.3390/molecules21020151.

Hammett, L. P., (1937). The effect of structure upon the reactions of organic compounds, benzene derivatives. *J. Am. Chem. Soc., 59*(1), 96–103.

Hansch, C., & Fujita, T., (1964). ρ-σ-π Analysis, a method for the correlation of biological activity and chemical structure. *J. Am. Chem. Soc., 86*, 1616–1626.

Hansch C., Leo A., Taft R. W., (1991). A survey of Hammett substituent constants and resonance and field parameters. *Chem. Rev. 91* (2), 165–195. doi:10.1021/cr00002a004

Hansen, O. R., (1962). Hammett series with biological activity. *Acta Chem. Scand., 16*, 1593–1600.

Hinchliffe, A., (2003). *Molecular Modeling for Beginners*. John Wiley & Sons Ltd., England.

Kubinyi, H., (1979). Lipophilicity and drug activity. *Prog. Drug Res., 23*, 97–198.

Kubinyi, H., (1993). 3D QSAR in drug design. *Theory, Methods, and Applications*. ESCOM, Science Publishers B.V., Leiden.

Hammett, L. P., (1935). Some relations between Reaction Rates and Equilibrium Constants. *Chem. Rev. 17* (1): 125–136. doi:10.1021/cr60056a010

Mezey, P. G., (2013). Molecular modeling: An open invitation for applied mathematics. *AIP Conference Proceedings, 1558*, 44–47.

Shorter, J., (1973). *Correlation Analysis in Organic Chemistry*. Clarendon, Oxford.

Taft, R. W., (1956). Separation of polar, steric, and resonance effects in reactivity. Chapter 13, In: *Newman*, M. S., (ed.), *Steric Effect in Organic Chemistry*. *Wiley, New York.*

Verloop, A., Hoogenstraaten, W., & Tipker, J., (1976). Development and application of new steric substituent parameters in drug design, In: Ariens, E. J., (ed.), *Drug Design* (Vol. 7, pp. 165–207). New York, Academic Press.

Verma, J., Khedkar, V. M., & Coutinho, E. C., (2010). 3D-QSAR in drug design–a review. *Current Topics in Medicinal Chemistry, 10*, 95–115.

Zahradnik, R., (1962). Influence of the structure of aliphatic substituents on the magnitude of the biological effect of substances. *Arch. Int. Pharmacodyn. Ther., 135*, 311–329.

CHAPTER 8

Coherent States

NICOLINA POP

Politehnica University of Timisoara, Department of Fundamental of Physics for Engineers, Vasile Pârvan Blv., 300223 Timisoara, Romania, E-mail: nicolina.pop@upt.ro

8.1 DEFINITION

The coherent states (CSs) form a complete set of states in the sense that any state of the mode can be expressed as a suitable sum taken over them:

$$|\alpha> = e^{-\frac{1}{2}|\alpha|^2} \sum_{n=0}^{\infty} \frac{\alpha^n}{n!} |n>,$$ where α is some complex number, $|n>$ express quantum-number states, $n = 0, 1, 2 \ldots$.

So they are equivalent to certain oscillator states defined by Schrödinger (1926) in his earliest discussions of wave functions.

8.2 HISTORICAL ORIGIN(S)

The coherent states show the correspondence between quantum oscillators and classical oscillators. They make it clear that even in fully quantum mechanical systems, not everything is quantized, not all waves are quantized, not all states are discrete.

Coherent states (CSs) [including Glauber states, *see* Glauber lectures (2008)] is very useful for understanding what goes on in oscillating systems, including harmonic oscillators (HO-1D, or 3D) as well as many kinds of waves.

The properties of CSs for HO-1D were generalized to the anharmonic oscillators potential. Therefore, the Morse oscillator (MO) has aroused interest also in the construction of CSs. The first works in this direction occurred more than three decades ago, based on the general treatment, *see*

Nieto (1979). Based on the Perelomov's definition of generalized CSs, *see* Perelomov (1986), other authors have obtained and studied the properties of such states for MO (Kais, 1990).

Cooper (1992) has shown that in the limit of small anharmonicity constant or equivalently, in the limit of large well depth, the minimum uncertainty CSs and annihilation operator CSs for the MO are equivalent. These states reduce, in the above limit, to the Perelomov's generalized CSs. An algebraic construction of the MO CSs, based on the supersymmetry quantum mechanics (SUSY QM), was published some years ago, *see* Benedict (1999), while in a series of papers are built the vibrational or molecular CSs of MO [Drăgănescu & Avram (1997, 1998, 2000)], also using a new analytic method for the calculation of creation and annihilation operators.

For anharmonic diatomic molecules were constructed the CSs, among it also the CSs for MO (Récamier, 2002). The group theoretical methods for obtaining the CSs of MO were also used in Dong, S.-H.(2002) and Popov et al. (2013). For MO were also constructed the CSs of Barut-Girardello kind (BG-CSs), *see* Fakhri and Chenaghlou (2003), of the Klauder-Perelomov (KP-CSs) kind or of the Gazeau–Klauder (KP-CSs) kind (Roy, 2002) and also the generalized and Gaussian CSs (Angelova and Hussin, 2008).

Generally, a CS must fulfill the defined by Klauder (1963) minimal set of conditions:

a) *continuity* in the complex variable $z = |z| \exp(i\varphi)$;
b) *normalization*;
c) *completeness*.

As a consequence, the general structure of a CS is:

$$| z;\lambda >= \left[\mathcal{N}\left(| z |^2 ;\lambda\right) \right]^{-\frac{1}{2}} \sum_{n=0}^{M} \frac{z^n}{\sqrt{\rho(n;\lambda)}} | n;\lambda > \qquad (1)$$

where λ is a real parameter, the positive quantities $\rho(n; \lambda)$ are called the *structure constants* of the CSs, while $\mathcal{N}(|z|; \lambda$ is the normalization function which plays an important role in the CSs formalism:

$$\mathcal{N}\left(| z |^2 ;\lambda\right) \equiv \sum_{n=0}^{M} \frac{(| z |^2)^n}{\rho(n;\lambda)} < \infty \qquad (2)$$

8.3 NANO-SCIENTIFIC DEVELOPMENT(S)

The properties of different nanostructures (quantum dots, wires, and wells) differ from those of the corresponding bulk structures. This fact leads to many applications in condensed matter, optoelectronic devices, sensors, electronic and light emitting components. So, it is interesting to examine the behavior of the quantum system (QS) using the coherent states (CSs) formalism, according to the sequence: nanoparticles → QS → CSs and statistical physics, *see* Kapsa (2011); Popov et al. (2014).

For the *pure CSs*, the corresponding *density operator* is just the projector onto the complex space of these states:

$$\rho^{(pure)}(z, z^*; \lambda) = | z, \lambda >< z, \lambda | \tag{3}$$

where $|z, \lambda >$, the general structure of a CS was defined as in Eq. (3)

For a *mixed state*, the associated density operator ρ in its diagonal form with respect to the CSs is:

$$\rho = \int d\mu(z, \lambda) | z, \lambda > P(| z |^2, \lambda) < z, \lambda | \tag{4}$$

where $P(|z|^2, \lambda)$ is the normalized P- quasi-distribution function and $d\mu(z; \lambda)$ is positive defined integration measure with weight function $h(|z|^2; \lambda)$:

$$d\mu(z; \lambda) = \frac{d^2 z}{\pi} h(| z |^2; \lambda) = \frac{d\varphi}{2\pi} d(| z |^2) h(| z |^2; \lambda) \tag{5}$$

Generally, the integration measure can be expressed through Meijer's G-function:

$$d\mu(z; \lambda) = \frac{d\varphi}{2\pi} d(| z |^2) \mathcal{N}(| z |^2; \lambda) G_{p,q}^{m,n}\left(| z |^2 | \ldots\right) \tag{6}$$

and must be unique which can be proved by using the Carleman's criterion (Popov, 2004)

As a representative example of the quantum system in the *mixed states* we consider a quantum oscillator's ideal gas (of HO-1D or of PHO) in thermodynamic equilibrium with a thermostat at temperature T. For such quantum systems, the *energy spectrum is linear* $E_n(\lambda) = an + b$, with $a = a(\lambda)$, $b = b(\lambda)$, and the mixed states are the *thermal states* characterized by the normalized density operator with the canonical probability distribution:

$$\rho = \frac{1}{Z(\beta)} \sum_{n=0}^{\infty} e^{-\beta E_n} | n; \lambda >< n; \lambda | = \frac{1}{\bar{n}+1} \sum_{n=0}^{\infty} \left(\frac{\bar{n}}{\bar{n}+1}\right)^n | n; \lambda >< n; \lambda | \tag{7}$$

Here the quantity $\bar{n} = \left(e^{\beta a} - 1\right)^{-1}$ is the mean number of photons (bosons) or Bose-Einstein distribution function, i.e., the mean photon number of thermal (blackbody) radiation at thermal equilibrium at a temperature $T = \left(k_B \cdot \beta\right)^{-1}$ where ω is the Boltzmann's constant.

The potential of the ***harmonic one-dimensional oscillator (HO-1D)*** with the mass m and the angular frequency ω in the variable x is:

$$V^{(HO)}(x) = \frac{m\omega^2}{2} x^2 \tag{8}$$

which allows the construction of three kinds of CSs defined in three manners:

a)   as the eigenvectors of the lowering operator a;
b)   as the result of the action of the displaced operator:

$$D(z) \equiv \exp(z\, a^+ - z^*\, a)$$

c)   as the states which minimize the uncertainty relations:

$$(\Delta x)_z^2 \, (\Delta p)_z^2 = \frac{\hbar}{2}$$

For HO-1D, all the three manners lead to the same ***canonical coherent state***.

All the three definitions of CSs are equivalent in the sense that they lead to the same expression for the CSs which are denoted by $|z>(\lambda=0)$.

So, we have:

$$|z> = e^{-\frac{1}{2}|z|^2} \sum_{n=0}^{\infty} \frac{z^n}{\sqrt{n!}} |n>, \quad d\mu^{(HO)}(z) = \frac{d^2 z}{\pi} = \frac{d\varphi}{2\pi} d(|z|^2) \tag{9}$$

The ***pseudoharmonic oscillator (PHO)*** (an intermediate oscillator between the HO-1D and the anharmonic oscillators (for example, Morse or Poschl-Teller oscillators) potential is a central field potential in the coordinate r:

$$V^{(PHO)}(r) = \frac{m\omega^2}{8} r_0^2 \left(\frac{r}{r_0} - \frac{r_0}{r}\right)^2 + \frac{\hbar^2}{2m} J(J+1) \frac{1}{r^2} \tag{10}$$

where m is the reduced mass of the quantum system described by the PHO potential (e.g., the diatomic molecule), r_0 is the equilibrium distance, and J is the rotational quantum number.

The quantum group associated with the PHO potential is SU(1,1) and, consequently, three different kinds of CSs can be built: a) the CSs of the

Barut-Girardello kind (BG-CSs); b) the CSs of the Klauder-Perelomov kind (KP-CSs), and c) the CSs of the Gazeau-Klauder kind (GK-CSs).

For PHO the three kinds of CSs have different expressions, and hence there is a variety of ways to calculate average values:

a) **BG-CSs for PHO** is defined as the eigenvectors of the lowering operator K_-:

$$K_- \mid z;k >_{BG} = z \mid z;k >_{BG} \tag{11}$$

and their development in the Fock-vectors basis is, *see* Popov (2001):

$$\mid z;k >_{BG} = \sqrt{\frac{\mid z \mid^{2k-1}}{I_{2k-1}(2\mid z\mid)}} \sum_{n=0}^{\infty} \frac{z^n}{\sqrt{n!\Gamma(n+2k)}} \mid n;k > \tag{12}$$

where $I_n(x)$ are modified Bessel functions of the first kind and the above integrals are of the kind (3.937.1) and (3.937.2) in the Gradshteyn and Ryshik table (2007) and $\Gamma(x)$ is the Euler's gamma function.

b) **KP-CSs for PHO** is defined by the action of the displaced operator $D[\alpha(z)] \equiv \exp(\alpha K_+ - \alpha^* K_-)$ on the ground (fiducial) state $\mid 0;k >$:

$$\mid z;k >= \exp(\alpha K_+ - \alpha^* K_-)\mid 0;k >= e^{zK_+} \, e^{\ln(1-\mid z\mid^2)K_3} \, e^{-z^*K_-} \mid 0;k >= \\ =(1-\mid z\mid^2)^k \, e^{zK_+} \tag{13}$$

where $\alpha = -\dfrac{\theta}{2}e^{-i\varphi}$, $z = \dfrac{\alpha}{\mid\alpha\mid}\tanh\mid\alpha\mid = -\tanh\dfrac{\theta}{2}e^{-i\varphi}$, $\theta \in (-\infty, +\infty)$,

$\varphi \in [0, 2\pi], \mid z\mid < 1$.

Their development in the Fock-vectors basis is as in Popov et al. (2006):

$$\mid z;k >_{KP} = (1-\mid z\mid^2)^k \sum_{n=0}^{\infty} \frac{z^n}{\sqrt{\dfrac{n!\Gamma(2k)}{\Gamma(n+2k)}}} \mid n;k > \tag{14}$$

c) **GK-CSs for PHO** have been built by performing some important modifications in the usual definition of the CSs (Gazeau and Klauder, 1999): the parameterization of the usual CSs in a single complex number z is extended by replacing z with two independent real numbers $J \geq 0$ and $-\infty < \lambda < +\infty$, namely $z = \sqrt{J}\exp(-i\varphi)$ and obtaining the CSs $\mid J; \gamma >$. So, the GK-CSs for the PHO could be expressed as:

$$\left| J; \gamma >= \frac{1}{\sqrt{{}_1F_1\left(1; k+1; \frac{J}{2}\right)}} \sum_{n=0}^{\infty} \frac{\left(\sqrt{\frac{J}{2}}\right)^n e^{-i\gamma e_n}}{\sqrt{(k+1)_n}} \left| n; k > \right. \tag{15}$$

where $e_n = 2(n + k)$ are the dimensionless energy eigenvalues of the PHO, ${}_1F_1(1; k + 1; x)$ is the confluent hypergeometric function and $(a)_n = \dfrac{\Gamma(a+n)}{\Gamma(a)}$ are the Pochhammer symbols.

Moreover it can be defined binomial states (BSs) [introduced by Stoler et al. (1985)], as a linear superposition of number states (NSs) $|n, \lambda >$ in an $(M+1)$-dimensional subspace of the infinite-dimensional Hilbert space of the Fock-vectors and, the excited binomial states (EBSs) using the pseudoharmonic oscillator (PHO) eigenvectors and the corresponding raising operator:

$$\left| z, M; \lambda; 0 \right) = \sum_{n=0}^{M} B_n^{(M;0)} z^n \left(1-|z|^2\right)^{\frac{M-n}{2}} |n; \lambda); \quad B_n^{(M;0)} = \binom{M}{n}^{\frac{1}{2}} \tag{16}$$

where $z \in C$, $|z|^2 < 1$, $M \in Z^+$ (*see* Popov & Pop, 2014). Even if it seems that the BSs (and implicitly EBSs) have the coherent state's behavior only for the harmonic limit, in fact, the EBSs behaves as the coherent states in all range for the complex variable $\left(|z|^2 < 1\right)$.

8.4   NANO-CHEMICAL APPLICATION: QUANTUM WELL MODEL

The quantum well model was used firstly as a theoretical model used for teaching purposes to familiarize students with the concept of energy quantization, but due to the development of the nanoscience and nanotechnology it has been reconsidered as a particular kind of heterostructure consisting of one thin "well" layer, surrounded by two "barrier" layers, of certain, practically of some finite height, but which, under certain conditions, can be considered as infinite.

In the following is presented the Barut-Girardello coherent states (BG-CSs) for a particle embedded in an infinite square quantum well (ISQW) (*see* Popov et al., 2014).

A quantum particle with effective mass m^* trapped inside an infinite square potential quantum well potential along the x-axis with width L and recall some well-known results regarding the motion of the particle. The corresponding Schrödinger equation for stationary states

$$H \Psi_n(x) = E_n \Psi_n(x), \qquad H = -\frac{\hbar^2}{2m^*}\frac{d^2}{dx^2} \tag{17}$$

with the boundary conditions: $\Psi_n(0) = \Psi_n(L) = 0$ leads to the following normalized eigenfunctions and corresponding energy Eigenvalues:

$$\Psi_n(x) = \sqrt{\frac{2}{L}}\sin\left(\frac{n\pi}{L}x\right), \quad E_n = \frac{\hbar^2\pi^2}{2m^*L^2}n^2 \equiv \hbar\omega n^2, \quad \omega \equiv \frac{\hbar\pi^2}{2m^*L^2} \tag{18}$$

The normalized eigenfunctions are real and by virtue of the general conditions of the orthogonality of wave functions:

$$\sum_{n=0}^{\infty}\Psi_n(x)\Psi_n(x') = \frac{2}{L}\sum_{n=0}^{\infty}\sin\left(\frac{n\pi}{L}x\right)\sin\left(\frac{n\pi}{L}x'\right) = \delta(x-x') \tag{19}$$

This equation could also be considered as a consequence of the boundary conditions for the canonical density operator in the coordinate representation

$$\rho(x,x';\beta) = \sum_{n=0}^{\infty}e^{-\beta E_n}\Psi_n(x)\Psi_n(x') \tag{20}$$

whose limit is

$$\lim_{\beta\to 0}\rho(x,x';\beta) = \sum_{n=0}^{\infty}\Psi_n(x)\Psi_n(x') = \delta(x-x') \tag{21}$$

In order to construct the coherent states, it is not less important to emphasize that the particle motion in an infinite quantum well is stationary and their energy is quantized, the difference between the energy levels is not equal, but increases with an increasing principal quantum number n, i.e., $\Delta E_n = E_{n+1} - E_n = \hbar\omega(2n+1)$. This means that the transition of a particle between the superior energy levels is more difficult than between lower levels. By considering an electron trapped inside an infinite quantum well of width $L = 10^{-2}$ m, the "variable quanta" $\hbar\omega$ is approximately $0.3\cdot10^{-38}$ J , while for the width of atomic dimension order $L = 10^{-9}$ m, this value is $0.3\cdot10^{-17}$ J.

The quantum group associated with this physical model is $SU(1,1)$ which allows building of these BG-CSs by choosing the Bargmann index $k = 1/2$, *see* Lemus & Frank (2003) and Thilagam & Lohe (2005).

The group generators act on the Fock basis vectors $|n;1/2 >\equiv|n>$ in the following manner:

$$K_+ \mid n >= (n+1) \mid n+1 >; \ K_- \mid n >= n \mid n-1 >; \ K_3 \mid n >= \left(n + \frac{1}{2} \right) \mid n > \qquad (22)$$

These generators ensure that the energy eigenvalues of a particle of mass m^* trapped inside infinite square quantum well (ISQW) assume a simple form:

$$E_n \equiv < n \mid H \mid n >= \frac{\hbar^2 \pi^2}{2m^* L^2} < n \mid K_+ K_- \mid n >= \hbar \omega n^2 \qquad (23)$$

It can be observed that the Hamiltonian operator H is expressed as the normal ordering product of the creation K_+ and annihilation K_- operators.

The coherent states for the infinite quantum well in the Barut-Girardello manner in the usual way, i.e., as the eigenvalues of the lowering operator K_- and we denote these states as BG-CSs (Popov, 2001):

$$K_- \mid z >= z \mid z > \qquad (24)$$

where $z = |z| \exp(i\varphi)$ is the complex variable labeling CSs.

There are some manners to perform realizations of the SU(1,1) generators, depending on the specific problems or quantum systems. Particularly, we consider such a realization of the SU(1,1) algebra in terms of a one-dimensional harmonic oscillator (HO-1D) creation and annihilation operators. So the BG-CSs for the ISQW are in fact a version of nonlinear CSs of the canonical CSs for the HO-1D.

Recalling that the Fock vectors $|n>$ satisfy the completeness relation

$$\sum_{n=0}^{\infty} \mid n >< n \mid = 1 \qquad (25)$$

the BG-CSs for the ISQW could be written as the superposition of the complete orthonormal Fock-vectors basis $|n>$:

$$\mid z >= \frac{1}{\sqrt{I_0(2 \mid z \mid)}} \sum_{n=0}^{\infty} \frac{z^n}{n!} \mid n > \qquad (26)$$

where $I_0(x)$ is the modified Bessel function of the first kind defined as:

$$I_0(2 \mid z \mid) = \sum_{n=0}^{\infty} \frac{1}{(n!)^2} \mid z \mid^{2n} \qquad (27)$$

As all CSs, the above-defined BG-CSs for the ISQW are also normalized but nonorthogonal, so the overlap of two such states is:

$$< z \,|\, z' > = \frac{I_0(2\sqrt{z^* z'})}{\sqrt{I_0(2\,|\,z\,|)\,I_0(2\,|\,z'\,|)}} \tag{28}$$

Moreover, a crucial property of any set of CSs is that they allow the decomposition (or resolution) of the unity operator, namely Antoine et al., (2001):

$$\int d\mu(z)\,|\,z > < z\,| = 1 \tag{29}$$

where $d\mu(z) = \dfrac{d\varphi}{2\pi} d(|\,z\,|^2)\,h(|\,z\,|)$ is the integration measure.

By substituting Eq. (28) and using Eq. (27) it obtained:

$$\sum_{n,m=0}^{\infty} \frac{|\,n > < m\,|}{n!\,m!} \int_0^{\infty} d(|\,z\,|^2)\,h(|\,z\,|)\frac{1}{I_0(2\,|\,z\,|)}\int_0^{2\pi}\frac{d\varphi}{2\pi}z^n\,(z^*)^m = 1 \tag{30}$$

The result of the angular integration is $(|\,z\,|^2)^n\,\delta_{nm}$, so that, in order to accomplish the completeness relation (29):

$$\int_0^{\infty} d(|\,z\,|^2)\,h(|\,z\,|)\frac{1}{I_0(2\,|\,z\,|)}(|\,z\,|^2)^n = (n!)^2 \tag{31}$$

If it performs the function change, as well as the exponent change:

$$g(|\,z\,|) \equiv h(|\,z\,|)\frac{1}{I_0(2\,|\,z\,|)}, \quad n = s - 1 \tag{32}$$

it obtains the Stietjes moment problem [6]

$$\int_0^{\infty} d(|\,z\,|^2)\,g(|\,z\,|)(|\,z\,|^2)^{s-1} = \left[\Gamma(n+1)\right]^2 \tag{33}$$

where $\Gamma(x)$ is the Euler's gamma function from Mathai and Saxena (1973).

The solution of this integral equation is $g(|\,z\,|) = 2K_0(2\,|\,z\,|)$, where $K_0(x)$ is the modified Bessel function of the second kind and, consequently, the integration measure becomes:

$$d\mu(z) = 2\frac{d\varphi}{2\pi}d(|\,z\,|^2)\,I_0(2\,|\,z\,|)\,K_0(2\,|\,z\,|) \tag{34}$$

The expectation value of a certain operator A, in the representation of the BG-CSs for the ISQW can be express as follows:

$$< z\,|\,A\,|\,z > = \frac{1}{I_0(2\,|\,z\,|)}\sum_{n=0}^{\infty}\frac{z^n\,(z^*)^m}{n!\,m!} < m\,|\,A\,|\,n > \tag{35}$$

So, the expectation values of particle number operator $N = K_3 - \dfrac{1}{2}$ in the BG-NCSs for the ISQW representation are

$$<z \mid N^s \mid z> = \frac{1}{I_0(2|z|)} \sum_{n=0}^{\infty} \frac{(|z|^2)^n}{(n!)^2} n^s = \frac{1}{2^s} \frac{1}{I_0(2|z|)} \left(|z| \frac{d}{d|z|} \right)^s I_0(2|z|) \qquad (36)$$

For the expectation value of the Hamiltonian operator in the BG-NCSs for the ISQW representation results:

$$E_{|z|^2} \equiv <z \mid H \mid z> = \frac{\hbar^2 \pi^2}{2m^* L^2} <z \mid K_+ K_- \mid z> = \frac{\hbar^2 \pi^2}{2m^* L^2} |z|^2 = \hbar \omega |z|^2 \qquad (37)$$

Now it is interesting to examine the behavior of the field of the BG-CSs for the ISQW and for this purpose it is useful to examine the values of the *Mandel parameter* as a function of the variable $|z|$. The Mandel parameter is defined by Walls and Milburn Gerard (1995) as:

$$Q_{|z|} = \frac{V_{|z|}}{<z \mid N \mid z>} - 1 \equiv \frac{<z \mid N^2 \mid z> - (<z \mid N \mid z>)^2}{<z \mid N \mid z>} - 1 \qquad (38)$$

Depending on the values of the Mandel parameter, the field of CSs is as follows:

$$Q_{|z|} \begin{cases} <0 & \text{sub-Poissonian field} & \rightarrow \text{sub-Poissonian statistics} \\ =0 & \text{Poissonian field} & \rightarrow \text{Poissonian statistics} \\ >0 & \text{super-Poissonian field} & \rightarrow \text{super-Poissonian statistics} \end{cases}$$

By using Eq. (36) the Mandel parameter BG-CSs for the ISQW can be express as:

$$Q_{|z|} = |z| \left[\frac{I_0(2|z|)}{I_1(2|z|)} - \frac{I_1(2|z|)}{I_0(2|z|)} \right] - 1 \qquad (39)$$

The Mandel parameter BG-CSs for the ISQW is always negative, i.e., the field of the BG-NCSs for the ISQW is always sub-Poissonian, and these states obey the sub-Poissonian statistics. This means that the variance $V_{|z|} = <z \mid N^2 \mid z> - (<z \mid N \mid z>)^2$ of these states is always less than the mean value in the BG-CSs for the ISQW representation $<z \mid N \mid z>$. In other words, the BG-NCSs for the ISQW have non-classical behavior. So, these states have received considerable attention in many modern fields of physics (nanotechnology, quantum optics, quantum communication, quantum information).

8.5 MULTI-/TRANS-DISCIPLINARY CONNECTION: QUANTUM INFORMATION

The coherent states formalism may play an important role in the quantum information theory, and the use of this representation is not only theoretical, but also of some practical relevance by its applications.

The fundamental unity of quantum information (qubit) can be express through the coherent states of the one-dimensional quantum oscillator. Density operator can also be expanded on the coherent states projectors. A sequence of qubits is a tensor product of coherent states. For the ensembles of qubits, it could be used the density operator, in order to describe the informational content of the ensemble.

A qubit $|\Psi>$ is a linear combination of two 2×1 basis vectors, Pop et al. (2013):

$$|\Psi> = \tilde{N}\left(\mu|0> + v|1>\right) = \tilde{N}\begin{pmatrix} \mu \\ v \end{pmatrix} \tag{40}$$

where μ and v are complex numbers and $|0> \equiv \begin{pmatrix} 1 \\ 0 \end{pmatrix}$, respectively $|1> \equiv \begin{pmatrix} 0 \\ 1 \end{pmatrix}$ are arbitrary base vectors from the state space.

As the application of the relationship qubit-coherent state it can be considered calculus of the standard fidelity of a transmission channel of this qubit through a long optical fiber, in order to achieve the transmission of quantum information. The optical fibers can be characterized, from the point of view of qubit transmission, as having an exponential energy loss $\exp(-\lambda L)$, where λ as loss coefficient of the fiber, and L is the transmission distance. High commercial fibers typically have $\lambda \approx 0.06/km$ (Vourdas, 2006).

The transmission channel randomly displaces an input pure coherent state according to the Gaussian distribution which results in a thermal state. The pure qubit input state can be express as $\rho_{in} = |\Psi><\Psi|$ and, because the transmission is imperfect, due to the noisy channel, the output state is a mixed state, described by the density operator ρ_{out}. A measure of the channel imperfect transmission is the channel standard fidelity (or quantum fidelity) F_{quant}, which can be defined as $F_{quant} = <\Psi|\rho_{out}|\Psi>$.

Generally, the fidelity is an important characteristic of quantum channels and a nontrivial quantity to study, because it evaluates how well the channel preserves the transmitted information. So the fidelity decrease with increasing of noise variance, i.e., when the channel is noisier, the transmitted state through the channel becomes more distorted.

KEYWORDS

- **density operator**
- **pseudo-harmonic oscillator**
- **pure and mixed states**
- **quantum system**
- **qubit**

REFERENCES AND FURTHER READING

Angelova, M., & Hussin, V., (2008). Generalized and Gaussian coherent states for the Morse potential. *J. Phys. A: Math. Theoret., 41*(30), 304016.

Antoine, J. P., Gazeau, J. P., Monceau, P., Klauder, J. R., & Penson, K. A., (2001). Temporally stable coherent states for infinite well and Pöschl-Teller potentials. *J. Math. Phys., 6,* 2349.

Avram, N. M., Draganescu, G. E., & Avram, C. N., (2000). Vibrational coherent states for Morse oscillator. *J. Opt. B: Quantum Semiclass. Opt., 2,* 214–220.

Benedict, M. G., & Molnár, B., (1999). Algebraic construction of the coherent states of the Morse potential based on super-symmetric quantum mechanics. *Phys. Rev A., 60,* R1737–R1740.

Cooper, I. L., (1992). A simple algebraic approach to coherent states for the Morse oscillator. *J. Phys. A: Math. Gen., 25*(6), 1671–1684.

Dong, S. H., (2002). The SU(2) realization for the Morse potential and its coherent states. *Can. J. Phys., 80*(2), 129–139.

Drăgănescu, G. E., & Avram, N. M., (1997). Molecular the coherent states for a generalized Morse oscillator. *Z. Phys. Chem., 200,* 51–56.

Draganescu, G. E., & Avram, N. M., (1998). Creation and annihilation operators for the Morse oscillator and the coherent states, *Can. J. Phys., 76*(4), 273–281.

Fakhri, H., & Chenaghlou, A., (2003). Barut–Girardello coherent states for the Morse potential. *Phys. Lett. A., 310*(1), 1–8.

Gazeau, J. P., & Klauder, J. R., (1999). Coherent states for systems with a discrete and continuous spectrum. *J. Phys. A: Math. Gen., 32,* 123–132.

Glauber, R. J., (2005). *One Hundred Years of Light Quanta.* Nobel Lecture.

Gradshteyn, I. S., & Ryshik, I. M., (2007). *Table of Integrals* (7th edn.). Series and Products, Academic Press, Amsterdam.

Kais, S., & Levine, R. D., (1990). Coherent states for the Morse oscillator,. *Phys. Rev. A., 41,* 2301–2305.

Kapsa, V., & Skála, L., (2011). Quantum mechanics, probabilities and mathematical statistics. *J. Comput. Nanosci., 8*(6), 998–1005.

Klauder, J. R., (1963). Continuous–representation theory. I. postulates of continuous–representation theory. *J. Math. Phys., 4,* 1055.

Lemus, R., & Frank, A., (2003). A realization of the dynamical group for the square-well potential and its coherent states. *J. Phys. A: Math. Gen., 36*, 4901.

Mathai, A. M., & Saxena, R. K., (1973). Generalized hypergeometric functions with applications in statistics and physical sciences. *Lect. Notes Math., 348*, Springer-Verlag, Berlin.

Nieto, M. M., & Simmons, L. M. Jr., (1979). Coherent states for general potentials. III. Nonconfining one-dimensional examples. *Phys. Rev. D., 20*, 1342–1350.

Perelomov, A., (1986). *Generalized Coherent States and Their Applications.* Springer-Verlag, Berlin.

Pop, N., Popov, D., & Davidovic, M., (2013). Density operator in terms of coherent states representation with the applications in the quantum information. *International Journal of Theoretical Physics, 52*(2), 332.

Popov, D., & Pop, N., (2014). Statistical properties of the excited binomial states for the pseudoharmonic oscillator. *Chinese Journal of Physics, 52*(2), 738–769.

Popov, D., (2001). Barut-Girardello coherent states of the pseudoharmonic oscillator. *J. Phys. A: Math. Gen., 34*, 5283–5296.

Popov, D., (2004). *Density Matrix–General Properties and Applications in Physics of Many-Particle Systems.* Politehnica Publisher, Timisoara (in Romanian).

Popov, D., Davidovic, D. M., Arsenovic, D., & Sajfert, V., (2006). P-function of the pseudoharmonic oscillator in terms of Klauder-Perelomov coherent states. *Acta Phys. Slovaca, 56*(4), 445–453.

Popov, D., Pop, N., Popov, M., & Simon, S., (2014). The information-theoretical entropy of some quantum oscillators. *AIP Conf. Proc., 1634*, 192.

Popov, D., Sajfert, V., Šetrajcic, J. P., & Pop, N., (2014). Coherent states formalism applied to the quantum well model. *Quantum Matter, 3*, 1–5.

Popov, D., Shi-Hai Dong, S. H., Pop, N., Sajfert, V., & Şimon, S., (2013). Construction of the Barut–Girardello quasi-coherent states for the Morse potential. *Annals of Physics, 339*, 122–134.

Récamier, J., García de León, P., Jáuregui, R., Frank, A., & Castaños, O., (2002). Coherent states for anharmonic diatomic molecules. *Int. J. Quant. Chem., 89*(6), 494–502.

Roy, B., & Roy, P., (2002). Gazeau–Klauder coherent state for the Morse potential and some of its properties. *Phys. Lett. A., 296*(4), 187–191.

Schrödinger, E., (1926). *Naturwissenschaften*, Der stetige Übergang von der Mikro- zur Makromechanik, *14*, 664.

Stoler, D., Saleh, B. E. A., & Teich, M. C., (1985). Binomial states of the quantized radiation field. *Opt. Acta., 32*, 345.

Thilagam, A., & Lohe, M. A., (2005). Coherent states polarons in quantum well. *Physica E., 25*, 625.

Vourdas, A., (2006). Analytic representations in quantum mechanics. *J. Phys. A: Math. Gen., 39*, R65.

Walls, D. F., & Milburn, G. J., (1995). *Quantum Optics*. Springer-Verlag, Berlin.

CHAPTER 9

Chemical Orthogonal Space

MIHAI V. PUTZ[1,2]

[1]*Laboratory of Structural and Computational Physical Chemistry for Nanosciences and QSAR, Biology-Chemistry Department, West University of Timisoara, Pestalozzi Street No. 44, Timisoara, RO-300115, Romania, Tel.: +40-256-592638, Fax: +40-256-592620, E-mail: mv_putz@yahoo.com, mihai.putz@e-uvt.ro*

[2]*Laboratory of Renewable Energies-Photovoltaics, R&D National Institute for Electrochemistry and Condensed Matter, Dr. A. Paunescu Podeanu Str. No. 144, Timisoara, RO-300569, Romania*

9.1 DEFINITION

The orthogonal space of chemical reactivity represents a chemical concept which can be explained with the aid of the reactivity indices electronegativity and chemical hardness, when one considers in chemical reactivity analysis the difference between the energetic level characterizing electronegativity and the energetic gap characterizing chemical hardness. Also, these two indices can "close" the chemical space of reactivity and bonding in two dimensions (2D).

9.2 HISTORICAL ORIGIN(S)

The electronegativity concept was first introduced by Pauling in 1932, being defined as the atom power to attract electrons to it (Pauling, 1932). Later, in 1934, Muliken states that for the atom valence state, the electronegativity can be defined as the average of the ionization potential (IP) and electron affinity (EA) (Muliken, 1934). Parr and his co-workers adopt a different approach, considering the electronegativity as the minus chemical potential for a multi-electronic system (Parr et al., 1978; Bartolotti et al., 1980; Parr and Bartolotti, 1983).

By performing on the ground state energy, E_N the finite difference approximation around the total electrons number N_0 (Parr, 1983; Parr and Yang, 1989; Kohn et al., 1996), the link between Parr definition and Muliken definition can be achieved. After determining the quantum formulation for electronegativity, chemical hardness was also introduced, as a parameter used in defining the maximum hardness (MH) and the hard-and-soft-acids-and-bases (HSAB), also known as the main chemical principles of reactivity (Parr and Pearson, 1983; Robles and Bartolotti, 1984; Gazquez and Ortiz, 1984; Sen and Mingos, 1993; Pearson, 1997; Putz, 2012).

9.3   NANO-SCIENTIFIC DEVELOPMENT(S)

The chemical orthogonal space is based on the idea of constructing an analogy between the (3+1) D model given by the physical laws and the chemical phenomena (Putz, 2013). This way, one can redefine both the velocity and the acceleration in terms of electronegativity (1) and chemical hardness (2), as follow:

$$\begin{cases} v = \partial r / \partial t \\ \partial E / \partial N = -\chi \cong \dfrac{E_{LUMO} + E_{HOMO}}{2} \end{cases} \tag{1}$$

and

$$\begin{cases} a = \partial^2 r / \partial t^2 \\ \partial^2 E / \partial N^2 = \eta \cong \dfrac{E_{LUMO} - E_{HOMO}}{2} \end{cases} \tag{2}$$

with the phenomenological equivalence (Putz and Putz, 2013):

$$E\left(\text{valence energy}\right) \leftrightarrow r\left(\text{coordinate}\right) \tag{3}$$

$$N\left(\text{no. valence e}^-\right) \leftrightarrow t\left(\text{time}\right) \tag{4}$$

where E represents the energy, N represents the total electron number, *HOMO* represents the highest occupied molecular orbital, and *LUMO* represents the lowest unoccupied molecular orbital (Putz, 2013).

Electronegativity describes the effect of donating electrons, being associated with the mid-level between HOMO-LUMO energy, while chemical hardness describes the effect of accepting electrons, being associated with the HOMO-LUMO gap (Figure 9.1), in this case, one can states that these two physicochemical indices describe a chemical orthogonal space. In this context, (de)formation mechanism and molecular reactivity can be easy analyzed is electronegativity and chemical hardness are considered the reaction orthogonal coordinate, through the orthogonal chemical bonding theory (OCB) (Putz et al., 2011a; Putz, 2013).

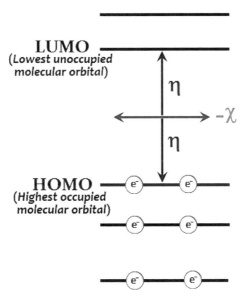

FIGURE 9.1 Schematic representation of a molecular orbital energy diagram with the chemical orthogonal space described by electronegativity and chemical hardness. Redrawn and adapted after Pearson (1987) and Putz (2008a, 2012).

By combining the electronegativity and the chemical hardness one can calculate the chemical power (Eq. 5), seen as a normalized/reduced electronegativity manifested by the system when its inertial hardness is considered (Putz and Dudas, 2013):

$$\pi = \frac{\chi}{2\eta} \tag{5}$$

and the electrophilicity (Eq. 6) which accounts the energy consumed by a system when its chemical power is manifested in a chemical orthogonal space (Putz and Dudas, 2013):

$$\omega = \chi \times \pi = \frac{\chi^2}{2\eta} \tag{6}$$

The reactivity indices can be used to analyze the molecular interaction mechanism in the case of a bonding complex which is chemically formed when a chemical, biological interaction occurs (Putz and Dudas, 2013).

On the conceptual (quantum fundamental) level, Putz (2011a, 2012) determined that the chemical structure is characterized by two different types of quantum manifestations whit their corresponding indices, in the balance between engaging or accepting electrons in bonding (Putz, 2012). In this context, electronegativity is considered a quantum observable which is spanned between the electronic attraction or release and the pure electronic attraction characters, as described by the lower and upper branches of the following equation:

$$\chi_\lambda = -\frac{E_0}{\rho_0} = -\mu = \begin{cases} \infty, & \rho_0 \to 0 \left(E_0 < 0 \right) \\ -E_0 = -\langle \psi_0 | H | \psi_0 \rangle, & \rho_0 \to 1 \end{cases} \qquad (7)$$

On the other hand, chemical hardness cannot be considered a quantum observable due to its absence of a definite or a non-zero value for any limit or electronic density realization, being usually associated with not definite in the sense of Eigenvalue in the case of fully occupied or empty states which can be inert or fully engaged respecting the chemical bonding, or with the chemical bonding open states in the case of fraction occupied states, as in first upper branch of the following equation (Putz, 2012):

$$\eta_\lambda = 0 \cdot E_0 \frac{1 + \lambda \rho_0}{\rho_0 \left(1 - \rho_0 \right)} = \begin{cases} 0, & \rho_0 \in (0.1) \\ 0 \cdot \infty = ?, & \rho_0 \to 0 \\ 0 \cdot \infty = ?, & \rho_0 \to 1 \end{cases} \qquad (8)$$

Even if the chemical hardness and aromaticity can virtually have the same observable character, its "metallic" behavior for the fractional states as seen on the top branch of the Eq. (8) indicates that the chemical hardness do not have any role in the electronic density range or occupancy, the system not being in the second order adjustment of the total energy phase, meaning that the systems electrons could move freely as influenced by the electronegativity according to Eq. (7) (Putz, 2012).

9.4   NANO-CHEMICAL APPLICATION(S)

In recent years, there are many studies which focus on explaining the increasing incidence of cancer, one of the most promising methods being based on the investigation of the carcinogenicity biological property through the computational and network interaction studies. In this context, the quantitative structure-activity relationship (QSAR) model represents a good alternative, having a defined endpoint and domain of applicability, an unambiguous algorithm, predictive power and statistical performance and a mechanistic interpretation (OECD, 2007; Putz et al., 2011b), however, its construction implies molecular representation, experimental data and fitting algorithms (Putz et al., 2011a).

In their work, Putz and his collaborators (2011a) propose two contributions on advanced residual-QSAR modeling of the genotoxic carcinogenesis (Putz, 2011b): they extend the residual QSAR method by considering, in structural alerts, the molecular fragment information, leading to a better understanding of the electrophilic theory of the chemical carcinogenesis, and investigate the influence of the electrophilic and reactivity parameters on the derived QSAR models of carcinogenesis in order to obtains mechanistic insight into mutagenesis (Putz et al., 2011a).

The alert-QSAR method proposed by the authors use in its implementation the following Hansch physicochemical parameters: polarizability (POL) associated with the molecule's vibrational motion in organism and, according to Millers theory, accounts for the potential electrophilic effects which activates cancer (Miller and Miller, 1977,1981; Putz, 2011b), hydrophobicity (LogP) which corresponds with the molecules translation motion and diffusion through trans-cellular membrane and optimal total energy (E_{tot}) which contains the steric information about the 3D structure of molecule[1] and represents the potential of the rotation motion of molecules which is triggered when interacting with the organism's receptor (Putz et al., 2011a). Additional reactivity indices were necessary in their QSAR study on chemical carcinogenicity, i.e., electronegativity (Eq. 9) and chemical hardnes (Eq. 10), which describe the donating electrons effect realized through HOMO levels and IP, and the accepting electrons effect realized through LUMO levels and EA, as following:

$$\chi = \frac{IP + EA}{2} \cong -\frac{\varepsilon_{LUMO} + \varepsilon_{HOMO}}{2} \tag{9}$$

$$\eta = \frac{IP - EA}{2} \cong \frac{\varepsilon_{LUMO} - \varepsilon_{HOMO}}{2} \tag{10}$$

In their QSAR analysis, Putz et al. (2011a) combine electronegativity and chemical hardness orthogonality with the associated two reactivity principles (Putz et al., 2008b). The first one, namely the electronegativity equalization

---

[1]HyperChem 7.01, Program package and documentation. Hypercube, Inc.: Gainesville, FL, USA, 2002.

principle (EE) was applied at the ligand-receptor binding level. First, the electronegativity is equalized with that of the receptor, which implies the receptor's pocket adjustment in order to fit with the electronegativity of the ligand, this stage being similar to electronegativity based docking, according to fundamental quantum EE principle, while the chemical hardness stabilized the induced interaction. The second one, namely the maximum hardness principle was constructed starting from the Pearson's observation, i.e., "the molecules arrange themselves to be as hard as possible" (Pearson, 1985), in other words, for a molecular sample, a stabilized interaction is associated with a maximum HOMO-LUMO gap (Putz et al., 2011a).

Both fundamental reactivity principles are referring to an intra-electrophilic/intramolecular electron transfer from the ligand molecule HOMO and LUMO and after that extra-electrophylic/intermolecular electron transfer occurs between the new molecule (SA)-HOMO* and receptor LUMO. As a result, the ligand HOMO* relaxation will determine a larger SA-HOMO-LUMO gap, being formally removed so that the gap which appears between the second order ligand's HOMO and the LUMO* increase, producing an overall electrophilic docking effect, as seen in Figure 9.2 (Putz et al., 2011a).

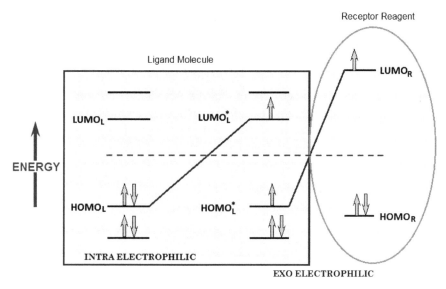

FIGURE 9.2 Schematic representation of the electrophilic docking structure-reactivity algorithm which correlates the chemical carcinogenesis with electronegativity and chemical hardness. Redrawn and adapted after Putz et al. (2011a).

In other words, the ligand-receptor docking mechanism can be described by electronegativity and chemical hardness through intra- and exo-electrophilic stages, this way, the theory of Millers referring to the direct electronic transfer between the molecular (SA)-HOMO and receptor LUMO being generalized (Miller and Miller, 1981; Putz et al., 2011a).

There are three operational stages associated with the carcinogenesis process: initiation, promotion and progression, each stage having a distinct mechanism and QSAR characteristics. In this context, the selection of appropriate molecular descriptors has an important role in a meaningful QSAR formulation (Putz et al., 2011a, Benigni et al., 2007). From their study, Putz et al., (2011a) determined that mutagenicity can be considered an interaction mechanism of the electrophilic ligand-receptor type which implies a covalent binding between the ligand molecule SA and receptor. They also determined that the residual QSAR analysis can be improved by the structural alert or molecular fragment analysis (Putz et al., 2011a).

9.5 MULTI-/TRANS- DISCIPLINARY CONNECTION(S)

The acids and bases have an important role in chemical bond and bonding theory, their characterization being a conceptual challenge for chemists for centuries (Bensaude-Vincent, 1992; Servos, 1990; Putz, 2008a), their first notable mention beginning in the XVII century with Lemery and Boyle Cartesian salt theory along with the associate reactivity principle which was driven by the "struggle between acids and alkali" (Putz, 2012). Years later, Volta, Gay-Lussac, and Liebig conducted several physicochemical experiments from which they determine that in order to be exchanged with a metal/radical of a different nature, the acids need to contain hydrogen, proposing the following principle (Putz, 2008a, 2012):

$$\text{acid} + \text{base} \leftrightarrow \text{salt} + \text{water} \tag{11}$$

Bronsted and Lowry gave a new insight, in 1923, by assuming the proton (which is never free) as the intermediate between the donor/acid and the acceptor/base when a chemical reaction occurs, their new acid-base interaction paradigm being of the form (Putz, 2012):

$$\text{acid } 1 + \text{base } 2 \leftrightarrow \text{ acid } 2 + \text{base } 1 \tag{12}$$

However, Lewis (1916) was the one who determined that the electronic pair can be used in defining the acid and bases along with their chemical bonding, the entire chemistry being reformulated in terms of Lewis compounds, in other words, the acid and the base represents the chemical species which are susceptible to accept, respectively to donate a pair of electrons. The acidic Lewis definition appears to covers general cases, while the Lewis base definition can be superimposed on the Bronsted-Lowry theory (Putz, 2008a, 2012).

The molecular orbital theory proposed by Pearson (1963, 1985, 1986, 1987, 1988, 1989, 1990, 1997) leads to a redefining of the acid-base theory. Starting from this, the chemical reactions and bonding have been considered in two steps: first is referring to the Coulombic interaction which are "solved" by electronegativity and the equalization principle associated, while in the second one the chemical hardness regulate the stability of the new chemical bond (Parr and Pearson, 1983; Zhou and Parr, 1990; Chattaraj and Parr, 1993; Parr and Gázquez, 1993; Putz, 2012).

In terms of molecular orbital theory, the acid and basis can be classified as soft ("s") and hard ("h") as follow: in a soft species, the electrons can be transferred easily in the vacant orbital (LUMO), the chemical reactions being more favorable when the HOMO level of one compound approaches the LUMO level vertically from the other compound involved in reaction (Putz, 2008a, 2012). In this context, the molecular systems can be seen as hard, and soft acid and bases (HSAB) structures, i.e., each molecule can be considered a hard-hard, soft-soft, hard-soft and soft-hard combination formed between the basic and acidic molecular regions. The chemical reactivity HSAB principle states that "soft acids prefer soft bases and hard acids prefer hard bases" (Putz, 2008a, 2012):

$$h_1 - s_1 + s_2 - h_2 \leftrightarrow h_1 - h_2 + s_1 - s_2 \tag{13}$$

In his work, Putz (2012) considers the electronegativity vs. chemical hardness approach to models the HSAB through the Koopmans's theorem (1934) with involving the valence vertical charge exchanges, or through the orbital relaxation effects. He starts from the general scheme of the acid-base reaction occurring between two acid-base complexes in gas-phase, where a base (X) is bonded to the hard acid (H^+) and removes the hard base (OH^-) from a soft acid (OH^+) as following:

$$H\text{-}X^+ + OH - OH \leftrightarrow HO\text{-}X^+ + H\text{-}OH \tag{14}$$

As results, Putz determined that the electronegativity returns positive and negative values, while the chemical hardness always returns positive values. On the other hand, by connecting the values of electronegativity and chemical hardness for neutral and ionic molecules in the compact finite difference computation scheme, they will form arches with curvatures of opposite arches, which proves once again that they are orthogonal quantum quantities, having different roles and effects in chemical reactivity (Putz, 2012).

9.6 OPEN ISSUES

Electronegativity equalization is a well-known concept in chemistry used in understanding the molecular charge distribution (Sanderson, 1983, 1951) its modern form, electronegativity equalization method (EEM) being used *in ab initio* atomic charges computation which utilized a set of electronegativity and chemical hardness parameters and the molecular geometry (Verstraelen and Bultinck, 2015; Mortier et al., 1985, 1986; Mortier, 1987). In other words, EEM can be seen as a semi-empirical form of density functional theory (DFT) which expand the system energy having a specific molecular geometry in a second order Taylor series of charges and model the molecular electron density using a minimum set of variables (Verstraelen and Bultinck, 2015; Parr and Yang, 1989; Kohn et al., 1996).

In their work, Verstraelen and Bultinck (2015) tested the EEM ability to reproduce the molecular dipole derivatives and the Cartesian Hessian, two important quantities of the infrared spectroscopy field on a set of 166 organic drug-like molecules (Bultinck et al., 2002a, 2002b; Smirnov and Bougeard, 2003). In this context, the EEM energy expression can be obtained by considering first the expression of molecular energy defined for a system with N atoms, which is approximated, in the atomic charge, to the second order, as follow (Verstraelen and Bultinck, 2015):

$$E(\{q\}) = \chi^T \mathbf{q} + \frac{1}{2} \mathbf{q}^T \overline{\overline{\eta}} \mathbf{q} \qquad (15)$$

with χ the column vector which has N atomic electronegativity parameters, $\mathbf{q}$ the column vector which has N atomic charges and $\overline{\overline{\eta}}$ the hardness matrix, a symmetric matrix $N \times N$ which contains all second-order coefficients (Verstraelen and Bultinck, 2015).

Moving forward, by minimizing this energy using a constraint on the total charge q_{mol} one can obtain the EEM ground state charges as follow:

$$E_{gs} = E(\{q\}_{gs}) = \min_{\{q\}, \mathbf{d}^T \mathbf{d} = q_{mol}} E(\{q\}) \qquad (16)$$

with $\mathbf{d}$ the column vector containing the N times number 1 (Verstraelen and Bultinck, 2015).

When using the EEM, one can model the molecule as a spherical atomic charge distribution superposition, which implies that the Cartesian components of the molecular dipole moment will have the form:

$$D_j = \sum_{i=1}^{N} R_{ij} q_{gs,i} \tag{17}$$

with q_i the atom i charge, R_{ij} the component j of the atom i Cartesian coordinates (Verstraelen and Bultinck, 2015).

After a series of DFT calculations, the authors determined that both Cartesian Hessian and dipole derivative cannot be described using only the EEM model, due to the fact that for computer simulation of infrared (IR) spectra is necessary a more advanced model of the atomic charge response to a nuclei displacement (Verstraelen and Bultinck, 2015).

KEYWORDS

- **chemical hardness**
- **chemical power**
- **electronegativity**
- **electronegativity equalization principle**
- **electrophilicity**
- **HOMO**
- **LUMO**
- **maximum hardness principle**

REFERENCES AND FURTHER READING

Bartolotti, L. J., Gadre, S. R., & Parr, R. G., (1980). Electronegativities of the elements from the simple X αtheory. *Journal of the American Chemical Society, 102*, 2945–2948.

Benigni, R., Netzeva, T. I., Benfenati, E., Bossa, C., Franke, R., Helma, C., et al., (2007). The expanding role of predictive toxicology: An update on the (Q)SAR models for mutagens and carcinogens. *Journal of Environmental Science and Health C., 25*, 53–97.

Bensaude-Vincent, B., (1992). *Histoire de la Chimie*. La Découverte: Paris.

Bultinck, P., Langenaeker, W., Lahorte, P., De Proft, F., Geerlings, P., Waroquier, M., & Tollenaere, J. P., (2002a). The electronegativity equalization method I: Parametrization and validation for atomic charge calculations. *Journal of Physical Chemistry A., 106*, 7887–7894.

Bultinck, P., Langenaeker, W., Lahorte, P., De Proft, F., Geerlings, P., VanAlsenoy, C., & Tollenaere, J. P., (2002b). The electronegativity equalization method II: Applicability of different atomic charge schemes. *Journal of Physical Chemistry A., 106*, 7895–7901.

Chattaraj, P. K., & Parr, R. G., (1993). Density functional theory of chemical hardness. *Structure and Bonding, 80*, 11–25.

Gázquez, J. L., & Ortiz, E., (1984). Electronegativities and hardness of open-shell atoms. *Journal of Chemical Physics, 81*, 2741–2748.

Kohn, W., Becke, A. D., & Parr, R. G., (1996). Density functional theory of electronic structure. *Journal of Physical Chemistry, 100*, 12974–12980.

Koopmans, T., (1934). Uber die zuordnung von wellen funktionen und eigenwerter zu den einzelnen elektronen eines atom. *Physica, 1*, 104–113.

Lewis, G. N., (1916). The atom and the molecule. *Journal of the American Chemical Society, 38*, 762–785.

Miller, E. C., & Miller, J. A., (1981). Searches for ultimate chemical carcinogens and their reactions with cellular macromolecules. *Cancer, 47*, 2327–2345.

Miller, J. A., & Miller, E., (1977). Ultimate Chemical Carcinogens as Reactive Mutagenic Electrophiles. In: Hiatt, H. H., Watson, J. D., & Winsten, J. A., (eds.), *Origins of Human Cancer, Cold Spring Harbor Laboratory* (pp. 605–628). Huntington, NY, USA.

Mortier, W. J., (1987). Electronegativity equalization and its applications. *Structure and Bonding, 66*, 125–143.

Mortier, W. J., Ghosh, S. K., & Shankar, S., (1986). Electronegativity-equalization method for the calculation of atomic charges in molecules. *Journal of the American Chemical Society, 108*, 4315–4320.

Mortier, W. J., Vangenechten, K., & Gasteiger, J., (1985). Electronegativity equalization: Application and parametrization. *Journal of the American Chemical Society, 107*, 829–835.

Mulliken, R. S., (1934). A new electro-affinity scale: Together with data on valence states and an ionization potential and electron affinities. *Journal of Chemical Physics, 2*, 782–793.

OECD principles. Guidance document on the validation of (quantitative) structure-activity relationship [(q)sar] models; OECD environment health and safety publications series on testing and assessment No. 69; Organization for Economic Co-operation and Development: Paris, France, 30 March 2007. Available online: https://www.oecd.org/env/guidance-document-on-the-validation-of-quantitative-structure-activity-relationship-q-sar-models-9789264085442-en.htm (accessed on 7 January 2020).

Parr, R. G., & Bartolotti, L. J., (1983). Some remarks on the density functional theory of few-electron systems. *Journal of Physical Chemistry, 87*, 2810–2815.

Parr, R. G., & Gázquez, J. L., (1993). Hardness functional. *Journal of Physical Chemistry, 97*, 3939–3940.

Parr, R. G., & Pearson, R. G., (1983). Absolute hardness: Companion parameter to absolute electronegativity. *Journal of the American Chemical Society, 105*, 7512–7516.

Parr, R. G., (1983). Density functional theory. *Annual Review of Physical Chemistry, 34*, 631–656.

Parr, R. G., Donnelly, R. A., Levy, M., & Palke, W. E., (1978). Electronegativity: The density functional viewpoint. *Journal of Chemical Physics, 68*, 3801–3808.

Parr, R., & Yang, W., (1989). *Density-Functional Theory of Atoms and Molecules*. Oxford Science Publications, Oxford, UK.

Pauling, L., (1932). The nature of the chemical bond IV. The energy of single bonds and the relative electronegativity of atoms. *Journal of the American Chemical Society, 54*, 3570–3582.

Pearson, R. G., (1963). Hard and soft acids and bases. *Journal of the American Chemical Society, 85*, 3533–3539.

Pearson, R. G., (1985). Absolute electronegativity and absolute hardness of Lewis acids and bases. *Journal of the American Chemical Society, 107*, 6801–6806.

Pearson, R. G., (1986). Absolute electronegativity and hardness correlated with molecular orbital theory. *Proceedings of the National Academy of Sciences, 83*, 8440–8441.

Pearson, R. G., (1987). Recent advances in the concept of hard and soft acids and bases. *Journal of Chemical Education, 64*, 561–567.

Pearson, R. G., (1988). Absolute electronegativity and hardness: Application to inorganic chemistry. *Inorganic Chemistry, 27*, 734–740.

Pearson, R. G., (1989). Absolute electronegativity and hardness: Application to organic chemistry. *Journal of Organic Chemistry, 54*, 1423–1430.

Pearson, R. G., (1990). Hard and soft acids and bases—the evolution of a chemical concept. *Coordination Chemistry Reviews, 100*, 403–425.

Pearson, R. G., (1997). *Chemical Hardness*. Wiley-VCH: Weinheim.

Putz, M. V., & Dudaş, N. A., (2013). Determining chemical reactivity driving biological activity from smiles transformations: The bonding mechanism of anti-HIV pyrimidines. *Molecules, 18*, 9061–9116.

Putz, M. V., & Putz, A. M., (2013). DFT chemical reactivity driven by biological activity: Applications for the toxicological fate of chlorinated PAHs. *Structure and Bonding, 150*, 181–232.

Putz, M. V., (2008a). Maximum hardness index of quantum acid-base bonding. *MATCH Communications in Mathematical and in Computer Chemistry, 60*(3), 845–868.

Putz, M. V., (2008b). *Absolute and Chemical Electronegativity and Hardness*. Nova Science Publishers Inc., New York, NY, USA.

Putz, M. V., (2011a). Electronegativity and chemical hardness: Different patterns in quantum chemistry. *Current Physical Chemistry, 1*(2), 111–139.

Putz, M. V., (2011b). Residual-QSAR. Implications for genotoxic carcinogenesis. *Chemistry Central Journal, 5*, 29(1–11).

Putz, M. V., (2012). *Chemical Orthogonal Spaces* (Vol. 14, pp. 240). Mathematical chemistry monographs, Kragujevac, Serbia.

Putz, M. V., (2013). Bonding in orthogonal space of a chemical structure: From in Cerebro to in silico. *International Journal of Chemical Modeling, 5*(4), 369–395.

Putz, M. V., Ionaşcu, C., Putz, A. M., & Ostafe, V., (2011). Alert-QSAR. Implications for an electrophilic theory of chemical carcinogenesis. *International Journal of Molecular Sciences, 12*, 5098–5134.

Putz, M. V., Putz, A. M., & Barou, R., (2011). Spectral-SAR realization of OECD-QSAR principles. *International Journal of Chemical Modeling, 3*(3), 173–188.

Robles, J., & Bartolotti, L. J., (1984). Electronegativities, electron affinities, ionization potentials, and hardnesses of the elements within spin-polarized density functional theory. *Journal of the American Chemical Society, 106*, 3723–3727.

Sanderson, R. T., (1951). An interpretation of bond lengths and a classification of bonds. *Science, 114*, 670–672.

Sanderson, R. T., (1983). *Polar Covalence.* Academic Press, New York.

Sen, K. D., & Mingos, D. M. P., (1993). *Chemical Hardness* (Vol. 80). Structure and Bonding, Springer, Berlin.

Servos, J. W., (1990). *Physical Chemistry from Ostwald to Pauling.* Princeton University Press, Princeton.

Smirnov, K. S., & Bougeard, D., (2003). Including the polarization in simulations of hydrated aluminosilicates. Model and application to water in silicalite. *Chemical Physics, 292,* 53–70.

Verstraelen, T., & Bultinck, P., (2015). Can the electronegativity equalization method predict spectroscopic properties? *Spectrochimica Acta Part A: Molecular and Biomolecular Spectroscopy, 136,* 76–80.

Zhou, Z., & Parr, R. G., (1990). Activation hardness: New index for describing the orientation of electrophilic aromatic substitution. *Journal of the American Chemical Society, 112,* 5720–5724.

CHAPTER 10

Electrochemical Impedance

MIRELA I. IORGA[1] and MIHAI V. PUTZ[1,2]

[1]*National Institute for Research and Development in Electrochemistry and Condensed Matter, Department of Applied Electrochemistry, Laboratory of Renewable Energies-Photovoltaics, 300569 Timisoara, Romania, E-mail: mirela_iorga@yahoo.com, mirela.iorga@incemc.ro*

[2]*West University of Timisoara, Biology-Chemistry Department, Laboratory of Structural and Computational Physical Chemistry for Nanosciences and QSAR, 300115 Timisoara, Romania, E-mail: mv_putz@yahoo.com, mihai.putz@e-uvt.ro*

10.1 DEFINITION

Electrochemical impedance represents the measurement of the circuit's ability to resist to electric current flow, similar to resistance, but with the difference that it is not limited to an ideal resistor. In electrochemistry, it could be measured by applying an alternating current potential to an electrochemical cell, and as a result, the current through the cell will be recorded. The main application of electrochemical impedance is one of the most important tools used in investigations and measurements, namely the electrochemical impedance spectroscopy (EIS).

10.2 HISTORICAL ORIGIN(S)

The concept of electrical impedance (Hunt, 2012) was discovered in the 1880s by Oliver Heaviside, an English mathematician. He invented the operational calculus which could be applied to convert differential equations describing the current flow in circuits containing resistors, capacitors, and inductors (Barett, 2008) into regular algebraic equations in Laplace-transformed independent and dependent variables (Calvert, 2002). These conversions laid the foundations of electrical engineering discipline (Nahin, 2002). The impedance term originates from the Latin word *impedire*, meaning *to hinder* (Bos, 2008). Heaviside invented the terms of impedance, admittance, and reactance; and he is known as the father of impedance spectroscopy (Wiener, 1993).

The operational impedance $Z(s)$ was defined as shown in Eq. (1):

$$Z(s) = \frac{\overline{V}(s)}{\overline{I}(s)} \tag{1}$$

where $\overline{V}(s)$ is the Laplace transform of the potential; $\overline{I}(s)$ is the Laplace transform of the current.

Then the Laplace space is transformed into Fourier space, following equation (2):

$$s = j\omega \tag{2}$$

Impedance is defined only in the case of steady-state systems, thus impedance as well admittance and reactance are defined only within linear systems theory (LST), which means that they must be linear, stable, causal and time-invariant (Macdonald, 2006).

In 1899, the impedance concept was applied by Warburg to electrochemical systems, following his original paper "Uber das Verhalten sogenannter unpolarisirbarer Elektroden gegen Wechselstrom" published in *Annalen der Physik und Chemie* (Warburg, 1899). This research work treated the diffusion transport impedance of an electroactive species to an electrode surface; this concept was named after him and is known as Warburg impedance (Bonciocat, 1996). The equation expressing impedance could be considered the most famous in electrochemistry (the oldest and frequently observed) (Macdonald, 2006):

$$Z(\omega) = \frac{\sigma}{\omega^{1/2}} - \frac{j\sigma}{\omega^{1/2}} \tag{3}$$

In the 1920s, Kramers-Konig transforms were developed, initially used only for optical data, and after many years for an independent check of impedance data validity (Kramers, 1929; Konig, 1926). These transforms derived from Cauchy's theorem and provides the scientific basis for causality (one of the LST constraints).

The earliest use of analogs to explain electrochemical data was done in 1940 by Dolin and Ershler (1940). They investigated the absorption of hydrogen on platinum and derived the equations describing the dependence of current frequency and electrode impedance similar to a parallel circuit of a resistance and a capacitance (Armstrong, 1978).

Then, in 1947, Randles took into consideration rapid metal/metal ion and redox reactions, where time-averaged concentrations are following the Nernst equation (Randles, 1947). In this case, could be considered that the impedance is given by a series circuit consisting of a resistance and a capacitance (Armstrong, 1978). Randles equivalent circuit made a dynamic simulation of the impedance characteristics, in a rapid charge transfer reaction in the case of a planar electrode. Based on the explicit expression of Warburg impedance, now appears the possibility to connect electrical elements with chemical parameters.

In the 1960s, deLevie (1967) developed the theory of porous electrodes, studying the effects of porous surfaces on electrochemical kinetics.

At the beginning of the 20th century, the primary use of impedance techniques was the study of electrified interfaces by measuring the electrode capacitance using reactive bridges at frequencies higher than 500 Hz (Lasia, 1999).

After the invention of the electronic potentiostat, at the beginning of the 1940s, and with the emergence of frequency response analyzers, in 1970s, the possibility to measure the impedance data at low frequencies (<1 mHz) arises (Macdonald, 2006). If initially, the primary applications were for determination of double-layer capacitance and alternating current polarography, this moment represented the beginning of an exponential development of electrochemical impedance spectroscopy applications (Lasia, 1999). The broad domains in which EIS will apply are heterogeneous reaction mechanisms, passivation, corrosion, conduction in solid (liquid) states, conduction (ionic and electronic) in polymers, etc.

10.3 NANO-SCIENTIFIC DEVELOPMENT(S)

Impedance represents the measurement of the circuit's ability to resist to electric current flow, similar to resistance, but with the difference that it is not limited to an ideal resistor. If an electrical perturbation is applied to an electrical circuit, it will cause a response following Ohm's Law:

- in the case of a single resistance R:

$$E(t) = i(t) \cdot R \tag{4}$$

- in the event of a series circuit of resistance, R, and a capacitance, C:

$$E(t) = i(t) R + \frac{Q(t)}{C} = i(t) R + \frac{1}{C} \int_0^t i(t) \, dt \tag{5}$$

According to Macdonald, the solution of Eq. (5) could be obtained by differential equations or by Laplace transform (Macdonald, 2006).

In the case of differential equations, for known $E(t)$, the following equation will be solved:

$$\frac{di(t)}{dt} + \frac{i(t)}{RC} = \frac{1}{R}\frac{dE(t)}{dt} \tag{6}$$

In the case of Laplace transform, the general form is:

$$\nabla\left[f(t)\right] = \bar{f}(s) = F(s) = \int_0^\infty f(t)\exp(-st)\,dt \tag{7}$$

$$\nabla\left(\int_0^t i(t)\,dt\right) = \frac{i(s)}{s} \tag{8}$$

where s represents the frequency.

If Laplace transform is applied to Eq. (5), the current will be expressed as the Eq. (9):

$$i(s) = \frac{E(s)}{R + \dfrac{1}{sC}} \tag{9}$$

and the expression of impedance, Z, will be:

$$Z(s) = \frac{E(s)}{i(s)} = R + \frac{1}{sC} \tag{10}$$

The inverse of impedance is named admittance, Y, and will have the expression:

$$Y(s) = \frac{1}{Z(s)} \tag{11}$$

Both $Z(s)$ and $Y(s)$ are transfer functions, and they have the property to transform one function into another.

The system impedance could be very facile determined if the Laplace transforms of the current and voltage are determined by numerical transform calculations (a much easier method).

Electrochemical impedance (Lasia, 1999) could be measured by applying an alternating current potential (a sinusoidal potential excitation) to an electrochemical cell. As a result, the current through the cell will be recorded (an AC signal, which could be analyzed as a sum of sinusoidal functions–a Fourier series).

In this case, the applied potential could be expressed as:

$$E_t = E_o \sin(\omega t) \tag{12}$$

$$\omega = 2\pi f \tag{13}$$

where: E_t represents the potential at time t; E_0 represents signal's amplitude; ω represents angular frequency; and f represents frequency.

The current response will have the expression:

$$I_t = I_0 \sin(\omega t + \varphi) \tag{14}$$

where I_t represents the current response; I_0 represents amplitude; and φ represents phase shift.

The expression for impedance will be:

$$Z_t = \frac{E_t}{I_t} = \frac{E_0 \sin(\omega t)}{I_0 \sin(\omega t + \varphi)} = Z_0 \frac{\sin(\omega t)}{\sin(\omega t + \varphi)} \tag{15}$$

where Z_0 in the magnitude.

The electrochemical impedance could be defined as the sum of real (noted Z_{Re} or Z') and imaginary (noted Z_{Im} or Z'') parts:

$$Z(\omega) = |Z_0|(\cos\varphi + j\sin\varphi) = |Z_0|\cos\varphi + j|Z_0|\sin\varphi = Z_{Re} + jZ_{Im} \tag{16}$$

Electrochemical impedance data (Bos, 2008) could be evaluated by two main types of plotting: Nyquist plot (also known as Cole-Cole plot or complex impedance plane plot) and Bode plot. For most of the data, they are analyzed by both graphics, Nyquist and Bode.

In the case of the Nyquist plot, the imaginary impedance component (Z'') is planned to the real impedance component (Z') at each excitation frequency (Cesulius et al., 2016). Each point on the Nyquist plot represents the impedance at one frequency. These are the most used in electrochemistry because it allows a simple prediction of the circuit parts and is easily related to an electrical pattern. The main disadvantage is that they do not contain all the necessary information, for example, frequencies are not obviously presented, and despite the fact that the ohmic resistance and the charge resistance could be read directly from the diagram, if one is necessary to obtain the electrode capacitance, the frequencies must be known.

A much better representation is realized by the Bode plot, which contains all the information needed. In this case, the impedance modulus and the phase shift are represented as a function of the logarithm of frequency. They are used particularly in circuit analysis, and this logarithmic representation allows a wide frequency range on the same chart and a more efficient data extrapolation.

To interpret the impedance data, two types of approaches exist: fundamental and phenomenological (Bos, 2008).

In the fundamental approach, a physical model for the process is necessary, based on which a transfer function (a mathematical dependence between the systems's input and output) could be expressed. This relation includes all the model's parameters (e.g., concentrations, diffusion coefficients, kinetic parameters, geometric features, hydrodynamics, etc.) and corresponds to the measured data. The objectives of physical models are to reproduce the process of interest and to characterize the mechanism of the interface process (Macdonald, 2006).

The phenomenological approach implies an interpretation based on electrical equivalent circuits (analogs), such as resistors, capacitors, and inductors, specially designed with similar impedance properties as the impedance data obtained by measurements; they reproduce the features of the system. The main electrical equivalent circuits and the impedance expressions are presented in Table 10.1 (Lvovich, 2012).

TABLE 10.1 The Expressions of Impedance vs. Electrical Equivalent Circuits (Lvovich, 2012)

Electrical equivalent circuit	Equation	Impedance expression
Resistor (R)	$E = I \times R$	$Z = R$
Capacitor (C)	$I = C \dfrac{dE}{dt}$	$Z = \dfrac{1}{j\omega C}$
Inductor (L)	$E = L \dfrac{di}{dt}$	$Z = j\omega L$

In the case of electrochemical cells, since they are very complex systems, sometimes this approach is preferred (Bos, 2008). Such a complex phenomenon made up of chemical, physical, electrical, and mechanical parts are represented concerning pure electricity. In this case, it is rarely used a single equivalent circuit element; both serial and parallel electrical circuits are favored.

The following equations give the equivalent impedance:

- Series circuit:

$$Z_{eq} = Z_1 + Z_2 + Z_3 \tag{17}$$

- Parallel circuit:

$$\frac{1}{Z_{eq}} = \frac{1}{Z_1} + \frac{1}{Z_2} + \frac{1}{Z_3} \qquad (18)$$

All the electrode processes, e.g., metal deposition, ions electro-reduction, the formation of oxide films, metal corrosion, etc., implies similar physico-electrochemical terms, such as electric double layer capacitance, electrolyte resistance, polarization resistance, charge transfer resistance, diffusion, coating capacitance and so on. Thus, equivalent electrical circuits could simulate these electrode processes, and if the model's equations are used, the electrochemical impedance could be calculated (Cesulius et al., 2016).

According to Cesiulis et al. (2016), the reactions that took place consequently are modeled by series circuits, while parallel circuits model the reactions that occur simultaneously.

Before modeling the experimental results, the impedance data must be checked to determine if the values are valid; this is done with a mathematical procedure introduced by Kramers and Kronig, and usually named KK integral transforms (Agarwal et al., 1993). The impedance data are valid if four criteria are met: linearity, causality, stability, and finiteness (Lasia, 1999).

10.4   NANO-CHEMICAL APPLICATION(S)

All of the electrochemical impedance techniques nowadays represent a stand-alone domain, known as electrochemical impedance spectroscopy (EIS), a powerful technique used in investigations and measurements. Different frequencies of the alternating current, on a large range ($0.2-10^5$ Hz) are used (Bonciocat, 1996).

This technique became widely used lately due to its ability to explore the electrochemical reactions mechanisms, to measure the dielectric and transport properties of materials, to study the properties of porous electrodes, and to investigate passive surfaces (Macdonald, 2006).

EIS is a non-destructive method, providing high experimental efficiency, and can give information about the electrode processes and complex interfaces. It can differentiate the electric and dielectric properties of the contributors from the investigated components.

Electrochemical impedance spectroscopy is applied in various domains, as follows: electrodeposition, corrosion, batteries, fuel cells, electrochemical capacitors, dielectric measurements, semiconductors, solid state devices, electrochromic materials, coatings, polymer films, industrial colloids, analytical chemistry, sensors, biomedical devices, imaging (Orazem and Triboullet, 2008). These application domains are schematically represented in Figure 10.1.

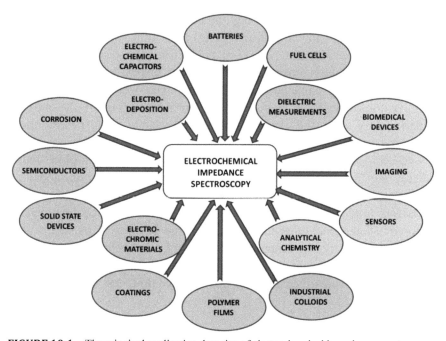

FIGURE 10.1   The principal application domains of electrochemical impedance spectroscopy.

For example, in electrodeposition, EIS is used to define the electrodeposition bath composition, for surface pretreatment, to control the deposition mechanism, and to characterize the obtained deposit. In electro-organic synthesis, it is used to study the reaction mechanism and the adsorption/desorption reactions.

In corrosion, EIS is applied for rate determination, passive layer investigations, inhibitor and coatings studies (i.e., dielectric measurements and corrosion protection). In batteries domain, the technique is used for materials selection, electrode design, and charge state evaluation.

In semiconductors field, EIS could be applied for photovoltaics and to study dopant distributions.

In his attempt to estimate the kinetic parameters and the concentrations of the electrochemically active species, Bonciocat (2005) demonstrated that the electrochemical impedance could be expressed using the first case of the Fredholm alternative.

In the case of a redox electrode reaction:

$$O + e^- \leftrightarrow R \tag{19}$$

The equation describing the kinetics is a Volterra type integral equation (representing a particular case of the Fredholm integral equation), namely:

$$i(t) = \lambda \int_0^t K(t,u) i(u) \, du + f(t) \tag{20}$$

where: $\lambda = -1$; $i(u)$ and $i(t)$ – faradic current densities at $u < t$ and t, respectively; $K(u,t)$ – integral equation kernel; and $f(t)$ have explicit expressions containing kinetic parameters of the electrode reaction, concentrations, diffusion coefficients of the electrochemically active species, times t and u, overvoltage at t time.

If only limitations of charge transfer and diffusion transfer are present, the main EIS method is the basic Faraday impedance method (applied in electrode's equilibrium state) (Bonciocat and Marian, 2006). In this process, a tiny (≤ 5 mV) alternating overvoltage is applied to the interface (in the equilibrium state), and the current density which passes the faradaic impedance, Z_F, of the interface is analyzed. Between the overvoltage, $\eta(t)$, and the current density, $i_F(t)$, a phase difference occurs, whose value depends on the reaction type (irreversible, reversible or quasi-reversible).

In the case of an irreversible equation, only limitations due to charge transfer are present, and Z_F is reduced to a charge transfer resistance, R_{ct}:

$$R_{ct} = \frac{RT}{Fi^0} \tag{21}$$

$\varphi = 0$, which means that $\eta(t)$ and $i_F(t)$ are in phase.

In the case of a reversible equation, when only limitations due to the diffusion transfer are present, a phase difference, φ, occurs between $\eta(t)$ and $i_F(t)$. This fact means that the diffusion limitations could not be characterized only by a diffusion resistance, R_W, and a "pseudo-capacitance," C_W, is needed; then the impedance is called Warburg impedance, Z_W.

In the case of a quasi-reversible equation, all three components (R_{ct}, R_W, C_W) are present, and φ depends on their relative contribution.

If the current density has a simplified expression:

$$i_F(t) = {}_1 I_F \sin \omega t \tag{22}$$

where $_1I_F$ refers to the amplitude of the first harmonic, the overvoltage could be deduced from Volterra type integral equation:

$$i_F(t) = \lambda \frac{i^0 N}{\pi^{1/2}} \int_0^t \frac{i_F(u)}{(t-u)^{1/2}} du - \frac{F}{RT} i^0 \eta(t) \tag{23}$$

So, the equation of the basic Faraday method for the first case of Fredholm alternative is:

$$_1I_F \sin \omega t = \lambda \frac{i^0 N}{\pi^{1/2}} \int_0^t \frac{_1I_F \sin \omega(t-u)}{u^{1/2}} du - \frac{F}{RT} i^0 \eta(t) \tag{24}$$

The combined overvoltage, given by charge transfer and diffusion transfer, will be expressed as follows:

$$\eta_{ct+d}(t) = -\sqrt{\left(\frac{RT}{Fi^0} + \frac{\sigma}{\omega^{1/2}}\right)^2 + \frac{\sigma^2}{\omega}} \; _1I_F \sin\left(\omega t - tg^{-1} \frac{\sigma/\omega^{1/2}}{\frac{RT}{Fi^0} + \frac{\sigma}{\omega^{1/2}}}\right) \tag{25}$$

where:

$$\sigma = \frac{RT}{\sqrt{2}F} N \tag{26}$$

The expressions for R_W and C_W will be:

$$R_W = \frac{\sigma}{\omega^{1/2}} \tag{27}$$

$$C_W = tg^{-1} \frac{\sigma/\omega^{1/2}}{\frac{RT}{Fi^0} + \frac{\sigma}{\omega^{1/2}}} \tag{28}$$

In this case, the expression for impedance is:

$$Z_F = (R_{ct} + R_W) - X_{c_W} j \tag{29}$$

where the capacitive reactance is:

$$X_{c_W} = \frac{1}{\omega C_W} \tag{30}$$

Since the basic Faraday impedance method determines the relation between charge transfer resistance (R_{ct}), Warburg resistance (R_W), Warburg capacitance (C_W), and the frequency of alternating current (ω), it could be defined as a spectroscopic method. The basic Faraday impedance method represents, in fact, the simplest method of stationary electrochemical impedance spectroscopy.

According to Eq. (16), the general expression for impedance is:

$$Z(\omega) = Z_{\text{Re}}(\omega) + jZ_{\text{Im}}(\omega) \tag{31}$$

where:

$$Z_{\text{Re}}(\omega) = R_{sol} + \frac{R_{ct} + \sigma\omega^{-1/2}}{\left(1 + \omega^{1/2}\sigma C_d\right)^2 + \omega^2 C_d^2 \left(R_{ct} + \sigma\omega^{-1/2}\right)^2} \tag{32}$$

and

$$-Z_{\text{Im}}(\omega) = \frac{\omega C_d \left(R_{ct} + \sigma\omega^{-1/2}\right)^2 + \sigma\omega^{-1/2}\left(1 + \omega^{1/2}\sigma C_d\right)}{\left(1 + \omega^{1/2}C_d\right)^2 + \omega^2 C_d^2 \left(R_{ct} + \sigma\omega^{-1/2}\right)^2} \tag{33}$$

where C_d represents the electrochemical double layer capacitance.

Expressions (32) and (33) represent the parametric equations of the impedance diagram $Z_{\text{Im}} = f(Z_{\text{Re}})$, where the parameter is the alternating current frequency.

The electrochemical impedance spectroscopy represents the whole methods based on impedance determination at various frequencies, and from these representations, electrochemical double layer capacitance, kinetic parameters of the redox reaction, and electrochemical active species concentrations could be estimated, as represented in Figure 10.2.

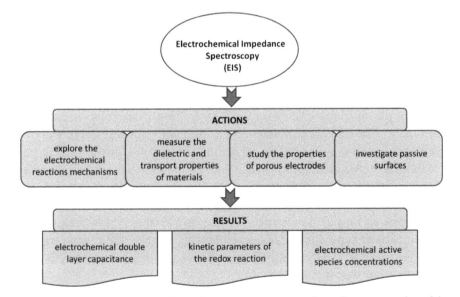

FIGURE 10.2 Electrochemical impedance spectroscopy–a schematic representation of the main actions and results.

Other authors (Saccocio et al., 2014) studied the application of regularized distribution of relaxation times (DRT) to characterize the electrochemical impedance spectroscopy data, to obtain relevant timescales in the case of electrochemical systems, using a Fredholm integral equation of the first kind. Nowadays, the analyzing of EIS data by DRT could be done by various methods, such as ridge regularization, preconditioned ridge regularization, Fourier filtering, and subsequent fitting, Monte Carlo methods, maximum entropy methods, genetic programming, etc. Saccocio et al. (2014) introduced two new test functions to identify the optimum regularization parameter by ridge regression, and subsequently benchmarked the results using the Least Absolute Shrinkage and Selection Operator (Lasso) regression.

By synthetic and real electrochemical experiments, in the case of exactly known DRT functions, the authors evaluated the performance of the proposed $Z_{Re}-Z_{Im}$ discrepancy and the $Z_{Re}-Z_{Im}$ cross-validation test and concluded that the minimization of $Z_{Re}-Z_{Im}$ cross-validation functions is a useful criterion to find the best regularization degree. The performance of the analysis is increased by the synergic use of both techniques (ridge and Lasso regression).

Munoz et al. (2015) used the DRT method to deconvolute the EIS data obtained for a solid oxide fuel cell. Based on the process noticed from the DRT spectra, a large equivalent circuit model was generated, which has been used to obtain the parameters characterizing the microstructure and the electrochemistry of the solid oxide fuel cell. The final purpose of these analyses is to gain the ability to predict performance decreasing in the case of solid oxide fuel cells and other components.

Song et al. (2006) used discrete Fourier transform (DFT) to deconvolute the penetrability coefficient distribution in the case of EIS data of porous electrodes, to study the microstructures of the electrodes and to improve the devices' performances. Such porous materials could be used as electrodes for example in electrochemistry storage devices (fuel cells, batteries or capacitors). The authors correlate the EIS data with a theoretical model by a Fredholm equation of the first kind. From the integral equation, taking into account the experimental impedance data and a known circuit model, they obtain the distribution of penetrability coefficient, from which subsequently the pore size distribution could be calculated.

Mertens et al. (2016) reported the use of EIS at very low frequencies, combined with time domain measurements to estimate the charge transfer resistance and the solid-state diffusion rate, and to determine the activation energies in the case of Li-Ion batteries.

According to Lvovich (2012), the applications in polymer field are related to electroactive polymer films, taking into account their outstanding properties: the capacity of mixing with raw polymeric materials, solution and melt flexible processing, stability, unique optical, and electric features. These particular properties determine the obtaining of new materials, such as electroactive inks, coatings, paintings, adhesives, electrochromic smart windows, supercapacitive materials, electroconductive films which protect from corrosion, enzyme-modified conductive polymers, and high-performance conductive fibers.

In the case of coatings, since most of them are supposed to degradation processes during the time, as a result of water penetration into the coating, a new interface liquid-metal is forming under the surface, where corrosion phenomena occur. EIS analyses are useful for the evaluation of failure degree of layers (the parameters analyzed are the double layer capacitance and the pore resistance). EIS could be applied for testing and assessment of industrial coatings and in the automotive industry (Kamarchik, 1993).

Cesiulis et al. (2016) described the practical applications of EIS in thin films, such as cathodic metals or alloys films deposition, characterization of oxide films and organic films deposited on metals, and in metal underpotential deposition. EIS data could be useful for several techniques in domains such as biosensors (Pires Carneiro, 2014) and biofuel cells (Cesiulis et al., 2016). From these techniques, one could mention biosensors surface characterization, biosensor action modeling, interpretation of bioanalytical signals, and evaluation of the bio-fuel cell efficiency.

EIS could be used to study the kinetics of the hydrogen underpotential deposition (HUPD) reaction to determine double layer capacitance, hydrogen adsorption pseudocapacitance, and charge transfer resistance of the HUPD reaction (Losiewicz et al., 2012).

Another field where impedance finds its applications in the industry of colloids and lubricants (Lvovich, 2012), due to their capacity to decrease the friction processes implied in industry and automotive. An important application domain is in medicine and biology, for biomedical sensors, cell suspensions, protein adsorption, and implantable biomedical devices. Other applications are for electro-rheological fluids, insulating films, and coatings, metal oxide films, and alloys, corrosion monitoring, etc.

10.5 MULTI-/TRANS-DISCIPLINARY CONNECTION(S)

The interpretation of electrochemical impedance data implies knowledge from various fields, such as chemistry, electrochemistry, physics, electricity, mechanics and, to obtain better results and to use this technique to its maximum potential, a high level of mathematical knowledge is needed.

Taking into account the large variety of application fields, a lot of domains where EIS finds its application could be mentioned as follows: biology, medicine, polymer technology, colloids and lubricants industry, etc.

10.6   OPEN ISSUES

By further development of mathematical methods, data analysis will be permanently a subject for improvement. Also with the growing up of the computer industry, the precision of measurement techniques will notably increase. The estimated result will be new applications, such as dielectric spectroscopy analysis of conduction in bulk polymers, analysis of biomedical implants, surface corrosion kinetics, etc.

According to Saccocio et al. (2014) in the future, a lot of problems related to DRT deconvolution must be solved, mainly speed and accuracy benchmarks for different methods, such as ridge regression, Lasso regression, Fourier filtering, Monte Carlo, maximum entropy, genetic programming, etc.

Macdonald (2006) considered that the analysis of EIS data would be facilitated if a comprehensive library of reaction mechanisms will be realized since the fundamental approach implying models is very appropriate for artificial neural network technology. In this case, new pattern recognition algorithms will contribute to scan the library and fit the best reaction mechanism.

ACKNOWLEDGMENT

This contribution is part of the Nucleus-Programme under the project "Deca-Nano-Graphenic Semiconductor: From Photoactive Structure to The Integrated Quantum Information Transport,"PN-18-36-02-01/2018, funded by the Romanian National Authority for Scientific Research and Innovation (ANCSI).

KEYWORDS

- **electrochemical impedance**
- **electrochemical impedance spectroscopy**

REFERENCES AND FURTHER READING

Agarwal, P., Orazem, M. E., & Garcia-Rubio, L. H., (1993). Application of the Kramers-Kronig relations in electrochemical impedance spectroscopy. In: Scully, J. R., Silverman, D. C., & Kendig, M. W., (eds.), *Electrochemical Impedance: Analysis and Interpretation* (pp. 115–119). American Society for Testing and Materials, Philadelphia.

Armstrong, B. D., Bell, M. F., & Metcalfe, A. A., (1978). *The A.C. Impedance of Complex Electrochemical Reactions* (pp. 244–263). A Specialist Periodical Report, Electrochemistry, The Chemical Society, Burlington House, London.

Barrett, T. W., (2008). *Topological Foundations of Electromagnetism*. World scientific series in contemporary chemical physics, World Scientific Publishing Co. USA.

Bonciocat, N., & Marian, I. O., (2006). *The Faraday Impedance Method and Its Alternatives* (in Romanian), Ed. Clujeana University Press.

Bonciocat, N., (1996). *Electrochemistry and Applications* (in Romanian), Dacia Europa-Nova, Timisoara, Romania.

Bonciocat, N., (2005). *Fredholm Alternative in Electrochemistry* (in Romanian), Mediamira, Cluj-Napoca, Romania.

Bos, W. M., (2008). *Prediction of Coating Durability. Early Detection Using Electrochemical Methods*. Gildeprint drukkerijen B.V., The Netherlands.

Calvert, J., (2002). *Heaviside, Laplace, and the Inversion Integral*, University of Denver.

Cesiulis, H., Tsyntsaru, N., Ramanavicius, A., & Ragoisha, G., (2016). The study of thin films by electrochemical impedance spectroscopy. In: Tiginyanu, I., et al., (eds.), *Nanostructures and Thin Films for Multifunctional Applications. Nanoscience and Technology*, Springer International Publishing Switzerland.

DeLevie, R., (1967). Electrochemical response of porous and rough electrodes. In: Delahay, P., (ed.), *Advances in Electrochemistry and Electrochemical Engineering* (Vol. 6, pp. 329–397). Wiley Interscience, New York.

Dolin, P. I., & Ershler, B. V., (1940). "Kinetics of Processes on the Platinium Electrode: 1. The Kinetics of the Ionization of Hydrogen Adsorbed on a Platinium Electrode," *Acta Physicochimica, URSS, 13,* 747–778.

Hunt, B. J., (2012). Oliver Heaviside: A first-rate oddity. *Physics Today, 65,* 41–48.

Kamarchik, P., (1993). Improved Coatings Testing and Evaluation Using Electrochemical Impedance Spectroscopy. In: Silverman, J. R., & Kendig, D. C., (eds.), *Electrochemical Impedance: Analysis and Interpretation* (pp. 463–474). Scully. American Society for Testing and Materials, Philadelphia.

Kramers, H. A., (1929). Die Dispersion und Absorption von Rontgenstrahlen. *Physikalische Zeitschrift, 30,* 522–523.

Kronig, R. De L., (1926). On the theory of dispersion of x-rays. *Journal of the Optical Society of America, 12,* 547–557.

Lasia, A., (1999). Electrochemical impedance spectroscopy, and its applications. In: Conway, B. E., Bockris, J., & White, R. E., (eds.), *Modern Aspects of Electrochemistry* (Vol. 32, pp. 143–248). Kluwer Academic/Plenum Publishers, New York.

Losiewicz, B., Jurczakowski, R., & Lasia, A., (2012). Kinetics of hydrogen under potential deposition at polycrystalline platinum in acidic solutions. *Electrochimica Acta, 80,* 292–301.

Lvovich, V. F., (2012). *Impedance Spectroscopy: Applications to Electrochemical and Dielectric Phenomena*. John Wiley & Sons, Inc.

Macdonald, D. D., (2006). Reflections on the history of electrochemical impedance spectroscopy, *Electrochemica Acta, 51*, 1376–1388.

Mertens, A., Vinke, I. C., Tempel, H., Kungl, H., De Haart, L. G. J., Eichel, R. A., & Granwehr, J., (2016). Quantitative analysis of time-domain supported electrochemical impedance spectroscopy data of Li-ion batteries: Reliable activation energy determination at low frequencies. *J. Electrochem. Soc., 163*, H521–H527.

Munoz, C. B., Pumiglia, D., McPhail, S. J., Montinaro, D., Comodi, G., Santori, G., Carlini, M., & Polonara, F., (2015). More accurate macro-models of solid oxide fuel cells through electrochemical and microstructural parameter estimation – Part I: Experimentation. *Journal of Power Sources, 294*, 658–668.

Nahin, P. J., (2002). *Oliver Heaviside: The Life, Work, and Times of an Electrical Genius of the Victorian Age*. Johns Hopkins University Press.

Orazem, M. E., & Tribollet, B., (2008). *Electrochemical Impedance Spectroscopy*, John Wiley & Sons.

Pires, C. L., (2014). *Development of an Electrochemical Biosensor Platform and a Suitable Low-Impedance Surface Modification Strategy*. KIT Scientific Publishing.

Randles, J. E. B., (1947). Kinetics of rapid electrode reactions. *Discussions of the Faraday Society, 1*, 11–19.

Saccoccio, M., Wan, T. H., Chen, C., & Ciucci, F., (2014). Optimal regularization in distribution of relaxation times applied to electrochemical impedance spectroscopy: Ridge and lasso regression methods–a theoretical and experimental study. *Electrochimica Acta, 147*, 470–482.

Song, H. K., Jang, J. H., Kim, J. J., & Oh, S. M., (2006). Electrochemical porosimetry: Deconvolution of distribution functions. *Electrochemistry Communications, 8*, 1191–1196.

Warburg, E., (1899). Uber das Verhalten sogenannter unpolarisirbarer Elektroden gegen Wechselstrom. *Annalen der Physik und Chemie, 67*, 493–499.

Wiener, N., (1993). *Invention: The Care and Feedings of Ideas* (pp. 70–75). MIT Press.

CHAPTER 11

Electromagnetic Signals in Nanosystems

CYNTHIA WHITNEY

Galilean Electrodynamics, 11660 239th Ave. NE, Redmond,
WA 98053-5613, USA, E-mail: Galilean_Electrodynamics@Comcast.net

11.1 DEFINITION

Throughout chemistry, atoms are bound together in ways that can generally be seen as electromagnetic (EM) in character. Atoms give or take electrons, become ions, and come together to make molecules and larger nanostructures. For the structures to then stay together, there must be continuing EM signals exchanged within them. Indeed, even within individual atoms, there have to exist EM signals between the electrons and the nucleus. In the 20th century, the subject of 'Signals' has been much developed within the disciplines of electrical engineering (EE) and applied mathematics (AM), and some of those developments can shed new light on chemistry, and indeed on physics as well. The reason why these opportunities exist is an accident of history. The most relevant EE and AM developments came in the middle of the 20th century. Claude Shannon launched information Theory (IT), and Norbert Wiener launched statistical communication theory (SCT). The accident is that, by then, both quantum mechanics (QM) and special relativity theory (SRT) had already been developed without the relevant IT or SCT concepts concerning 'signals.' Signals travel at some finite speed, but in QM, distant correlations occur. Signals are pulse-like, but in SRT, signals are more like infinite plane waves of light. The following short essay revisits these issues, and shows how a more realistic signal model can make a difference to our understanding of atoms and molecules, since all such nanosystems contain electrons and ions, and EM signals fly between them.

11.2   HISTORICAL ORIGINS

We presently have a hint of a crisis emerging in science: the two main pillars of 20th-century physics seem somewhat incompatible with each other. In QM, there exist distant correlations, instantaneously enforced, without clear evidence of communication via traveling signals. In SRT, particle speeds are limited to values less than light speed c, and signal speeds are equal to exactly c, specified relative to the receiver of the signal. So instant correlations between distant particles are inconceivable in SRT.

User communities in chemistry, biochemistry, nanotechnology, and so on, cannot be sure of the reliability of conflicting conclusions drawn from QM *vs*. SRT. This is a situation in need of remedy.

How did the present situation arise? Its development has something to do with the sometimes-erratic path of history. Both QM and SRT emerged early in the 20th century, with two separate works of Albert Einstein (1905, 1905). At that time, some relevant science about EM signals was not yet available. Shannon's IT grew out of technological developments – telegraphs, telephones, etc., through the mid 20th century (*see*, Brillouin (2013) for more details on IT). Wiener's SCT emerged soon after that (*see*, Lee (1960) for more details on SCT).

IT clarifies an important property that signals have to possess in order to convey any information whatsoever: they have to be somewhat pulse-like. The 'infinite plane wave,' so often referenced in SRT, cannot convey any information at all! That is because the infinite plane wave has no marker on it to distinguish 'before' *vs*. 'after' its 'arrival.'

IT has been available in the early 20th century. Einstein might have considered finite-energy pulses, not only for the 'photon' that (to his dismay) launched QM, but also for the 'signal' that figures in SRT.

Instead, SRT relied a lot on the infinite plane wave, and its phase speed, c. About this, Einstein made his famous Second Postulate, that the speed of light is always the same number c with respect to all observers in all inertial reference frames.

If, instead, a finite-energy pulse signal had been considered for SRT, some interesting behavior would have emerged: over time, pulses spread, and develop into wavelets. This development, and its implication, is discussed next.

11.3   NANO-SCIENTIFIC DEVELOPMENT

In the late 19th century, there existed a well-developed approach for dealing with any problem that possesses differential equations, known families of solutions to the differential equations, and problem-specific boundary conditions.

The signal problem has all those standard attributes. The differential equations should be Maxwell's four coupled field equations. The solutions can be wavelets developed from initial smooth pulses. The boundary conditions can be: (1) no energy leakage behind the source, and (2) no energy overflow beyond the receiver. Those conditions can be assured by putting additional phantom sources before and after the propagation path to cancel the electric fields at the boundaries.

All this is all very standard, and no reason has ever been given for not at least trying the standard approach. So let us try it here. Maxwell's four coupled field equations go:

$$\nabla \cdot \mathbf{B} = 0, \ \nabla \cdot \mathbf{D} = 4\pi\rho, \ \nabla \times \mathbf{E} + \frac{1}{c}\partial \mathbf{B} / \partial t = 0, \ \nabla \times \mathbf{H} - \frac{1}{c}\partial \mathbf{D} / \partial t = \frac{4\pi}{c}\mathbf{J} \qquad (1)$$

Here $\mathbf{B} = \mu_0 \mathbf{H}$, and μ_0 is magnetic permeability; $\mathbf{D} = \varepsilon_0 \mathbf{E}$, and ε_0 is electric permittivity; ρ is charge density and $\mathbf{J}$ is current density–the driving sources. *See*, for example, Jackson (1975).

Inserting two of the first-order coupled field equations into the other two reduces the system to two uncoupled second-order wave equations, in which there appears the product $\varepsilon_0\mu_0$ that we routinely identify as the inverse square of light speed c. In a source-free region of space we have:

$$\partial^2 \mathbf{E} / c^2 \partial t^2 - \nabla^2 \mathbf{E} = 0 \ \text{ and } \ \partial^2 \mathbf{B} / c^2 \partial t^2 - \nabla^2 \mathbf{B} = 0 \qquad (2)$$

What can be said about solutions to Eq. (1) *vs.* Eq. (2) as signal models? For one thing, the two uncoupled wave equations are less restrictive than the four coupled field equations are. For example, we could imagine a plane wave of all $\mathbf{E}$ without any $\mathbf{B}$. That wave solution cannot represent a signal because it cannot travel.

So it is important to take account here of an idea from modern Set Theory (ST). The solutions to the four first-order coupled field equations (1) are a *subset* of the solutions to the two uncoupled second-order wave equations (1). ST was already available in the late 19th century, but it seems not to have been invoked here.

The coupling through (1) means that a problem that starts with a signal consisting of a simple pulse in $\mathbf{E}$, and a similar pulse in $\mathbf{B}$ to permit travel, will evolve into a problem with an extended wavelet in $\mathbf{E}$, and a corresponding extended wavelet in $\mathbf{B}$. Over time, this wavelet will grow longer and longer in extent, and it will shrink lower and lower in amplitude.

To model this evolution, let us begin with the problem of modeling a pulse. We need a convenient function. Some candidate functions are available from statistical mechanics. There, one finds distribution functions for random variables of various types–discrete, continuous, one-sided, and two-sided. Here are some examples:

	discrete variable n	continuous variable x
one-sided function	Poisson $\propto \exp(-\lambda n)$	Exponential $\propto \exp(-\lambda x)$
two-sided function	Binomial $\propto (N-n)! n! / N!$	Gaussian $\propto \exp(-x^2 / 2\sigma^2)$

All of these functions maximize 'entropy' subject to the conditions described. For modeling a signal pulse, the most appropriate of the example functions appears to be the continuous and two-sided Gaussian. About the Gaussian: besides maximizing entropy, it also minimizes 'uncertainty.' This statement sounds peculiar, inasmuch as both words, 'entropy' and 'uncertainty,' are about information not possessed. The explanation involves the operation known as 'Fourier transformation.' A function of a variable x can be re-expressed as a function of another variable k through Fourier transformation. Entropy is about the situation in one space or the other, x or k, while uncertainty is about the situation between the two spaces, x and k, together. The Gaussian function is unique in that its Fourier transform is also a Gaussian.

In full detail, the Gaussian function is.

$$G(x) = \exp\left(-x^2 / 2\sigma^2\right)\Big/\sqrt{2\pi}\,\sigma \tag{3}$$

where σ means 'standard deviation,' or $\sqrt{<x^2>}$, where '<>' means 'expected value.' The Gaussian serves even if we deal, not with a single pulse, but with a string of pulses, like a telegraph message. That is where SCT takes over: it deals, not with single pulses, but with ongoing strings of them, by introducing the autocorrelation function of the string, which can again be a single Gaussian.

The result of applying derivatives to a Gaussian function is to generate Hermite polynomials multiplying the original Gaussian function. Hermite polynomials are defined in detail in math handbooks, such as Abramowitz and Stegun (1964), but the detail doesn't matter here. The main point is that each derivative applied multiplies the generating Gaussian with a higher-order Hermite polynomial, creating more zero crossings and more peaks, all evenly spaced, thereby creating a wavelet that is spread out along its propagation direction.

Note that the evolution of the Gaussian signal pulse into an extended wavelet means that speaking of the speed of light c is an over-simplification. Wavelet shape evolution means that light speed cannot be just a number; it has to be a function, depending on location along the propagation path, and on time into the propagation process.

This means Einstein's second postulate was too strong. We stand reminded that there is always a risk attendant to injecting a new postulate into a working mathematical system. The risk is that of creating, either redundancy or else conflict, with one or more of the original Postulates. The system can then generate paradoxes. And SRT certainly did generate Paradoxes–there exists a vast literature about them. The most famous ones have to do with time. Ordinary Newtonian time does not work with SRT. Every observer has his own clock, and each one sees the clocks of all other observers moving relative to himself to be running slow.

Going deep into history, we see that Euclid established a good discipline for all of us to follow: set out all axioms (Greek) or postulates (Latin) at the beginning, and never inject any more of them. The developers of the several modern non-Euclidean geometries followed this discipline, and took care to relax, or remove, a Euclidean postulate before injecting a new postulate and starting their own new developments.

So what might now replace the Einstein postulate? An opportunity to capture some dependence on location and time variables comes with the so-called 'reference' for light speed c. In the early twentieth century, there were two proposals about the reference for light speed: Einstein (1905) chose the receiver as his reference, and Ritz (1909) proposed the source as his reference. The Ritz proposal did not work in astronomical situations. The Einstein proposal has been judged to work, but only with the many SRT paradoxes being tolerated. Therefore, the 'reference' seems to invite some new definition that better captures the whole waveform development scenario.

Fortunately, there exist several ways to describe a new and plausible reference for light speed. For example, one could make the reference the spatial mid-point between the source and the receiver, or one could make the reference switch from the source to the receiver at the temporal midpoint of the scenario, or one could make the reference move continuously from the source and to the receiver through the scenario. All of these are equivalent to the path-averaged speed c relative to the scenario spatial mid-point.

This new reference for c produces new results for the most telling of all problems. This problem concerns the potentials and fields created by a source charge and sensed by a distant receiver charge; namely, the famous Liénard-Wiechert (1898, 1901) potentials and fields. These results predated Einstein, are consistent with Einstein, and apparently still in modern textbooks. They are the reason why QM had to begin with the injection of yet another new postulate – the famous quantum hypothesis; namely, that the hydrogen atom simply is stable, despite the radiation expected from its orbiting, and hence accelerating, electron.

The new reference for c changes the expression of 'retardation' (i.e., allowance for propagation time) in the Liénard-Wiechert (LW) potentials and fields. The standard LW potentials were:

$$\Phi(\mathbf{r},t) = e\left[1/\kappa R\right]_{\text{retarded}} \text{ and } \mathbf{A}(\mathbf{r},t) = e\left[\boldsymbol{\beta}/\kappa R\right]_{\text{retarded}} \qquad (4)$$

where $\kappa = 1 - \mathbf{n}\cdot\boldsymbol{\beta}, \mathbf{n}$ is propagation direction, $\boldsymbol{\beta} = \mathbf{v}/c$, where is $\mathbf{v}$ source velocity, with magnitude v, limited to the number c, and 'retarded' means evaluated at the earlier time that allows for propagation to the receiver at speed c relative to the receiver. These potentials gave the fields:

$$\mathbf{E}(\mathbf{r},t) = e\left\{(\mathbf{n}-\boldsymbol{\beta})(1-\beta^2)/\kappa^3 R^2 + \mathbf{n}\times\left[(\mathbf{n}-\boldsymbol{\beta})\times(d\boldsymbol{\beta}/dt)\right]/c\kappa^3 R\right\}_{\text{retarded}} \qquad (5)$$

$$\text{and } \mathbf{B}(\mathbf{r},t) = \mathbf{n}_{\text{retarded}} \times \mathbf{E}(\mathbf{r},t)$$

The $1/R^2$ fields were called 'near fields,' or 'Coulomb-Ampere fields,' and the $1/R$ fields were called 'far fields,' or 'radiation fields.' But the distinction is actually moot, because radiation is really described with the Poynting vector $\mathbf{P} \propto \mathbf{E}\times\mathbf{B}$, and in the far field $\mathbf{P}$ becomes $\propto 1/R^2$ just like the near fields.

The LW results are suspect because the $1/R^2$ part of the $\mathbf{E}$ vector lies in the direction of $(\mathbf{n}-\boldsymbol{\beta})_{\text{retarded}} \approx \mathbf{n}_{\text{retarded}} - \boldsymbol{\beta} \approx \mathbf{n}_{\text{present}}$, but the $1/R^2$ part of the $\mathbf{P}$ vector lies in the direction of $\mathbf{n}_{\text{retarded}}$ – a different direction. That is, the Coulomb tug and the radiation torch disagree. Physically and logically, these directions ought to agree.

The disagreement can be resolved with the new interpretation for c as path-averaged speed, and the new reference for c as the scenario's spatial mid-point. This new interpretation suggests that the relevant items of information – the path length and the propagation direction – are not those items measured from the source at emission to the receiver at absorption, but rather those items measured from the source to the receiver at the scenario midpoint. That is, we should be doing half-retardation.

With this change to half-retardation, the $\mathbf{P}$ vector lies along $\mathbf{n}_{\text{half-retarded}}$, and the Coulomb $\mathbf{E}$ vector lies along a direction $\mathbf{n}_{\text{retarded}} - \boldsymbol{\beta}/2$; i.e., also $\mathbf{n}_{\text{half-retarded}}$. That is, the Coulomb tug and the radiation torch come into agreement. This makes sense.

11.4 NANO-CHEMICAL APPLICATION(S)

With the fields between charges better formulated, we can restart QM without yet another new postulate. With the hydrogen atom less mysteriously explained, we can better understand all atoms, ions, and molecules, and do chemistry more easily, using just algebra, and then we can make new advances in nano-chemical applications.

KEYWORDS

- **information theory**
- **quantum mechanics**
- **special relativity theory**

REFERENCES AND FURTHER READING

Abramowitz, M., & Stegun, A., (1964). *Handbook of Mathematical Functions with Formulas, Graphs, and Mathematical Tables*. U.S. Department of Commerce, 10th printing, with corrections (1972).

Brillouin, L., (1962). *Science and Information Theory* (2nd edn.) Dover republication (2013).

Einstein, A., (1905). On a heuristic point of view concerning the production and transformation of light. *Annalen der Physik, 17*, 132–148; English translation in *Collected Papers of Albert Einstein, vol. 2, The Swiss Years, 1900–1909,* Princeton University Press (1989).

Einstein, A., (1905). On the electrodynamics of moving bodies. *Annen der Physik, 17*, 891–921; English translation in *Collected Papers of Albert Einstein, vol. 2, The Swiss Years, 1900–1909,* Princeton University Press (1989).

Jackson, J. D., (1975). *Classical Electrodynamics* (2nd edn., p. 218). John Wiley, New York, USA.

Lee, Y. W., (1960). *Statistical Theory of Communication*. John Wiley & Sons, Inc.

Liènard, A., (1898). "Champ Electriqueet Magnétiqueproduit par une Charge Electrique Concentrée en un Point et Animée d'un Movement Quelconque," *L'Eclairage Electrique, XVI(27)* 5–14, (*28*) 53–59, (*29*) 106–112.

Ritz, W., (1908). Recherches critiques surl'electrodynamiquegenerale, *Ann. Chim. Phys., 13*, 145–256.

Wiechert, E., (1901). Elektrodynamische Elementargesetze, *Archives Néerlandesises des Sciences Exacteset Naturellessérie II, Tome IV*.

CHAPTER 12

Entropy in Physics and Chemistry

CYNTHIA WHITNEY

Galilean Electrodynamics, 11660 239th Ave. NE, Redmond, WA 98053-5613, USA, E-mail: Galilean_Electrodynamics@Comcast.net

12.1   DEFINITION

'Entropy' is a foundational concept in the subjects of thermodynamics and statistical mechanics, which, in turn, are foundational for the subject of chemistry, the oldest of the nanosciences. The idea of entropy captures the observation that nature always seems to run toward greater disorder. Entropy constitutes a quantitative way of characterizing the disorder. That is to say, natural processes tend to increase entropy.

12.2   HISTORICAL ORIGIN(S)

Entropy has been a part of science for a long time, but even so, it is not yet very clear how entropy arises physically, or how, or even *if*, it always grows over time. After all, life develops, and it looks contrary to the disorder commonly associated with entropy.

The uncertain situation is largely the result of accidents of history. Modern mathematical physics really started with Newton's derivation of idealized equations for the mechanics of idealized point particles. Those equations do not necessarily contain any messy energy loss mechanism, such as friction. In principle, Newton's equations can be reversible in time.

The most that one can say about irreversibility in Newton's equations is something practical. For just two bodies, Newton's equations can be solved exactly, in closed form. But for more than two bodies, they cannot be solved in closed form. Nevertheless, the trajectories can be computed step-by-step, and in reverse too. The only thing that can degrade the results is that small differences at the beginning could morph into larger differences after a long time. This is the phenomenon of 'chaos.' But in principle, the math is still locally reversible in time. So Newton's equations do not seem to mandate the notion of entropy.

So where does entropy really enter into physics? The big idealization in Newtonian mechanics is the instantaneous communication among particles. This idealization goes away with the advent of Maxwell's equations. Electromagnetic signals have to comport with Maxwell's equations. So can Maxwell's equations be the portal for entropy?

This author believes so. Here is how entropy can enter the story. Maxwell's equations are four first-order coupled field equations. It is always possible to convert them into two uncoupled wave equations. The reader can find this done in many textbooks, along with the assertion that two uncoupled wave equations are entirely equivalent to four coupled field equations. One example is the text by Jackson (1975).

But the assertion of equivalence between four coupled field equations and two wave equations is not quite true. The uncoupled wave equations always admit some solutions that the coupled field equations do not admit. That is to say, the solutions to the coupled field equations form a special subset of the set of solutions to the uncoupled wave equations.

Considering just the two uncoupled wave equations, one can easily imagine a solution that does not work for the coupled field equations. For example, consider a compact electromagnetic pulse. With only the wave equations to satisfy, it is possible for this electromagnetic pulse to travel without evolving in shape. But with the full set of four coupled field equations, the electromagnetic pulse will not only travel, but over time it will also evolve in shape, becoming a wavelet extended in length. Observe that this spreading-out of energy over a greater and greater amount of space is equivalent to an increase of entropy.

12.3　NANO-SCIENTIFIC DEVELOPMENT(S)

One line of development that flows from the idea that Maxwell is the portal through which entropy comes into physics is represented by some other entries in this volume. It begins with electromagnetic signals in nanosystems (EMS), progresses to hydrogen – the first atom (H-FA), leads to algebraic chemistry (AC), ionization potentials (IPs), ionic configurations (ICs), and various application areas, such as hydrocarbons (HCs) and new energy technology (NET).

Here is one more development along that line: we can use the concept of ever-increasing entropy to figure out more about how chemical catalysis works. In Whitney (2013), the author revisited a catalysis problem from an old textbook, to study the explanation given there, and see if the purported set of catalysis steps really could have occurred, and if not, what different set of catalysis steps actually must have occurred.

The reaction in question was:

$$2Cr^{3+} + 6(SO_4)^- + 7(H_2O) \rightarrow (Cr_2O_7)^{2-} + 6(SO_4)^{2-} + 14H^+ \tag{1}$$

The ions Cr^{3+} and H^+ can be evaluated directly. The most likely (lowest energy) ionic configurations of the larger radicals and molecules are:

$$(SO_4)^- = S^{3+} + 4O^- \tag{2}$$

$$H_2O = 2H^- + O^{2+} \tag{3}$$

$$(Cr_2O_7)^{2-} = 2Cr^{4+} + 4O^- + 3O^{2-} \tag{4}$$

$$(SO_4)^{2-} = S^{4+} + 2O^- + 2O^{2-} \tag{5}$$

So in terms of ions we have:

$$\begin{aligned} 2Cr^{3+} + 6(S^{3+} + 4O^-) + 7(2H^- + O^{2+}) \\ \rightarrow 2Cr^{4+} + 4O^- + 3O^{2-}) + 6(S^{4+} + 2O^- + 2O^{2-}) + 14H^+ \end{aligned} \tag{6}$$

The energy requirements for the individual species are:

$$2Cr^{3+} : 2 \times 20.0845 = 40.169 \text{ eV},$$

$$6(S^{3+} + 4O^-) : 6 \times [25.1739 + 4 \times (-12.4354)] = -147.4062 \text{ eV},$$

$$7(2H^- + O^{2+}) : 7 \times [2 \times (-90.6769) + 9.6214] = -1202.1268 \text{ eV},$$

$$(2Cr^{4+} + 4O^- + 3O^{2-}) : [2 \times 26.2142 + 4 \times (-12.4354) + 3 \times (-27.3788)] =$$
$$(52.4284 - 49.7416 - 82.1364) = -79.4496 \text{ eV}$$

$$6(S^{4+} + 2O^- + 2O^{2-}) : 6 \times [30.9895 + 2 \times (-12.4354) + 2 \times (-27.3788)] =$$
$$6 \times (30.9895 - 24.8708 - 54.7576) = -291.8334 \text{ eV}$$

$$14 \times H^+ : 14 \times 14.250 = 199.5 \text{ eV}$$

The total energy requirement on the left is:

$$40.169 - 147.4062 - 1202.1268 = -1309.364 \text{ eV} \tag{7}$$

The total energy requirement on the right is:

$$-79.4496 - 291.8334 = -371.283 \text{ eV} \tag{8}$$

So the reaction takes:

$$-371.283 - (-1309.364) = 938.081 \text{ eV} \tag{9}$$

This is a huge positive energy. So this reaction definitely does *not* transpire without external assistance.

The addition of silver ions was known to make the reaction possible. The believed explanation involved two steps:

$$2SO_4^- + Ag^+ \rightarrow 2SO_4^{2-} + Ag^{3+} \text{ (3 times over)} \tag{10}$$

followed by

$$2Cr^{3+} + 3Ag^{3+} + 7H_2O \rightarrow Cr_2O_7^{2-} + 3Ag^+ + 14H^+ \tag{11}$$

The assumption there was that the silver ions just go from charge +1 to +3 in one step, and then go back from +3 to +1 in the other step. This proposal cannot solve the energy problem. It cannot be the right explanation for how the catalysis works.

Reaction steps there may be, but these two reaction steps cannot do the job. What is necessary is at least one step, and possibly several steps, involving just silver ions by themselves. Note that the silver atoms admit a variety of ionization states, all with different energies. There can be all sorts of reactions that do nothing but change the ionization states of participating silver atoms.

The principle of entropy maximization implies that the silver ions have to end up distributed in a Gaussian way over all their possible ionization/energy states. The governing parameter is the ambient temperature T. The average energy of a silver ion has to be kT, where k is Boltzmann's constant.

Whatever happens with the chromium, sulfur, and water, the resulting population of silver ions has to interact with itself in order to revert to its Gaussian distribution over energy states, and conform to the requirement for average energy kT.

The involvement of temperature here recalls the very first chemistry, and the very first catalysis, ever practiced by humans; namely, cooking!

12.4   NANO-CHEMICAL APPLICATION(S)

If we can understand the phenomenon of catalysis a little better, we can probably improve on our exploitation of it. This can be important in various engineering applications; for example, in designing catalytic converters for cleaning up emissions from motor vehicles or power plants.

12.5   MULTI-/TRANS-DISCIPLINARY CONNECTION(S)

There is a political dimension to all environmental issues.

KEYWORDS

- **catalysis**
- **chemical reactions**
- **electromagnetic signals in nanosystems**
- **ionic configurations**
- **ionization potentials**

REFERENCES AND FURTHER READING

Jackson, J. D., (1975). *Classical Electrodynamics* (2nd edn., p. 659). John Wiley & Sons: New York, USA.

Whitney, C., (2013). *Algebraic Chemistry Applications and Origins*. Chapter 7; NOVA Publishers: New York, USA.

CHAPTER 13

Excitons

NICOLINA POP

Politehnica University of Timisoara, Department of Fundamental of Physics for Engineers, Vasile Pârvan Blv., 300223 Timisoara, Romania, E-mail: nicolina.pop@upt.ro

13.1 DEFINITION

Optical excitations in condensed matter are called excitons. An exciton could be considered as an entity consisting of an electron which has been excited so much as to leave its own atom and the positive charge that it has left behind. There are two types of excitons. One of them appears in molecular crystals, and they are called Frenkel excitons. The excitons appearing semiconductors are called Mott-Wannier excitons. Their main characteristics are that both are bounded particle-hole pairs. In molecular crystals (anthracene, naphthacene, naphthalene, etc.), electron and hole remain at the molecule, and these excitons are called "low radius excitons." The electron wave functions of neighbor molecules have a weak covering.

In semiconductors, hole is in valence zone while particle (electron) is in the conductive zone. The connection between elements of this pair remains stable. The electron-hole distance is relatively high, and these excitons are called "high" radius excitons.

Our study will be concerned to Frenkel excitons, mainly and in special, to the problem of short one-dimensional macromolecule whose number of excitons is not conserved.

13.2 HISTORICAL ORIGIN(S)

The concept of excitons was first proposed by Yakov Frenkel in 1931, when he presented the excitation of atoms in a lattice of insulators. He proposed that the excited state could travel in a particle-like fashion through the lattice without the net transfer of charge. There are two types of Frenkel excitons. One type is caused by electron subsystem excitations of an isolated molecule, while second is caused by excitations of internal molecular oscillations. The last is called vibrons, and their Hamiltonian is expressed in Boze operators. The excitons which arise due to excitations of electron subsystem have mixed statistics of operators which create and annihilate excitons. At one lattice point, they have Fermi statistics, while for different lattice points they are more close to bozons. The question about kinematical and dynamical interactions of excitons became actually after the appearance of lasers. For two-level molecular excitations, exciton operators are Pauli operators. For the multilevel scheme of molecular excitations, exciton operators have more complicated kinematics than Paulson's ones.

The excitons founded in semiconductor crystals with small energy gaps and high dielectric constants were named for Gregory Wannier and Nevill Francis Mott. Wannier-Mott excitons have also been identified in liquids, such as liquid xenon. They are also known as large excitons.

An intermediate case between Frenkel and Wannier excitons, charge-transfer excitons (called simply CT excitons) occur when the electron and the hole occupy adjacent molecules. They occur primarily in ionic crystals. Unlike Frenkel and Wannier excitons, they display a static electric dipole moment.

13.3 NANO-SCIENTIFIC DEVELOPMENT(S)

The exciton interactions were studied by nonlinear optics (Ahmanov., Hohlov, 1966; Mills, 1991). The problems of nonlinear optics are the appearance of higher harmonics, multiphoton absorption (Tošić, 1967) and problem of non-conservation of a number of excitons. These problems become more complicated in nanostructures. So, first of all, was considered a linear chain containing N molecules where $N \in (3,...,30)$ molecules that will be analyzed by ASQ (Approximate Second Quantization) method.

So, only quadratic part of the exciton Hamiltonian expressed in Bose operators B^+ and B will be analyzed. The non-conservation of excitons will be taken into account, i.e., the fact that $\left[H, \sum_n B_n^+ B_n \right] \neq 0$. Consequently, the Hamiltonian of the chain, is given by

$$H = \sum_{n=0}^{N} \Delta B_n^+ B_n + \sum_{n=0}^{N} \left(X_{n,n+1} + X_{n,n-1} \right) B_n^+ B_n + \sum_{n=0}^{N} B_n^+ \left(Y_{n+1} B_{n+1} + Y_{n-1} B_{n-1} \right) +$$

$$+ \frac{1}{2} \sum_{n=0}^{N} B_n^+ \left(Z_{n,n+1} B_{n+1}^+ + Z_{n-1} B_{n-1}^+ \right) + B_n \left(Z_{n,n+1} B_{n+1} + Z_{n,n-1} B_{n-1} \right) \tag{1}$$

In this Hamiltonian $\Delta \sim 3-5 eV$ is excitation of an isolated molecule while X, Y, and Z are matrix elements of dipole-dipole interactions and they are about 100 times less than Δ.

Since a number of molecules N are small the boundary conditions must be taken into account:

$$F_{0,-1} = F_{N,N+1} = 0; \; F_{n,n\pm 1} = F;$$

and the Hamiltonian (1) has the form

$$H = H_1 + H_2 \tag{2}$$

where

$$H_1 = \sum_{n=0}^{N} \Delta B_0^+ B_0 + X B_0^+ B_0 + Y B_0^+ B_1 + \Delta B_N^+ B_N + X B_N^+ B_N + Y B_N^+ B_{N+1} +$$

$$+ \Delta \sum_{n=0}^{N} B_n^+ B_n + 2X \sum_{n=0}^{N} B_n^+ B_n + Y \sum_{n=0}^{N} B_n^+ \left(B_{n+1} + B_{n-1} \right) \tag{3}$$

$$H_2 = \frac{1}{2} Z \left[B_0^+ B_1 + B_1^+ \left(B_2^+ + B_0^+ \right) + B_0 B_1 + B_1 \left(B_2 + B_0 \right) \right] +$$

$$+ \frac{1}{2} Z \left[B_N^+ B_{N-1} + B_{N-1}^+ \left(B_N^+ + B_{N-2}^+ \right) + B_N B_{N-1} + B_{N-1} \left(B_N + B_{N-2} \right) \right] +$$

$$+ \frac{1}{2} Z \sum_{n=0}^{N} \left[B_n^+ \left(B_{n+1}^+ + B_{n-1}^+ \right) + B_n \left(B_{n+1} + B_{n-1} \right) \right] \tag{4}$$

The exciton Hamiltonian will be investigated by means of Green's functions (Sajfert, 2011)

$$G_{n,m}(t) = \Theta(t) < \left[B_n(t), B_m^+(0) \right] > \tag{5}$$

$$D_{n,m}(t) = \Theta(t) < \left[B_n^+(t), B_m^+(0) \right] > \tag{6}$$

where $\Theta(t)$ is Heaviside step function.

The energy of the elementary excitations in the chain is given by:

$$E_v = \sqrt{(\Delta+2X)^2 + 4(Y^2 - Z^2)\cos\varphi_v \cos\chi_v + 2(\Delta+2X)\left[(Y+Z)\cos\varphi_v + (Y-Z)\cos\chi_v\right]} \text{ with}$$

$$\sin(n+2)\varphi_v + \frac{2X}{Y+Z}\sin(n+1)\varphi_v + \left(\frac{X}{Y+Z}\right)^2 \sin n\varphi_v = 0 \qquad (7)$$

$$\sin(n+2)\chi_v + \frac{2X}{Y-Z}\sin(n+1)\chi_v + \left(\frac{X}{Y-Z}\right)^2 \sin n\chi_v = 0$$

The energy of excitations, as well as concentrations of excitons and the pairs, are the functions of two arguments, φ and χ, and in the coordinate system φ, χ and E (or $< B_n^+ B_n >$, $< B_n^+ B_n^+ >$ and E) represent surfaces. This essentially differs from the behavior of the same characteristics for an infinite chain where E, $< B^+B >$ and $< B^+B >$ are represented as lines in function of the angle $\gamma = ak$.

In infinite chain energy is given by

$$E_\gamma = \sqrt{(\Delta + 2X + 2Y\cos\gamma)^2 - 4Z^2 \cos^2\gamma} \qquad (8)$$

The comparison between dispersion laws of the finite and infinite chain is shown by dispersion law diagrams from Figures 13.1a and 13.1b.

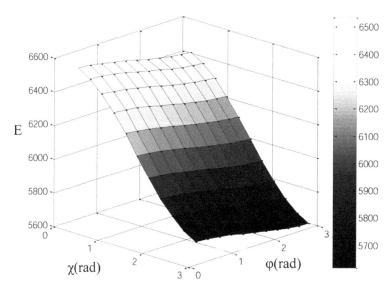

FIGURE 13.1a Dispersion law of finite chain with effective negative mass for a set of parameters $X = 50$, $Y = 400$, $Z = 300$.

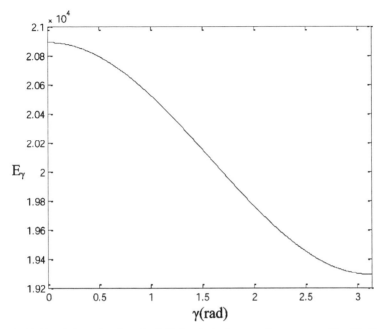

FIGURE 13.1b   Dispersion law for infinite chain for set of parameters $X = 50$, $Y = 400$, $Z = 300$.

The analysis of non-conserving excitons in short one-dimensional chain has shown that it fundamentally differs from the corresponding infinite linear chain.

The physical characteristics of a finite chain depend on two independent parameters and set of energies as well as a set of concentration points forms a discrete surface-like diagram of points. In the same time, the mentioned characteristics of the infinite chain are curves depending on one angle $\gamma = ak$. The physical explanation of this fact is not a simple problem. It comes due to the presence of usual excitons and exciton pairs which are behaving in some sense independently. For another comparison, *see* Sajfert (2005).

13.4   NANO-CHEMICAL APPLICATION(S): EXCITON WAVE FUNCTION OF POLYETHYLENE CHAIN

Assuming that polyethylene linear chain has N monomers (N is 30–50 monomers) after the introduction of boundary conditions the energies of excitons in the polymer chain is given by:

$$E_v = \Delta + 2X - 2Y \cos \varphi_v \qquad (9)$$

Using this formula, it can be found the transitions whose wavelengths correspond to experimentally obtained wavelengths.

Besides these wavelengths, we shall determine the probabilities of the finding of exciton at the given energy level, as well as the probabilities of exciton transition between two energy levels. These probabilities can be found by means of exciton one particle wave function

$$\left| \psi_v \right\rangle = \sum_{v=1}^{N+1} A_v(n) B_n^+ \left| 0 \right\rangle \qquad (10)$$

The average values $< B_n^+ B_n >$ are dependent on position n. We determined seven maximal values of $< B_n^+ B_n >$ for values $\Delta = 3$ eV D $= 2.5$ eV; J $= 2.7$ eV and they are given on Table 13.1.

n	16	0	22	26	8	3	14
$<B_n^+ B_n>$	0.01393	0.00973	0.00860	0.00842	0.00824	0.00798	0.00779

Taking that initial exciton concentration of macrochain is 10^{-8} and micro-chain initial concentration was taken as $\dfrac{1}{32}$ it found that internal energy of microchain U_{micro} comparing to 10^{-7} U_{macro} (macrochain has 10^7 microchain) we found that $\dfrac{U_{micro}}{10^{-7}U_{macro}} = 1.55 \cdot 10^8$. This result is optimistic: polyethylene folie leads to the absorption which is 10^8 times higher.

13.5   MULTI-/TRANS-DISCIPLINARY CONNECTION(S)

Macromolecules whose number of excitons is not conserved are very important for biophysical processes.

13.6   OPEN ISSUES

- Concise presentation of the limits of the concept within nanosciences, in both conceptual and applicative sides.
- Presents the potential risks in nanotechnology, the possible improvements, doubts, critics, etc.

KEYWORDS

- **exciton wave function of polyethylene chain**
- **finite chain**
- **non-conservation of excitons**

REFERENCES AND FURTHER READING

Ahmanov, S. A., & Hohlov, R. V., (1966). "The Problems of Nonlinear Optics" (in Russian) *UFN, 88*, 439.

Sajfert, V., Jaćimovski, S., Šetrajčić, J., Mašković, L., Bednar, N., Pop, N., & Tošić, B., (2011). Optical Properties of Nanostructures, *Journal of Computational and Theoretical Nanoscience 8*(11), 2285–2290.

Sajfert, V., Šetrajčić, J., Popov, D., & Tošić, B., (2005). Statistical Mechanics and its Applications, *Physica A 353C*, 217–234.

Tošić, B. S., (1967). *FTT, 9*, 1773.

CHAPTER 14

Electron Localization Function

MIHAI V. PUTZ[1,2]

[1]*Laboratory of Structural and Computational Physical Chemistry for Nanosciences and QSAR, Biology-Chemistry Department, West University of Timisoara, Pestalozzi Street No. 44, Timisoara, RO-300115, Romania, Tel.: +40-256-592638, Fax: +40-256-592620, E-mail: mv_putz@yahoo.com, mihai.putz@e-uvt.ro*

[2]*Laboratory of Renewable Energies-Photovoltaics, R&D National Institute for Electrochemistry and Condensed Matter, Dr. A. Paunescu Podeanu Str. No. 144, Timisoara, RO-300569, Romania*

14.1   DEFINITION

Electron localization function is a term used to measure the Pauli repulsion, i.e., which is the probability for an electron to be in the neighborhood of another electron which has the same spin.

14.2   HISTORICAL ORIGIN(S)

The concept of spatial localized electrons has been in the center of attention for scientists. From the Hartree-Fock theory is known that the canonical orbitals appear to be delocalized through the space of a molecule, but then is also known that the Hartree-Fock total energy remains unchanged when equivalent localized orbitals are generated by canonical orbitals unitary transformations. On the other hand, orbital-independent descriptors of electron localization were also studied; Bader and his collaborators (Bader and Essen, 1984; Bader et al., 1984) use the topography of electronic density Laplacian as a mean to determine the electron pairs or atomic shell structure. However, the Laplacian cannot be used in analyzing the shell structure in the case of heavy atomic systems (Sagar et al., 1988; Shi and Boyd, 1988). They (Bader and Stephens,

1975; Bader et al., 1988) proposed that the electron localization can be related with the Fermi hole function and the parallel spin pair probability, from which one can determine the effects of Pauli exchange repulsion. A formulation similar with the ELF concept was introduced by Artmann (Artmann and Nururforsch, 1946, p. 426), starting from the quantum mechanics principle according to which for N-particle, the sum of squares of their wave function is corresponding to a probability density. He determined that the maximum of this quantity can occur in bonds (Savin et al., 1997). Based on the Hartree-Fock pair probability, Becke and Edgecombe developed in 1990 a new orbital independent measure of electron localization, named the "electron localization function" (ELF), dependent on its gradient, the kinetic energy and the total electronic density (Becke and Edgecombe, 1990).

14.3 NANO-SCIENTIFIC DEVELOPMENT(S)

The electron localization information can be extracted from the Taylor expansion of the probability of the spherical averaged pair, in other words, when the reference electron is highly localized there is a smaller probability of finding a second like-spin electron in the proximity of the reference point (Becke and Edgecombe, 1990), respecting the minimum value of the following expression:

$$D_\sigma = \tau_\sigma - \frac{1}{4}\frac{(\nabla\rho_\sigma)^2}{\rho_\sigma} \tag{1}$$

Starting from this equation, Becke and Edgecombe introduce the electron delocalization function (ELF) with the expression:

$$ELF = \left(1 + \chi_\sigma^2\right)^{-1} \tag{2}$$

with

$$\chi_\sigma = D_\sigma / D_\sigma^0 \tag{3}$$

$$D_\sigma^0 = \frac{3}{5}\left(6\pi^2\right)^{2/3}\rho_\sigma^{5/3} \tag{4}$$

where D_σ^0 is corresponding to a uniform electron gas which has the spin density equal with the local value of $\rho(r)$ and χ_σ represent dimensionless localization index (Becke and Edgecombe, 1990).

Putz propose a new method of defining ELF, by taking the general Markovian ELFs shape and particularizing it using Becke–Edgecombe formulation (Putz, 2005):

$$\eta^{MBE1}(r) = \frac{1}{\left\{1+\left[\frac{g(r)}{h(r)}\right]^2\right\}^{3/2}} = \left[\eta^{BE}(r)\right]^{3/2} \tag{5}$$

its exponential form:

$$\eta^{MEXP1}(r) = \exp\left\{-\frac{3}{2}\left[\frac{g(r)}{h(r)}\right]^2\right\} \tag{6}$$

and its hyperbolic form:

$$\eta^{MSECH1}(r) = \left\{\sec h\left[\sqrt{2}\frac{g(r)}{h(r)}\right]\right\}^{3/2} \tag{7}$$

The electron localization function can also be analyzed by examining the spatial region, i.e., the *f*-localized domain, where is valid the inequality ELF$\geq f$ with *f* a constant. When *f* has a large value each of the *f*-localization domain contains only one attractor, when ELF values are smaller than *f* means that there are more than one attractor in the *f*-localization domain, and at a specific *f*-value (smaller), different *f*-localization domains are in contact at an ELF saddle point (Silvi and Savin, 1994; Mezey, 1990; Savin et al., 1996). Another approach (Savin et al., 1997) assume that theoretically, ELF can also be derived from the experimental data, i.e., one can obtain ELF values only from the electron density (Tal and Bader, 1978; Dreizler and Gross, 1990). In this case, by ignoring the Pauli principle determines that the orbital of the lowest energy will be occupied by all electrons, leading to the following equation:

$$\Delta T = \int\left[\frac{1}{2}\sum_i|\nabla\varphi_i|^2 - \sum_i\left|\nabla\sqrt{\rho(\vec{r})/N}\right|^2\right]d^3r \tag{8}$$

where $\varphi_i = \sqrt{\rho(\vec{r})/N}$, N being the electron number.

The same local value as obtained for the ELF expression proposed by Becke and Edgecombe is obtained when is applied the same scaling, such as the electron gas interval *(0,1)* (Savin et al., 1997).

14.4 NANO-CHEMICAL APPLICATION(S)

Even if aromaticity studies have been made starting with the 16th century, a theoretical characterization of the correspondence between the electrons and the aromatic character of a molecule has yet to be achieved. In this context, Hou et al. (2014) were trying to explain the aromaticity using the Electron Localization Function (ELF) (Silvi, 2004; Fuster et al., 2000), which based on the electron delocalization property, by dividing ELF in two components: ELF_σ and ELF_π for analyzing the σ and π character. In order to define ELF, one can make use of the excess of local kinetic energy density (Pauli exclusion principle) $T(\rho(r))$ along with the Thomas-Fermi kinetic energy density $T_h(\rho(r))$ (Santos et al., 2004), which will lead to the formula (Hou et al., 2014):

$$ELF = \left[1+\left(\frac{T(r)}{T_h(r)}\right)^2\right]^{-1} \qquad (9)$$

By considering the ELF value in the (3, −1) critical points (CPs) of the ELF basins (relative bifurcation points), one can determine the values for ELF_σ, ELF_π, respectively after the formula (Hou et al., 2014):

$$ELF_\sigma = \frac{\sum_{n=1}^{n_\sigma} ELF_\sigma}{n_\sigma}, \; ELF_\pi = \frac{\sum_{n=1}^{n_\pi} ELF_\pi}{n_\pi} \qquad (10)$$

with n_σ as the total numbers of σ orbital CPs from the molecule and n_π as the total numbers of π orbital CPs from the molecule.

Both ELF_σ and ELF were constructed starting from the premise that the interaction between adjoining ELF domain is measured by the ELF value at the bifurcation point. When the ELF value is higher, the electrons between these domains will have a higher level of delocalization and will give the aromatic character to a molecule (Hou et al., 2014).

The solids present different bonding types, as a result of numerous atom-atom contacts and high coordination index, their characterization being more difficult compared to the molecular class. Based on the theoretical approach, there is not yet made a rigorous quantitative classification for these bonds (Mori-Sanchez et al., 2002). In this context, Bader (1990) and Pendas et al., (1997) develop the topological analysis of the electron density (AIM) and states that in the stable structure of a solid, the existence of a bond implies a bond path. Also, the nature of the bond can be analyzed using the electron localization function (ELF) (Gatti, 2005).

Starting from the premise that the ELF formalism cannot be applied in high pressure condensed matter due to the computational limitation derived from periodicity (Gatti, 2005; Silvi and Gatti, 2000; Savin, 2004), Contreras-Garcia and his co-workers (2008) developed a computational code which can be used to visualize different chemical entities, in evaluating the volumes and charges of basins and localizing the ELF critical points. This way, the gap between the ELF topological analyses of bonding in both crystalline and molecular domains is fulfilled (Contreras-Garcia et al., 2008).

The newly developed program (Contreras-Garcia et al., 2008) have as input an all-electron crystalline wave function, and performs three different tasks related to ELF. In the beginning, the complete set of critical points is searched inside the cell, by constructing the irreducible Wigner Seitz with topological properties, these properties being further used in recursive algorithms, followed by a spherical search as an intermediary step, at the shell radii distance around the nuclei. The advantage of analyzing ELF profile is that, by identifying the relationship between basins, one can characterize the week interaction by means of localizing the bond points situated between basins (Contreras-Garcia et al., 2008). The next step starts from the identification of attractors which lead the program to locate which is the boundary limit of their associated space region. At this point, the program can also locate the basins and can calculate the integrals for any provided density operator. The final step consists in constructing of one-, two- and three-dimensional grids in order to visualize the net of critical points (Contreras-Garcia et al., 2008).

Applied to covalent solids, like diamonds, they will display a continuous 3D network representing the carbons tetrahedrally coordinated, which are bound by single bound basins B(C–C), the compound's perfect covalency implying the existence of the perfect electron pairs. In the case of ionic compounds, such as LiF, they present closed-shell basins (K and L, C(Li), C (F) and V(F)) with charge correspondent with the one for a complete transfer and shape close to a sphere (Contreras-Garcia et al., 2008).

The authors (Contreras-Garcia et al., 2008) consider in their study the crystalline phases of the elements of the periodic table from the first two rows, and use a method of calculating the electronic structure the DFT implemented in the CRYSTAL code (Saunders et al., 1998). They calculate several pressure (p)–volume (v) points for each compound, the new code being used in the topological analysis of ELF, i.e., to evaluate the rate at which occur a decrease of the basins volume as a result of the pressure action (Contreras-Garcia et al., 2008).

There were found two different types of attractors, bonding and atomic shells, and when one considers the analogy with Jellium model, the electrons which came from the last shell is correspondent with valence and the inner shells with the core. Results show that, in crystalline systems, the valence is responsible for bulk compression, the cores being periodically untouched (Contreras-Garcia et al., 2008).

14.5   MULTI-/TRANS-DISCIPLINARY CONNECTION(S)

In molecular systems, an alternative information measure of the electron localization, the so-called information-theoretic ELF (IT-ELF) was proposed by Nalewajski and his co-workers (2005) in order to interpret in terms of the Fisher, or local, a measure of information the key ingredient of ELF (Fisher, 1925; Frieden, 2000). They (Nalewajski et al., 2005) start by constructing the original ELF with the "squared" reciprocity relation as follow:

$$\mathrm{ELF}_\sigma(\mathbf{r}) = \left(1 + \chi_\sigma\left[\rho_\sigma;\mathbf{r}\right]^2\right)^{-1} \tag{11}$$

$$\chi_\sigma\left[\rho_\sigma;\mathbf{r}\right] = D_\sigma\left[\rho_\sigma;\mathbf{r}\right] / D_\sigma^0(\mathbf{r}) \tag{12}$$

with the specification that the local density approximation value (LAD) is respected:

$$D_\sigma^0(r) = D_\sigma^0\left[\rho_\sigma;r\right] = \left\lfloor 3\left(6\pi\right)^{2/3} / 5\right\rfloor \rho_\sigma(r)^{5/3} \tag{13}$$

Using this approach, where the function is invariant and dimensionless still respecting the unitary transformation of orbitals, one can determine the electron localization relative to a specific LDA. Its working interval range between 0 and 1, for a delocalized electron gas (homogenous) ELF take the value ½, the perfect localization being obtained at ELF = 1 (Nalewajski et al., 2005).

The authors also define the overall shape factor in terms of the unitary normalized probability distributions, as in equation:

$$p_\sigma(\mathbf{r}) = \rho_\sigma(\mathbf{r}) / N_\sigma \tag{14}$$

$$\int p_\sigma(\mathbf{r})\, d\mathbf{r} = 1 \tag{15}$$

with $N_\sigma = \int \rho_\sigma(\mathbf{r}) d\mathbf{r}$ as the total number of electrons of the spin variety σ.

In the same manner, the probability of orbital systems in molecules is determined from the following expressions, considering that the electron occupies a specific i^{th} molecular orbital (MO) (Nalewajski et al., 2005):

$$\rho_{i\sigma}(\mathbf{r}) = p_{i\sigma}(\mathbf{r}) / P_{i\sigma} \equiv \pi_{i\sigma}(\mathbf{r}) \tag{16}$$

$$P_{i\sigma} = \int p_{i\sigma}(\mathbf{r})\, d\mathbf{r} = 1 / N_\sigma \equiv P_\sigma \tag{17}$$

$$\sum_i^\sigma p_{i\sigma}(\mathbf{r}) = p_\sigma(\mathbf{r}) \tag{18}$$

$$\sum_i^\sigma \int p_{i\sigma}(\mathbf{r})\, d\mathbf{r} = \int \pi_{i\sigma}(\mathbf{r})\, d\mathbf{r} = \sum_i^\sigma P_{i\sigma} = 1 \qquad (19)$$

With $p_{i\sigma}(\mathbf{r})$ as the joint probability of simultaneous events of an electro originates from spin-orbital $\phi_{i\sigma}$ being at the point r, and $P_\sigma = \{P_{i\sigma} = 1/N_\sigma\}$ groups as the condense probabilities of finding on a specific MO an electron with σspin (Nalewajski et al., 2005).

Using these two parameters (the probability of orbital systems in molecules and the overall shape factor), one can express the MO information-theoretic quantities, the LDA reference function being expressed using the molecular probability distribution as follow (Nalewajski et al., 2005):

$$d_\sigma^0\big[p_\sigma(\mathbf{r})\big] = \Big[3(6\pi)^{2/3}/5\Big]\big[\rho_\sigma(\mathbf{r})/N_\sigma\big]^{5/3} = D_\sigma^0\big[\rho_\sigma;\mathbf{r}\big]/N_\sigma^{5/3} \qquad (20)$$

The modified ELF ratio is here of the form:

$$\chi_\sigma\big[p_\sigma;\mathbf{r}\big] = d_\sigma\big[p_\sigma;\mathbf{r}\big]/d_\sigma^0\big(p_\sigma(\mathbf{r})\big) = N_\sigma^{2/3} D_\sigma(\mathbf{r})/D_\sigma^0(\mathbf{r}) = N_\sigma^{2/3}\big[\rho_\sigma;\mathbf{r}\big] \qquad (21)$$

from which one can also construct the IT-ELF.

Moving forward, the ordinary "inverse" relationship is used and is obtained the following expression (Nalewajski et al., 2005):

$$\mathrm{elf}_\sigma(\mathbf{r}) = N_\sigma^{2/3}/\Big(N_\sigma^{2/3} + \chi_\sigma\big[p_\sigma;\mathbf{r}\big]\Big) = \big(1 + \chi_\sigma\big[\rho_\sigma;\mathbf{r}\big]\big)^{-1} \qquad (22)$$

In this case, the assumed normalization of Becke and Edgecombe is maintained, such as when $\chi_\sigma(\rho_\sigma;r) = 0$, elf(r) = 1 in the case of the perfectly localized case and when $\chi_\sigma[\rho_\sigma;r] = 1$, elf(r) = ½ in the case of homogenous, perfect delocalized electron gas (Nalewajski et al., 2005).

This study determined that the ELF (Becke and Edgecombe, 1990) defined as a conditional two-electron probability function can be used to measures the nonradiative part of the Fisher information density in the MO resolution for the locality. Starting from this considerate and based on the first power inverse relationship, one can construct the modified IT-ELF which appear to be comparable with the original ELF (Nalewajski et al., 2005).

14.6   OPEN ISSUES

The ELF can be analyzed using its graphical representation such as 2D-cut, due to the fact that is not an easy task to visualize a 3D function. This method implies displaying several plots, the condense information being determined by determining the salient feature. Starting from Bader theory (Bader 1990), one can use stationary point (Silvi and Savin, 1994) in order to characterize the functions, in other words, for determining the position in the space of the maximum value of ELF and for using in these specific points the ELF value. Even if there are many application domains of ELF maximum value (Savin et al., 1992a, 1992b, 1992c, 1992d), studies in this area determined that ELF do not have an absolute scale (Savin, 2005). Another limitation of ELF came from the amount of its interpretations (even if the procedures are simple), when many-body wave function came in question, because only for independent particles one can use the same interpretation in analyzing. Also, a limitation is given by the local nature of ELF into which were projected non-local feature, as well as the Heisenberg uncertainty principle which characterize the space separation in regions with electrons in a pair (Savin, 2005).

ACKNOWLEDGMENT

This contribution is part of the research project "Fullerene-Based Double-Graphene Photovoltaics" (FG2PV), PED/123/2017, within the Romanian National program "Exploratory Research Projects" by UEFISCDI Agency of Romania.

KEYWORDS

- **Fermi function**
- **Hartree-Fock total energy**
- **information-theoretic ELF**
- **molecular orbitals**

REFERENCES AND FURTHER READING

Artmann, Von K., (1946). Zur quantentheorie der gewinkelten Valenz, I. Mitteilung: Eigenfunction und Valenzbetätigung des Zentralatoms. *Zeitschrift für Naturforschung, 1,* 426–432.

Bader, R. F. W., (1990). *Atoms in Molecules: A. Quantum Theory.* Oxford University Press: Oxford, UK.

Bader, R. F. W., & Essen, H., (1984). The characterization of atomic interaction. *Journal of Chemical Physics, 80*, 1943–1960.

Bader, R. F. W., & Stephens, M. E., (1975). Spatial localization of the electronic pair and number distributions in molecules. *Journal of the American Chemical Society, 97*(26), 7391–7399.

Bader, R. F. W., Gillespie, R. J., & MacDougall, P. J., (1988). A physical basis for the VSEPR model of molecular geometry. *Journal of the American Chemical Society, 110*(22), 7329–7336.

Bader, R. F. W., MacDougall, P. J., & Lau, C. D. H., (1984). Bonded and nonbonded charge concentrations and their relation to molecular geometry and reactivity. *Journal of the American Chemical Society, 106*, 1594–1605.

Becke, A. D., & Edgecombe, K. E., (1990). A simple measure of electron localization in atomic and molecular systems. *Journal of Chemical Physics, 92*, 5397–5403.

Contreras-Garcia, J., Pendas, A. M., Silvi, B. J., & Recio, M., (2008). Useful applications of the electron localization function in high-pressure crystal chemistry. *Journal of Physics and Chemistry of Solids, 69*, 2204–2207.

Dreizler, R. M., & Gross, E. K. U., (1990). Basic formalism for stationary non-relativistic systems. In: *Density Functional Theory: An Approach to the Quantum Many-Body Problem* (pp. 15). Springer: Berlin.

Fisher, R. A., (1925). Theory of statistical estimation. *Proceedings of the Cambridge Philosophical Society, 22*, 700–725.

Frieden, B. R., (2000). *Physics from Fisher Information: A Unification*. Cambridge University Press: Cambridge.

Fuster, F., Sevin, A., & Silvi, B., (2000). Topological analysis of the electron localization function (ELF) applied to the electrophilic aromatic substitution. *Journal of Physical Chemistry, A104*, 852–858.

Hou, B., Yi, P., Wang, Z., Zhang, S., Zhao, J., Mancera, R. L., Cheng, Y., & Zuo, Z., (2014). Assignment of aromaticity of the classic heterobenzenes by three aromatic criteria. *Computational and Theoretical Chemistry, 1046*, 20–24.

Mezey, P., (1990). In: Lipkowitz, K. B., & Boyd, D. B., (eds.), *Reviews in Computational Chemistry* (Vol. 1). VCH: Weinheim.

Mori-Sanchez, P., Martin, P. A., & Luana, V., (2002). A classification of covalent, ionic, and metallic solids based on the electron density. *Journal of the American Chemical Society, 124*(49), 14721–14723.

Nalewajski, R. F., Ko1ster, A. M., & Escalante, S., (2005). Electron localization function as information measure. *Journal of Physical Chemistry, A109*, 10038–10043.

Pendas, A. M., Costales, A., & Luana, V., (1997). Ions in crystals: The topology of the electron density in ionic materials. *I. Fundamentals, Physical Review, B55*, 4275–4284.

Putz, M. V., (2005). Markovian approach of the electron localization functions. *International Journal of Quantum Chemistry, 105*, 1–11.

Sagar, Jr., P., Ku, A. C. T., Smith, Jr., V. H., & Simas, A. M., (1988). The Laplacian of the charge density and its relationship to the shell structure of atoms and ions. *Journal of Chemical Physics, 88*, 4367–4374.

Saunders, V. R., Dovesi, R., Roetti, C., Causa, M., Harrison, N. M., Orlando, R., & Zicovich-Wilson, C. M., (1998). *CRYSTAL 98 User's Manual*. University of Torino: Torino.

Savin, A., (2004). Phase transition in iodine: a chemical picture. *Journal of Physics and Chemistry of Solids, 65*, 2025–2029.

Savin, A., (2005). The electron localization function (ELF) and its relatives: Interpretations and difficulties. *Journal of Molecular Structure: THEOCHEM727*, 127–131.

Savin, A., Flad, H. J., Flad, J., Preuss, H., & Von Schnering, H. G., (1992a). Zur Bindung in Carbosilanen. *Angewandte Chemie, 104*(2), 185–186.

Savin, A., Flad, H. J., Flad, J., Preuss, H., & Von Schnering, H. G., (1992b). On the Bonding in Carbosilanes. *Angewandte Chemie International Edition in English, 31*, 185.

Savin, A., Jepsen, O., Flad, J., Andersen, O. K., Preuss, H., & Von Schnering, H. G., (1992c). Die Elektronenlokalisierung in den Festkörperstrukturen der Elemente: die Diamantstruktur. *Angewandte Chemie, 104*, 187–188.

Savin, A., Jepsen, O., Flad, J., Andersen, O. K., Preuss, H., & Von Schnering, H. G., (1992d). Electron localization in solid-state structures of the elements: The diamond structure. *Angewandte Chemie International Edition in English, 31*, 187–188.

Savin, A., Nesper, R., Wengert, S., & Fassler, T. E., (1997). ELF: The electron localization function. *Angewandte Chemie International Edition in English, 36*, 1808–1832.

Savin, A., Silvi, B., & Colonna, F., (1996). Topological analysis of the electron localization function applied to delocalized bonds. *Canadian Journal of Chemistry, 74*(6), 1088–1096.

Shi, Z., & Boyd, R. J., (1988). The shell structure of atoms and the Laplacian of the charge density. *Journal of Chemical Physics, 88*, 4375–4376.

Silvi, B., (2004). How topological partitions of the electron distributions reveal delocalization. *Physical Chemistry Chemical Physics, 6*, 256–260.

Silvi, B., & Gatti, C., (2000). Direct space representation of the metallic bond. *Journal of Physical Chemistry A., 104*, 947–953.

Silvi, B., & Savin, A., (1994). Classification of chemical bonds based on topological analysis of electron localization functions. *Nature (London), 371*, 683–686.

Tal, Y., & Bader, R. W. F., (1978). Studies of the energy density functional approach. I. Kinetic energy. *International Journal of Quantum Chemistry, Supplement: Proceedings of the International Symposium on Atomic, Molecular, and Solid-state Theory, Collision Phenomena, and Computational Methods, S12*, 153–168.

CHAPTER 15

Families of Molecular Descriptors

LORENTZ JÄNTSCHI[1] and SORANA D. BOLBOACĂ[2]

[1]*Technical University of Cluj-Napoca, Romania,*
Tel.: +40-264-401775, E-mail: lorentz.jantschi@gmail.com

[2]*Iuliu Haţieganu University of Medicine and Pharmacy Cluj-Napoca,*
Romania

15.1   DEFINITION

Families of molecular descriptors were designed and used to relate the structural information with measured properties and activities for different series of chemical compounds. Since a whole pool of experimental designs are devised for conducting measurements (preferably for series of compounds) of which outcomes depend on a variety of factors (including environmental ones) it is unlikely to expect a unique pattern of relating the structure of the compounds with their measurements. On the other hand, by taking into account that from one series of measurements to another, some settings may have small variations, it is expected that the series of the measurements to be related one to the other as well to be both related with the structure of the compounds. This general idea may have different materializations, one of them being the creation and the use of related structural descriptors–families of molecular descriptors.

The strategy to materialize a family of molecular descriptors is to use a genetic code embedding in it a series of different operators (the genome of the family) applying on the structure of the chemical compound (preferably on both topological and geometrical information) to produce a parameterized numerical outcome.

Several families of molecular descriptors were developed to date, FPIF (from Fragmental Property Index Family), MDF (from Molecular Descriptors Family), MDFV (from Molecular Descriptors Family–Vertex), SAPF (from Structural Atomic Property Family), SMPI (from Szeged Matrix

Property Indices) and its extension FMPI (from Fragmental Matrix Property Indices) as well as ChPE, which is an extension of the Characteristic Polynomial intended to be used on molecules.

15.2　HISTORICAL ORIGIN(S)

First steps to the molecular models are recorded in 1861 (Loschmidt, 1861). Today molecular modeling involves theoretical methods and computational techniques for pushing further (*see* Rhinehardt et al., 2015) the knowledge about the molecular structure.

When a series of compounds are involved, then the expected result of a model is to provide a function or a relation between the structure and observed macroscopic behavior of the molecules. Strategies like docking (Taha et al., 2015), assaying (Peng et al., 2015) and mapping (Radwan and Abdel-Mageed, 2014) are involved in exploiting the feature of the systematic experimental observation better, but nevertheless, strategies to develop families of molecular descriptors (FMDs) began to attract concerns (*see* Kihl et al., 2015).

Modeling the molecular structure is the way of understanding of the microscopic level and its expression at the macroscopic one level. The accessing of the microscopic level is via measurements (*see* Figure 15.1).

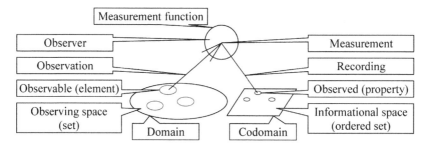

FIGURE 15.1　Encoding the information from measurements.

In regard to the measurements, there are many ways of expressing the encoded information, differing one from each other by the quality of the representation.

Thus, the primary measurement scale is binomial which encodes (in the informational space) logical values having as allowed operations equality ("=") and negation ("!") providing a structure of Boolean algebra

(Boole, 1854). Mode and Fisher exact (Fisher, 1922) are the allowable statistics on, and examples of measurements associated with the encoded values are distinguishing between dead and alive, and looking for occurrences of the sides of a coin.

(Multi)nomi(n)al scale uses a finite and known series of unordered values to record the observations, being a discrete scale and having allowed the test for equality ("=") providing a structure of a standard set. One statistic has a clear meaning on the values measured on this scale–mode–and comparisons between a series of measurements using this scale can be conducted with a Chi-square test (Pearson, 1900). Examples of measurements expressed with this scale include 'ABO' blood group system, but also the classification of living organisms.

An ordinal scale is encoding discrete values, and the allowed operations include the test for equality ("=") and (strict) inequality ("<") providing a structure of commutative algebra (Krull, 1935). The allowable statistic is the median and on the information collected with this scale is possible the ranking. An example of information collected using this scale is the number of atoms in molecules.

Interval scale provides continuous values implicitly falling into an interval or domain. As operations are possible to do comparisons using inequality operator ("≤") as well as to do subtractions. It provides a structure of one-dimensional affine space (Berwald, 1918), having allowed calculating the mean, standard deviation, correlation, regression, and ANOVA. Examples include measurements of temperature, distance, time, and energy.

Ratio scale provides too continuous values on the non-negative domain having as allowed operations inequality ("≤"), subtraction ("–") and multiplication ("*"). It provides a structure of a one-dimensional vector space (Bolzano, 1804) having allowed the most comprehensive list of statistics including geometric and harmonic means, the coefficient of variation, doing of logarithms (Napier, 1614), and examples include chemical and biological measurements such as pH and sweetness relative to sucrose.

Molecular modeling requires and is feed with measurements. If on the one hand stays the measured values (*see* Figure 15.1), on the other hand, stays the chemical structure (*see* Figure 15.2). If the Universe is seen as the whole observing space (*see* Figure 15.2), then radiant energy differentiates as having a velocity comparable with light velocity (relativistic velocity) grouping radiations such as β, γ, being differentiated through properties.

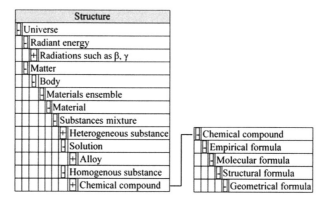

FIGURE 15.2   To the layers of the chemical structure.

The other main group contains the matter (Figure 15.2) is seen as the whole non-relativistic observing space in which the body is seen as having the velocity much less than the velocity of light. It contains materials ensemble with possibly variable and discontinue (chemical) composition. Going deeper in the classification, on the next layer stays materials with variable and continue (chemical) composition which generally groups mixtures of substances possessing well-defined chemical composition from which homogenous substances have constant (chemical) composition, to finally arrive at the chemical compound concept with well defined and unique chemical composition. From this point on, we may start to discuss about the chemical structure, and an empirical formula provides the ratio between the atoms in the compound, the molecular formula provides further the number of atoms from each type in the molecule, the structural formula reveals the structural groups in the molecule, and finally geometrical formula defines the relative arrangement of the atoms in the molecule. Although it is the last refinement level, sometimes (actually quite often), the geometrical formula may degenerate to being well known the geometrical isomerism (*see* Warder, 1890). Namely, knowing the distances between the atoms and the angles between them we still don't have enough knowledge to define a unique chemical structure, which in some cases may be problematic.

Modeling the molecular structure is a prerequisite for structure-activity inference analysis. The building of a three-dimensional model (3D) is necessary when the calculated descriptors on the structure use the geometry of the molecule. Obtaining the 3D model can be achieved using a molecular modeling program (*see* Table 15.1).

TABLE 15.1   Molecular Modeling Software

Name	Provider website
Abalone	http://biomolecular-modeling.com
ADF	http://scm.com
ChemBioOffice	http://cambridgesoft.com
Gaussian	http://gaussian.com
HyperChem	http://hyper.com
Materials Studio	http://accelrys.com
Q-Chem	http://q-chem.com
Spartan	http://wavefun.com

When certain software (as given above) is used, sometimes conversions between different formats storing the chemical information are useful, as well as it helps some software for visualizing (only) of the obtained models (*see* Table 15.2).

TABLE 15.2   Molecular Modeling Auxiliary Software

Name	Intend
GLmol	Browser-based visualization
Jmol	Java applet for visualization
MDL Chime	Browser plugin for visualization
Open Babel	Conversions
PyMOL	Python application for visualization
RasMol	GNU GPL application for visualization
WebQC	Conversions

Obtaining of the 3D model of the molecule involves a series of steps, as given below:

- Constructing of the topology, namely specification of the atoms by atom type and of the bonds by bond order.
- Building of a 3D arrangement, when typical routines possibly including molecular mechanics force fields, such as are CHARM (Brooks et al., 1983), AMBER (Cornell et al., 1995), MMFF94 (Halgren, 1996), and OPLS (Jorgensen and Tirado-Rives, 1998).
- Refining of the 3D arrangement may involve semi-empirical methods, such as are AM1 (Dewar et al., 1985), PM3 (Stewart, 1989), RM1 (Rocha et al., 2006), and PM6 (Stewart, 2007).

- Further refining of the geometry with DFT (density functional theory) approaches including HF (Hartree-Fock, *see* Hartree, 1928; Fock, 1930), post-HF–such as are perturbation theory (Møller and Plesset, 1934), coupled cluster (Purvis and Bartlett, 1982), configuration interaction (Maurice and Head-Gordon, 1999), and composite methods (Ohlinger et al., 2009) and KS (Kohn-Sham, *see* Kohn and Sham, 1965), such as LDA (Parr and Yang, 1994), GGA (Perdew et al., 1992), and PBE (Perdew et al., 1996).

Special precautions at building and of refining of the 3D model should be given to the structures with geometrical isomers, because during the geometrical optimization the passing from one geometrical conformation to another is quite often encountered.

One of the outcomes of the molecular modeling is the charge distribution over the atoms in the molecule, or partial charges. There are different approaches are available:

- Born (*see* Born and Goppert-Mayer, 1931);
- Callen (*see* Callen, 1949);
- Szigeti (*see* Szigeti, 1949);
- Mulliken (*see* Mulliken, 1955 and thereafter);
- Coulson (*see* Coulson et al., 1962);
- Politzer (*see* Politzer, 1968)
- Löwdin (*see* Löwdin, 1970);
- Hirshfeld (*see* Hirshfeld, 1977);
- Cioslowski (*see* Cioslowski, 1989);
- Bader (*see* Bader, 1990);
- An optimization method based on electrostatic potentials (*see* for instance Wang and Ford, 1994).

Along with the partial charges, the outcome of the molecular modeling includes the (relative) coordinates of the atoms (usually given in Å), the bonds and their types (*see* Table 15.3).

TABLE 15.3 Typical Information from Molecular Modeling

The list of the atoms			
Label	Type	Coordinates (x, y, z)	Partial charge(s)
The list of the bonds			
Atom Label		Atom Label	Bond type or order

Usually, the methodology for relating the structure with the experimental measurements in a series of compounds uses the molecular structure in which the hydrogen atoms are neglected (deleted). Some of the reasons are given below:

- Biological activities determined in vivo have as the environment (medium) aqueous solutions in which processes of (partial) dissociation in which the hydrogen atoms pass in the form of protons in solution, leaving the place occupied in the molecular structure.
- Hydrogen atoms can form a single bond; if they are deleted, excepting their geometrical position information can always be rebuilt.
- Because form a single bond, the hydrogen atoms do not contribute to the complexity of molecular (not create chains and branches are just terminals for the structure).
- Deleting of the hydrogen atoms reduces the amount of calculations for a certain structure; considering only an alkane of the general formula C_nH_{2n+2}, removing the hydrogen atoms reduces the complexity of the topology to $1/9$ (a topological matrix records values for each pair of atoms and the atoms are about one third less).

15.3 NANO-SCIENTIFIC DEVELOPMENT(S)

A genetic code can formally describe the breeding of the molecular descriptors of a family (*see* Figure 15.3). The outcome of the breeding process is the population of phenotypes, in which the individuals are labeled with letters encoding their genomic code ($L_n...L_1$ in Figure 15.3).

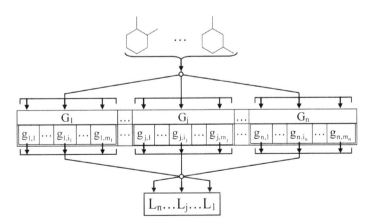

FIGURE 15.3 Encoding the information from structure with FMDs.

The process of generating the population of descriptors is to be applied for each molecule of the entry dataset in part (*see* Figure 15.3) and basically is a far each procedure (*see* Figure 15.4).

```
Input data: series of the molecules ({Mol₁, ..., Molₖ})
For each w from 1 to k //for Molᵥᵥ
        For each i₁ from 1 to m₁ //for G₁
            ... //for the rest of the genes
                    For each iⱼ from 1 to mⱼ //for Gⱼ
                        ... //for the rest of the genes
                                For each iₙ from 1 to mₙ //for Gₙ
                                    Lₙ...Lⱼ...L₁(Molᵥᵥ) ← Calculate the descriptor using
                                                          [Molᵥᵥ, g₁,i₁, ..., g₁,iⱼ, ..., g₁,iₙ]
                                                          current values
                                End for
                            ...
                    End for
                ...
        End for
End for
Output data: Series of descriptors values for each molecule ({Lₙ...L₁(Molᵥᵥ), 1 ≤ w ≤ k})
```

FIGURE 15.4 Obtaining the descriptors of a FMD for a set of molecules.

As can be deduced, always a FMD generates exactly the same number of descriptors for any molecule (in the general case, *see* Figures 15.3 and 15.4) the number is $m_1 \cdot \ldots \cdot m_j \cdot \ldots \cdot m_n$.

FPIF (from Fragmental Property Index Family; Jäntschi and Diudea, 2000; *see* Table 15.4) is a matrix-based method, in which the matrices collects properties derived from structure for fragments obtained for each pair of atoms.

TABLE 15.4 Code of FPIF Descriptors

Gene	I_M	D_M	A_P	P_D	F_C	S_M	M_I	L_O
	R	T	M	_p_	si	S	P_	I
	D	G	E	_d_	se	P	P2	R
			C	_1/p_	ji	A	E_	L
Genome			Q	_1/d_	je	G	E2	
				_p*d_	fi	H		
				_p/d_	fe			
				_p/d2				
				p2/d2				

FPIF = $L_O \times I_M \times D_M \times A_P \times P_D \times F_C \times S_M \times M_I$. Examples, IRGseCp2/d2SE2, RDGjeP_p/d2GP_

FPIF uses $d_M(a,b)$ – the topological distance in structure M from atom a to atom b; $\delta_M(a,b)$ – the topological detour – i.e., longest path – in structure M from atom a to atom b; $W_M(a,b)$ – set of walks; $P_M(a,b)$ – set of paths; $D_M(a,b)$ – set of distances – i.e., shortest paths; $\Delta_M(a,b)$ – set of detours – i.e., longest paths; M/p – substructure derived from structure M when the atoms inside the path p are removed from M together with their connections. Sets (one or more) of atoms of a molecule M for every pair of atoms (a,b) are calculated for every of the following (six) set collecting criteria (called F_C – fragmentation criteria):

- With $Sz_{a,b,d/\delta} = \{c \in M \mid d/\delta_M(c,a) < d/\delta_M(c,b)\}$:
  - $F_C = si$, $SzDi_{a,b} = \{c\ M \mid d_M(c,a) < d_M(c,b)\}$;
  - $F_C = se$, $SzDe_{a,b} = \{c \in M \mid \delta_M(c,a) < \delta_M(c,b)\}$;
- With $CJ_{a,b,p} = \{c \in M \mid d_M(c,a) < d_M(c,b) \text{ and } \exists w \in W_M(c,a) \mid \{a\} = w \cap p\}$:
  - $F_C = ji$, $Cj_{a,b,p}$ when $p \in D_M(a,b)$;
  - $F_C = je$, $Cj_{a,b,p}$ when $p \in \Delta_M(a,b)$;
- With $Cf_{a,b,p} = \{c \in M \mid d_{G\backslash p}(c,a) < d_{G\backslash p}(c,b)\}$:
  - $F_C = fi$, $CfDi_{a,b} = Cf_{a,b,p}$ when $p \in D_M(a,b)$;
  - $F_C = fe$, $CfDe_{a,b} = Cf_{a,b,p}$ when $p \in \Delta_M(a,b)$;

Four atomic properties (A_P an atomic property) are taken into calculation, M ($A_P = M$) as relative atomic mass; E ($A_P = E$) as electronegativity (Sanderson scale, Sanderson, 1983); C ($A_P = C$) as (set) cardinality; P ($A_P = P$) as partial charge (class I, *see* Cramer, 2002, from Mulliken population analysis, Mulliken, 1955). Eight property descriptor (P_D a property descriptor) expressions account atomic properties, p ($P_D = p$) – atomic property; d ($P_D = d$) – distance; $P_D = 1/p$; $P_D = 1/d$; $P_D = pd$; $P_D = p/d$; $P_D = p/d^2$; $P_D = p^2/d^2$. Five overlapping methods (S_M – superposing method) overlap atomic properties to provide the fragmental property, S ($S_M = S$) – sum; P ($S_M = P$) – multiplication; A ($S_M = A$) – arithmetic mean; G ($S_M = G$) – geometric mean; H ($S_M = H$) – harmonic mean. Two models of interaction give transform in a vector a descriptor (I_M – interaction model), R ($I_M = R$) – rare (uses the assumption that the property of all atoms are approximately located in the fragment center of property, of which position is consequently obtained and used to express the descriptor vector); D ($I_M = D$) – dense (the effect of each atom are superposed using vector summation). Two distance metrics (D_M – metric of distance) provides the distance for expressing the descriptor values, T ($D_M = T$) – topological (from connectivity) and D ($D_M = D$) – topographical (from

3D model of the molecule obtained from different levels of theory – for a discussion about, *see* Clark, 1985). Four square-matrix based indices (M_I– matrix index) collects overall molecular property, P_ (M_I = P_) – half-sum of matrix elements; P2 (M_I = P2) – half-sum of squared matrix elements; E_ (M_I = E_) – half-sum of Hadamard product of matrix with adjacency matrix; E2 (M_I = E2) – half-sum of squared Hadamard product of matrix with adjacency matrix. Finally, a molecular descriptor is obtained via a linearization operator (L_O – linearization operator) meant to transform nonlinearities to linearity at relationships, *I* (L_O = I) – identity function; *R* (L_O = R) – reciprocal function (f(x) = 1/x); L (L_O = L) – logarithm function (f(x) = ln(x)). Thus, FPIF family of molecular descriptors puts together a total number of individuals equal with the number of all multiplications described above (2·2·4·8·6·5·4·3–46080) – *see* Table 15.4.

MDF (from Molecular Descriptors Family; Jäntschi, 2004; Jäntschi, 2005; *see* Table 15.5) is a method based on molecular fragments obtained for pairs of atoms.

TABLE 15.5 Code of MDF Descriptors

Gene	D_M	A_P	I_D				I_M	F_C	S_M				L_O
Genome	t	C	D	Q	L	F	r	m	m	A	G	H	I
	g	H	d	q	l	f	R	M	M	a	g	h	i
		M	O	J	V	S	m	D	n	B	F	I	A
		E	o	j	E	s	M	P	N	b	f	i	a
		G	P	K	W	T	d		S	P	s		L
		Q	p	k	w	t	D						l

MDF = $D_M \times A_P \times P_D \times I_M \times F_C \times S_F \times L_O$. Examples, lsPRLGg, IhDDDCt.

Similarly with FPIF, MDF it uses two distance operators (D_O), topological (t) and geometrical (g), six atomic properties (A_P), cardinality (C), number of directly connected hydrogen atoms (H), relative atomic mass (M), electronegativity (E – Sanderson scale, group electronegativity (G – Diudea and Silaghi, 1989), partial atomic charge (Q – Mulliken, and twenty-four interaction descriptors (I_D) as follows, D(d), d(1/d), O(p_1), o(1/p_1), P(p_1p_2), p(1/p_1p_2), Q($\sqrt{p_1p_2}$), q(1/$\sqrt{p_1p_2}$), J(p_1d), j(1/p_1d), K(p_1p_2d), k(1/p_1p_2d), L(d$\sqrt{p_1p_2}$), l(1/d$\sqrt{p_1p_2}$), V(p_1/d), E(p_1/d_2), W(p_1^2/d), w(p_1p_2/d), F(p_1^2/d^2), f(p_1p_2/d^2), S(p_1^2/d^3), s(p_1p_2/d^3), T(p_1^2/d^4), t(p_1p_2/d^4). Interaction were modeled (I_M) using six functions, *R* and *r* – being rare, M and m – being medium, and D and d being dense – the upper letter encoded one having as reference the first atom of the fragment (a in the notation given at defining of FPIF) and lower letter nominating the reference on the probe atom (b in the notation given at defining of FPIF). Fragmentation is driven by one fragmentation criterion (F_C), m (F_C = m) – defines smallest fragment containing atom a; M (F_C = M) – defines largest fragment not containing atom b; D (F_C = D) – defines so called Szeged fragments (closer to atom a than to atom b), P (F_C = P) – Cluj path based fragments (*see* FPIF definition for the definition of Cluj path based fragments – $CF_{a,b,p}$, $p \in D_M(a,b)$), nineteen overlapping strategies for fragments interaction (S_F – superposing formula), m (S_F = m) – smallest value; M (S_F = M) – biggest value; n (S_F = n) – smallest absolute value; N (S_F = n) – biggest absolute value; S (S_F = S) – sum of; A (S_F = A) – S divided to number of fragments possessing real value of descriptor; a (S_F = a) – S divided to total number of fragments; B (S_F = B) – S divided to number of atoms; b (S_F = b) – S divided to number of bonds; P (S_F = P) – product of; G geometric mean rooted P as S is divided for A (S_F = A); g (S_F = g) – rooted P as S divided for a (S_F = a); F (S_F = F) – rooted P as S divided for B (S_F = B); f (S_F = f) – rooted P as S divided for b (S_F = b); s (S_F = s) – harmonic sum; H (S_F = H), h (S_F = h), I (S_F = I), i (S_F = i) harmonic means following same procedure from s as G (S_F = G), g (S_F = g), F (S_F = F), f (S_F = f) were derived as geometric means from P and same procedure as for A (S_F = A), a (S_F = a), B (S_F = B), b (S_F = b) derived as arithmetic means from S. Six linearization operators (L_O) being, I (L_O = I) – identity(f(x) = x); i (L_O = i) – inverse (f(x) = 1/x), A (L_O = A) – absolute of (f(x) = |x|), a (L_O = a) inverse of absolute of (f(x) = 1/|x|), L (L_O = L) – logarithm of (f(x) = ln(x)) and l (L_O = l) – logarithm of absolute of (f(x) = ln(|x|)). Thus, MDF puts together a total number of individuals equal with the number of all multiplications ($2 \cdot 6 \cdot 6 \cdot 24 \cdot 4 \cdot 19 \cdot 6$ = 787968) – *see* Table 15.5.

MDFV (from Molecular Descriptors Family – Vertex; Bolboacă and Jäntschi, 2009; *see* Table 15.6) uses atoms in place of pairs of atoms (as FPIF and MDF uses). It implements two distance metrics (D_O), t (topological) and g(geometrical), seven atomic properties (A_P), C (cardinality), H (hydrogen's), M (mass), E (electronegativity, Sanderson scale), Q (partial charge, Mulliken population analysis), L (melting point under normal temperature and pressure conditions), A (electronic affinity), 58 interaction descriptors (I_D, *see* Table 15.6).

TABLE 15.6 Code of MDFV Descriptors

Gene	D_O	A_P	I_D							S_F	S_M	I_T	E_U	L_O
T	C	J	R	N	Z	V	I	D	A	A	f	D	I
G	H	j	r	n	z	v	i	d	a	a	F	d	R
		M	O	K	W	S	F	A	0	I	I	c		L
		E	o	k	w	s	f	a	1	i	i	C		
		Q	P	L	X	T	G	B	2	F	F	p		
		L	p	l	x	t	g	b	3	P	P	P		
		A	Q	M	Y	U	H	C	4	C	C	a		
			q	m	y	u	h	c	5			A		
									6			i		
									7			I		

(Left margin label: *Genome*)

$MDFV = D_O \times A_P \times I_D \times S_F \times S_M \times I_T \times E_U \times L_O$. Examples, GA0AAaDI, TQzPPPdR.

Atoms (or vertices in graph theory naming) are cut, and fragments (connected atoms) are collected. It is calculated first the fragmental property using one out of ten strategies (I_T – interaction type):

- $I_T = f$ – fragment's field – superposes (adds) axial projections of I_D for all pairs of atoms (b,c) from fragment ((b,c) $\in$ Fr(a)) taken once – giving interactions in the fragment independent of atom cut).
- $I_T = F$ – field of the fragment in the cut – superposes (adds) axial projections of I_D for all pairs of atoms (a,b) with one atom in the fragment (b$\in$Fr(a)) – giving interaction of the fragment in the cut.
- $I_T = c$ – fragment's descriptor center – computes coordinates of the center of the descriptor using once every pair of atoms of the fragment (b,c)$\in$Fr(a).
- $I_T = C$ – fragmentation descriptor center – computes coordinates of the center of the descriptor using all pairs of atoms (a,b) with one atom in the fragment (b$\in$Fr(a)) – giving the weight of the fragment in the cut.
- $I_T = p$ – fragment's potential – uses all pairs (b,c)$\in$Fr(a) to obtain the average direction (average of the directions) of the field; uses all pairs (b,c)$\in$Fr(a) to obtain the cumulated value (sums of the effects); gives the intrinsic potential of the fragment.
- $I_T = P$ – potential of the fragment relative to the cut – uses all pairs of atoms (a,b) with one atom in the fragment (b$\in$Fr(a)) for giving the extrinsic potential of the fragment at the cut.

- I_T = a – select highest descriptor present in the fragment (from all pairs (b,c)∈Fr(a) of atoms present in the fragment); give strongest interaction in the fragment.
- I_T = A – select highest descriptor of the fragment with the cut (from all pairs (a,b) with b∈Fr(a)); give strongest interaction in the cut.
- I_T = m – select lowest descriptor present in the fragment (from all pairs (b,c)∈Fr(a) of atoms present in the fragment); give weakest interaction in the fragment.
- I_T = M – select highest descriptor of the fragment with the cut (from all pairs (a,b) with b∈Fr(a)); give weakest interaction in the cut.

In general, for a vertex cut, more than one fragment may occur. Thus, this fact are accounted using superposing of the descriptors interaction at fragments (between fragments of same cut) level by the superposing at fragment (S_F) formula. When operates in the Minkowski space (using absolute values) two superposing derives, a $(S_F = a)$ – standing for max $(|(\cdot_x|+|\cdot_y|+|\cdot_z|)$ and i $(S_F = i)$ -standing for min$(|(\cdot_x|+|\cdot_y|+|\cdot_z|)$. When operates in the Euclidian space (using square values and after squared root of) other two superposing derives, A $(S_F = A)$ – standing for max$(\sqrt{(\cdot_x^2+\cdot_y^2+\cdot_z^2)})$ and I$(S_F = I)$ – standing for min$(\sqrt{(\cdot_x^2+\cdot_y^2+\cdot_z^2)})$. When the effects of two or more fragments are superposed, we can superpose it as vectors, and then S_F takes value of F $(S_F = F)$, we can superpose only their directions (and add their values), and then S_F takes the value of P $(S_F = P)$ or weighing their effect, and then S_F takes the value of C $(S_F = C)$. Finally, superposing is conducted at molecular level from all cuts using same procedure described above at superposing at fragments of a cut. Thus, S_M superposes as minimum absolute (when S_M = i), as maximum absolute (when S_M = a), as minimum in Euclidean space (when S_M = I), as maximum in Euclidean space (when S_M = A), weighting effects (when S_M = C), superposing directions (when S_M = P) or vectorial superposing (when S_M = F). All values of the descriptors at molecular level obtained using the procedure described above possess two things, a value and a reference (a coordinate of its position). Thus, we can express as molecular descriptor the value of it (and then E_U = D) or a reference of it (a distance, and then E_U = d) where E_U is the expressing unit). A linearization operator (L_O) serves for linear regression designing of the analysis with MDFV family of descriptors, and it takes three values, *I* (standing for identity with), *R* (standing for reciprocal or inverse of) and *L* (standing for logarithm of). Thus, MDFV family of molecular descriptors puts together a total number of individuals equal with the number of all multiplications described above $(2·7·58·7·7·10·2·3 = 2387280)$ – *see* Table 15.6.

Transforming of MDF to a more complex and large family (as MDFV is) does not provided expected significant improvement of QSAR (quantitative structure-activity relationships) models (with MDFV) as were obtained (with MDF), another approach were developed, SAPF (*see* Table 15.7).

SAPF (from Structural Atomic Property Family; Sestraş et al., 2012; *see* Table 15.7; calculation details given in Jäntschi, 2012) cumulates atomic properties at molecular level. It locates the molecular centre using one (out of three methods, C_F) for this task involving a metric (out of two, M_D) for the distance, a atomic property (out of eight defined till date, A_P), a rising power for the distance (D_P, seven cases), a rising power for the property (P_P, same seven cases). At molecular level one of two sorts of operators (O_M, mean type or sum type) build the molecular property as generalized mean or sum (*see* O_M) of descriptor's values rising it at a power (G_M, again one out of same seven cases) and the result are subject to linearization (L_O, one out of seven cases).

TABLE 15.7 Code of SAPF Descriptors

Gene	C_F	D_O	A_P	D_P	P_P	O_M	M_P	L_O
	D	T	C	I	I	S	I	I
	P	G	H	E	E	M	E	A
Genome	C		M	H	H		H	S
			E	G	G		G	T
			A	A	A		A	Q
				Q	Q		Q	R
				S	S		S	L

$SAPF = L_O \times M_P \times O_M \times P_P \times D_P \times A_P \times D_O \times C_F$. Examples, SISHQEGC, TESHIMGP.

Thus, SAPF family of molecular descriptors puts together a total number of individuals equal with the number of all multiplications described above $(7 \cdot 7 \cdot 2 \cdot 7 \cdot 7 \cdot 9(5) \cdot 2 \cdot 3 = 259308$ – with nine atomic properties; 144060 with five atomic properties, *see* Table 15.7).

Szeged matrix property indices (SMPI; Bolboacă and Jäntschi, 2016; *See* Table 15.8) has an online interface and it is free to be used (Jäntschi, 2014).

TABLE 15.8 Code of SMPI Descriptors

Gene	A_P	D_M	I_D	M_O	L_O
Genome	A	T	E	m	I
	B	G	U	M	R
	C	U	D	I	L
	D		P	J	
	E			E	
	F			F	
	G				

SMPI = $L_O \times M_O \times I_D \times D_M \times A_P$. Examples, ImETA, LFPUG.

For SMPI distance matrix are calculated, and then for each pair of (distinct) atoms the atoms closer to the first than to the second atom of the pair are collected into (these are fragments; are exactly one fragment associated to a pair of atoms by this way) a matrix (similarly to the unsymmetrical Szeged matrix on paths, but containing sets of atoms in place of their number; for [USzp] matrix definition, *see* Diudea et al., 2001). To each fragment it is assigned an atomic property A_P = A, Atomic mass (a.u.), as sum of; A_P = B, Atomic number (Z), as harmonic sum of; A_P = C, Cardinality (= 1), as sum of; A_P = D, Solid state density (kg/m^3), as harmonic mean of; A_P = E, Electronegativity (revised Pauling; for Pauling electronegativity, *see* Pauling, 1932; for revised Pauling electronegativity, *see* Allred, 1961), as geometrical mean of; A_P = F, First ionization energy (kJ/mol), as average of; A_P = G, Melting point temperature (K), as Euler (PM(p), p = 2) mean of. A distance matrix is calculated using three alternatives–D_M = T, Topological distance (bonds); D_M = G, Geometrical distance (Å); D_M = U, Weighted topological distance (as reversed bond order). An interaction descriptor produces the interaction effects matrix operating on the properties and on the distances matrices–I_D = E, $E_{i,j} = P_{i,j}*D_{i,j}$; I_D = U, $U_{i,j} = P_{i,j}/D_{i,j}$; I_D = D, $D_{i,j} = 1*D_{i,j}$; I_D = P, $P_{i,j} = P_{i,j}*1$. On the resulted interaction effects matrix a molecular level operator calculates a value–M_O = m, min; M_O = M, max; M_O = I, half-sum($M_{i,j}$); M_O = J, half-sum($M_{i,j}*M_{j,i}$); M_O = E, half-sum($M_{i,j}*Ad_{i,j}$); M_O = F, half-sum($M_{i,j}*M_{j,i}*Ad_{i,j}$). Finally the calculated value is subject to a linearization – L_O = I, I(x) = x; L_O = R, R(x) = 1/x; L_O = L, L(x) = Ln(x). A total number of 1512 (7·3·4·6·3) descriptors reflects the molecular structure of a molecule from (slightly) different (from one to another) perspectives (Table 15.8).

- FMPI (from Fragmental Matrix Property Indices; *see* Table 15.9) is an improvement were made to SMPI, by extending the principle applied for Szeged fragments (assigned letter, S) to other two matrices collecting fragments from molecule for pairs of atoms, namely to maximal fragments (assigned letter, M) – the largest set containing the first atom of the pair along with all its connected atoms after removal of the second atom of the pair from molecule and to complements of the maximal fragments (assigned letter, N) – the set containing the second atom of the pair along with the rest of the atoms lost from the molecule when maximal fragments were extracted.

TABLE 15.9 Code of FMPI Descriptors

Gene	F_C	A_P	D_M	I_D	M_O	L_O
Genome	S	A	T	E	m	I
	M	B	G	U	M	R
	N	C	U	D	I	L
		D		P	J	
		E			E	
		F			F	
		G				

FMPI = $L_O \times M_O \times I_D \times D_M \times A_P \times F_C$. Examples, ImETAS, LFPUGN.

Therefore, the gene sequence of FMPI is increased from SMPI with one gene, and the number of descriptors is multiplied with 3 (arriving at 4536; *see* Table 15.9).

ChPE (from Characteristic Polynomial Extend; *see* Table 15.10) is an extension was recently made to the Characteristic Polynomial in order to be used to produce a family of molecular descriptors (Jäntschi and Bolboacă, 2017). The calculations for the extended characteristic polynomial ChPE = $|\lambda \cdot I_A - C_M|$ were conducted diversifying the (atom's) identities (I_A) using atomic properties in 12 levels ('A' – atomic mass, /294.01; 'B' – boiling point, /3915, 'C' – count – always 1, ESP; 'D' – solid state density, in kg/m^3, /30,000; 'E' – electronegativity (revised Pauling scale, /4.00; 'F' – first ionization potential, in kJ/mol, /1312.0; 'G' – melting point temperature, in K, /3820.0; 'H' – attached hydrogen atoms, /4, 'I,' 'J,' and 'K' – partial charges from electrostatic, Mulliken and natural approaches, 'L' – spin), diversifying the (molecule's) connectivity (C_M) in 2·4 levels – two, by adjacencies ('g,' 'c,' 'b' and 't') and by distances ('G,' 'C,''B' and 'T') and three, diversifying the distance metric ('g' and 'G' – inverse of the geometrical distance; 'c' and 'C' – inverse of the bond order weighted topological distances; 't' and 'T' – inverse of the topological distances) as given in Table 15.10.

TABLE 15.10 Code of ChPE Descriptors

Gene		L_O	I_A	C_M	S_A	$\overline{d_1 d_2 d_3 d_4}$
		I	A	G	N	0000..1000
		R	B	C	P	
	Genome	L	C	T		
			D	g		
			E	c		
			F	t		
			G			
			H			

ChPE $= L_O \times I_A \times C_M \times S_A \times d_1 d_2 d_3 d_4$. Examples, IAGN1000 – first, LHtP1000 – last.

Using the selected alternatives for atom's identities (I_A) and for molecule's connectivity the polynomial ChPE $= |\lambda \cdot I_A - C_M|$ is evaluated in 2001 points by giving to the argument (λ) a sign (S_A, 'N'–negative or 'P'–positive) and a value ($\overline{d_1 d_2 d_3 d_4}$, from 0000 to 1000), as $\lambda = \pm \overline{d_1 d_2 d_3 d_4} / 1000$. The result of the evaluation is finally is subject to a linearization–$L_O = I$, $I(x) = x$; $L_O = R$, $R(x) = 1/x$; $L_O = L$, $L(x) = Ln(x)$. A total number of 288144 ($3 \cdot 8 \cdot 6 \cdot 2001$) descriptors reflects the molecular structure of a molecule from (slightly) different (from one to another) perspectives (*see* Table 15.10).

15.4 NANO-CHEMICAL APPLICATION(S)

Not implicitly all FMDs are suitable for nano-chemical applications. Namely, a FMD is suitable to be applied at nanolevel if the complexity of the calculation did not surpass a polynomial one (e.g., it not involves solving of a hard problem – for hard problems, *see* Falkenauer, 1998). From this point of view, for generating of the fragments, FPIF require the construction of the list of all paths between two atoms, which goes into a hard problem for cycles containing structures. A small modification to the MDF (MDF →MDF2004), namely by replacing Cluj path based fragments with the complement of the maximal fragments makes MDF (MDF→MDF2015) suitable to be used to the nanolevel, because the complexity of all others calculations did not surpass $O(N^4)$, where N is the number of the atoms in a molecule. For SMPI and FMPI, the complexity of the calculation is $O(N^4)$ too, while for MDFV, SAPF, and ChPE is $O(N^3)$.

SAPF and ChPE are implemented as very fast versions in Free Pascal (for Free Pascal, *see* Codère et al., 1998–2015), and the rest of the families suitable for nanolevel (MDF, MDFV, SMPI, FMPI) were implemented using PHP language (for PHP language, *see* Lerdorf, 1994; Gutmans and Suraski, 1994–2004).

All implementations were parallelized for exploiting the multi-core feature of the modern computers. Parallelization is portable when is implemented to pass to different tasks different molecules from a set. Thus, Table 15.11 contains these two implementations (in PHP and in Free Pascal).

Initially, the applications were designed to work with a database and to save the descriptors as well as the later conducted regression analysis on a database; the applications were revised to produce text-based human-readable files. The following working plan is to be used for an analysis conducted with FMDs (as the ones described in the previous section):

- **Stage 0. Preliminary requirements:** This stage is to be applied after a procedure which assumes that the geometry of the molecules is obtained and is saved in '*.hin'–HyperChem format and the partial charges are calculated. Much convenient is to optimize the structures with software which have possibility to parallelize the calculation, such as is Spartan. If it is the case, then conversions from Spartan ('input' and 'output' files) to HyperChem are required. A program (spartan_hin_convert_qsar.php) was designed to do this (and it requires '*.spinput' files to be placed in a directory, '*.txt' Spartan output files to be placed in other one, as well as it requires that the Spartan calculations to be conducted with 'verbose log' in order to contain the partial charges too; then, in a new directory the HyperChem files are generated).
- **Stage 1. Generation of the descriptors:** A folder containing the structure files is the input data for all programs providing the descriptors in a single file (e.g., mdf2004_a_generate.php → mdf2004.txt; it also applies for mdf2015_a_generate.php, mdfv2008_a_generate.php, sapf2011.exe, smpi2014.php, fmpi2015.php, and chpf2015.exe). The output files contain matrix-based data, with molecules in columns and descriptors in lines. The values are expressed with four significant digits as numbers in general form (in which are expressed with the smallest number of characters).
- **Stage 2. Filtering of the descriptors:** This step is intended only to remove the duplicates – it is possible for simple molecules to have two different descriptors with exactly the same series of values for all molecules in the dataset. Also, it is possible that at given precision that the values to be different only in order of magnitude; thus, the values of the descriptors should be expressed relatively to the order

TABLE 15.11 Code of Parallelization for Processing of Sets of Molecules (PHP and Free Pascal)

PHP	Free Pascal
Initialization code:	Initialization code:
define("max_splits,"8); define("f_CPU,"0);	uses SysUtils, Classes, Process, Windows;
define("a_CPU,""NUMBER_OF_PROCESSORS");	const f_CPUs = 0; m_CPUs = 8;
$srv = &$_SERVER; $invoked = $srv["argv"];	var splits,c_split,i:byte; s:string[255];
$me = &$srv["PHP_SELF"];	SystemInfo:SYSTEM_INFO; AProcess:TProcess;
if(array_key_exists("1,"$srv["argv"]))	begin
$splits = $srv["argv"][1]; else $splits = 1;	splits: = 1; current_split: = 0; //parallelization default state
if(array_key_exists("2,"$srv["argv"]))	if(paramcount = 0)then begin
$c_split = $srv["argv"][2]; else $c_split = 0;	GetSystemInfo(SystemInfo);
if((!is_numeric($splits)) || (!is_numeric($c_split)) ||	splits: = SystemInfo.dwNumberOfProcessors-f_CPUs;
($splits>max_splits)){die("wrong execution!\r\n"); }	Str(splits,s); AProcess: = TProcess.Create(NIL);
if($splits>1){ echo("$me. in parralel! Task ");	AProcess.InheritHandles: = FALSE;
echo(($c_split+1)." of ".$splits.."\r\n");	for i: = 1 to GetEnvironmentVariableCount do
}else{ $n_CPU = 0;	AProcess.Environment.Add(GetEnvironmentString(i));
if(array_key_exists(c_CPU,$srv))	AProcess.Executable : = Paramstr(0);
$n_CPU = $srv[c_CPU];	AProcess.Parameters.Add(s);
if($n_CPU ==0){	for c_split: = splits-1 downto 0 do begin
if($c_split ==0){ echo("Run a single task!\r\n"); }	Str(c_split,s); AProcess.Parameters.Add(s);
else die();	AProcess.Execute; AProcess.Parameters.Delete(1);
}else{ $splits = $n_CPU-f_CPU;	writeln('Starting ', c_split,' child');
echo("n-procs: $n_CPU n-splits: $splits."\r\n");	end;

TABLE 15.11 *(Continued)*

PHP	Free Pascal
```for($i = 0; $i<$splits; $i++){```	```AProcess.Free; writeln('Started ',splits,' childs.'); halt;```
```$run_s = "START php ."$me." ."$splits." ."$i;```	```end else begin```
```$run[$i] = popen($run_s,"r");```	```if(paramcount<>2)then halt; //wrong call```
```pclose($run[$i]);```	```val(paramstr(1),splits,i); if(i<>0)then halt;```
```} die("All tasks started.\r\n");```	```val(paramstr(2),current_split,i); if(i<>0)then halt;```
```}}```	```if((c_split>splits)or(splits>m_CPUs))then halt;```
```if($splits>1)```	```end;```
```echo("Split ."($c_split+1)." of ."$splits."\r\n");```	```writeln('Splits = ',splits,' Split = ',c_split);```
In the program body:	In the program body:
```for($i = 0; $i<$n_hin; $i++){//for the list of molecules```	```for i: = m1 downto 0 do begin //for the list of molecules```
```if($splits>1){if($i % $splits <> $c_split) continue; }```	```if(splits>1)then if(i mod splits <> c_split)then continue;```
```... //do the job for the molecule $i```	```... //do the job for the molecule i```
```}```	```end;```

of magnitude of the highest (absolute value). A program compact the outputs (v2_mdf_x_compactize.php for 'mdf*' families and v2_others_compactize.php for the rest); the output files are created as in the following example, mdf2004.txt → mdf2004_r.asc. The 1_sort_all.php program is feed with '*_r.asc' files to produce sorted and distinct series of values (for the descriptors & for the molecules) as '*_t.asc' files as in the following example, mdf2004_r.asc → mdf2004_t.asc.

- **Stage 3. Building of structure-property files:** The properties and/or activities are collected in a file ('properties.asc'), keeping the association with the structure (from '*.hin' files) with the first line having the names of the files containing the structures of the molecules for which the property (or properties) have that value(s). The first column contains the name of the property/activity. The generate_property_files_v2.php program generates files for each property, family_name + "_" + property_name + ".txt."

- **Stage 4. Regression analysis:** From this point on any software may be feed with the data to conduct the regression analysis. A program (_r1v_all.exe) was designed to provide ("r1_" + input_filename) simple linear regressions and other (r2f_all_v2.exe) to account for additive and multiplicative effects with two descriptors.

  - The use of the FMDs is exemplified in the modeling of C_{42} fullerene isomers continuum elasticity expressed as Total Strain Energy (TSE, in eV), molecules included in the Bolboacă and Jäntschi (2016) study. The geometries were already available from Tománek (2015) and here the partial charges were obtained applying HF 6-31G* calculations, and extracted as ground-state Mulliken net atomic charges (from the analysis of SCF wave function). The analysis of TSE was conducted by joining the pool of all FMDs described above (except FPIF, for the reason given above).

  - The methodology included the generation of the FMDs descriptors, scaling of the values relative to the highest one, filtering, sorting, joining, and filtering and sorting again.

  - The size of the valid descriptors from families was as follows, 183050 in ChPE, 1176 in FMPI, 40485 in MDF2004, 188195 in MDF2015, 83187 in MDFV, 6748 in SAPF, 673 in SMPI and 475073 in the joined pool.

  - The regression analysis included simple linear regression, and multiple linear regression with two structure-based descriptors selected systematically from the whole pool.

The molecules along with TSE values are given in Table 15.12.

TABLE 15.12 All Isomers of C_{42} Fullerene

TSE, 31.06 eV	TSE, 30.537 eV	TSE, 29.791 eV	TSE, 29.805 eV	TSE, 30.618 eV
01_C42_C2	02_C42_C1	03_C42_C1	04_C42_C1	05_C42_C2
TSE, 29.85 eV	TSE, 30.608 eV	TSE, 29.782 eV	TSE, 28.527 eV	TSE, 29.393 eV
06_C42_C2v	07_C42_C2	08_C42_C1	09_C42_C1	10_C42_C1
TSE, 29.475 eV	TSE, 28.34 eV	TSE, 28.157 eV	TSE, 27.147 eV	TSE, 29.955 eV
11_C42_Cs	12_C42_Cs	13_C42_C2v	14_C42_C1	15_C42_C1
TSE, 28.175 eV	TSE, 28.276 eV	TSE, 29.474 eV	TSE, 27.408 eV	TSE, 28.175 eV
16_C42_C2v	17_C42_C1	18_C42_C1	19_C42_Cs	20_C42_C1
TSE, 27.283 eV	TSE, 29.14 eV	TSE, 28.765 eV	TSE, 27.743 eV	TSE, 27.487 eV
21_C42_C2v	22_C42_Cs	23_C42_C2	24_C42_C1	25_C42_C1
TSE, 28.353 eV	TSE, 28.014 eV	TSE, 29.051 eV	TSE, 27.489 eV	TSE, 28.972 eV
26_C42_C1	27_C42_C2	28_C42_C2	29_C42_C1	30_C42_C1

TABLE 15.12 *(Continued)*

TSE, 27.484 eV	TSE, 26.657 eV	TSE, 26.639 eV	TSE, 27.371 eV	TSE, 26.554 eV
31_C42_C2	32_C42_C1	33_C42_C1	34_C42_C1	35_C42_Cs
TSE, 27.973 eV	TSE, 29.764 eV	TSE, 31.101 eV	TSE, 26.639 eV	TSE, 27.501 eV
36_C42_C1	37_C42_C1	38_C42_C2	39_C42_C1	40_C42_C2
TSE, 26.672 eV	TSE, 28.665 eV	TSE, 28.284 eV	TSE, 26.737 eV	TSE, 25.661 eV
41_C42_C2	42_C42_Cs	43_C42_C2	44_C42_C1	45_C42_D3

The simple linear regression with the highest explanatory power selected a MDF2015 descriptor (INVRcTu·10^1) with an explained determination of $r^2 = 0.9886$. Also, all other models involving two descriptors (additive, multiplicative and additive-multiplicative) selected MDF2015 descriptors exclusively, this being an important result for MDF2015 in regard to the use of it in nanochemistry (*see* Table 15.13).

TABLE 15.13 Best Selected Additive, Multiplicative and Additive-Multiplicative Effects

Model	Equation	r^2
2+1·1v	$\hat{Y} = 0_{t=0} + AMDrgTs_{MDF2015} \cdot 10^6 \cdot 42.53_{t=39} - IDPRgJj_{MDF2015} \cdot 133.8_{t=2.38}$ $+ AMDrgTs_{MDF2015} \cdot 10^6 \cdot IDPRgJj_{MDF2015} \cdot 42.53_{t=74}$	99.78
1·1v	$\hat{Y} = 195.1_{t=163} - IDBRgKs_{MDF2015} \cdot aMTrgTs_{MDF2015} \cdot 10^{-7} \cdot 16.23_{t=140}$	99.79
2v	$\hat{Y} = 88.15_{t=12} + AMDrgTs_{MDF2015} \cdot 10^6 \cdot 88.1_{t=97} - IDERgkS_{MDF2015} \cdot 31.49_{t=27}$	99.81

The results from Table 15.13 suggest that an additive model with descriptors from MDF (the 2015 version) provides the highest explanatory power of the association of the TSE with the structure of the C$_{42}$ isomers.

15.5 MULTI-/TRANS- DISCIPLINARY CONNECTION(S)

The using of FMDs to estimate and predict biological activities proved constantly better results when was compared with other alternatives (e.g., *see* Bolboacă and Jäntschi, 2010).

15.6 OPEN ISSUES

There is no ideal recipe on about how to design FMDs. Obtained results have shown that the refining of the FMDs leads to improvements of the estimation power.

KEYWORDS

- characteristic polynomial extensive
- fragmental matrix property indices
- fragmental property index family
- molecular descriptors family
- molecular descriptors family–vertex
- structural atomic property family
- Szeged matrix property indices

REFERENCES AND FURTHER READING

Allred, A. L., (1961). Electronegativity values from thermochemical data. *Journal of Inorganic and Nuclear Chemistry, 17*(3/4), 215–221.

Bader, R. F. W., (1990). *Atoms in Molecules, a Quantum Theory* (p. 438). Oxford, Oxford University Press.

Berwald, L., (1918). About affine geometry XXVII. Liesche F2, affine normal and mean affine curvature (in German). *Math Z. 1*(1), 63–78.

Bolboacă, S. D., & Jäntschi, L., (2009). Comparison of QSAR performances on carboquinone derivatives. *The Scientific World Journal, 9*(10), 1148–1166.

Bolboacă, S. D., & Jäntschi, L., (2010). From molecular structure to molecular design through the molecular descriptors family methodology. In: Castro, E. A., (ed.), *QSPR-QSAR Studies on Desired Properties for Drug Design* (pp. 117–166). Kerala, Research Signpost.

Bolboacă, S. D., & Jäntschi, L., (2016). Nanoquantitative structure-property relationship modeling on C_{42} fullerene isomers. *Journal of Chemistry.* 1791756 (8p.). DOI: 10.1155/2016/1791756.

Bolzano, B., (1804). Considerations about some items of elementary geometry (in German) (p. 65). Prague, Karl Barth.

Boole, G., (1854). *An Investigation of the Laws of Thought* (p. 430). (Reprinted 2003 as Laws of Thought. New York, Prometheus Books).

Born, M., & Goppert-Mayer, M., (1931). Dynamic lattice theory of crystals (in German). *Handb Phys., 24*(2), 623–794.

Brooks, B. R., Bruccoleri, R. E., Olafson, B. D., States, D. J., Swaminathan, S., & Karplus, M., (1983). CHARMM, A program for macromolecular energy, minimization, and dynamics calculations. *J. Comp. Chem., 4*(2), 187–217.

Callen, H. B., (1949). Electric breakdown in ionic crystals. *Phys. Rev., 76*, 1394–1402.

Cioslowski, J., (1989). General and unique partitioning of molecular electronic properties into atomic contributions. *Phys. Rev. Lett., 62*(13), 1469–1471.

Clark, T., (1985). *A Handbook of Computational Chemistry* (p. 332). New York, Wiley.

Codère, C. E., Mantione, D., Klämpfl, F., Maebe, J., Canneyt, Van M., Vreman, P., et al., (1998). *Free Pascal, Open Source Compiler for Pascal and Object Pascal* (v.0.99.10); 2000, FreePascal v.1.0; 2005, FreePascal v.2.0; 2009, FreePascal v.2.2.4; 2011, FreePascal v.2.4.2; 2012, FreePascal v.2.6.0.2015, FreePascal v.3.0.0.

Cornell, W. D., Cieplak, P., Bayly, C. I., Gould, I. R., Merz, K. M. Jr., Ferguson, D. M., et al., (1995). A second generation force field for the simulation of proteins, nucleic acids, and organic molecules. *J. Am. Chem. Soc., 117*, 5179–5197.

Coulson, C. A., Redei, L. B., & Stocker, D., (1962). The electronic properties of tetrahedral intermetallic compounds. *I. Charge Distribution, Proceedings of the Royal Society of London. Series A, Mathematical and Physical Sciences, 270*(1342), 357–372.

Cramer, C. J., (2002). *Essentials of Computational Chemistry, Theories and Models* (p. 542). New York, J. Wiley.

Dewar, M. J. S., Zoebisch, E. G., Healy, E. F., & Stewart, J. J. P., (1985). Development and use of quantum molecular models. 75. Comparative tests of theoretical procedures for studying chemical reactions. *The Journal of the American Chemical Society, 107*(13), 3902–3909.

Diudea, M. V., & Silaghi-Dumitrescu, I., (1989). Valence group electronegativity as a vertex discriminator. *Revue Roumaine de Chimie, 34*(5), 1175–1182.

Diudea, M. V., Gutman, I., & Jäntschi, L., (2001). *Molecular Topology* (p. 335). New York, Nova Science.

Falkenauer, E., (1998). *Genetic Algorithms and Grouping Problems* (p. 220). New York, Wiley.

Fisher, R. A., (1922). On the interpretation of $\chi2$ from contingency tables, and the calculation of P. *Journal of the Royal Statistical Society, 85*(1), 87–94. doi: 10.2307/2340521.

Fock, V. A., (1930). "Self-consistent field" with exchange for sodium (In German). *Zeitschrift für Physik, 62*(11/12), 795–805.

Fock, V. A., (1930). An approximation method for solving the quantum mechanical many-body problem (In German). *Zeitschrift für Physik, 61*(1/2), 126–148.

Gutmans, A., & Suraski, Z., (1997). *PHP3 (Open Source Online Since 1998; PHP v.3.0).*

Gutmans, A., & Suraski, Z., (1998). PHP4 (with 'Zend' engine; open source online since 1999; PHP v.4.0).

Gutmans, A., & Suraski, Z., (2004). PHP5 (with 'Zend' engine 2.0; open source online since 1999; PHP v.5.0).

Halgren, T. A., (1996). Merck molecular force field. I. Basis, form, scope, parameterization, and performance of MMFF94. *J. Comp. Chem., 17*(5/6), 490–519.

Hartree, D. R., (1928). The wave mechanics of an atom with a non-coulomb central field, Part I. *Theory and Methods. Math. Proc. Cambridge, 24*(1), 89–110.

Hartree, D. R., (1928). The wave mechanics of an atom with a non-coulomb central field, Part II. *Some Results and Discussion, Math. Proc. Cambridge, 24*(1), 111–132.

Hirshfeld, F. L., (1977). Bonded-atom fragments for describing molecular charge densities. *Theoret. Chim. Acta, 44*, 129–138.

Jäntschi, L., (2000). Prediction of the Physic-chemical and biological properties with the help of mathematical descriptors (in Romanian). PhD Thesis (Chemistry) – advisor Prof. Diudea MV. Universitatea "Babeş-Bolyai" din Cluj-Napoca, Cluj-Napoca, Ian. 2000.

Jäntschi, L., (2004). MDF–A new QSAR/QSPR molecular descriptors family. *Leonardo Journal of Sciences, 3*(4), 68–85.

Jäntschi, L., (2005). Molecular descriptors family on structure-activity relationships, 1 Review of the methodology. *Leonardo Electronic Journal of Practices and Technologies, 4*(6), 76–98.

Jäntschi, L., (2012). *Structure vs. Properties, Algorithms and Models. Habilitation thesis in Chemistry* (p. 175). Bucharest, CNATDCU (defended in 2013 in Cluj-Napoca, Babeş-Bolyai University).

Jäntschi, L., (2014). *Szeged Matrix Property Indices*. Online calculation software (release version). URL, http://l.academicdirect.org/Chemistry/SARs/SMPI/

Jäntschi, L., & Bolboacă, S. D., (2017). *Extending Characteristic Polynomial for Characterization of C_{20} Fullerene Congeners*. RoPM & AM 2017 presentation.

Jorgensen, W. L., & Tirado-Rives, J., (1988). The OPLS force field for proteins; energy minimizations for crystals of cyclic peptides and crambin. *J. Am. Chem. Soc., 110*(6), 1657–1666.

Kihl, O., Picard, D., & Gosselin, P. H., (2015). A unified framework for local visual descriptors evaluation. *Pattern Recognition, 48*(4), 1170–1180.

Kohn, W., & Sham, L. J., (1965). Self-consistent equations including exchange and correlation effects. *Physical Review, 140*(4A), A1133–A1138.

Krull, W., (1935). Idealtheorie. As, Ergebnisse der Mathematik, und ihrer Grenzgebiete, *4*(3), 152. Berlin, Julius Springer.

Lerdorf, R., (1994). *PHP/FI–Initially from Personal Home Page Tools (Open Source Online Since 1995; PHP v.1.0).*

Loschmidt, J., (1861). Ideal theory (In German). Ergeb Math Grenzg. As, Ostwalds Klassiker der exacten wissenschaften Nr. 190. Leipzig, Richard Anschütz. p. 384.

Löwdin, P. O., (1970). On the nonorthogonality problem. *Advances in Quantum Chemistry, 5*, 185–199.

Maurice, D., & Head-Gordon, M., (1999). Analytical second derivatives for excited electronic states using the single excitation configuration interaction method, theory and application to benzo[a]pyrene and chalcone. *Molecular Physics, 96*(10), 1533–1541.

Møller, C., & Plesset, M. S., (1934). Note on an approximation treatment form many-electron systems. *Physical Review, 46*(7), 618–622.

Mulliken, R. S., (1955). Electronic population analysis on LCAO–MO molecular wave functions-I. *J. Chem. Phys., 23*, 1833–1840.

Napier, J., (1614). *Ideal construction of the logarithms (In Latin)* (p. 155). Edinburgh, Andrew Hart.

Ohlinger, W. S., Klunzinger, P. E., Deppmeier, B. J., & Hehre W. J., (2009). Efficient calculation of heats of formation. *The Journal of Physical Chemistry A., 113*(10), 2165–2175.

Parr, R. G., & Yang, W., (1994). *Density-Functional Theory of Atoms and Molecules* (p. 333). Oxford, Oxford University Press.

Pauling, L., (1932). The nature of the chemical bond. IV, the energy of single bonds and the relative electronegativity of atoms. *Journal of the American Chemical Society, 54*(9), 3570–3582.

Pearson, K., (1900). On the criterion that a given system of deviations from the probable in the case of a correlated system of variables is such that it can be reasonably supposed to have arisen from random sampling. *Philosophical Magazine Series, 5, 50*(302), 157–175. doi: 10.1080/14786440009463897.

Peng, X., Wang, Q., Mishra, Y., Xu, J., Reichert, D. E., Malik, M., Taylor, M., Luedtke, R. R., & Mach, R. H., (2015). Synthesis, pharmacological evaluation and molecular modeling studies of triazole containing dopamine D3 receptor ligands. *Bioorganic and Medicinal Chemistry Letters, 25*(3), 519–523.

Perdew, J. P., Burke, K., & Ernzerhof, M., (1996). Generalized gradient approximation made simple. *Phys. Rev. Lett., 77*(18), 3865–3868.

Perdew, J. P., Chevary, J. A., Vosko, S. H., Jackson, K. A., Pederson, M. R., Singh, D. J., & Fiolhais, C., (1992). Atoms, molecules, solids, and surfaces, applications of the generalized gradient approximation for exchange and correlation. *Physical Review B., 46*(11), 6671–6687.

Politzer, P., (1968). Electron affinities of atoms. *Transactions of the Faraday Society, 64*, 2241–2246.

Purvis, D. G., & Bartlett, R. J., (1982). A full coupled-cluster singles and doubles model. The inclusion of disconnected triples. *The Journal of Chemical Physics, 76*(4), 1910–1919.

Radwan, A. A., & Abdel-Mageed, W. M., (2014). In silico studies of quinoxaline-2-carboxamide 1,4-di-N-oxide derivatives as antimycobacterial agents. *Molecules, 19*(2), 2247–2260.

Rhinehardt, K. L., Mohan, R. V., & Srinivas, G., (2015). Computational modeling of peptide-aptamer binding. *Methods in Molecular Biology, 1268*, 313–333.

Rocha, R. B., Freire, R. O., Simas, A. M., & Stewart, J. J. P., (2006). RM1, A reparameterization of AM1 for H, C, N, O, P, S, F, Cl, Br, and I. *The Journal of Computational Chemistry, 27*(10), 1101–1111.

Sanderson, R. T., (1983). Electronegativity and bond energy. *Journal of the American Chemical Society, 105*(8), 2259–2261.

Sestraş, R. E., Jäntschi, L., & Bolboacă, S. D., (2012). Quantum mechanics study on a series of steroids relating separation with structure. JPC. *Journal of Planar Chromatography-Modern TLC, 25*(6), 528–533.

Stewart, J. J. P., (1989). Optimization of parameters for semiempirical methods I. method. *The Journal of Computational Chemistry, 10*(2), 209–220.

Stewart, J. J. P., (2007). Optimization of parameters for semiempirical methods V, modification of NDDO approximations and application to 70 elements. *The Journal of Molecular Modeling, 13*(12), 1173–1213.

Szigeti, B., (1949). Polarisability and dielectric constant of ionic crystals. *Trans. Faraday Soc., 45*, 155–166.

Taha, M., Ismail, N. H., Imran, S., Selvaraj, M., Rahim, A., Ali, M., et al., (2015). Synthesis of novel benzohydrazone-oxadiazole hybrids as β-glucuronidase inhibitors and molecular modeling studies. *Bioorganic and Medicinal Chemistry, 23*(23), 7394–7404.

Tománek, D., (2015). C_{42} Isomers. In: *Guide through the Nanocarbon Jungle, Buckyballs, Nanotubes, Graphene, and Beyond.* http://nanotube.msu.edu/fullerene/fullerene.php?C = 42.

Wang, B., & Ford, G. P., (1994). Atomic charges derived from a fast and accurate method for electrostatic potentials based on modified AM1 calculations. *J. Comput. Chem., 15*, 200–207.

Warder, R. B., (1890). Recent theories of geometrical isomerism. *Science, 16*(400), 188–188.

CHAPTER 16

Fractal Dimension

PABLO M. BLANCO[1], SERGIO MADURGA[1], ADRIANA ISVORAN[2,3],
LAURA PITULICE[2,3], and FRANCESC MAS[1]

[1]*Department of Material Science and Physical Chemistry and Research Institute of Theoretical and Computational Chemistry (IQTCUB) of Barcelona University, C/Martí I Franquès, 1, 08028-Barcelona (Spain), E-mail: fmas@ub.edu*

[2]*Department of Biology-Chemistry, West University of Timisoara, 16 Pestalozzi, Timisoara, Romania*

[3]*Advanced Environmental Research Laboratory, West University of Timisoara, 4 Oituz, Timisoara, Romania*

16.1 DEFINITION

In order to properly understand the concept of fractal dimension, a suitable definition of a fractal object is needed. A fractal object (or set) is one that fulfills some or all of the following properties:

- It has a fine structure (i.e., it has details in all the observed scales).
- It is irregular enough to avoid being described using the classic geometry.
- It has self-similarity (i.e., it has the same behavior at different scales).
- It has a Fractal Dimension which is greater than its topological dimension.

In this context, the Fractal Dimension is a real number which characterizes the fractal set and gives an insight of its irregularity and complexity (Figure 16.1).

16.2 HISTORICAL ORIGINS

In natural systems, there are many complex structures that cannot be described in terms of classic geometry and analysis. The Polish mathematician Benoît Mandelbrot is considered the father of Fractal Geometry. His work about the geometric properties of natural systems (e.g., the continental coast, the tree branches, or the cloud surface) can be considered the origin of Fractal Geometry (Mandelbrot, 1982). Benoît Mandelbrot defined the word "fractal" using the Latin word *fractus* which means irregular or fragmented. This new approach allows the study of systems that have complex geometry and behavior but present self-similarity.

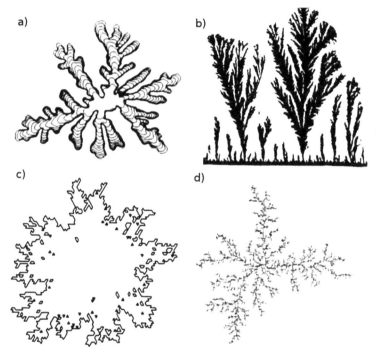

FIGURE 16.1 Examples of complex nanostructures which are observed in different experiments: (a) Growing of NH_4Cl crystals by precipitation (Reprinted with permission from Ohta and Honjo, 1988. © American Physical Society). (b) Formation of quasi-2D deposits of $ZnSO_4$ using electrochemical deposition (Adapted from Mach et al., 1994). (c) Model of the interface between two immiscible fluids with different viscosity (Reprinted with permission from Blunt and King, 1988. © American Physical Society). (d) Electric discharge in a dielectric material (Reprinted with permission from Niemeyer, et al., 1984. © American Physical Society).

The first mathematical definition of fractal dimension (FD) was proposed by the German mathematician Hausdorff and by the Russian mathematician Besicovitch (Hausdorff, 1919; Besicovitch, 1935) before the work of Mandelbrot. Although their definition of FD is valid in any set, its calculation is usually too complex. Therefore, alternative definitions of FD as the box-counting dimension (Kolmogorov and Tihomirov, 1959; Rényi, 1978) or the empiric fractal dimension were developed in order to simplify the FD calculation.

16.3 NANO-SCIENTIFIC DEVELOPMENTS

In order to illustrate the need of the FD concept, one should analyze two classical fractal sets: the Cantor set and the von Koch curve.

The Cantor set (C) was developed by Georg Cantor (a German mathematician) in 1877. This fractal set is generated slicing a unit segment in three equal parts and removing the central one. Then, the two ending parts are sliced as well in three equal parts, and the central part is removed, and this process is infinitely repeated (Figure 16.2). Using this rule, in the k iteration there are 2^k segments of $(1/3)^k$ length (Figure 16.3).

FIGURE 16.2 Generation of the Cantor set. It is only shown until the third iteration. (Adapted from Mas et al., 1996).

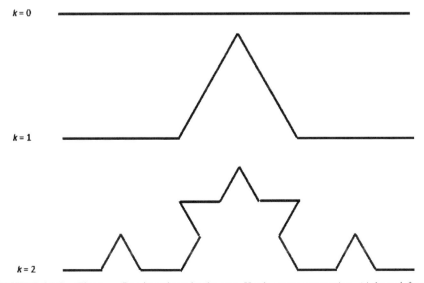

FIGURE 16.3 The two first iterations in the von Koch curve generation. (Adapted from Mas et al., 1996).

Although in the infinite limit this set has an infinite number of elements, it can be easily seen that the total length is null.

$$L(k) = 2^k \left(\frac{1}{3}\right)^k = \left(\frac{2}{3}\right)^k ; L = \lim_{k \to \infty} L(k) \to 0 \tag{1}$$

The von Koch curve (K) was introduced by the Swedish mathematician Helge von Koch in 1904. In order to generate this set, a unit segment is again divided in three equal parts. The central one is replaced by two new segments of the same length which are placed forming a triangle as shown in Figure 16.3. Thus, in the k iteration, there are 4^k segments of $(1/3)^k$ length. When this curve is infinitely iterated, one generates an object with an infinite length.

$$L(k) = 4^k \left(\frac{1}{3}\right)^k = \left(\frac{4}{3}\right)^k ; L = \lim_{k \to \infty} L(k) \to \infty \tag{2}$$

This result points out that the von Koch curve is, in fact, a two-dimensional object. However, if the area occupied by the von Koch curve is calculated using squares of $(1/3)^k$ placed in each segment one founds that it has a null total area.

$$A(k) = 4^k \left(\frac{1}{3}\right)^{2k} = \left(\frac{4}{9}\right)^k ; A = \lim_{k \to \infty} A(k) \to 0 \tag{3}$$

These two examples show one of the most important characteristics of the fractal objects iteratively generated. If d is the Euclidean dimension of the object and $\mu(d)$ is the d-dimensional measure, the properties $\mu(d) = 0$ and $\mu(d-1) = \infty$ are fulfilled. This leads towards the need of a fractionate dimension to characterize each fractal set, called Fractal Dimension. It is easy to deduce that for the Cantor set the FD has to be within the $[0,1]$ interval, and for the Koch curve it has to be within the $[1,2]$ interval.

Various definitions of the FD will be introduced in the following subsections. It is important to notice that all these definitions are equal and they are valid for any set (i.e., it does not matter if it is a fractal set or not).

16.3.1   *FRACTAL DIMENSION: HAUSDORFF-BESICOVITCH MEASURE*

The oldest and probably the most important definition of the FD is the one proposed by Hausdorff and Besicovitch. On the one hand, this FD has the advantage to be valid in any set, but on the other hand, it is hard to be calculated using computational methods.

Consider F as a set ($F \in \mathbb{R}^n$) in which the fractal dimension can be calculated, and U, a non-empty subset of the n-dimensional Euclidean space ($\mathbb{R}^n$) with a diameter defined as $|U| = \sup\{|x-y| : x,y \in U\}$. If $|U|$ is a countable collection (finite) of subsets with a maximum diameter of ϵ which cover the F set, U_i is called to be ϵ-coverer of F. Since F is a subset of $\mathbb{R}^n$, it can be defined for any $s > 0$ and $\epsilon > 0$ ($s,\epsilon \in \mathbb{R}^n$)

$$\mathcal{H}_\varepsilon^s(F) = \inf\left\{\sum_i |U_i|^s : |U_i| \text{ is a } \varepsilon - \text{coverer of } F\right\} \tag{4}$$

where $\mathcal{H}_\varepsilon^s$ is the so-called Hausdorff-Besicovitch (HB) measure of the F set. Therefore, one has to minimize the sum of subsets with ϵ diameter of power s. If ϵ decreases, the possible coverings of F also decrease. Thus, the HB measure can be written as:

$$\mathcal{H}^s(F) = \lim_{\varepsilon \to 0} \mathcal{H}_\varepsilon^s(F) \tag{5}$$

Then, the FD of the set F is defined as:

$$FD \equiv \dim_{HB} F \equiv \inf\left\{s : \mathcal{H}^s(F) = 0\right\} = \sup\left\{s : \mathcal{H}^s(F) = \infty\right\} \tag{6}$$

In other words, given a covering for the F set with U_i subsets, the unique s exponent able to change the behavior of $\mathcal{H}^s(F)$ from 0 to ∞ is the FD of F.

As a practical example of the use of this definition, it has been applied to the Cantor and von Koch sets. In order to use Eq. (4) a collection of subsets U_i must be chosen. In the Cantor set case, the covering able to minimize the HB measure is the U_i subset which covers each of the segments of these set. Therefore, the diameter of each subset is $|U_i| = (1/3)^k$. Since there are 2^k subsets, the HB measure (Eq. (4)) for the Cantor set is

$$\mathcal{H}_\varepsilon^s(C) = \sum_i |U_i|^s = 2^k\left[\left(\frac{1}{3}\right)^k\right]^s = \left(\frac{2}{3^s}\right)^k \tag{7}$$

Taking into account that $\epsilon = (1/3)^k$, the limit $\epsilon \to 0$ is equivalent to the limit $k \to \infty$. If one focuses on Eq. (6) and takes the limit $k \to \infty$ is easy to find that

$$\mathcal{H}^s(C) = \infty; \text{ if } \frac{2}{3^s} > 1 \tag{8}$$

$$\mathcal{H}^s(C) = 0; \text{ if } \frac{2}{3^s} < 1 \tag{9}$$

The FD is the s exponent which changes the behavior of $\mathcal{H}^s(C)$ from 0 to ∞. Thus, the FD is the value of s which fulfills

$$\frac{2}{3^s} = 1 \rightarrow s = \frac{\log 2}{\log 3} \rightarrow FD(C) = \frac{\log 2}{\log 3} = 0.6309\ldots \tag{10}$$

The same procedure can be applied to the von Koch curve. In this set there are 4^k subsets (U_i) of diameter $|U_i| = (1/3)^k$. Therefore, Eq. (4) is

$$\mathcal{H}_\varepsilon^s(K) = \sum_i |U_i|^s = 4^k \left[\left(\frac{1}{3}\right)^k \right]^s = \left(\frac{4}{3^s}\right)^k \tag{11}$$

Which in the limit $k \rightarrow \infty$ becomes

$$\mathcal{H}^s(K) = \infty; \text{ if } \frac{4}{3^s} > 1 \tag{12}$$

$$\mathcal{H}^s(K) = 0; \text{ if } \frac{4}{3^s} < 1 \tag{13}$$

Therefore, the FD of the von Koch curve is

$$\frac{2}{3^s} = 1 \rightarrow s = \frac{\log 4}{\log 3} \rightarrow FD(K) = \frac{\log 4}{\log 3} = 1.2618\ldots \tag{14}$$

16.3.2 BOX-COUNTING DIMENSION

It is one of the FD definitions mostly used because it can be easily calculated. It is commonly referred to as: Kolmogorov entropy, Entropy Dimension, Capacity Dimension, Logarithmic Dimension or Dimension of Information.

Consider F a non-empty subset of $\mathbb{R}^n$ and $N_\varepsilon(F)$ the minimum number of subsets of ϵ diameter needed to cover F (Falconer, 1990). If all these subsets have an equal diameter of ϵ, the Eq. (4) can be rewritten as:

$$\mathcal{H}_\varepsilon^s(F) = \sum_i^{N_\varepsilon(F)} \varepsilon^s = N_\varepsilon(F) \varepsilon^s \tag{15}$$

The s exponent which makes the HB measure finite and non-null is the FD of HB (Eq. (6)). This exponent can be isolated using logarithms, which leads towards

$$FD \equiv \dim_{BC} F \equiv \lim_{\varepsilon \to 0} \frac{\log N_\varepsilon(F)}{-\log \varepsilon} \tag{16}$$

In order to calculate the Box-Counting FD, it is necessary to do several coverings of F with different-sized boxes (i.e., different ϵ) and extrapolate the result to $\epsilon \rightarrow 0$.

Once again, this definition can be applied to the Cantor set and the von Koch curve in order to illustrate the concept (Mach et al., 1994). In the k iteration of the Cantor set, one needs 2^k one-dimensional boxes (segments) of $(1/3)^k$ length. Therefore, using Eq. (16) the FD of the Cantor set is

$$FD(C) = \lim_{k \to \infty} \frac{\log 2^k}{-\log\left(\dfrac{1}{3}\right)^k} = \frac{\log 2}{\log 3} = 0.6309... \tag{17}$$

In the von Koch curve case, in the k iteration, $4k$ square boxes of $(1/3)^k$ length are necessary to cover the set. Thus, applying Eq. (16), the FD of the von Koch curve can be calculated as:

$$FD(K) = \lim_{k \to \infty} \frac{\log 4^k}{-\log\left(\dfrac{1}{3}\right)^k} = \frac{\log 4}{\log 3} = 1.2618... \tag{18}$$

16.3.3 EMPIRIC FRACTAL DIMENSION

The main idea of this definition is "to measure in the ϵ scale" (Vicsek, 1989). The Euclidean measure of F with a unit of measurement with ϵ diameter is $M_\epsilon(F)$. This measure is the d-dimensional volume ($d \in \mathbb{N}$) multiplied by the d-dimensional boxes necessary to cover the F set

$$M_\varepsilon(F) = \varepsilon^d N_\varepsilon(F) \tag{19}$$

If one isolates $N_\epsilon(F)$ and replaces it in Eq. (15), the HB measure can be rewritten as:

$$\mathcal{H}_\varepsilon^s(F) = M_{\mathring{a}}(F) \varepsilon^s \varepsilon^{-d} \tag{20}$$

Following the same procedure as in the Box-counting FD, the empiric FD is obtained by

$$FD \equiv \dim_E F \equiv d + \lim_{\varepsilon \to 0} \frac{\log M_\varepsilon(F)}{-\log \varepsilon} \tag{21}$$

In order to calculate the empiric FD one needs to measure the F set with a unit of measure of ϵ diameter. Then, a logarithmic representation is done of the measure of the F set as a function of the unit of measure. The slope of this representation gives an estimation of the empiric FD.

The Cantor set and the von Koch curve provide a useful example of the application of this definition again. In the Cantor set, 2^k units of $(1/3)^k$ length can be measured (Mach, 1995). The Euclidean measure is $d = 1$ because we are measuring a length, so the FD of the Cantor set is:

$$FD(C) = 1 + \lim_{k \to \infty} \frac{\log\left(2 \cdot \frac{1}{3}\right)^k}{-\log\left(\frac{1}{3}\right)^k} = 1 + \lim_{k \to \infty} \frac{\log\left(\frac{2}{3}\right)}{\log 3} = \frac{\log 2}{\log 3} = 0.6309\ldots \tag{22}$$

In the von Koch curve case, one can measure 4^k units of $(1/3)^k$ length. Therefore, the FD of this fractal set is:

$$FD(K) = 1 + \lim_{k \to \infty} \frac{\log\left(4 \cdot \frac{1}{3}\right)^k}{-\log\left(\frac{1}{3}\right)^k} = 1 + \lim_{k \to \infty} \frac{\log\left(\frac{4}{3}\right)}{\log 3} = \frac{\log 4}{\log 3} = 1.2618\ldots \tag{23}$$

16.3.4   MULTIFRACTAL ANALYSIS

There are complex sets that cannot be described by means of a single FD, but they need an infinite number of FD. To study this kind of sets in terms of Fractal Geometry, they are divided into several subsets. Each subset is composed of points with the same measure, and therefore a FD can be calculated in all them (Mach, 1995; Mach et al., 1995).

Nowadays, the characterization of a measure distributed on a support can be estimated by using elements of multifractal analysis (Vicsek, 1989). A procedure used as a basis in the multifractal analysis is to make a profit of the expression (Falconer, 1990)

$$\Phi(q, \tau) = E\left(\sum_{i=1}^{N} p_i^q \varepsilon_i^{-\tau}\right) \tag{24}$$

where q and τ are real numbers and p_1 and ε_1 are random quantities that represent the measure factor and the size factor, respectively. The sum is extended over the N separated parts composing the object and E holds for the average value. The function Φ, therefore, is the coupled q-moment of the measure and the τ-moment of the size.

When one wants to characterize an experimental measure, Eq. (24) is resolved considering p_i or ε_i constant and it is assumed to collapse into a constant value function $\Phi(q,\tau)$ (Mach et al., 1995). On the one hand, if ε_i is constant (i.e., we work in the q-representation), the size factor ε can be taken out of the sum and Eq. (24) can be rewritten as:

$$\varepsilon \sim cnt \rightarrow E\left(\sum_{i=1}^{N(e)} p_i^q\right) \sim \varepsilon^\tau \tag{25}$$

where $N(\varepsilon)$ is the number of parts with size ε needed to cover the whole support where the measure is defined. On the other hand, if the averaged quantity in Eq. (24) is weighted with the probability distribution p_i then

$$p^{(q-1)}\varepsilon^{-\tau}{}_p \sim cnt \tag{26}$$

If p is considered constant (i.e., we work in the p-representation), Eq. (26) becomes

$$p \sim cnt \rightarrow \varepsilon^{-\tau}{}_p \sim p^{1-q} \tag{27}$$

where $\langle\ \rangle_p$ is the average value computed according to the probability distribution p_i. Therefore, there are two alternative representations of the multifractal indices, taking either q or τ as the fundamental moment. In this context, the multifractal indices used to characterize a non-uniform measure are:

- The set of generalized fractal dimensions $D_q = \dfrac{\tau(q)}{q-1}$ (Grasssberger, 1983; Hentschel and Procaccia, 1983).
- The so-called Hölder exponent α which satisfies $\alpha = d\tau(q)/dq$.
- The Legendre transformed function $f(\alpha)$ which is defined as $f(\alpha) = q\alpha - \tau$.

If the measure is uniformly distributed the set of generalized fractal dimensions collapses in one single FD and $D_q = D_0$. In general, when a measure is not uniformly distributed there are a spectra of infinite generalized FD D_q. In particular, D_0 is the FD of the measure, D_1 is its information dimension, and D_2 is its correlation dimension. These three properties indicate the three conjugated moments of the spectra. The Hölder exponent α is a conjugated variable to q (e.g., like entropy Legendre transformations in thermodynamics). This implies that $\tau(q)$ is a kind of energy of the measure and $f(\alpha)$ is its free-energy.

In Table 16.1, the behavior of the different indices in q-representation is summarized. The different properties of the multifractal indices depending on the measure that is uniformly distributed or it follows a multifractal distribution can be consulted in Table 16.2. The equivalent information in τ-representation can be consulted in the work of Mach (1995).

TABLE 16.1   The Most Important Properties of the Multifractal Indices in q-Representation

q	Characteristic values
$-\infty$	$D_{-\infty} = \alpha_{-\infty} = \alpha_{max} > 0$
	$D_f \geq f(\alpha_{-\infty}) \geq 0$
0	$D_0 = D_f$
	$\tau_{0=-}D_f$
	$f(\alpha_0) = f_{max} = D_f$
1	$\tau_1 = 0$
	$D_1 = \alpha_1 = f(\alpha_1)$
$+\infty$	$D_{+\infty} = \alpha_{+\infty} = \alpha_{min} > 0$
	$D_f \geq f(\alpha_{+\infty}) \geq 0$

TABLE 16.2   Different Behavior for the Multifractal Indices in q-Representation Depending on the Distribution of the Measure

Multifractal index	Uniform measure	Multifractal measure
$q(\tau)$	$q(\tau) = \tau/(D_f + 1)$ straight line	Strictly increasing function
$\alpha(q)$	Constant $= D_f$	Strictly decreasing function
$f(\alpha(q))$	Constant $= D_f$	Concave function with a maximum equal to D_f when $q = 0$
D_q	Constant $= D_f$	Decreasing function, $D_0 = D_f$

An example of the use of multifractal analysis is the study of the electro-chemical deposition of Zn (Mach et al., 1995). The study of this phenomenon shows that the electrodeposits of Zn has a uniform mass distribution, but its growth probability follows a multifractal distribution with a FD of $D_0 = 1.53$ (Figure 16.4).

16.4   NANO-CHEMICAL APPLICATIONS

The Fractal Geometry and the concept of Fractal Dimension are of crucial importance for the characterization of complex nanostructures. Some examples are the study of crystal growing in a supersaturated solution (Ohta and Honjo, 1988), the viscous fingering (Blunt and King, 1988) which is formed when a fluid is injected inside another less viscous and immiscible fluid or the electrodeposition (Sagués et al., 1990; Mas and Sagués, 1992; Mach et al., 1994) of metal ions in an electrode (Figure 16.4).

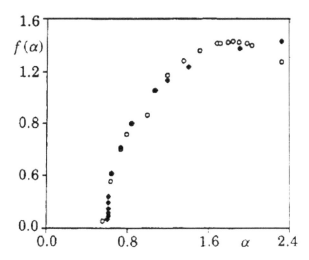

FIGURE 16.4 Multifractal scaling analysis of the growth probability distribution of $ZnSO_4$ electrodeposits with an open-fractal morphology. Empty circles hold for normal velocities distribution and filled circles for a harmonic (Laplacian) measure (Adapted from Mach et al., 1995).

The application of fractal theory for the characterization of complex nanostructures is efficiently illustrated by the study of structural properties of proteins in correlations with their biological functions. From this point of view, the surface roughness of a protein is quantitatively characterized by the fractal dimension of its surface, and also the spatial structure of a protein is quantitatively described by the fractal dimension of its backbone and/or by its mass fractal dimension.

The backbone of a protein reflects two fractal dimensions, one associated to the local folding (concerning an interval of about 10 residues) and the other one associated to the global folding (Craciun et al., 2009). In order to describe proteins in terms of mass fractals, the scaling law between the radius of gyration (R_g) and the number of amino acids (N) (or the molecular weight, M) is considered

$$N \sim M \sim R_g^{MD} \tag{28}$$

and this equation allows computing the mass fractal dimension, MD as the slope of the line when plotting the *log(N)* versus *log(R$_g$)* (Abad-Zaparetto and Lin, 1990; Dewey, 1997; Oleschko et al., 2000). The mass fractal dimension of proteins ranges between 2.30 and 2.83 and its value allows distinguishing between structural classes of proteins (Abad-Zaparetto and Lin, 1990; Dewey, 1997; Oleschko et al., 2000; Stawiski et al., 2000; Stawiski et al., 2002; Pitulice et al., 2008; Pitulice et al., 2009)

It is well known that the shape and chemical properties of the protein surface are important features for many biological and chemical phenomena: diffusion, absorption, and binding of small molecules, protein-protein interactions, protein-nucleic acids interactions, heterogeneous catalysis, etc. The surface of a protein is rough presenting pockets, cavities and channels in various shapes and sizes (Laskovski et al., 1996).

There are multiple possibilities to define and compute the surface area of a protein, which are recently reviewed (Craciun et al., 2009). The computation of the surface fractal dimension of a protein depends on the way of defining the surface, and this is a generic property of the protein. When the contact surface (CS), defined as the surface containing all contact points of the protein and a test sphere particle of radius R, is used the surface fractal dimension (SD) is obtained from the following power law (Connoly, 1983):

$$CS \sim R^{2-SD} \tag{29}$$

The literature contains many results reflecting the fractal nature of proteins surface with fractal dimensions values lying between 2.10 and 2.80. (Craciun et al., 2009, 2010, 2012; Kaczor et al., 2012; Banerji, 2013a; Craciun and Isvoran, 2014; Tscheliessnig and Pusztai, 2016).

The fractality of a protein surface manifests itself through distinctive global and local features of the surface roughness, resulting in dissimilar local and global fractal surface dimensions. Usually, the cavities of the active site reflect higher roughness than the global protein surface (Craciun et al., 2009; Craciun et al., 2012; Kaczor et al., 2012).

Also, the fractal nature of the channels and protrusions found in the protein interior has been emphasized (Banerji and Gosh, 2011; Banerji, 2013b).

16.5   MULTI-/TRANS-DISCIPLINARY CONNECTIONS

The concept of fractal dimension was developed in mathematics, but it has applications in many other fields, from decorative art to science and technology. It is not the aim of this paper to describe in detail the applications of the concept of fractal dimension in different fields. However, some applications will be shortly illustrated.

Fractal dimension can be used to characterize the complex geometry of natural macroscopic systems such as coasts, clouds or snowflakes with direct applicability in geography, paleontology, geology and many other scientific areas.

The trajectory of diffusing particles in crowded media is fractal, and its fractal dimension is correlated with the properties of the environment: concentration, mobility, and size of particles (Pitulice et al., 2016). It illustrates the applicability of this concept in studying diffusion process in many systems of interest in physics and chemistry, including biological systems. In Material Science, it can be used to study the transport of particles in disordered systems.

In medicine, the fractal dimension can be used to describe the branched structure of nerves and capillaries, the respiratory tract, the renal system, etc. Also, computation of the fractal dimension allows to characterize the morphological structure of organs quantitatively and to distinguish between their physiological and pathological functioning (Uahabi and Atounti, 2015).

There are also applications of the fractal dimension in the art as fractal shapes may capture the details of irregularities of natural and artificial objects. A well-known example consists in the use of fractal dimension computation as a way of authentication of Pollock's work by identifying fractal patterns of his paintings (Coddington et al., 2008). Also, it has been found that the fractal dimension of an artistic work is important in appreciating the beauty in visual art, especially for abstract and natural pictures (Forshyte et al., 2011). Not, at last, the fractal dimension is also applicable in music for its characterization. It reflects that music possesses self-similar incidence of the frequency melodic intervals, of the changes of acoustic frequency or/and of the note durations (Hsü and Hsü, 1991).

16.6 OPEN ISSUES

Fractal technology may lead to advances in photovoltaic electric generating systems, membranes for fuel cells or gas separation, soft-touch polymers, unique thermal properties, and high-resolution touch screens.

Recent discoveries in superconductivity suggest an improvement when the microstructure is most "connected." Therefore, it is possible to trace a path with the same nanostructure (i.e., by oxygen atoms) over a large distance. The microstructure, in this case, is fractal-like because if the material structure is enlarged its appearance, at increasing levels of magnification, remains unaltered.

Nowadays, fractal nanostructures have been discovered on the surface of cancer cells. These fractal nanostructures confer to cancer cells a higher fractal dimension than normal cells, and this fractal dimension can be further used to distinguish cancer cells from normal cells. This important finding inspired us to explore whether fractal nanostructures can be utilized to program interfacial materials for cancer cell recognition. If this holds true, it would be of great importance for clinical diagnosis.

A notable trend in the actual literature is that a wide range of disordered systems (e.g., linear and branched polymers, biopolymers, epoxy resins, and percolation clusters) can be characterized by the fractal nature over a microscopic correlation length. This is in agreement with the fact that energy transformations are permitted on a small scale or nanoscale.

Recent research on retinal implants is devoted to nanoclusters deposition which can create artificial fractal interactions. Despite this promising approach, their possible toxicity must be reduced by choosing non-oxidable and non-toxic metals.

KEYWORDS

- **fractal geometry**
- **Hausdorff-Besicovitch measure**
- **multifractal analysis**
- **self-similarity**

REFERENCES AND FURTHER READING

Abad-Zappareto, C., & Lin, C. T., (1990). Statistical descriptors for the size and shape of globular proteins. *Biopolymers, 29,* 1745–1754.

Banerji, A., (2013a). Characterization of protein-protein interfaces, considering surface-roughness and local shape. *Fractal Symmetry of Protein Exterior,* 17–31.

Banerji, A., (2013b). Fractal symmetry of protein interior. *Springer Briefs in Biochemistry and Molecular Biology,* 19–84.

Banerji, A., & Ghosh, I., (2011). Fractal symmetry of protein interior: What have we learned? *Cellular and Molecular Life Sciences, 68*(16), 2711–2737.

Besicovitch, A. S., (1935). On the sum of digits of real numbers represented in the dyadic system. *Mathematische Annalen., 110*, 321–330.

Blunt, M., & King, P., (1988). Scaling structure of viscous fingering. *Physical Review A., 37*, 3935.

Coddington, J., Elton, J., Rockmore, D., & Wang, Y., (2008). Multifractal analysis and authentication of Jackson Pollock paintings. In: David, G. S., & Jim, C., (eds.), *Computer Image Analysis in the Study of Art/Proceedings of SPIE–The International Society for Optical Engineering 29 February 2008* (Vol. 6810, p. 12).

Connolly, M. L., (1983). Analytical molecular surface calculation. *Journal of Applied Crystallography, 16*, 151–176.

Craciun, D., & Isvoran, A., (2014). Computational analysis of surface properties of EF-hand calcium-binding proteins. *Romanian Journal of Physics, 59*(3/4), 339–345.

Craciun, D., Isvoran, A., Reisz, R. D., & Avram, N. M., (2010). Distinct fractal characteristics of monomer and multimer proteins. *Fractals, 18*(2), 207–214.

Craciun, D., Pitulice, L., & Isvoran, A., (2009). Fractal features of proteins structure and dynamics. Part I–structure. *International Journal of Chemical Modeling (IJCHEMO), 2*(1), 33–48.

Craciun, D., Pitulice, L., Ciorsac, A., & Isvoran, A., (2012). Proteins surface roughness analysis. Comparison of crystallographic and NMR structures. *Romanian Reports in Physics, 64*(1), 116–126.

Dewey, T. G., (1997). *Fractals in Molecular Biophysics*. Oxford University Press: New York.

Falconer, K. J., (1990). *Fractal Geometry: Mathematical Foundations and Applications*. John Willey: New York.

Forsythe, A., Nadal, M., Sheehy, N., Cela-Conde, C. J., & Sawey, M., (2011). Predicting beauty: Fractal dimension and visual complexity in art. *British Journal of Psychology, 102*(1), 49–70.

Grassberger, P., (1983). Generalized dimensions of strange attractors. *Physics Letters, A97*, 227–230.

Hausdorff, F., (1919). Der wertvorrat einer bilinearform. *Mathematische Annalen, 79*, 157.

Hentschel, H. G. E., & Procaccia, I., (1983). The infinite number of generalized dimensions of fractals and strange attractors. *Physica D., 8*, 435–444.

Hsü, K. J., & Hsü, A., (1991). Self-similarity of the "1/f noise" called music. *Proceedings of the National Academy of Sciences, 88*(8), 3507–3509.

Kaczor, A. A., Guixà-González, R., Carrió, P., Obiol-Pardo, C., Pastor, M., & Selent, J., (2012). Fractal dimension as a measure of surface roughness of G protein-coupled receptors: Implications for structure and function. *Journal of Molecular Modeling, 18*(9), 4465–4475.

Kolmogorov, A. M., & Ihomirov, N. M., (1959). ε-entropy and ε-capacity of sets in function spaces. *Uspekhi Matematicheskikh Nauk, 13*(2), 3–86.

Laskowski, R. A., Luscombe, N. M., Swindells, M. B., & Thornton, J. M., (1996). Protein clefts in molecular recognition and function. *Protein Science, 5*, 2438–2452.

Mach, J., (1995). *Anàlisi Multifractal: Algorismes Computacionals i Aplicació a l'Electrodeposició*. Doctoral thesis, University of Barcelona, Barcelona.

Mach, J., Mas, F., & Sagués, F., (1994). Laplacian multifractality of the growth probability distribution in electrodeposition. *Europhysics Letters, 25*(4), 271–276.

Mach, J., Mas, F., & Sagués, F., (1995). Two representations in the multifractal analysis. *Journal of Physics A: Mathematical and General, 28*(19), 5607–5622.

Mandelbrot, B., (1982). *Fractal Geometry of Nature*. Freeman: San Francisco.

Mas, F., & Sagués, F., (1992). Scaling properties of the growth probability distribution in electrochemical deposition. *Europhysics Letters, 17*(6), 541–546.

Mas, F., Mach, J., Triguerus, P. P., & Sagués, F., (1996). Creixement fractal, Als limits de la modelittació. In: Casassac, E., & Estesan, M., (eds.), Modelització macroscópica en ciéncies experimentals Institut Fitudin Catalans: Barieldona, Chapter 6.

Niemeyer, L., Pietronero, L., & Wiesmann, H. J., (1984). Fractal dimension of dielectric breakdown. *Physical Review Letters, 52*(12), 1033–1036.

Ohta, S., & Honjo, H., (1988). Growth probability distribution in irregular fractal-like crystal growth of ammonium chloride. *Physical Review Letters, 60*(7), 611–614.

Oleschko, K., Figueroa, B. S., Miranda, M. E., Vuelvas, V. A., & Salleiro, E. R., (2000). Mass fractal dimensions and some selected physical properties of contrasting soil and sediments of Mexico. *Soil and Tillage Research, 55*(1/2), 43–61.

Pettersen, E. F., Goddard, T. D., Huang, C. C., Couch, G. S., Greenblatt, D. M., Meng, E. C., & Ferrin T. E., (2004). UCSF chimera: A visualization system for exploratory research and analysis. *Journal of Computational Chemistry, 25*(13), 1605–1612.

Pitulice, L., Craciun, D., Vilaseca, E., Madurga, S., Pastor, I., Mas, F., & Isvoran, A., (2016). Fractal dimension of the trajectory of a single particle diffusing in crowded media. *Romanian Journal of Physics, 61*(7/8), 1276–1286.

Pitulice, L., Isvoran, A., & Chiriac, A., (2008). Structural features of proteins reflected by statistical scaling laws. *Journal of the Serbian Chemical Society, 73*(8/9), 805–813.

Pitulice, L., Isvoran, A., Craescu, C. T., & Chiriac, A., (2009). Scaling properties of the radius of gyration and surface area for E–F hand calcium-binding proteins. *Chaos, Solitons & Fractals, 40*(2), 684–690.

Rényi, A., (1978). *Probability Theory.* North-Holland Publishing Company: Amsterdam.

Sagués, F., Mas, F., Vilarrasa, M., & Costa, J. M., (1990). Fractal electrodeposits of zinc and copper. *Journal of Electroanalytical Chemistry and Interfacial Electrochemistry, 278*(1/2), 351–360.

Stawiski, E. W., Baucom, A. E., Lohr, S. C., & Gregoret, L. M., (2000). Predicting protein functions from structure: Unique structural feature of proteases. *Proceedings of the National Academy of Sciences, 97*(8), 3954–3958.

Stawiski, E. W., Mandel-Gutfreund, Y., Lowenthal, A. C., & Gregoret, L. M., (2002). Progress in predicting protein function from structure: Unique features of O-glycosidases. *Pacific Symposium on Biocomputing, 7*, 637–648.

Tscheliessnig, R., & Pusztai, L., (2016). Proteins in solution: Fractal surfaces in solutions. *Condensed Matter Physics, 19*(1), 13803.

Uahabi, K. L., & Atounti, M., (2015). Applications of fractals in medicine. *Annals of the University of Craiova, Mathematics and Computer Science Series, 42*(1), 167–174.

Vicsek, T., (1989). *Fractal Growth Phenomena* (2nd ed.). World Scientific: Singapore.

CHAPTER 17

Fractal Kinetics

FRANCESC MAS[1], LAURA PITULICE[2,3], SERGIO MADURGA[1],
JOSEP LLUÍS GARCÉS[4], EUDALD VILASECA[1], and ADRIANA ISVORAN[2,3]

[1]Department of Material Science and Physical Chemistry and Research Institute of Theoretical and Computational Chemistry (IQTCUB), University of Barcelona, C/Martíi Franquès, 1, 08028-Barcelona (Spain), E-mail: fmas@ub.edu

[2]Department of Biology-Chemistry, West University of Timisoara, 16 Pestalozzi, Timisoara, Romania

[3]Advanced Environmental Research Laboratory, West University of Timisoara, 4 Oituz, Timisoara, Romania

[4]Department of Chemistry, University of Lleida (UdL), Av. Alcalde Rovira Roure 191, 25198-Lleida, Spain

17.1 DEFINITION

Fractal kinetics (FK) is a generalization of the classical kinetic reaction scheme in which the kinetic constant is no longer time independent. The new kinetic coefficient becomes time-dependent, scaling with time as t^{-h}, with an exponent that reflects the characteristic of the medium. This behavior occurs when reactants are not well mixed after each reaction step mainly due to the fact that reactions are diffusion-limited either in low-dimensional geometries (e.g., interfaces, polymeric chains, and pores), or in a heterogeneous environment, where there are regions of disconnected topologies modeled by fractal geometries, or in crowded environments.

17.2 HISTORICAL ORIGINS

Classical reaction kinetics gives us a rate law that considers the change of the reactants/products of a reaction mechanism with time. This rate law depends on a time-independent kinetic constant and on the nature of reactants and medium where the reaction is embedded. When reactions take place in heterogeneous environments, the classical Smoluchowski-law of diffusion-controlled reaction kinetics, $k \propto D$, may be no longer valid, where D is the relative diffusion coefficient of the two reactants in a bimolecular reaction (Avnir, 1989). This behavior can be determined by different situations for which the assumption of well-mixed reactants after each reaction step does not hold. Such examples consist of reactions taking place at interfaces or linear geometries, as polymeric chains and pores, or in a heterogeneous environment, with disconnected topologies in which reactants are only allowed to diffuse over fractal trajectories, or reactions embedded in crowded environments.

R. Kopelman, from the Chemistry Department of University of Michigan (USA), pioneered the introduction of the concept of kinetic coefficient as a power-law depending on time, $k \propto t^{-h}$, a generalization of the time-independent kinetic constant k, (Anacker & Kopelman, 1984; Kopelman, 1986; Kopelman, 1988), based on preliminary ideas about the scaling approach of the kinetics of a recombination process of two different kind of species (Kang & Render, 1984).

17.3 NANO-SCIENTIFIC DEVELOPMENTS

A diffusion process of a solute in a dilute solution can be described with the well-known Einstein-Smoluchowski equation for Brownian motion:

$$r^2(t) = (2d)Dt \tag{1}$$

where d is the topological dimension of the medium where the process is embedded, and D is its diffusion coefficient (Ben-Avraham & Havlin, 2000).

In crowded media (Minton, 2001) the existence of different macromolecular species hinders the diffusion process. In these cases, Eq. (1) must be generalized to a more complex process known as anomalous diffusion (Bouchaud & Georges, 1990; Banks & Fradin, 2005; Pastor et al., 2010; Vilaseca et al., 2011), which can be described by:

$$r^2(t) = (2d)\Gamma t^\alpha \tag{2}$$

where α is defined as the anomalous exponent (where $0 < \alpha < 1$ is the case of subdiffusion and $\alpha > 1$ holds for the case of superdiffusion) and Γ is a generalized transport coefficient, also known as anomalous diffusion coefficient.

The motion of a single particle can be described using the concept of a random walk because the particle movement consists of a succession of random steps. The trajectory of a random walk is obtained by connecting the visited sites, and it has interesting geometric properties described in terms of fractal geometry (Ben-Avraham & Havlin, 2000). Also, the fractal geometry has been proven to be useful in explaining features of diffusion in crowded media (Banks & Fradin, 2005; Pastor et al., 2010; Vilaseca et al., 2011). The random walk motion is quantitatively described by the fractal exponent through the following relationship (Ben-Avraham & Havlin, 2000; Pitulice et al., 2016):

$$d_\omega = \frac{2}{\alpha} \tag{3}$$

where d_ω is the fractal dimension of the random walk with $d_\omega = 2$ for normal diffusion and $d_\omega > 2$ for anomalous diffusion.

In terms of statistical mechanics, it is useful to define the *fracton* as a random walker that diffuses over a fractal environment (Alexander & Orbach, 1982; Avnir, 1989), which has a spectral dimension, d_s, related to the probability of the random walker to come back to its initial position, and it is related to the fractal dimension of the random walk as follows:

$$d_s = \frac{2d_f}{d_\omega} \tag{4}$$

where d_f is the fractal dimension of the fractal environment where the random walker diffuses. Normally, $d_s < d_f < d$. For a free classical particle that diffuses on a bulk solution ($d = 3$), $d_f = d$, and $d_\omega = 2$, then $d_s = \frac{1}{2}$, as predicted by the classical Maxwell-Boltzmann distribution law.

In classical kinetics, a bimolecular reaction that takes place in batch conditions (well-mixed bulk environment) follows the classical rate law with a second order for the reaction:

$$A + A \rightarrow poducts \Rightarrow \frac{d[A]}{dt} = R - 2k[A]^2 \tag{5a}$$

$$A + B \rightarrow products \Rightarrow \frac{d[A]}{dt} = R - k[A][B] \tag{5b}$$

where R is related to a possible flux of reactants.

When the reaction is diffusion-controlled, the kinetic constant depends on the relative diffusion coefficient given by the Einstein-Smoluchowski equation (1), and it is time independent:

$$k_d = 4\pi L D_{AB} r_{AB} \qquad (6)$$

where L is the Avogadro constant, D_{AB} is the relative diffusion coefficient, and r_{AB} is the minimum approaching distance between reactants. This equation is only valid for bulk systems ($d = 3$) while for low-dimensional geometries it does not always hold.

In fractal kinetics, Kopelman (1986, 1988) generalizes the rate laws (5) with a power-law expression:

$$k(t) = k_0' \, t^{-h} \qquad (7)$$

where $k(t)$ is the reaction coefficient, k_0' is an instantaneous reaction constant and h, called the fractal parameter, is an exponent ($0 \le h \le 1$) which depends on the topological dimensionality of the medium in which the reaction occurs. If one considers the schemes shown in Eq. (5) and follows the universal scaling laws for recombination processes (Kang & Render, 1984), it becomes possible to relate this fractal parameter with the spectral dimension d_s of the reactants as random walkers (Kopelman, 1986):

$$A + A \rightarrow products \Rightarrow h = 1 - \frac{d_s}{2} \qquad (8a)$$

$$A + B \rightarrow products \Rightarrow h = 1 - \frac{d_s}{4} \qquad (8b)$$

For random fractals, embedded in any topological dimension d, as the percolation or the limited diffusion aggregation (DLA) clusters (Avnir, 1989), $d_s \approx 4/3$, which yields a value for the fractal parameter of $h \approx 1/3$, for the reaction scheme shown by Eq. (5a).

If we work in stationary state conditions, using computer simulations, the flux of reactants follows a power law with an anomalous reaction order, x (Anacker & Kopelman, 1984):

$$R = 2k[A]^x \qquad (9)$$

where $x = 1 + \dfrac{2}{d_s}$, yielding to the following relationship with the fractal parameter, h:

$$x = \frac{2 - h}{1 - h} \qquad (10)$$

Special cases are reactions in pores ($d = 1$), which gives $h = \dfrac{1}{2}$ and $x = 3$, and in random fractals, which gives $h \approx 2.5$.

In general:

$$d = 3 \Rightarrow d_s = 2 \Rightarrow h = 0 \Rightarrow x = 2 \; (classical \; kinetics) \tag{11a}$$

$$d < 3 \Rightarrow d_s < 2 \Rightarrow h > 0 \Rightarrow x > 2 \; (fractal \; kinetics) \tag{11b}$$

Another interesting feature is that this kind of bimolecular reactions (5b) that take place in heterogeneous low dimensional media ($d < 2$) can yield segregation (Avnir, 1989).

It is important to summarize these features shown in fractal kinetics with a simple question: what is the meaning of fractal kinetics? To answer this question, we can suppose the following reaction scheme of a dimerization reaction followed by the disappearance of the product:

$$A + A \rightarrow A_2 \uparrow \tag{12}$$

with a value of the fractal parameter, $h = \dfrac{1}{3}$. For this particular reaction scheme, two different experiments which have the same instantaneous concentration of reactant species at different times as, for example, $[A]^I_{t=8s} = [A]^{II}_{t=1s}$, yields to $\{k_0'(t=8s)\}^I = \dfrac{1}{2}\{k_0'(t=1s)\}^{II}$, i.e., the system reacts differently for the same concentration. This important fact implies that the system depends on the history of the reaction. A feasible explanation of this feature is that the particle distribution of the reactants is different because the time-independent 'at random' distribution for classical kinetics does not hold for fractal kinetics, for which the distribution evolves to a more ordered and less "at random." The fractal kinetics seems to be more rational because it is not necessary to generate a self-mixture of the reactants after every reactive collision.

The relationship between anomalous diffusion exponent, α, and the fractal parameter, h, is given, for some particular cases. In the Zeldovich regime (Argyrakis et al., 1993; Lin et al., 1996), subdiffusive reactants with an anomalous diffusion coefficient α on a homogeneous substrate with $d_f = d$ yield, from Eq. (8b), $h = 1 - \alpha d/4$, for $d < 4$.

The first example of fractal kinetics applied to fractal surface reactions of physiological interest was done by T.G. Dewey, who studied the proton exchange kinetics in lysozyme, a protein, which can be modeled with a fractal surface where the protons are exchanged (Dewey, 1994, 1997). This process is modeled as a Michaelis-Menten mechanism:

$$E + S \underset{k_{-1}}{\overset{k_1}{\rightleftharpoons}} ES \overset{k_2}{\rightarrow} E + P \tag{13}$$

where k_i is the kinetic constants of the different steps, being the first two diffusion-controlled and the last one, reaction-controlled. However, due to the fractal characterization of the protein surface, the two first-rate constants must be seen as kinetic coefficients that follow a time-dependence power-law given by Kopelman's equation Eq. (7). In stationary conditions, the observed forward rate, k_f, is given by

$$k_f = \frac{k_1(t)k_2}{k_{-1}(t) + k_2} = \frac{k_0'k_2t^{-h}}{(\frac{k_0'}{K_1})t^{-h} + k_2} \tag{14}$$

where $K_1 = \frac{k_1(t)}{k_{-1}(t)}$ is the equilibrium constant for the formation of the complex ES that can be time independent. When the protein is in excess in respect to the tritium isotope, the limiting behavior of the rate of exchange tritium isotopes becomes:

$$[S] \propto exp(k_0't^{1-h}) = exp\left[\left(\frac{t}{\tau}\right)^{\alpha'}\right] \tag{15}$$

where τ is the corresponding relaxation time and $\alpha' = 1-h$. The α' can be calculated for large time scales (de Gennes, 1982) assuming normal diffusion, $d_\omega = 2$ as $\alpha' = 1 - h = \frac{d_f}{4}$, being d_f the fractal dimension of the surface of the protein.

Savageau (1995) generalized the stationary state of Michaelis-Menten mechanism for fractal kinetics following its biochemical systems theory, which uses fractional kinetic orders similarly to the stationary fractal kinetics developed by Kopelman et al. [Eq. (9)]. It is important to note that both Kopelman and Savageau refer to their mathematical formalisms as 'fractal-like kinetics' because they have both evolved for modeling reactions in fractal environments. However, despite the common name the two approaches are mathematically quite distinct (Schnell & Turner, 2004; Bajzer et al., 2006, 2008).

Many enzymatic reactions have been modeled by the well-known Michaelis-Menten reaction mechanism, which illustrates the fractal kinetic behavior when the reaction occurs in a crowded environment, as it happens in the interior of cells (Aon & Cortassa, 2015). Monte Carlo simulations have been used to test the time dependence of rate coefficients when the reactants (substrate and enzyme) react in a crowded environment (Berry, 2002; Schnell & Turner, 2004; Isvoran et al., 2006, 2008; Pitulice et al., 2014). Schnell & Turner generalized the power-law introduced by Kopelman, Eq. (7), taking into account the initial behavior of the reaction, as $t \to 0$, and using a Zipf-Mandelbrot distribution, which will further enable the modeling and analysis of biochemical reactions occurring in crowded intracellular environments (Schnell & Turner, 2004):

$$k(t) = \frac{k_0^{''}}{(t+\tau)^h} ; (0 \leq h \leq 1) \tag{16}$$

where τ is a positive constant and its physical meaning is the time after which the reaction 'feels' the crowding effects. However, both $k_0^{'}$ and $k_0^{''}$ are instantaneous reaction constants, which do not have the usual dimension of a second order rate constant.

Recently, the Zipf-Mandelbrot distribution has been generalized (Pitulice et al., 2014) in order to take into account the correct dimensions of the parameters of the distribution:

$$k(t) = k_0 \left(1 + \frac{t}{\tau}\right)^{-h} ; (0 \leq h \leq 1) \tag{17}$$

with the correct limiting behaviors:

$$\lim_{t \to 0} k(t) = k_0; \quad \lim_{t \gg \tau} k(t) = k_0 \tau^h t^{-h} = k_0^{'} t^{-h} \Rightarrow k_0^{'} = k_0 \tau^h \tag{18}$$

This kind of equation has also been taken into account by Neff et al. (2011). The authors also generalized different approximations used in the literature in order to understand which kind of formalism better describes the biochemical reactions in a crowded environment, and used them in an experimental study to test different models for reaction kinetics in intracellular environments.

It can be demonstrated that in 3D, the initial rate constant, k_0, can be seen as a rate constant of a diffusion-controlled reaction within a confined volume, before the reactants 'feel' that they are in the crowded environment, and it can be broken down as follows (Pitulice et al., 2014):

$$k_0 = k_d \left(\frac{r_{cv}}{r_{cv} - r_{ES}}\right) \tag{19}$$

where k_d is the diffusion-limited reaction constant, given in Eq. (7), and r_{cv} is the radius of the averaged confined volume, which depends on the excluded volume of the system. From this equation, it can be observed that k_0 is always greater than k_d.

This behavior can be seen in a representation of the diffusion-controlled bimolecular reaction coefficient, k_1, of the Michaelis-Menten scheme, Eq. (13), *versus* time (Figure 17.1). It should be noted here that the proposed Eq. (17), not only describes the two limit linear regions, Eq. (18), of the plot but also the transition curve between them. The log-log plot allows the determination of the crossover time value, τ, which corresponds to the passage from one limiting expression to another. It may also be estimated as the time value related to the intersection of the two linear fittings of the linear regions.

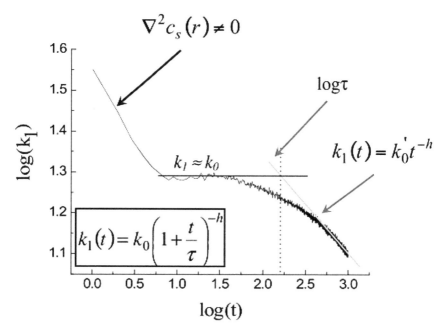

FIGURE 17.1   The $log(k_1)$ versus $log(t)$ plot for a particular case of simulation Michaelis-Menten scheme in crowded media.

Finally, recent Monte-Carlo simulations (Mourao et al., 2014a) show that both first-order and second-order rate coefficients, as well as the fractal-kinetic parameter, are influenced by the density and size of obstacles. The authors also propose a new form of anomalous kinetics that includes diffusion to explain lattice gas automata simulation results. In addition, they investigate the effects of particle rotation and weak force interactions in the reaction kinetics. They find that both particle rotation and weak force interactions shift the balance from diffusion-limited to reaction-limited kinetics. This fact significantly reduces fractal-like kinetics, suggesting that in lattice-based models, fractal-like kinetics generally appears under more restrictive conditions than previously believed.

17.4 NANO-CHEMICAL APPLICATIONS

This kind of non-classical kinetics represents a new insight, which may be the key to understanding heterogeneous transport and chemical kinetics in low-dimensional media (Kopelman, 1986). It is worth mentioning some examples of heterogeneous reactions such as industrial catalysis (involving powders, porous materials, etc.), enzymatic reactions, geochemical porous rock, atmospheric dust and pollutants, electrochemical rough electrodes (Trigueros et al., 1993) and even interstellar formation of nucleic acids on grains of ice and/or dust. Moreover, there are experimental evidences in some heterogeneous photophysical reactions such as electron-hole recombination on surfaces and in amorphous materials, photoelectron trapping, and, finally, exciton trapping an exciton annihilation (Kopelman, 1986).

Many functions of living cells involve complex biochemical reactions of which rates must be as fast as possible to allow a wide range of processes taking place. To study biochemical reactions, one must take into account that cellular media are not homogenous but highly compartmented being characterized by a high total macromolecular content known as macromolecular crowding (Zhou et al., 2008; Pastor et al., 2014; Balcells et al., 2015). Schnell & Turner (2004) excellently reviewed a few non-classical approaches regarding biochemical reactions developing in crowded media. They are mainly divided into deterministic approaches, particularly comprising fractal kinetic approaches (Kopelman, 1986, 1988), kinetics based on fractional reaction orders (Savageau, 1995; Bajzer et al., 2006, 2008) and stochastic ones (Turner et al., 2004).

There are different physiological fields of interest where fractal kinetics can be applied:

a) in signal transduction, as for example preconditioning – a process of subjecting a cell or a tissue to a stimulus – can be studied through fractal-like kinetics (Vasilescu et al., 2011), or the endotoxin tolerance due to a particular case of negative preconditioning (Vasilescu et al., 2013).

b) in pharmacokinetics, as the importance of heterogeneity in drug distribution processes in the body (Kosmidis et al., 2003, 2004; Pereira, 2010).

c) in different physiological environments, due to the anomalous diffusion of the intracellular biochemical reactants (Hiroi et al., 2011; Varjú et al., 2015).

17.5 MULTI-/TRANS-DISCIPLINARY CONNECTIONS

Heterogeneous reaction kinetics poses a challenge to several fields of science such as chemical kinetics, surface science, solid-state physics, physiology, biochemistry, biotechnology, and biomedicine, among others. This is a good example of a problem that requires interdisciplinary study and research (Kopelman, 1986; West, 1990; Brouers & Sotolonga-Costa, 2006).

17.6 OPEN ISSUES

Fractal kinetics is one approximate way to study reaction kinetics mechanisms in complex media. It is generally known that the reaction kinetics can be influenced by the transport properties of the reactants (Bénichou et al., 2010). The concept of 'geometry-controlled kinetics' is an emergent concept that introduces the initial distance between reactants as a key parameter to characterize the kinetics of diffusion-limited reactions. This concept is related to the spatial organization of reactants in complex systems (e.g., genes in transcription kinetics).

However, much less attention has been given to the effects of macromolecular crowding on cell physiology. Recent findings that shed some light on the role of crowding in various cellular processes, such as reduction of biochemical activities, structural reorganization of the cytoplasm, cytoplasm fluidity, and cellular dormancy are well reviewed (Mourao et al., 2014b), and conclusions withdrawn by presenting some unresolved problems requiring the attention of biophysicists, biochemists, and cell physiologists. Although it is still underappreciated, macromolecular crowding plays a critical role in life, as we already know it.

KEYWORDS

- **diffusion-limited reactions**
- **enzymatic reactions**
- **fractal geometry**
- **heterogeneous environment**
- **macromolecular crowding**
- **reaction kinetics**

REFERENCES AND FURTHER READING

Alexander, S., & Orbach, R., (1982). Density of states on fractals: Fractions. *Journal of Physics (Paris) Letters, 43*, L625–L631.

Anacker, L. W., & Kopelman, R., (1984). Fractal chemical kinetics: Simulations and experiments. *Journal of Chemical Physics, 81*, 6402–6403.

Aon, M. A., & Cortassa, S., (2015). Function of metabolic and organelle networks in crowded and organized media. *Frontiers in Physiology, 5*, 523.

Argyrakis, P., Kopelman, R., & Lindenberg, K., (1993). Diffusion-limited binary reactions: The hierarchy of nonclassical regimes for random initial conditions. *Chemical Physics, 177*, 693.

Avnir, D., (1989). *The Fractal Approach to Heterogeneous Chemistry*, John Wiley & Sons: New York.

Bajzer, Z., Huzak, M., Neff, K. L., & Prendergast F. G., (2006). Reaction kinetics in intracellular environments: The two proposed models yield qualitatively different predictions. *Croatia Chimica Acta, 79*, 437–444.

Bajzer, Z., Huzak, M., Neff, K. L., & Prendergast, F. G., (2008). Mathematical analysis of models for reaction kinetics in intracellular environments. *Mathematical Biosciences, 215*, 35–47.

Balcells, C., Pastor, I., Pitulice, L., Hernández, C., Via, M., Garcés, J. L., et al., (2015). Macromolecular crowding upon *in vivo-like* enzyme-kinetics: Effect of enzyme-obstacle size. *New Frontiers in Chemistry, 24*, 3–16.

Banks, D. S., & Fradin, C., (2005). Anomalous diffusion of proteins due to molecular crowding. *Biophysical Journal, 89*, 2960–2971.

Ben-Avraham, D., & Havlin, S., (2000). *Diffusion and Reactions in Fractals and Disordered Systems*. Cambridge University Press: New York.

Bénichou, O., Chevalier, C., Klafter, J., Meyer, B., & Voituriez, R., (2010). Geometry-controlled kinetics. *Nature Chemistry, 2*, 472–477.

Berry, H., (2002). Monte Carlo simulations of enzyme reactions in two dimensions: Fractal kinetics and spatial segregation. *Biophysical Journal, 83*(4), 1891–1901.

Bouchaud, J. P., & Georges, A., (1990). Anomalous diffusion in disordered media: Statistical mechanisms, models and physical applications. *Physics. Reports, 195*, 127–293.

Brouers, F., & Sotolongo-Costa, O., (2006). Generalized fractal kinetics in complex systems (application to biophysics and biotechnology). *Physica, A368*, 165–175.

De Gennes, P. G., (1982). Kinetics of diffusion-controlled processes in dense polymer systems. I. Nonentangled regimes. *The Journal of Chemical Physics, 76*(6), 3316–3321.

Dewey, T. G., (1994). Fractal analysis of proton exchange kinetics in lysozyme. *Proceedings of the Natural Academy of Sciences USA, 91*, 12101–12104.

Dewey, T. G., (1997). *Fractals in Molecular Biophysics*. Oxford University Press: New York.

Hiroi, N., Lu, J., Iba, K., Tabura, A., Yamashita, S., Okada, Y., et al., (2011). Physiological environment induces quick response-slow exhaustion reactions. *Frontiers in Physiology, 2*, 50.

Isvoran, A., Vilaseca, E., Garcés, J. L., Unipan, L., & Mas, F., (2008). Nonclassical kinetics for enzymatic reactions in two-dimensional media. *Romanian Biophysical Journal, 18*, 283–291.

Isvoran, A., Vilaseca, E., Ortega, F., Cascante, M., & Mas, F., (2006). About implementing a Monte Carlo simulation algorithm for enzymatic reactions in crowded media. *Journal Serbian Chemical Society, 71*, 75–86.

Kang, K., & Redner, S., (1986). Scaling approach for the kinetics of recombination processes. *Physical Review Letters, 52*, 955–958.

Kopelman, R., (1986). Rate processes on fractals: Theory, simulations, and experiments. *Journal of Statistical Physics, 42*, 185–200.

Kopelman, R., (1988). Fractal reaction kinetics. *Science, 241*(4873), 1620–1626.

Kosmidis, K., Argyrakis, P., & Macheras, P., (2003). Fractal kinetics in drug release from finite fractal matrices. *Journal of Chemical Physics, 119*, 6373–6377.

Kosmidis, K., Karalis, V., Argyrakis, P., & Macheras, P., (2004). Michaelis-Menten kinetics under spatially constrained conditions: Application to mibefradil pharmacokinetics. *Biophysical Journal, 87*, 1498–1506.

Lin, A., Kopelman, R., & Argyrakis, P., (1996). Nonclassical kinetics in three dimensions: Simulations of elementary A+B and A+A reactions. *Physical Review, E53*, 1502–1509.

Minton, A. P., (2001). The influence of macromolecular crowding and macromolecular confinement on biochemical reactions in physiological media. *Journal of Biological Chemistry, 276*(14), 10577–10580.

Mourao, M., Hakim, J. B., & Schnell, S., (2014b). Connecting the dots: The effects of macromolecular crowding on cell physiology. *Biophysical Journal, 107*(12), 2761–2766.

Mourao, M., Kreitman, D., & Schnell, S., (2014a). Unraveling the impact of obstacles in diffusion and kinetics of an enzyme catalyzed reaction. *Physical Chemistry Chemical Physics, 16*, 4492–4503.

Neff, K. L., Offord, C. P., Claride, A. J., Strehler, E. E., Prendergast, F. G., & Bajzer, Z., (2011). Validation of fractal-like kinetic models by time-resolved binding kinetics of dansylamide and carbonic anhydrase in crowded media. *Biophysical Journal, 100*, 2495–2503.

Pastor, I., Pitulice, L., Balcells, C., Vilaseca, E., Madurga, S., Isvoran, A., Cascante, M., & Mas, F., (2014). Effect of crowding by dextrans in enzymatic reactions. *Biophysical Chemistry, 185*, 8–13.

Pastor, I., Vilaseca, E., Madurga, S., Garcés, J. L., Cascante, M., & Mas, F., (2010). Diffusion of alpha-chymotrypsin in solution-crowded media: A fluorescence after photobleaching recovery study. *The Journal of Physical Chemistry B., 114*, 4028–4034. Erratum, ibid, 114, 12182.

Pereira, L. M., (2010). Fractal pharmacokinetics. *Computational and Mathematical Methods in Medicine, 11*(2), 161–184.

Pitulice, L., Craciun, D., Vilaseca, E., Madurga, S., Pastor, I., Mas, F., & Isvoran. A., (2016). Fractal dimension of the trajectory of a single particle diffusing in crowded media. *Romanian Journal of Physics, 61*(7/8), 1276–1286.

Pitulice, L., Vilaseca, E., Pastor, I., Madurga, S., Garcés, J. L., Isvoran, A., & Mas, F., (2014). Monte Carlo simulations of enzymatic reactions in crowded media. Effect of the enzyme-obstacle relative size. *Mathematical Biosciences, 251*, 72–82.

Savageau, M. A., (1995). Michaelis-Menten mechanism reconsidered: Implications of fractal kinetics. *Journal of Theoretical Biology, 176*(1), 115–124.

Schnell, S., & Turner, T. E., (2004). Reaction kinetics in intracellular environments with macromolecular crowding: Simulations and rate laws. *Progress in Biophysics & Molecular Biology, 85*, 235–260.

Trigueros, P. P., Mas, F., Claret, J., Sagués, F., Galceran, J., & Puy, J., (1993). Monte-Carlo simulation of diffusion-controlled response functions at 2D experimental rough electrodes. *Journal of Electroanalytical Chemistry and Interfacial Electrochemistry, 348*, 221–246.

Turner, T. E., Schnell, S., & Burrage, K., (2004). Stochastic approaches for modeling *in vivo* reactions. *Computational Biological Chemistry, 28*, 165–178.

Varjú, I., Tenekedjiev, K., Keresztes, Z., Pap, A. E., Szábo, L., Thelwell, C., Longstaff, C., Machovich, R., & Kolec, K., (2014). Fractal kinetic behavior of plasmin on the surface of fibrin meshwork. *Biochemistry, 53*, 6348–6356.

Vasilescu, C., Olteanu, M., & Flondor, P., (2008). Description of the preconditioning effect in biology by means of fractal-like enzyme kinetics of intracellular reactions. *Revue Roumaine de Chimie, 56*(7), 751–757.

Vasilescu, C., Olteanu, M., Flondor, P., & Calin, G. A., (2013). Fractal-like kinetics of intracellular enzymatic reactions: A chemical framework of endotoxin tolerance and a possible non-specific contribution of macromolecular crowding to cross-tolerance. *Theoretical Biology and Mathematical Modeling, 10*, 55.

Vilaseca, E., Isvoran, A., Madurga, S., Pastor, I., Garcés, J. L., & Mas, F., (2011). New insights into diffusion in 3D crowded media by Monte Carlo simulations: Effect of size, mobility and spatial distribution of obstacles. *Physical Chemistry Chemical Physics, 13*(16), 7396–7407.

West, B. J., (2012). *Fractal Physiology and Chaos in Medicine* (2nd ed.). World Scientific: Singapore.

Zhou, H. X., Rivas, G., & Minton, A. P., (2008). Macromolecular crowding and confinement: Biochemical, biophysical and potential physiological consequences. *Annual Review of Biophysics, 37*, 375–397.

CHAPTER 18

Fredholm Integral Equation in Electrochemistry

MIRELA I. IORGA[1] and MIHAI V. PUTZ[1,2]

[1]*National Institute for Research and Development in Electrochemistry and Condensed Matter, Department of Applied Electrochemistry, Laboratory of Renewable Energies-Photovoltaics, 300569 Timisoara, Romania, E-mail: mirela_iorga@yahoo.com, mirela.iorga@incemc.ro*

[2]*West University of Timisoara, Biology-Chemistry Department, Laboratory of Structural and Computational Physical Chemistry for Nanosciences and QSAR, 300115 Timisoara, Romania, E-mail: mv_putz@yahoo.com, mihai.putz@e-uvt.ro*

18.1 DEFINITION

Fredholm integral equation represents in mathematics an integral equation whose solutions have to give rise to Fredholm theory, Fredholm nuclei study, and Fredholm operators. Fredholm integral equation was the basis of Fredholm's theorems, one of them being known as Fredholm alternative. The applicability of Fredholm integral equation was demonstrated in several domains, among them one of great importance is electrochemistry because the solutions of Fredholm integral equations facilitate the determination of process conditions, parameters, and so on.

18.2 HISTORICAL ORIGIN(S)

The concept of Fredholm integral equation was introduced in mathematics in 1900 by Erik Ivar Fredholm with his presentation *"Sur une nouvelle méthode pour la résolution du problème de Dirichlet"* (Fredholm, 1900) at the Congress of Mathematics at Stockholm, and in his works published in the following years he solved this equation (Fredholm, 1903). His scientific papers had a major influence on the subsequent development of integral equation theory. His studies covered Fredholm determinants and Fredholm's theorems demonstration.

In the 1960s, Tikhonov (1963a, 1963b) and Philips (1962) established independently a method which transforms first kind Fredholm integral equations to second kind Fredholm equations, known as the regularization method.

A new approach of Fredholm alternative, applicable in electrochemistry, was presented by Bonciocat in his works (Bonciocat, 1997). Thus, through Fredholm integral equations, the equations which describe various electro-chemical methods, such as potentiostatic method, galvanostatic methods, cyclic or direct voltammetry, electrochemical impedance spectroscopy methods, etc., were determined (Bonciocat, 2005).

18.3   NANO-SCIENTIFIC DEVELOPMENT(S)

The Fredholm alternative can be expressed in various ways, i.e., as a linear algebraic theorem, as a theorem of integral equations or as theorem based on Fredholm operators (Fredholm, 1900). Part of the results states that a complex number different from zero from a spectrum of a compact operator is an eigenvalue (Fredholm, 1903).

According to Abadi & Rezazadeh (2010), the linear algebraic expression states as follows: if V is an n-dimensional space vector and T is a linear transformation, one of the following is sustained:

a)   for every v vector in V, there is an u vector in V so that $T(u) = v$; in other words, T is surjective (and also bijective, meanwhile V is dimensional finite);
b)   dim $(Ker(T))>0$.

In matrix terms a more simple formulation of the concerned theorem states as follows: if a mn matrix A and a $m1$ column vector b are given, one of the following situations will occur:

a) $Ax = b$ has an x solution, or
b) $A^T y = 0$ has a y solution with $y^T b \neq 0$.

In other words, $Ax = b$ has a solution if and only if for every $A^T y = 0$, $y^T b = 0$.

If it is formulated as a theorem of integral equations (Bonciocat, 2005; Rahman, 2007), the general expression of such Fredholm type equations, in the domain $a \leq x \leq b$ will be as follows:

- the nonhomogenous Fredholm linear integral equation of the second kind, as presented in equation (1):

$$y(x) = \lambda \int_a^b K(x,u) y(u) du + f(x) \tag{1}$$

- the Fredholm linear integral equation of the first kind, as presented in Eq. (2):

$$f(x) = \int_a^b K(x,u) y(u) du \tag{2}$$

- the Fredholm nonlinear integral equation of the second kind, as presented in Eq. (3):

$$y(x) = \lambda \int_a^b K(x,u) \{y(u)\}^2 du + f(x) \tag{3}$$

where $K(x,u)$ represents the integral equation kernel, defined in the rectangle R, for which $a \leq x \leq b$ and $a \leq u \leq b$.

The following different situations could exist:

- if the kernel $K = (x,u)$ is expressed as presented in Eq. (4) and has a finite value, it is named a regular kernel, and the corresponding integral equation is a regular integral equation:

$$\int_a^b \int_a^b |K(x,u)|^2 dxdu < \infty \tag{4}$$

- if the kernel $K = (x,u)$ expressed as presented in Eq. (5) the corresponding integral equation is named a weakly singular integral equation:

$$K(x,u) = \frac{H(x,u)}{|x-u|^\alpha} \tag{5}$$

- if the kernel $K = (x,u)$ expressed as presented in Eq. (6) the corresponding integral equation is named a singular integral equation with Cauchy kernel:

$$K(x,u) = \frac{H(x,u)}{x-u} \tag{6}$$

- if the kernel $K = (x,u)$ expressed as presented in Eq. (7) the corresponding integral equation is named a hypersingular integral equation with Cauchy kernel:

$$K(x,u) = \frac{H(x,u)}{(x-u)^2} \tag{7}$$

- if the kernel $K = (x,u) \equiv 0$, $b = x$, $x < u \le b$, the equation becomes an integral equation Volterra type.

The theorems describing the properties of the integral equation of the second kind such as Eq. (1) are known as Fredholm's theorems (Bonciocat, 2005) and states as follows:

Theorem 1. The following alternative presented: whatever is $f(x)$ function from an enough wide class, equation (1) – integral equation linear and non-homogenous Fredholm type has a unique solution or the appropriate homogenous integral equation (8):

$$y(x) = \lambda \int_a^b K(x,u) y(u) du \tag{8}$$

has at least one non-zero solution (which is not reducing identically to zero).

Theorem 2. If for integral equation (8) the first case of the alternative took place, then, whatever is $f(x)$ function from an enough wide class, both for the adjunct non-homogenous integral equation (9):

$$z(x) = \lambda \int_a^b K(u,x) z(u) du + f'(x) \tag{9}$$

the first case of the alternative will take place.

Theorem 3. If for integral equation (1) the second case of the alternative occurs, the homogenous integral equation (8) has the same number of non-zero solutions linear independent as the adjunct non-homogenous integral equation (10):

$$z(x) = \lambda \int_a^b K(u,x) z(u) du \tag{10}$$

and this number is finite.

Theorem 4. The necessary and sufficient condition that in the second case of the Fredholm alternative, solutions of the non-homogenous integral equation (1) exist, is that the *f(x)* function to be orthogonal on all the linear independent solutions of the integral equation (10):

$$\int_a^b f(x)z_i(x)dx = 0 \qquad (i = 1, 2, ..., k) \tag{11}$$

If these conditions are fulfilled, the non-homogenous integral equation (1) in fact has an infinity of solutions.

Fredholm theorems demonstration was made by many authors. In the case of formulation based on Fredholm's operators (Ramm, 2001), the following could be stated:

Definition 1. An A operator is referred Fredholm if and only if

$$\dim N(A) = \dim N(A^*) := n < \infty \tag{12}$$

and

$$R(A) = \overline{R(A)}, \ R(A^*) = \overline{R(A^*)} \tag{13}$$

where the superior line means closing.

Considering the equations:

$$Au = f \tag{14}$$

$$Au_0 = 0 \tag{15}$$

$$A^*v = g \tag{16}$$

$$A^*v_0 = 0 \tag{17}$$

The Fredholm alternative could be formulated as follows:

Theorem 1. If B is an isomorphism and F is a finite rank operator, then $A = B + F$ is a Fredholm operator. For every A Fredholm operator, the following Fredholm alternative could be stated:

a) either equation (15) has only the trivial solution $u_0 = 0$, and in this case, equation (17) has only the trivial solution, and (14) and (16) equations have a unique solution for any f or g from the right side, or

b) equation (15) has $n > 0$ nonlinear independent solutions $\{\phi_j\}$, for $1 \leq j \leq n$ and then equation (17) has also n linear independent solutions $\{\psi_j\}$ for $1 \leq j \leq n$, Eqs. (14) and (16) are solvable if and only if $(f,\psi_j) = 0$ for $1 \leq j \leq n$ and correspondingly $(g,\phi_j) = 0$ for $1 \leq j \leq n$.

If these are solvable, their solutions are not unique and their general solutions are as follows:

$$u = u_p + \sum_{j=1}^{n} a_j \phi_j \tag{18}$$

respectively:

$$v = v_p + \sum_{j=1}^{n} b_j \psi_j \tag{19}$$

where a_j and b_j are arbitrary constants and u_p and v_p are some specific solutions of Eqs. (14) and (16), respectively.

Theorem 2. A bounded linear operator A is a Fredholm operator if and only if $A = B + F$, where B is an isomorphism and F has a finite rank.

The results obtained for the Fredholm operator determines their generalization for vector spaces of infinite dimensions, Banach spaces.

According to Berkani (2013), an extended version of Fredholm alternative was studied by including the B-Fredholm case. Thus, if T is a B-Fredholm operator which acts on a Banach space, type $T(x) = y$ equation was investigated, for x, y ranging from $R(T^n)$ to T, for $n \in N$. Also, the B-Fredholm unbounded operators for Banach spaces were defined.

Other authors (Shestopalov & Smirnov, 2002) presented fundamental concepts and statements regarding the theory of linear and integral operators. Fredholm theory is presented both in Banach spaces and Hilbert spaces. The elements of the harmonic functions theory are presented, including the problem of limit value reduction in the case of integral equations, several methods for integral equation solving, as well as some methods with numerical solutions.

A lot of solution techniques are nowadays available to solve the Fredholm integral equations (Rahman, 2007). Among them could be mentioned the main traditional methods, such as the method of successive approximations and the method of successive substitutions, the series method, the direct computational method, and more recent methods such as Adomian decomposition method (ADM), the modified decomposition method, and the numerical techniques.

Zemyan (2012) have demonstrated that the Fredholm integral equation of the second kind with a general kernel will have a unique solution, only if the product of λ parameter and the kernel "size" is small. Afterward, the validity of Fredholm theorems for an unlimited value of λ and a general kernel is demonstrated, introducing numerical methods to give an approximation of a Fredholm integral equation solution.

In the last years, to solve such type of equations, more attention was given to numerical methods (Darania & Shal, 2015). The numerical solutions of Volterra-Fredholm non-linear integral equation were studied, by product integration method. Thus, the equations could be transformed in a non-linear algebraic system, which could be solved by classical methods. The authors studied both the applicability and the precision of the method for non-linear equation solving by several numerical experiments. A considerable advantage of the method is that the unknown coefficients are readily determined using software programs.

According to Jerri (1999), to solve the Fredholm integral equations some precisely, approximate, and numerical methods were proposed. An outstanding condition, named regularity condition, for solution development of the Fredholm integral equation is that the kernel $K(x,u)$ to be square integrable in x and u, in $\{(x,t): a \leq x \leq b, a \leq u \leq b\}$ square.

Ziyaee & Tari (2014) proposed a development of the regularization method to linear and non-linear two-dimensional Fredholm integral equations of the first kind, finding a simple and reliable method to solve multiple integral equations of the first kind. Generally speaking, the regularization method transforms the two-dimensional Fredholm integral equation of the first kind (20) into a two-dimensional Fredholm integral equation of the second kind (21), as follows:

$$f(x,t) = \int_c^d \int_a^b K(x,t,y,z)u(y,z)dydz \qquad (20)$$

$$\alpha u_\alpha(x,t) = f(x,t) - \int_c^d \int_a^b K(x,t,y,z)u_\alpha(y,z)dydz \qquad (21)$$

Fredholm integral equation of the second kind is widespread in many areas of engineering and applied mathematics. Among the variety of numerical solutions for this equation, the quadrature and modified quadrature methods stand out, contributing to the decreasing of the computational complexity of the method. Thus, some authors (Rahbar & Hashemizadeh, 2008) used Mathematica software to establish and verify the algorithms by implementing the quadrature method and its modified version, have proved that the proposed algorithms lead to appropriate results.

Alturk (2016) reported the application of the regularization-homotopy method, which represents a combination between regularization method and the homotopy perturbation method (HPM), to find a solution to the two-dimensional linear and non-linear Fredholm integral equations of the first kind, and demonstrated the validity and the applicability of this approach.

Elbeleze et al. (2012) presented in their paper the application of numerical methods such as HPM and variational iteration method (VIM) to linear and non-linear initial-boundary conditions for Fredholm integrodifferential equation of fractional order and illustrated the accuracy of those methods.

According to Molabahrami (2013), any physical transformation could be modeled by differential equations, integral equations, integrodifferential equations or a system of these. Thus, he applied the integral mean value method to the general nonlinear Fredholm integrodifferential equations under the mixed conditions and concluded that this represents a valid method for this type of problems.

Saha Ray & Sahu (2013) reviewed various numerical methods to solve both linear and non-linear Fredholm integral equations of the second kind. They considered that a computational approach to solving Fredholm integral equations is crucial. In the case of linear Fredholm integral equations of the second kind could be mentioned the B-spline wavelet method, the method of moments based on B-spline wavelets, and the VIM. Since in the case of linear Fredholm integral equations system of the second kind were mentioned rationalized Haar functions method, block-pulse functions, Taylor series expansion method, and Haar wavelet method with operational matrices of integration. For non-linear Fredholm integral equations of the second kind could be enumerated quadrature method, B-spline wavelet method, wavelet Galerkin method, ADM, VIM, and HPM.

18.4 NANO-CHEMICAL APPLICATION(S)

The applicability of the Fredholm integral equation was demonstrated in several domains, such as applied mathematics, physics, chemistry and engineering, as shown in Figure 18.1. In many applications from different scientific areas, non-linear phenomena occur and could be modeled by non-linear integrodifferential equations. Electrochemistry represents such a domain where Fredholm integral equation finds applicability since the solutions of this equation contribute to facilitating the determination of processes conditions and parameters. There is a multitude of technical devices based on such electrochemical phenomena, as follows: accumulators, galvanic cells, electronic devices (capacitors, electronic displays), electrochemical sensors, electrofilter devices (e.g., in the case of artificial kidneys or for water desalination), metal galvanization reactors, substances synthesis reactors, etc.

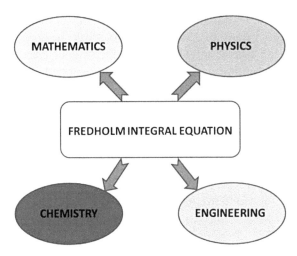

FIGURE 18.1 The principal application areas of Fredholm integral equation.

Theoretical modeling of transient experiments in analytical electrochemistry could be done by a particular mathematical approach, known as the integral equation method (Bieniasz, 2015). The analytical electrochemistry is the discipline that studies the physicochemical processes occurring both in the vicinity and at the interfaces between electronic conductors (metals, semiconductors, conductor polymers, etc.) and/or ionic conductors (liquid electrolytic solutions, solid electrolytes, electrolyte gels, etc.), and at liquid/liquid interfaces. High interests are the charge transfer processes by interfaces, where the electric charge could be transferred by electron or ions. If a sequence of the phases is constructed in such a manner to form a closed circuit, this gives rise to an electric current flow. The electroanalytical studies are of great scientific and practical interest, because the interfaces through electrical current passes and these types of processes occur almost everywhere in nature, including inorganic matter and living organisms.

An important feature of the integral equation method is the profound mathematical vision onto the models, resulted from the need to give partial analytical solutions. The integral equation method could facilitate the obtaining of analytical formulas for some limit cases of those models (e.g., concerning steady-state solutions).

Mirkin and Bard (1992a, 1992b) reported the applicability of multidimensional integral equations in the case of different types of microelectrode systems (microelectrode disk, microband electrode, etc.). They found that every solution of those equations was suitable in the situation of an electrode reaction for any values of kinetic parameters of heterogeneous electron transfer and in the case of any electrode applied signal shape; for other reaction mechanisms, similar equations could be developed. The algorithms for equations solving is quite similar.

In his works, Bonciocat (1997, 2005) proposed the idea that the equation describing the kinetics of a redox electrode reaction:

$$O + ne^- \leftrightarrow R \tag{22}$$

is a Volterra type integral equation (representing a particular case of the Fredholm integral equation), namely:

$$i(t) = \lambda \int_0^t K(t,u) i(u) \, du + f(t) \tag{23}$$

where: $\lambda = -1$; $i(u)$, $i(t)$ – faradic current densities at time $u = t$, respectively t; $K(t,u)$ – integral equation kernel.

$K(t,u)$ and $f(t)$ have explicit expressions containing kinetic parameters of the electrode reaction, concentrations, diffusion coefficients of the electrochemically active species, times t and u, overvoltage at t time.

Depending on how the overvoltage is applied, different forms of the equation (23) are obtained; by solving these equations, various forms of the studied interface response occurred. These equations have unique solutions for any λ so the first case of Fredholm alternative it's applying. By unsteady techniques, which study the reaction (22), kinetic parameters of the redox electrode reaction and diffusion coefficients (or concentrations) of the two-electrochemical active species are determined.

The conclusion is that any electrochemical technique could be expressed starting from the basic equation (23), which is an integral equation (Volterra type). All the equations describing unsteady electrochemical techniques could be deduced through the same mathematical procedure since all implies solving the same kind of integral equations. Figure 18.2 shows a simplified layout illustrating the process.

Further are exemplified several unsteady methods in continuous or alternative current for kinetic parameters determined according to Bonciocat (2005), in the case of inert metal/redox electrolyte electrodes and inert semiconductor/redox electrolyte electrodes, theoretically expressed by a non-homogenous integral equation Fredholm type with a unique solution.

The overvoltage applied to the interface is a continuous function on $[0,a]$ closed interval, and both functions $K(t)$ and $f(t)$ are continuous uniform functions for $t \in [0,a]$.

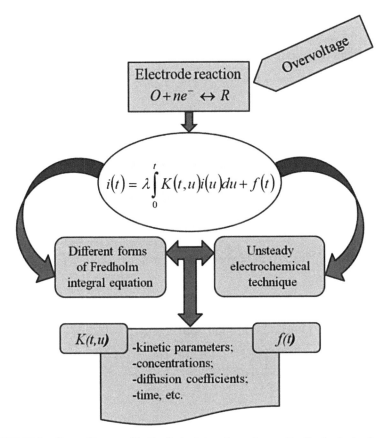

FIGURE 18.2   General layout of the Fredholm integral equation application in electrochemistry.

18.4.1   POTENTIOSTATIC METHOD

In the case of inert metal/redox electrolyte electrodes, after applying a constant voltage, the electric response of the electrode is registered, which coincide with the faradic current density, and Eq. (23) becomes:

$$i(t) = \lambda \frac{i^0 N}{\pi^{1/2}} \int_0^t \frac{i(u)}{(t-u)^{1/2}} du - \frac{F}{RT} i^0 \eta \tag{24}$$

Finally, the equation of the potentiostatic method for small overvoltages and times is obtained:

$$i(t) = -\frac{F}{RT} i^0 \eta \left(1 - \frac{2i^0 N}{\pi^{1/2}} t^{1/2}\right)$$ (25)

The obtained equations facilitate the determination of the change current density i^0 and the charge transfer reaction kinetic parameters i^{00} and β.

$$\ln i^0 = \ln i^{00} c_R^\beta + (1-\beta) \ln c_0$$ (26)

In the case of some inert semiconductor or redox electrolyte electrodes, $i_F(t)$ represents the total faradic current density, consisting of two energy bands contributions (conduction and valence), according to equation (27):

$$i_F(t) = i_{nF}(t) + i_{pF}(t)$$ (27)

If the charge transfer is only through the conduction band, i.e., $i_p = 0$, Eqs. (25) and (26) become:

$$i(t) = -\frac{F}{RT} i_n^0 \eta \left(1 - \frac{2i_n^0 N}{\pi^{1/2}} t^{1/2}\right)$$ (28)

$$\ln i_n^0 = \ln i_n^{00} c_R^{\alpha_n} + (1-\alpha_n) \ln c_0$$ (29)

If the charge transfer is through both energy bands, Eq. (25) becomes:

$$i(t) = -\frac{F}{RT} \left(i_n^0 + i_p^0\right) P \left(1 - \frac{2\left(i_n^0 + i_p^0\right) N}{\pi^{1/2}} t^{1/2}\right)$$ (30)

These equations enable to calculate the sum of the current exchange standard densities through both energy bands.

18.4.2   GALVANOSTATIC METHOD

In the case of some inert metal or redox electrolyte electrodes, the current density is constant in the external circuit. The sum of faradic and capacitive components remains constant, although their value is changing with time. In this case, Eq. (23) becomes:

$$i - i_C(t) = \lambda \frac{i^0 N}{\pi^{1/2}} \int_0^t \frac{du}{(t-u)^{1/2}} - \frac{F}{RT} i^0 \eta$$ (31)

Finally, the galvanostatic method equation for small overvoltage is obtained, with two cases:

- if the capacitive current is neglect:

$$\eta(t) = -\frac{RT}{Fi^0}i\left(1+\frac{2i^0 N}{\pi^{1/2}}t^{1/2}\right)$$

(32)

- if the capacitive current is taken into account:

$$\eta(t) = -\frac{RT}{Fi^0}i\left(1+\frac{2i^0 N}{\pi^{1/2}}t^{1/2}-\frac{RT}{F}N^2 C_d i^0\right)$$

(33)

The equations facilitate the determination of charge transfer reaction kinetic parameters i^{00} and β.

In the case of some inert semiconductor or redox electrolyte electrodes, if the charge transfer is for example only through the conduction band, i.e., $i_p = 0$, Eq. (33) becomes:

$$\eta(t) = -\frac{RT}{Fi_n^0}i\left(1+\frac{2i_n^0 N}{\pi^{1/2}}t^{1/2}-\frac{RT}{F}N^2 C_d i_n^0\right)$$

(34)

The equation facilitates the determination of the specific capacity of the electrochemical double layer.

If the charge transfer is through both energy bands, Eq. (33) becomes:

$$P(t) = -\frac{RT}{F\left(i_n^0+i_p^0\right)}i\left(1+\frac{2\left(i_n^0+i_p^0\right)N}{\pi^{1/2}}t^{1/2}-\frac{RT}{F}N^2 C_d\left(i_n^0+i_p^0\right)\right)$$

(35)

where P is the polarization.

Through a successive approximation process is possible to estimate all the kinetic parameters: $\beta, \beta^*, \alpha_n, \alpha_p, i_n^{00}, i_p^{00}, i_n^0, i_p^0$.

Inert semiconductor/redox electrolyte electrodes with charge transfer through a single energy band are used to produce photo-electrochemical cells which convert solar energy into electric energy.

18.4.3   *VOLTAMMETRY WITH POTENTIAL LINEAR SWEEP*

In the case of inert metal/redox electrolyte electrodes the overvoltage applied to the electrode depends on time and scan rate, according to the equation (36):

$$\eta(t) = vt \tag{36}$$

In this case, if the capacitive current is neglected, Eq. (23) becomes:

$$i = \lambda \frac{i^0 N}{\pi^{1/2}} \int_0^t \frac{i(u)}{(t-u)^{1/2}} du - \frac{Fi^0}{RT} vt \tag{37}$$

and, if the capacitive current could not be neglected, Eq. (23) becomes:

$$i = \lambda \frac{i^0 N}{\pi^{1/2}} \int_0^t \frac{i(u)}{(t-u)^{1/2}} du - vC_d - \frac{2i^0 N}{\pi^{1/2}} vC_d \sqrt{t} - \frac{Fi^0}{RT} vt \tag{38}$$

Finally, the equation of voltammetry with the potential linear sweep for small overvoltages and times is obtained, and there are two cases:

• if the capacitive current is neglect:

$$i(t) = -\frac{Fi^0}{RT} vt \left(1 - \frac{4i^0 N}{3\pi^{1/2}} t^{1/2} \right) \tag{39}$$

• if the capacitive current is taken into account:

$$i(t) = -\frac{Fi^0}{RT} vt \left(1 - \frac{4i^0 N}{3\pi^{1/2}} t^{1/2} \right) - vC_d \tag{40}$$

The obtained equations facilitate the determination of change current standard density and specific capacity of the electrochemical double layer.

18.4.4 BASIC FARADAY IMPEDANCE METHOD

In the case of inert metal/redox electrolyte electrodes, a slight overvoltage is applied to the steady-state interface, and the current density that passes through the interface faradaic impedance (Z_F) is analyzed. Between overvoltage and current density, a phase gap occurs, depending on the electrode redox reaction character. If the reaction is irreversible, limitations are determined by the charge transfer. Z_F depends only on the charge transfer resistance (R_{ts}), and both parameters are in phase ($\varphi = 0$). If the reaction is reversible, the limitations are given by the mass transfer, more specifically by diffusion, Warburg impedance Z_W is provided by both components R_W and C_W, and the two parameters are phase shifted with φ. In the case of quasi-reversible equations, all three components occur (R_{ts}, R_W, C_W), and the value φ of the phase shift depends on their relative contribution.

In this case Eq. (23) becomes:

$$i_F = I_F \sin \omega t = \lambda \frac{i^0 N}{\pi^{1/2}} \int_0^t \frac{I_F \sin \omega (t-u)}{u^{1/2}} du - \frac{F i^0}{RT} \eta(t) \qquad (41)$$

According to Bonciocat & Marian (2006), basic Faraday impedance method is the simplest electrochemical impedance spectroscopic method and determines de dependence between R_s and C_W on the alternative current frequency/pulsation.

Electrochemical impedance spectroscopy (EIS) represents all the methods based on the impedance determination at different frequencies. By its interpretation through Fredholm alternative could be determined: electrochemical active species concentrations, redox reaction kinetic parameters, electrochemical double layer capacity, etc.

By Fredholm integral equations many problems occurring in electrocatalysis or impedance spectroscopy could be described, and regularization techniques could solve such problems.

Saccocio et al. (2014) used Fredholm integral equation of the first kind for studies of the application of regularized distribution of relaxation times (DRT) to characterize the EIS data, to obtain relevant timescales in the case of electrochemical systems.

Hansen et al. (2015) studied the application of Fredholm integral equation of the first kind in the case of the inverse problem associated with measurements of electrochemical impedance spectra to obtain a model of the underlying physical reaction mechanism that could be applied in fuel cells analysis.

Lopes et al. (2013) analyzed the dynamic behavior of hydrogen fuel cells using the distribution of relaxation times (DRT). To determine the DRT from impedance data, a solution of a Fredholm integral equation was needed. The approach proposed was a combination between a general fractional-order Legendre basis for the representation of EIS data and the linearity of the Fredholm integral equations. Finally, the method was tested with real EIS data obtained for hydrogen polymer electrolyte fuel cell under different operating conditions.

According to Dion & Lasia (1998) to determine the regularization parameters a software program was developed. It was further applied to determine the adsorption energy distribution function in electrocatalysis, for Temkin and Frumkin isotherms, for hydrogen underpotential deposition (HUPD) on polycrystalline platinum, and time constants distribution in alternative current impedance for constant phase elements and fractal models. Appling regularization techniques allow distribution functions determination. The instruments used for regularization suggest appropriate values for λ parameter and permit the determination of distribution functions in the presence of a moderate instrumental noise.

Must be mentioned the fact that the alternative of modeling based on integral equations is closely connected with the technique of experimental data analysis, developed for electrochemical experiments with controlled potential, known as "convolutive analysis of the data" (Bieniasz, 2015). The method is based on the fact that in the interactions at the concentration-flow interface, concentrations are often expressed as flow convolution integrals or as faradic currents depending on the flow. This fact suggests that by analyzing of a convolution integral suitable with the experimental faradic current, instead of current investigate, the relevant electrochemical information could be extracted more directly from experimental data. Mainly, by current convolution, the interface concentrations could be considered directly observable.

Schmuck and Bazant (2015) derived the effective Poisson–Nernst–Planck (PNP) equations for ion transport in charged porous media under forced convection by an asymptotic multiscale expansion with drift. They appealed to Fredholm alternative to solve the periodic reference cell problem in the case of major modifications of the equations for the ionic diffusivities and mobilities that become tensorial coefficients in connection with the microstructure. These equations could be applied to electrochemical and biological systems where electrodiffusion in charged porous media is present.

Other authors (Velasquez-Arcos et al., 2013) studied Fredholm integral application in electromagnetic waves field.

Wesse (1992) proposed an improved method based on Tikhonov regularization to calculate a solution of a Fredholm integral equation of the first kind from noisy experimental data in the case of chemical and physical applications.

Another use was reported by Gauden and Al (2003), as a new "adsorption stochastic algorithm" to solve the linear Fredholm integral equation of the first kind. In this program, several procedures for one-dimensional relative minimum estimation were tested. The proposed algorithm was then applied to the reconstruction of a distribution function of pore dimensions (monomodal and multimodal) and experimentally verified, proving the fact that by this method similar result with that obtained by Tikhonov regularization method was obtained. This method is very useful in the evaluation of pore dimension distribution from experimental data.

18.5   MULTI-/TRANS-DISCIPLINARY CONNECTION(S)

Fredholm integral equation was discovered in mathematics and consist the basis for the further development of the integral equation theory. It was expressed by theorems belonging to linear algebra or mathematical analysis (integral equation or operators). It is considered one of the most important integral equations.

According to Saha Ray & Sahu (2013) integral equations found applications in several domains, such as: continuum mechanics, fluid mechanics, fracture mechanics and quantum mechanics, potential theory, electricity, magnetism, radiation, acoustics, steady state heat conduction, radiative heat transfer, kinetic theory of gases, mass transfer in astrophysics and reactor theory, genetics, hereditary phenomena in physics and biology, medicine, optimization, optimal control systems, mathematical economics, communication theory, etc.

In the case of Fredholm integral equation, the high application degree was initially demonstrated in the principal areas, such as applied mathematics, physics, chemistry and engineering.

In particularly, Fredholm integral equation is frequently used in electrochemistry, mechanics, geophysics, electricity, magnetism, kinetic theory of gases, quantum mechanics, quantum chemistry, hereditary phenomena in biology, medicine, ecology, etc., the solutions of this equation lead to a more readily determination of processes parameters.

On the other side, just from the beginning, the Fredholm equation was used in economics, mathematical statistic, etc.

18.6   OPEN ISSUES

Taking into account the vast areas in which Fredholm integral equation could be successfully applied, it must be considered the fact that from Fredholm theory occur the possibility to set new conditions for novel nanomaterials obtaining and analyzing, for most real fields, such as medicine, energy (photovoltaics), environment protection, etc.

ACKNOWLEDGMENT

This contribution is part of the Nucleus-Programme under the project "Deca-Nano-Graphenic Semiconductor: From Photoactive Structure to the Integrated Quantum Information Transport," PN-18-36-02-01/2018, funded by the Romanian National Authority for Scientific Research and Innovation (ANCSI).

KEYWORDS

- **Fredholm alternative**
- **Fredholm integral equation**
- **Fredholm integral equation in electrochemistry**
- **Fredholm operators**
- **Fredholm theory**

REFERENCES AND FURTHER READING

Abadi, A. R. K., & Rezazadeh, H. R., (2010). *About the Proof of the Fredholm Alternative Theorems* (pp. 1–3). arXiv:1011.4480 [math. FA].

Alturk, A., (2016). The regularization-homotopy method for the two-dimensional Fredholm integral equations of the first kind. *Mathematical and Computational Applications, 21*, 1–10, doi: 10.3390/mca21020009.

Berkani, M., (2013). On the B-Fredholm alternative. *Mediterr. J. Math. 10*, 1487–1496, doi: 10.1007/s00009-013-0260.

Bieniasz, L. K., (2015). *Modeling Electroanalytical Experiments by the Integral Equation Method*, Springer Verlag Berlin.

Bonciocat, N., (1996). *Electrochimie și aplicatii* (In English: Electrochemistry and Applications), Dacia Europa-Nova, Timisoara, Romania.

Bonciocat, N., (2005). *Alternativa Fredholm în Electrochimie* (in English: Fredholm Alternative in Electrochemistry), Mediamira, Cluj-Napoca, România.

Bonciocat, N., & Marian, I. O., (2006). *Metoda impedantei Faraday si variantele sale* (In English: Faraday Impedance Method and Its Alternatives), Presa Universitara Clujeana, Cluj-Napoca, România.

Darania, P., & Shali, J. A., (2015). Convergence analysis of product integration method for nonlinear weakly singular volterra-Fredholm integral equations. *Sahand Communications in Mathematical Analysis (SCMA), 2*, 57–69.

Dion, F., & Lasia, A., (1998). Applications of regularization techniques in electrochemistry. In: Van Zee, J. W., Fuller, T. F., Foller, P. C., & Hine, F., (eds.), *Proceedings of the Symposium on Advances in Mathematical Modeling and Simulation of Electrochemical Processes and Oxygen Depolarized Cathodes and Activated Cathodes for Chlor-Alkali and Chlorate Processes* (pp. 84–94). The Electrochemical Society, Inc., Pennington, NJ, USA. doi: 10.1109/ICEAA2013.6632268.

Elbeleze, A. A., Kilicman, A., & Taib, B. M., (2012). Application of homotopy perturbation and variational iteration methods for Fredholm integrodifferential equation of fractional order. *Abstract and Applied Analysis*, ID 763139, 1–14, doi: 10.1155/2012/763139.

Fredholm, E. I., (1900). Sur une nouvelle méthode pour la résolution du problème de Dirichlet. *Stockh. Öfv., 57*, 39–46.

Fredholm, E. I., (1903). Sur uneclassed'equationsfonctionnelles, *Acta Math., 27*, 365–390.

Gauden, P. A., Kowalczyk, P., & Terzyk, A. P., (2003). Toward solving the unstable linear Fredholm equation of the first kind: A new procedure called the adsorption stochastic algorithm (ASA) and its properties. *Langmuir, 19*, 4253–4268.

Hansen, J., Hogue, J., Sander, G., Renaut, R. A., & Popa, S. C., (2015). Non-negatively constrained least squares and parameter choice by the residual periodogram for the inversion of electrochemical impedance spectroscopy, *Journal of Computational and Applied Mathematics, 278*, 52–74.

Jerri, A. J., (1999). *Introduction to Integral Equations with Applications.* John Wiley & Sons, Inc., New York.

Lopes, V. V., Rangel, C. M., & Novais, A. Q., (2013). Modeling and identification of the dominant phenomena in hydrogen fuel-cells by the application of DRT analysis. *Computer-Aided Chemical Engineering, 32*, 283–288.

Mirkin, M. V., & Bard A. J., (1992a). Multidimensional integral equations. Part. 1. A new approach to solving microelectrode diffusion problems, *J. Electroanal. Chem., 323*, 1–27.

Mirkin, M. V., & Bard A. J., (1992b). Multidimensional integral equations. Part. 2. Applications to microband electrodes and the scanning electrochemical microscope. *J. Electroanal. Chem., 323*, 29–51.

Molabahrami, A., (2013). Integral mean value method for solving a general nonlinear Fredholm integrodifferential equation under the mixed conditions. *Communications in Numerical Analysis*, 1–15, doi: 10.5899/2013/cna-00146.

Philips, D. S., (1962). A technique for the numerical solution of certain integral equation of the first kind. *Journal of the Association for Computing Machinery, 9*, 84–96.

Rahbar, S., & Hashemizadeh, E., (2008). A computational approach to the Fredholm integral equation of the second kind. In: *Proceedings of the World Congress on Engineering*. London, U.K.

Rahman, M., (2007). *Integral Equations and their Applications*, WIT Press, Southampton, UK.

Ramm, A. G., (2001). A simple proof of the Fredholm alternative and a characterization of the Fredholm operators. *American Mathematical Monthly, 108*, 855–860.

Saccoccio, M., Wan, T. H., Chen, C., & Ciucci, F., (2014). Optimal regularization in distribution of relaxation times applied to electrochemical impedance spectroscopy: Ridge and lasso regression methods–a theoretical and experimental study. *Electrochimica Acta, 147*, 470–482.

Schmuck, M., & Bazant, M. Z., (2015). Homogenization of the Poisson–Nernst–Planck equations for ion transport in charged porous media. *J. Appl. Math., 75*, 1369–1401.

Shestopalov, Y. V., & Smirnov, Y. G., (2002). *Integral Equations. A compendium.* Karlstad University, Division for Engineering Science, Physics and Mathematics.

Tikhonov, A. N., (1963a). On the solution of incorrectly posed problem and the method of regularization, *Soviet Mathematics, 4*, 1305–1308.

Tikhonov, A. N., (1963b). Regularization of incorrectly posed problems. *Soviet Mathematics, 4*, 1624 1627.

Velasquez-Arcos, J. M., Grandos-Samaniego, J., & Vargas, C. A., (2013). The confinement of electromagnetic waves and Fredholm's alternative. *Electromagnetics in Advanced Applications*.

Weese, J., (1992). A reliable and fast method for the solution of Fredholm integral equations of the first kind based on Tikhonov regularization. *Computer Physics Communications, 69,* 99–111.

Zemyan, S. M., (2012). *The Classical Theory of Integral Equations: A Concise Treatment,* doi: 10.1007/978-0-8176-8349-8 2, Springer Science & Business Media, Basel.

Ziyaee, F., & Tari, A., (2014). Regularization method for the two-dimensional Fredholm integral equations of the first kind. *International Journal of Nonlinear Science, 18,* 189–194.

CHAPTER 19

Hydrogen—The First Atom

CYNTHIA WHITNEY

Galilean Electrodynamics, 11660 239th Ave. NE, Redmond,
WA 98053-5613, USA, E-mail: Galilean_Electrodynamics@Comcast.net

19.1 DEFINITION

The hydrogen atom is the gateway atom for the development of the quantum mechanics (QM) of all atoms, and hence all molecules, and hence all nano-structures. In the early twentieth century, the understanding for the hydrogen atom taken from classical physics (CP) clearly did not work; indeed, it indicated that the hydrogen atom ought to self-destruct. That was the conclusion because the orbiting electron should, by its on-going acceleration, create radiation, and so lose orbit energy, and so spiral into the nucleus. Therefore, a different development path was taken. Planck's constant was known to form various experimental situations, and it was incorporated into the founding Postulates from which QM was developed. But in retrospect, that the conclusion about classical physics was drawn, and that development path for QM was taken, largely because certain mathematical and engineering analysis developments of the mid-twentieth century were not yet in place. We did not yet have information theory (IT), or the practical techniques of statistical communication theory (SCT). If we now put those tools in place, we can see how the observed phenomena of atoms can be better understood, and how further developments in areas such as nanotechnology may be achieved more easily.

19.2 HISTORICAL ORIGINS

Like SRT, QM also embedded a postulate about signal speed. But in QM, the signal speed was not declared to be the number c; instead, it was presumed to tend to infinity. However, since QM directs our attention away from specific trajectories of individual particles, and toward the statistics of populations of particles, that problem does not matter so much to QM *per se*.

The one place where QM is seriously affected by the signal speed is in the foundational puzzle about the immortality of the hydrogen atom: Why does it not radiate to death? The solution to this particular problem sets the stage for the study of all atoms, all ionization potentials, and hence for algebraic chemistry.

The origin of the problem is the radiation phenomenon. From classical physics (CP) [*See*, for example, Jackson (1975)], we have the classical Larmor formula for radiation power generated by a particle with charge e and mass m_e, accelerating at rate a_e:

$$P_{radiation} = \left(2e^2/3c^3\right)(a_e)^2 \tag{1}$$

For an electron orbiting a stationary proton at the radius r_e, the acceleration $a = e^2/m_e(r_e)^2$, so the radiation power output is:

$$P_{radiation} = 2e^6/3c^3(m_e)^2(r_e)^4 \tag{2}$$

This is the only mechanism acknowledged in either CP or in special relativity theory (SRT), so it implies that the hydrogen atom ought to self-destruct. This problem was the big motivation for the early development of the QM or atoms, using as a postulate the fact that the hydrogen atom does not self-destruct.

19.3   NANO-SCIENTIFIC DEVELOPMENTS

A more conservative analysis of the hydrogen atom just identifies the approximations that caused the classical analysis to go badly, and removes those approximations. The first of the approximations is that the mass of the proton, m_p, is essentially infinite, in which case the orbit radius of the proton, r_p, is essentially zero, and the speed of the proton, v_p, is essentially zero. Removing those assumptions, the radiation power becomes:

$$P_{radiation} = (2e^2/3c^3)\left[(a_e)^2 + (a_p)^2\right] = \frac{2e^6}{3c^3(r_e + r_p)^4}\left[1/(m_e)^2 + 1/(m_p)^2\right] \tag{3}$$

Clearly, using the real m_p and the real r_p does not change the radiation P_{rad} very much. The numerically significant feature was, and remains, the power 4 of its decline with separation. It is a steep decline. The conceptually significant feature is: there are two particles moving. Both particles orbit the center of mass of the hydrogen atom system.

With two particles moving, a second physical process transpires. The new physical process is caused by the finite propagation delay of the signals exchanged between the electron and the proton. The delay causes tiny tangential components of force on each of the orbiting particles, of order $F = e^2 / (r_e + r_p)^2$. The atomic system overall then has a tiny torque on it, and that constitutes a source of power input to the system. Its magnitude is approximately

$$P_{\text{tourque}} = (e^4 / m_p c) / (r_e + r_p)^3 \qquad (4)$$

Here m_p is the mass of the proton, which is about 1800 times larger than the mass of the electron, and r_p is the orbit radius of the proton, which is about 1800 times smaller than the orbit radius of the electron, r_e. Observe that the P_{torque} declines with power 3 of separation $r_e + r_p$. So it crosses the P_{rad} line at some value of separation. That means the hydrogen atom can have a balance between energy loss and energy gain.

At the experimentally estimated value of $r_e \approx r_e + r_p$, the torquing power (3) is quite large compared to the radiation power (2). This discrepancy suggests that there exists yet another phenomenon involved in the problem of the hydrogen atom. This third phenomenon is something very interesting that wasn't discovered until 1927, long after the 1905 launch of SRT, and well into the development of QM. It is the natural rotation of systems driven in a circular orbit that was discovered by Thomas (1927).

The context for the discovery of Thomas rotation was an investigation of the sequential application of non-co-linear Lorentz transformations (LT's). The purpose was to use LT's, not just in the historic passive sense, for relating different observers, but also in a new active sense, for modeling on-going physical forces.

Because of its context in the study of LT's, the fact of accumulated Thomas rotation was long understood as a consequence of SRT. But actually, it has no special connection to SRT. Thomas rotation also emerges with sequential non-co-linear Galilean transformations, or any other kind of sequential non-co-linear transformations. It is a fact of Geometry, not of Physics.

The fact is this: when a system is driven around a circle at an angular rate Ω, the system itself will rotate internally at the angular rate $\Omega/2$. Or conversely, when a system internally drives itself around a circle at rate Ω, the outside world sees effects like radiation power delivery compatible with rotation at rate 2Ω.

The circumstance in the hydrogen atom is this: the big difference between orbit radii r_e and r_p means that the electron can exert a much larger tangential force on the proton than the proton can exert on the electron. The tangential forces do not balance, and radial forces must not balance either. Without instantaneous forces, the atom is just not a Newtonian system. Its center of mass is not necessarily stationary. Indeed, its center of mass is driven around a circle.

The consequence is this: more radiation delivered to the outside world:

$$P_{radiation} \rightarrow P_{enhanced\ radiation} = 2^4 \times 2e^6/3c^3 m_e^2 r_e^4 = 2^5 \times e^6/3c^3\ m_e^2 r_e^4 \qquad (5)$$

The new factor of $2^5 = 32$ makes the radiation much more significant.

Now, consider the possible balance between $P_{tourque}$ and $P_{enhanced\ radiation}$. The value of $r_e + r_p \approx r_e$ for which is:

$$r_e + r_p \approx r_e \approx 32 m_p e^2 / 3 m_c^2 c^2 \approx 5.5 \times 10^{-9}\ \text{cm} \qquad (6)$$

This result is quite close to the presently accepted value for the first orbit radius of the hydrogen atom, which is 5.29×10^{-9} cm. So this result is promising.

19.4 NANO-CHEMICAL APPLICATIONS

The basic approach to the hydrogen atom that is illustrated above can be characterized as follows: take account of all processes that can be identified, and look for a natural balance between them. There exist other research areas awaiting further application of this approach. They include explaining the observed behaviors of ionization potentials (*see* Chapter 26), visualizing interactions among electrons within atoms, and among atoms in molecules and nanostructures, and providing a new interpretation of some facts from spectroscopy.

19.4.1 *IONIZATION POTENTIALS*

The next element after hydrogen is helium, and already there is something to puzzle about. The ionization potential of helium is very high–the highest among all elements. It invites the interpretation that electrons very much like to form a pair. Electrons form pairs despite their natural repulsion due to having the same charge. Presently, the behavior is explained in terms of the concept of electron spin. But an additional explanation could help with elements beyond helium.

19.4.2   INTERACTIONS BETWEEN MORE THAN TWO ELECTRONS

Electron pairs are only the beginning. Every peak that occurs in the empirical plot of IPs indicates the end of some natural family of electrons. They come in families of 3, 5, and 7. If more elements can be synthesized in the future, we may see families of 9.

19.4.3   SPECTROSCOPY

Hydrogen atoms, and all other atoms as well, are presently envisioned as having infinitely many 'excited states' that are described via the multiple solutions of Schrödinger's differential equation for the atom. These excited states are presently understood as conditions that an individual atom possesses. That may not be the most straightforward way to think about excitation. The term 'excited state' may more aptly describe a situation where several atoms are involved, not in making a molecule, but rather in making a temporary structure that holds energy in its physical arrangement, and upon the collapse of that arrangement, dumps the energy in the form of photons.

One common factor that all these situations allow is the glue that electromagnetic signals can provide. Obviously, such signals can make a positive and a negative charge stay together. But there is much more to say. The author have advocated a signal model that modernizes what was used at the foundation of special relativity theory (SRT) (*see*, signals between ions (SBI)). That means the author also advocate against the speed limit c that comes from SRT. Superluminal speed is not impossible, and may be quite a common inside of atoms. Within an electron family of 2, 3, etc., the speeds could be superluminal. That would allow signals to arrive 'out of phase,' such that repulsion would change sign and become an attraction, and thereby keep the family together.

KEYWORDS

- **information theory**
- **periodic arch**
- **quantum mechanics**
- **special relativity theory**

REFERENCES AND FURTHER READING

Jackson, J. D., (1975). *Classical Electrodynamics* (2nd edn., p. 659). John Wiley & Sons: New York, USA.

Thomas, L. H., (1927). The kinematics of an electron with an axis. *Phil. Mag. Series 7, 3*(13), 1–22.

Whitney, C. K., (2013). *Algebraic Chemistry* (p. 659). Nova Science Publishers: New York, USA.

CHAPTER 20

Helium

CYNTHIA WHITNEY

Galilean Electrodynamics, 11660 239th Ave. NE, Redmond,
WA 98053-5613, USA, E-mail: Galilean_Electrodynamics@Comcast.net

20.1 DEFINITION

Helium is the first multi-electron atom, the first noble gas, and the first element with truly amazing properties. It is the first, and most noble, of the noble gasses. It has the highest first ionization potential of any element. It has a second ionization potential that is higher than one might expect, based just on the first ionization potential of hydrogen (*see* Chapter 26). In the liquid state, helium exhibits amazing behaviors: climbing up walls, spilling over edges, and escaping containment. This behavior suggests that helium forms a sheet-like macro-structure. Such a structure would be similar to the so-called 'exclusion zone' water (Pollack, 2013). The latter subject is discussed in the volume entry for ionic configurations (*see* Chapter 25). Let us here do the same for helium.

20.2 HISTORICAL ORIGIN(S)

The extraordinary behavior of helium in the liquid state has been a topic of study for decades. There hasn't been a simple explanation for it. A study of helium from the electrical viewpoint looks very useful for this purpose.

20.3 NANO-SCIENTIFIC DEVELOPMENT(S)

A very short list of model data is involved in the analysis of helium. We need just:

Hydrogen: $Z = 1$, $M = 1.008$; model $IP_{1,1} = 14.250$ eV, $\Delta IP_{1,1} = 0$ eV

Helium: $Z = 2$, $M = 4.003$; model $IP_{1,2} = 49.875$ eV, $\Delta IP_{1,2} = 35.625$ eV

Lithium: $Z = 3$, $M = 6.941$; model $IP_{1,3} = 12.469$ eV, $\Delta IP_{1,3} = 1.781$ eV

Beryllium: $Z = 4$, $M = 9.012$; model $IP_{1,4} = 23.327$ eV, $\Delta IP_{1,4} = 9.077$ eV

The formulae involved are developed in general in the volume entry on algebraic chemistry (AC). Eq. (5) evaluated for $Z = 2$, and with $\Delta IP_{1,1} = 0$, gives the energy requirement:

$$_2\text{He} \rightarrow {_2\text{He}^+} : \left(IP_{1,1} \times 2 + \Delta IP_{1,2} \times 2 \right)/M_2 . \tag{1}$$

Similarly, Eq. (9), with, $\Delta IP_{1,0} = 0$, gives the energy requirement

$$_2\text{He} \rightarrow {_2\text{He}^{2+}} : \left[IP_{1,1}\left(2 + \sqrt{2 \times 1} \right) + \Delta IP_{1,2} \times 2 \right]/M_2 . \tag{2}$$

Next, Eq. (13) gives the energy requirement

$$_2\text{He} \rightarrow {_2\text{He}^-} : -\left(IP_{1,1}\sqrt{2 \times 3} \right)/M_2 + \left(\Delta IP_{1,2} \times 2 - \Delta IP_{1,3} \times 3 \right)/M_2 . \tag{3}$$

and Eq. (16) gives the energy requirement

$$_2\text{He} \rightarrow {_2\text{He}^{2-}} : -\left[IP_{1,1} \times \left(\sqrt{2 \times 3} + \sqrt{2 \times 4} \right) \right]/M_2 + \left(\Delta IP_{1,2} \times 2 - \Delta IP_{1,4} \times 4 \right)/M_2 . \tag{4}$$

For numerical evaluations, we have:

$$\begin{aligned}
&_2\text{He} \rightarrow {_2\text{He}^+} : \\
&\left(14.250 \times 2 + 35.625 \times 2 \right)/4.003 = \\
&\left(28.5 + 71.250 \right)/4.003 = 24.919 \text{ eV}
\end{aligned} \tag{5}$$

$$\begin{aligned}
&_2\text{He} \rightarrow {_2\text{He}^{2+}} : \\
&\left(14.250 \times 3.4142 + 35.625 \times 2 \right)/4.003 = \\
&\left(48.6524 + 71.25 \right)/4.003 = 29.9532 \text{ eV}
\end{aligned} \tag{6}$$

$$\begin{aligned}
&_2\text{He} \rightarrow {_2\text{He}^-} : \\
&-14.250 \times 2.4495/4.003 + \left(35.625 \times 2 + 1.781 \times 3 \right)/4.003 = \\
&-34.9054/4.003 + \left(71.250 + 5.345 \right)/4.003 = \\
&41.6876/4.003 = 10.4141 \text{ eV}
\end{aligned} \tag{7}$$

$$\begin{aligned}
&_2\text{He} \rightarrow {_2\text{He}^{2-}} : \\
&-14.250 \times \left(2.4495 + 2.8284 \right)/4.003 + \left(35.625 \times 2 - 9.077 \times 4 \right)/4.003 = \\
&-18.788 + \left(71.25 - 36.308 \right)/4.003 = \\
&-18.788 + 8.729 = -10.059 \text{ eV}
\end{aligned} \tag{8}$$

Most of these energy requirements for making helium ions are positive. The one that is negative is not negative enough to balance with any combination of positive ones. This all means that Helium atoms resist making any ions with balancing changes of electron counts. That is a reason why Helium is the noblest of all noble gasses. It is also a reason why liquid Helium tends to spread out: it likes to form a mono-atomic plane, rather than a three-dimensional volume, because a plane gives it just three Helium neighbors with which to avoid any interaction, whereas a volume would present even more of the tiresome neighbors to avoid.

20.4 MULTI-/TRANS-DISCIPLINARY CONNECTION(S)

People study liquid helium because it presents puzzles for our understanding of physics, especially in the area of thermodynamics (*see* Tisza, 1966). Within thermodynamics, the most central and mysterious concept is that of entropy (*see* Chapters 11 and 12).

20.5 OPEN ISSUES

Liquid helium has two named sub-states, called helium I and helium II. Helium I is not very conspicuously different from ordinary helium gas, since it has low refraction, viscosity, and density, making its surface hard to notice. Helium II is the one with the interesting behavior: climbing container walls and such.

What makes the difference between these two sub-states of liquid helium? Might it have to do with the population, if any, of ions? That is to say, Helium I might have a few ions within it, whereas helium II might have none at all.

KEYWORDS

- **algebraic chemistry**
- **ionic configurations**
- **ionization potentials**

REFERENCES AND FURTHER READING

Pollack, G., (2013). *The Fourth Phase of Water, Beyond Solid Liquid Vapor*. Chapter 4, Ebner & Sons, Seattle, USA.

Tisza, L., (1966). *Generalized Thermodynamics*. Especially Paper 5, The M.I.T. Press: Cambridge, MA, USA.

Whitney, C., (2013). *Algebraic Chemistry Applications and Origins*. Chapter 5, Nova Publishers: New York, USA.

Whitney, C., (2014). EZ water: A new quantitative approach applied–many numerical results obtained. *Water, 5*, 105–120.

CHAPTER 21

Helical Symmetry

ATTILA BENDE

Molecular and Biomolecular Physics Department, National Institute
for Research and Development of Isotopic and Molecular Technologies,
Donat Street, No. 67-103, RO-400293 Cluj-Napoca, Romania,
Tel: 0040-264-584037, Fax: 0040-264-420042, E-mail: bende@itim-cj.ro

21.1 DEFINITION

The helical symmetry operation is defined as a simultaneous translation and rotation of a given molecular system (considered as a unit cell) along the same symmetry axis. It can be used as a special periodic boundary condition for one-dimensional infinite long polymer chain. Its advantage is that instead of a long translational cell which contains a molecular fragment which covers to the whole rotation (rotation with 360°) one can use a much smaller unit cell which corresponds for a rotation with a certain angle. This means that, for example, in the case of the DNA polymer chain, instead of a unit cell with ten base pairs one can consider only a single base pair. It can be demonstrated that in a one-dimensional periodic system the symmetry group of a combined symmetry operation is isomorphic with the group of translation and therefore the Hamiltonian of a system is invariant under the applied helical symmetry operation. It can be applied in the case of one-dimensional polymer chains, like single- or double-strand DNA, helical (3-10, α or π helix) proteins or carbon nanotubes.

21.2 HISTORICAL ORIGIN(S)

It was always a need for understanding how small "building blocks" can make up larger complex structures and show new collective effects. Several theoretical models and experimental techniques were developed in order to find the relationship between the macroscopic properties and microscopic picture of matter. One of the possibilities is when we take into account the symmetry property of "building blocks." In some special cases, these properties are not simple translation-like operations, but more complicated process of translations and rotations called helical operations. Considering the helical symmetry operations one can build one-dimensional infinite long polymer chains with special electronic and mechanical properties. Several helical polymers are of central importance to biological systems. The famous double helix of the DNA stores our genetic code; the alpha-helix is one of the most common secondary structures in proteins, while some biologically relevant polysaccharides can also take different helical shapes. Accordingly, there is a need for an adequate theoretical framework which is able to describe in details the electronic structure of such molecular materials.

21.3 NANO-SCIENTIFIC DEVELOPMENT(S)

The functionality of various hierarchically structured materials, like solids, biomolecules or supramolecular assemblies is assured by the particular geometry arrangement of their building blocks (Chung et al., 2011). They possess in many cases highly ordered microscopic structures, like crystals (Hoddeson et al., 1992), co-crystals (Braga et al., 2009), dendrimers (Buhleier et al., 1978; Yates et al., 2005) or coiled peptide chains (Pauling et al., 1951) which often exhibit periodic structures by repeating itself in different directions in the space. The discovery in 1953 of the double helix structure of the DNA by James Watson and Francis Crick (Watson et al., 1953) is considered as the starting point of the modern molecular biology. The functionality of biomolecular structures is given by their three-dimensional shapes, therefore knowing their specific physicochemical properties we can elucidate how these specific functionalities are related to the structure of biomolecules.

21.3.1 ELECTRONIC STRUCTURE OF THE ONE-DIMENSIONAL INFINITE HELICAL CHAIN

The ordered arrangement of atoms, ions or molecules builds up macroscopic systems like polymers, solids or crystalline liquids. The structure of such kind of material can be described in terms of its unit cell, which repeating in different directions generates the whole crystal structure. Assuming that the nuclei are stationary with respect to the electron motions (Born-Oppenheimer approximation) the electronic structure of the system can be described by the TISE. Applying the variational principle for the TISE and considering the molecular orbital theory approximation (linear combination of the atomic orbitals, LCAO) one can derive a set of N-coupled equations for the N spin-orbitals, called the Hartree-Fock equation (Szabo et al., 1996):

$$\mathbf{FC} = \varepsilon \mathbf{SC} \tag{1}$$

where $\mathbf{F}$ is the Fock operator, $\mathbf{C}$ the matrix of the orbital LCAO coefficients, $\mathbf{S}$ the overlap matrix and ε the orbital energies.

Introduce Born-von Karman periodic boundary conditions, the whole F and S matrixes can be decomposed in the block form, where the blocks are defined by the unit cells:

$$
\begin{bmatrix}
\begin{array}{|c|c|c|}\hline 0 & 1 & 2 \\\hline\end{array} & \cdots & \begin{array}{|c|}\hline 2N \\\hline\end{array} \\
\begin{array}{|c|c|c|c|}\hline -1 & 0 & 1 & 2 \\\hline\end{array} & \cdots & \begin{array}{|c|}\hline 2N\text{-}1 \\\hline\end{array} \\
\vdots & & \\
\begin{array}{|c|}\hline -2N \\\hline\end{array} \cdots & \begin{array}{|c|c|}\hline -1 & 0 \\\hline\end{array}
\end{bmatrix} \tag{2}
$$

Making some basic matrix operation, like block-diagonalization and unitary transformation as well as moving to the reciprocal (or momentum) space defined by the k lattice vector, one can write the Fock operator form explicitly expressed with the help of the k-vector (Ladik et al., 1999):

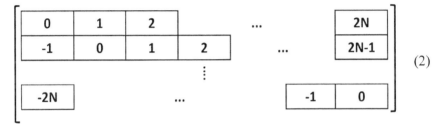

$$
\begin{aligned}
F_{k,l}^{i,j} =\ & \langle \chi_k^i \left| -\frac{1}{2}\Delta - \sum_{q_1}\sum_{\alpha=1}^{M} \frac{Z_\alpha}{|\mathbf{r}-\mathbf{R}_\alpha^{q_1}|} \right| \chi_l^j \rangle + \\
& + \sum_{\substack{q_1,q_2=\\=-N^3}}^{N^3}\sum_{u,v=1}^{m} P_{u,v}(q_1-q_2)\, \langle \chi_k^i(\mathbf{r}_1)\chi_u^{q_1}(\mathbf{r}_2)\left| \frac{1}{r_{12}}\left(1-\frac{\hat{P}_{1\leftrightarrow2}}{2}\right)\right| \chi_l^j(\mathbf{r}_1)\chi_v^{q_2}(\mathbf{r}_2) \rangle
\end{aligned} \tag{3}
$$

where

$$
P_{u,v}(q_1-q_2) = \frac{1}{\omega}\sum_{i=1}^{n^*}\int_\omega 2d(\boldsymbol{k})_{i,u}\, d(\boldsymbol{k})_{i,v} \times \exp\left(i\boldsymbol{k}\cdot\left(\mathbf{R}_{q_1}-\mathbf{R}_{q_2}\right)\right) d\boldsymbol{k} \tag{4}
$$

Eqs. (1), (3), and (4) define the so-called HFCO method where the analytical form of the crystal-orbital can be written

$$\psi_k^i(\mathbf{r}) = \sum_{l=1}^{m} \sum_{q} \exp\left(ik\mathbf{R}_q\right) d(k)_{i,l} \, \chi_l^q \tag{5}$$

as a linear combination of the Bloch functions.

Since **F** is a non-linear matrix operator (it depends on its own solution through the orbitals), Eq. (1) can be solved iteratively using the so-called self-consistent-field procedure.

Eq. (1) can be solved for any translational symmetry case, but it was shown (André et al., 1967; Blumen et al., 1977; Del Re et al., 1967) that the theory is also valid for the helical symmetry case (translation in one direction and rotation along this translation axis). The helix operator in its general form can be written as:

$$\hat{S}(\alpha,\tau) = \hat{D}(\alpha) + \tau \tag{6}$$

where the operator $\dot{D}(\alpha)$ performs a rotation around the helix axis by an angle α and τ means a translation with $|\tau|$ along the helix axis (*see* Figure 21.1).

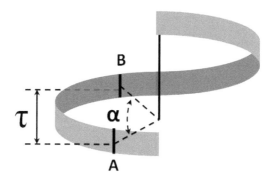

FIGURE 21.1   The schematic figure of the helical operation.

Considering the helical symmetry defined by Eq. (6) the analytical form of the crystal orbitals can be written as:

$$\psi_k^i(\mathbf{r}) = \frac{1}{(2N+1)^{1/2}} \sum_{q=-N}^{N} \sum_{i=1}^{m} \exp\left(\frac{2\pi ikq}{2N+1}\right) \times d(k)_{k,i} \, \chi_i \left\{ \hat{D}(q\alpha) \left[\mathbf{r} - \hat{S}^{-q}\left(\mathbf{R}_q^{S_A}\right) \right] \right\} \tag{7}$$

The helical operator can be used as the periodic boundary condition for the Hartree-Fock equitation defined by Eq. (1) where the block matrix given by the scheme (2) will have the block configuration characteristic for this helical-type symmetry. After several matrix operations and applying the self-consistent-field procedure, one can easily obtain the band structure at each *k*-point of the first Brillouin zone.

21.4 NANO-CHEMICAL APPLICATION(S)

21.4.1 HELICAL STRUCTURE OF DNA

The DNA double helix is not only the carrier of the original genetic code but also the protector of it. Compact packing of DNA is essential to improve the efficiency of gene delivery, achieved through the winding of the DNA double helix into a higher order helix form (Luger et al., 1997; Richmond et al., 2003); thus it occupies a smaller footprint and blocks the possibility of copying the genetic code in an uncontrolled manner. Its structure consists in 147 base pair long B-DNA sequence which is wrapped around the nucleosome core particle built by eight histone proteins (*see* Figure 21.2).

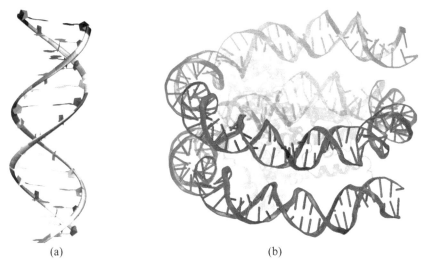

(a)   (b)

FIGURE 21.2   The B-DNA double helix in a one-dimensional infinite chain (a) and the B-DNA superhelix wrapped around the nucleosome core particle (b).

Using the HFCO method, defined by the Eqs. (3) and (4), the band structure (*see* Figure 21.3) of an infinite long double-stranded B-DNA chain has been calculated for the four homopolynucleotide sequence cases (Ladik et al., 2007, 2008).

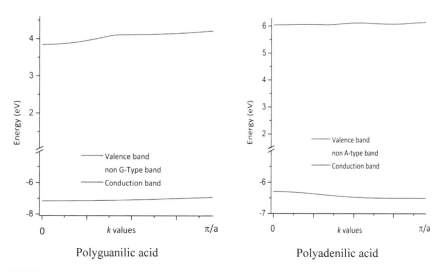

Polyguanilic acid   Polyadenilic acid

FIGURE 21.3   The dispersion curves of the valence and conduction bands as well as the impurity bands (non-base-type bands) in the fundamental gap of the polyguanilic and polyadenylic acids in the presence of water and Na+ ions (Reprinted with permission from Ladik et al., 2008).

The DNA chain is left-handed and has 1.67 superhelical turns inside the nucleosome core particle. Furthermore, the superhelix is strongly screwed around the nucleohistone protein fixed by several salt-bridges (Schalchet al., 2005) between the negatively charged PO_4 groups of the DNA backbone and the positively charged (protonated) lysine and arginine amino acids of the protein chains. It was shown (Ladiket al., 1960) that in the case DNA base pairs, the one-dimensional helix symmetry can induce similar structures of electronic bands as in the case of a three-dimensional solid with translational symmetry. The semiconductor behavior of a double-stranded DNA chain was for the first time experimentally proven by Eley & Spivey (1962). Since then, a huge number of scientific papers have been published considering both theoretical and experimental techniques.

The most complex theoretical studies consist in large scale DNA band structure calculations based on quantum mechanical theories were done by Endres and Ladik (Endres et al., 2004; Ladik et al., 2008). The subject of the conduction behavior of the DNA chain is still under debate. There are different opinions about whether the conduction mechanism is a long-range coherent charge transport or the thermally induced hopping phenomena.

21.4.2 HELICAL STRUCTURE OF PROTEINS

The biochemical process through a protein structure attains its functional shape or conformation is called protein folding (Dill et al., 2012). The most plausible explanation is the distance-dependence behavior of the conduction, where in the case of shorter homogeneous base sequences the charge transport would be a coherent process, while at longer distances over the aperiodic base sequences the thermally induced hopping is more likely (Giese et al., 2001; Genereux et al., 2011). On the other hand, this special electron structure of the DNA chain could play an important role in detecting DNA damages (Sontz et al., 2012) or could be used as a very efficient spin filter at room temperature for spintronic applications (Göhler et al., 2011). This process is mainly guided by hydrophobic interactions, the formation of intramolecular hydrogen bonds, and van der Waals forces, and it is opposed by conformational entropy (Miller et al., 2014). As a result of this protein folding the most abundant class of secondary structures is the helical configuration, especially the so-called α-helix. The helical structure of the polypeptide chains was for the first time experimentally observed by Pauling et al. (1951), and different forms of the helical configuration can be seen in Figure 21.4.

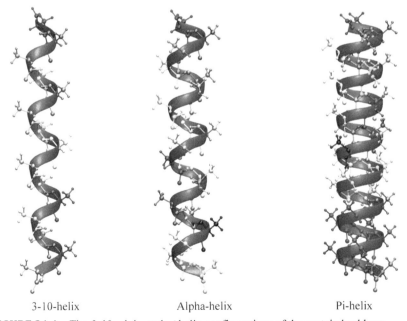

| 3-10-helix | Alpha-helix | Pi-helix |

FIGURE 21.4 The *3-10, alpha* and *pi* helix configurations of the protein backbone.

In the case of proteins, the helical structure provides mainly the hardness of the protein structures. Their electronic structures do not change significantly in the sense that due to the saturated chemical bonds from the polypeptide chain there are no any electronic bands formed as the overlap of the atomic orbitals of the neighboring unit cells. On the other hand, the helical symmetry also has the role of the right orientation of the amino acid side chains which will determine later the tertiary structure of a protein.

21.4.3   HELICAL STRUCTURES OF SINGLE-WALL NANOTUBES

Due to their extraordinary thermal, mechanical and electrical properties, cylindrical nanostructures, called nanotubes, are the subject of one of the most important areas of research in nanotechnology. A certain group of these nanotubes also show helical symmetry (*see* Figure 21.5) and their electronic structures were studied in details.

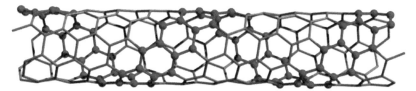

FIGURE 21.5   The helical structure the chiral nanotubes.

Accordingly, the chiral nanotubes based on carbon were investigated by White and coworkers (1993), those built by silicon atoms by Baňacký et al. (2013) and the bit more complicated structure of TiS_2 by Teich et al. (2011). As a general conclusion, it was found, that in all these cases the electronic band structure does not explicitly depend on helical parameter but on the structure of the unit cell of the system. More precisely, it depended on the diameter of the nanotube and based on this value we have systems with metallic or small-gap semiconductor-like behavior.

This is not the case when one deals with the more complicated structure of the helical carbon nanocoils or superhelices (Amelinckx et al., 1994) when instead of a linear chain configuration we have a helical turn of the nanotube similar to the DNA turns in the nucleosome (Figure 21.2b). Analyzing the electronic structure of these carbon nanocoils (Akagi et al., 1995) a semimetal feature was often observed, which was not found for the case of straight carbon nanotubes. Furthermore, it was demonstrated (Ben et al., 2015) that these kinds of helical nanosprings show huge nonlinearities and the Hooke's law for the elastic range may not be correct for some special cases of them.

The model can also be used in cases of nanospirals (Avdoshenko et al., 2013), peptide nucleic acid (Nielsen et al., 1991), and orhelicenes (Collins et al., 2006).

First principle investigations using the special PBC of the helical symmetry can be used for studying electronic band structures (Bogár et al., 2001, 2004; Ferrari et al., 2010), charge carrier mobilities and charge conductivity as well as elastic properties of an infinite long 1D polymer chain (Bende et al., 2013; Bogár et al., 2014).

The helical PBC software module is implemented in the commercially available CRYSTAL software package (Dovesi et al., 2014) and in some "home-made" softwares developed by F. Bogár(University of Szeged) or by J. W. Mintmire (Oklahoma State University) (Mintmire et al., 1991).

21.5   MULTI-/TRANS-DISCIPLINARY CONNECTION(S)

First principle investigations using the special PBC of the helical symmetry has a great deal of applications in fields of condensed matter physics, polymer science or theoretical biophysics. Using this method one can study, for example, DNA conductibility, polymer chain stability or electrical properties of different materials.

21.6   OPEN ISSUES

One of the major limits of the concept is that it can be applied for systems build up by homogeneous blocks of the unit cell. In this sense, solutions given by the present model are a bit far from the real case of the DNA base sequence where the four base types follow each other without any property of the periodicity. On the other hand, it seems highly probable (Bogár et al., 2014), that only the homogenous or quasi-homogenous DNA base sequences show band structure-like electronic configurations. Moreover, excepting the DNA-like helical structures with strong $\pi-\pi$ interaction between the stacks, there are not many cases where the electronic configuration of the system show band structure form and this fact makes some limitation of the applicability of the present model.

KEYWORDS

- **one-dimensional polymer chain**
- **periodic boundary condition**
- **rotation**
- **translation**

REFERENCES AND FURTHER READING

Akagi, K., Tamura, R., Tsukada, M., Itoh, S., & Ihara, S., (1995). Electronic structure of helically coiled cage of graphitic carbon. *Phys. Rev. Lett., 74*, 2307–2310.

Amelinckx, S., Zhang, X. B., Bernaerts, D., Zhang, X. F., Ivanov, V., & Nagy, J. B., (1994). A formation mechanism for catalytically grown helix-shaped graphite nanotubes. *Science, 265*, 635–639.

André, J. M., Gouverneur, L., & Leroy, G., (1967). L'Etude Théorique des Systèmes Périodiques. I. La Méthode LCAO–HCO. *Int. J. Quant. Chem., 1*, 427–450.

André, J. M., Gouverneur, L., & Leroy, G., (1967). L'Étude Théorique des Systèmes Périodiques. II. La Méthode LCAO–SCF–CO. *Int. J. Quant. Chem., 1*, 451–461.

Avdoshenko, S. M., Koskinen, P., Sevinçli, H., Popov, A. A., & Rocha, C. G., (2013). Topological signatures in the electronic structure of graphene spirals. *Nature Sci. Rep., 3*, 1632.

Baňacký, P., Noga, J., & Szöcs, V., (2013). Electronic structure of single-wall silicon nanotubes and silicon nanoribbons: Helical symmetry treatment and effect of dimensionality. *Adv. Cond. Matter. Phys.*, 374371.

Ben, S., Zhao, J., & Rabczuk T., (2015). Does Hooke's law work in helical nanosprings? *Phys. Chem. Chem. Phys., 17*, 20990–20997.

Bende, A., Bogár, F., & Ladik, J., (2013). Hole mobilities of periodic models of DNA double helices in the nucleosomes at different temperatures. *Chem. Phys. Lett., 565*, 128–131.

Blumen, A., & Merkel, C., (1977). Energy band calculations on helical systems. *Phys Status Solidi, 83*, 425–431.

Bogár, F., & Ladik, J., (2004). Correlation corrected band structures of homopolypeptides. Part V. B3LYP band structures of 19 homopolypeptides. *Int. J. Quantum Chem., 99*, 47–52.

Bogár, F., Bende, A., & Ladik, J., (2014). Influence of the sequence on the *Ab initio* band structures of single and double-stranded DNA models. *Phys. Lett., A378*, 2157–2162.

Bogár, F., Van Doren, V., & Ladik, J., (2001). Correlation corrected band structures of different homopolypeptides. Part II. quasi-particle band structures of seven homopolypeptides with aliphatic side chains. *Phys. Chem. Chem. Phys., 3*, 5426–5429.

Braga, D., Grepioni, F., Maini, L., & Polito, M., (2009). Crystal polymorphism and multiple crystal forms. In: Hosseini, M. W., (ed.), *Molecular Networks, Structure and Bonding* (Vol. 132. pp. 87–95). Springer-Verlag Berlin Heidelberg.

Buhleier, E., Wehner, W., & Vögtle F., (1978). "Cascade"- and "nonskid-chain-like" syntheses of molecular cavity topologies. *Synthesis, 2*, 155–158.

Chung, W. J., Oh, J. W., Kwak, K., Lee, B. Y., Meyer, J., Wang, E., Hexemer, A., & Lee, S. W., (2011). Biomimetic self-templating supramolecular structures. *Nature, 478*, 364–368.

Collins, S. K., Grandbois, A., Vachon, M. P., & Côté, J., (2006). Preparation of helicenes through olefin metathesis. *Angew. Chem. Int. Ed., 45*(18), 2923–2926.

Del Re, G., Ladik, J., & Biczó, G., (1967). Self-consistent-field tight-binding treatment of polymers. I. infinite three-dimensional case. *Phys. Rev., 155*, 997–1003.

Dill, K. A., & MacCallum, J. L., (2012). The protein-folding problem, 50 years on. *Science, 338*(6110), 1042–1046.

Dovesi, R., Orlando, R., Erba, A., Zicovich-Wilson, C. M., Civalleri, B., Casassa, S., et al., (2014). CRYSTAL14, a program for the *Ab initio* investigation of crystalline solids. *Int. J. Quantum Chem., 114*, 1287–1317. URL: http://www.crystal.unito.it.

Eley, D. D., & Spivey, D. I., (1962). Semiconductivity of organic substances. Part 9. –nucleic acid in the dry state. *Trans. Faraday Soc., 58*, 411–415.

Endres, R. G., Cox, D. L., & Singh, R. R. P., (2004). Colloquium: The quest for high-conductance DNA. *Rev. Mod. Phys., 76*, 195–214.

Ferrari, A. M., Civalleri, B., & Dovesi, R., (2010). *Ab initio* periodic study of the conformational behavior of glycine helical homopeptides. *J. Comput. Chem., 31*, 1777–1784.

Genereux, J. C., Wuerth, S. M., & Barton J. K., (2011). Single-step charge transport through DNA over long distances. *J. Am. Chem. Soc., 133*, 3863–3868.

Giese, B., Amaudrut, J., Köhler, A. K., Spormann, M., & Wessely, S., (2001). Direct observation of hole transfer through DNA by hopping between adenine bases and by tunneling. *Nature, 412*, 318–320.

Göhler, B., Hamelbeck, V., Markus, T. Z., Kettner, M., Hanne, G. F., Vager, Z., Naaman, R., & Zacharias, H., (2011). Spin selectivity in electron transmission through self-assembled monolayers of double-stranded DNA. *Science, 331*, 894–896.

Hoddeson, L., Braun, E., Teichmann, J., & Weart, S., (1992). *Out of the Crystal Maze: Chapters from the History of Solid State Physics.* Oxford University Press.

Ladik, J. J., (1999). Polymers as solids: A quantum mechanical treatment. *Phys. Report, 313*, 171–235.

Ladik, J. J., Bende, A., & Bogár, F., (2007). Calculation of the band structure of polyguanilic acid in the presence of water and Na^+ ions. *J. Chem. Phys., 127*, 055102.

Ladik, J. J., Bende, A., & Bogár, F., (2008). The electronic structure of the four nucleotide bases in DNA, of their stacks and of their homopolynucleotides in the absence and presence of water. *J. Chem. Phys., 128*, 105101.

Ladik, J., (1960). Investigation of the electronic structure of deoxyribonucleic acid. I. *Acta Phys. Hung., 11*, 239.

Luger, K., Mäder, W., Richmond, R. K., Sargent, D. F., & Richmond, T. J., (1997). Crystal structure of the nucleosome core particle at 2. 8 Å resolution. *Nature, 389*, 251–260.

Miller, S. E., Watkins, A. M., Kallenbach, N. R., & Arora, P. S., (2014). Effects of side chains in helix nucleation differ from helix propagation. *Proc. Natl. Acad. Sci., USA, 111*(18), 6636–6641.

Mintmire, J. W., (1991). *Local-Density Functional Electronic Structure of Helical Chain Polymers* (pp. 125–137). New York. Springer-Verlag.

Nielsen, P. E., Egholm, M., Berg, R. H., & Buchardt, O., (1991). Sequence-selective recognition of DNA by strand displacement with a thymine-substituted polyamide. *Science, 254*(5037), 1497–1500.

Pauling, L., Corey, R. B., & Branson, H. R., (1951). The structure of proteins: Two hydrogen-bonded helical configurations of the polypeptide chain. *Proc. Natl. Acad. Sci. USA, 37*, 205–211.

Richmond, T. J., & Davey, C. A., (2003). The structure of DNA in the nucleosome core. *Nature, 423,* 145–150.

Schalch, T., Duda, S., Sargent, D. F., & Richmond, T. J., (2005). X-ray structure of a tetranucleosome and its implications for the chromatin fiber. *Nature, 436,* 138–141.

Sontz, P. A., Mui, T. P., Fuss, J. O., Tainer, J. A., & Barton, J. K., (2012). DNA charge transport as a first step in coordinating the detection of lesions by repair proteins. *Proc. Natl. Acad. Sci. USA, 109,* 1856–1861.

Szabo, A., & Ostlund, N. S., (1996). *Modern Quantum Chemistry.* Mineola, New York: Dover Publishing.

Teich, D., Lorenz, T., Joswig, J. O., & Seifert, G., (2011). Structural and electronic properties of helical TiS2 nanotubes studied with objective molecular dynamics. *J. Phys. Chem. C., 115,* 6392–6396.

Terrones, M., (2003). Science and technology of the twenty-first century: Synthesis, properties, and applications of carbon nanotubes. *Ann. Rev. Mater. Res., 33,* 419–501.

Watson, J. D., & Crick, F., (1953). A structure for deoxyribose nucleic acid. *Nature, 171*(4356), 737–738.

White, C. T., Robertson, D. H., & Mintmire, J. W., (1993). Helical and rotational symmetries of nanoscale graphitic tubules. *Phys. Rev. B., 47*(9), 5485–5488.

Yates, C. R., & Hayes, W., (2005). Utilization of dendritic architectures in molecular recognition and self-assembly processes. *Mini-Reviews in Organic Chemistry, 2,* 1–22.

CHAPTER 22

Hydrocarbon Burning

CYNTHIA WHITNEY

*Galilean Electrodynamics, 11660 239th Ave. NE, Redmond,
WA 98053-5613, USA, E-mail: Galilean_Electrodynamics@Comcast.net*

22.1   DEFINITION

Hydrocarbon burning (HCB) is the oxidation of any molecule that contains mainly Hydrogen and carbon. Such reactions have long been used to produce energy. Concomitantly, they also produce residual molecules, such as carbon dioxide (CO_2), carbon monoxide (CO), and water (H_2O). HCB is currently a subject of great interest because its energy production is vital to our current industrial civilization, but some of its residual waste molecules can present problems. It is to be hoped that a novel approach to the study of HCB can help and lead us to identify novel alternatives to present-day HCB. This chapter investigates the use of algebraic chemistry (AC) (*see* Chapter 1), to study and characterize HCB, and to identify candidate alternative ideas.

22.2   HISTORICAL ORIGINS

Hydrocarbon burning, is the primary source of energy for our current level of industrial civilization. But HCB presents problems. It generates carbon dioxide, CO_2, and that may affect our climate adversely. It generates carbon monoxide CO, and that can poison us. Only the water, H_2O, that it produces seems harmless. So we are certainly interested in alternatives to HCB.

But to arrive at any such discussion of alternatives, we have begun first from where we are now. Let us look at some reactions characteristic of HCB. Consider the simple hydrocarbon chain molecules. The general formula for combustion of a carbon chain molecule of length N is:

$$2\left(C_N H_{2N+2}\right)+\left[4+3(N-1)\right]\left(O_2\right) \rightarrow 2N\left(CO_2\right)+2(N+1)\left(H_2O\right) \qquad (1)$$

The simplest case is the case of $N = 1$, methane combustion:

$$2\left(CH_4\right)+4\left(O_2\right) \rightarrow 2\left(CO_2\right)+4\left(H_2O\right) \qquad (2)$$

The next case is $N = 2$, ethane combustion:

$$2\left(C_2 H_6\right)+7\left(O_2\right) \rightarrow 4\left(CO_2\right)+6\left(H_2O\right) \qquad (3)$$

And so it goes from there on, with an even number of O_2 oxygen molecules for an odd N and an odd number of O_2 oxygen molecules for an even N (For odd N, we could cancel an overall factor of 2 in (2), but for the sake of consistency between even and odd N, it is better not to do so.)

The end products of this, and other HCBs are always carbon dioxide, CO_2, and water, H_2O. The water is innocuous, but the CO_2 is a greenhouse gas. It occurs naturally, of course, but it also occurs significantly because of HCB. We worry about CO_2 as a possible cause for climate warming, ice melting, and seas rising.

What further can we learn about HCB? It can certainly be studied by any technique that addresses chemical reactions, so let us try out algebraic chemistry (AC) on HCB.

22.3 NANO-SCIENTIFIC DEVELOPMENTS

For this short essay, consider the simplest case of HCB, combustion of methane, $N = 1$ (2). Methane CH_4 allows two possible ICs. The conventional IC is $C^{4-} + 4H^+$, and the alternative IC is $C^{4+} + 4H^-$. AC allows us to compare these ICs on the basis of energy. The conventional IC, $C^{4-} + 4H^+$, has slightly positive energy: $-54.6304 + 4 \times 14.1369 = +1.9172$ eV. The alternative IC, $C^{4+} + 4H^-$, has very negative energy: $22.187 + 4 \times (-90.6769) = -340.5206$ eV.

If energy were the only consideration, then the alternative IC would dominate the methane-burning scenario. But there is another consideration as well: geometry. An H^- ion has two electrons, and that makes it much larger than an H^+ ion, which has no electrons at all, and just amounts to one lone proton.

It is geometrically difficult to make $C^{4+} + 4H^-$ fit together. So the conventional IC, $C^{4-} + 4H^+$ actually dominates this HCB scenario. (The very low-energy of the alternative IC, $C^{4+} + 4H^-$, invites interpretation as the strange form of methane that has in recent times been discovered trapped at high pressure under the sea. It is called 'methane hydrate.' If it escapes from under the sea, it can become a terrible climate disruptor. It can absorb energy from sunlight, and turn into ordinary methane $C^{4-} + 4H^+$. Then, since ordinary methane is a greenhouse gas, it can trap heat escaping from planet Earth. If methane hydrate starts to escape, we will absolutely need to capture and burn it.)

In Eq. (2) for methane combustion, the $2(CH_4)$ in the IC $C^{4-} + 4H^+$ takes

$$2 \times 1.9172 = 3.8344 \text{ eV}. \tag{4}$$

Like methane CH_4, oxygen O_2 also allows two plausible ICs: $O^+ + O^-$ and $O^{2+} + O^{2-}$. These two ICs are not very different in energy: forming $O^+ + O^-$ takes $7.9399 - 12.4354 = -4.4955$ eV, while forming $O^{2+} + O^{2-}$ takes $17.5613 - 27.3788 = -9.8175$ eV.

These two energies are so similar that both ICs must co-exist. That means the left side of (2) could take a number of different values, within a limited range. We can at best characterize the range. So in Eq. (2) for methane combustion, the $4(O_2)$ takes:

$$4 \times (-4.4955 \text{ to} - 9.8175) = -17.982 \text{ to} -39.2700 \text{ eV} \tag{5}$$

Therefore, Eq. (2) says that the methane combustion takes

$$3.8344 + (-17.982 \text{ to} -39.2700) = -14.1476 \text{ to} -35.4356 \text{ eV} \tag{6}$$

That is to say, the methane combustion can deliver $+14.1476$ to $+35.4356$ eV.

But there is more to say about the delivery of energy. The overall hydrocarbon combustion reaction (2) generally occurs in steps. For our methane example, the steps go as follows:

$$2(CH_4) + 2(O_2) \rightarrow 2(H_2CO) + 2(H_2O) \tag{7}$$

The first product listed, H_2CO, is formaldehyde. This is a little bit nasty. The second reaction product, H_2O, is harmless enough though. Next:

$$2(H_2CO) \rightarrow 2(HCO) + H_2 \tag{8}$$

The HCO is a formyl radical, which might be a little bit nasty. The H_2 is a hydrogen gas molecule. This is definitely explosive. Next:

$$2(HCO) + H_2 + O_2 \rightarrow 2(CO) + 2(H_2O) \tag{9}$$

The first product is carbon monoxide. This is poisonous. Next:

$$2(CO) + O_2 \rightarrow 2(CO_2) \tag{10}$$

The final product is carbon dioxide CO_2, the greenhouse property of which is problematic for our climate.

Wherever a reaction occurs in steps, is always possible that circumstances will occur such that one or the another step fails to complete, and some intermediate product is left unconsumed. In HCB, such a result is clearly undesirable.

22.4   NANO-CHEMICAL APPLICATION(S)

The main issues are exposed with the above analysis of methane, even though the scenarios where the issues play out involve much bigger hydrocarbon molecules, such as heptane and octane, used in present-day vehicles.

One thing we see from the methane analysis is that the value of hydro-carbon fuels lies, not so much in the positive energy that their own ICs hold, but more significantly in the negative energy of the IC of the normal water that the burning of the hydrocarbons produces. This is a potentially useful way to look at the fuel-burning situation. We can analyze alternative fuels in a different way, focusing more on the water produced in burning, and less on the fuels *per se*. Anything that can produce water when burned can be a useful fuel.

Another thing we can see from the methane analysis is that reaction steps are important. Too many reaction steps can mean too many opportunities for something toxic or dangerous to be left behind, unconsumed.

22.5   MULTI-/TRANS-DISCIPLINARY CONNECTION(S)

If the real story about HCB is the water output, and not the hydrocarbon fuel input, then carbon may not even be necessary, and so CO_2 may not be an inevitable byproduct.

Consider reviewing the periodic arch (PA) topic discussed in Chapter 35. Carbon is in a keystone position, which befits its important role in fueling our civilization. Just above carbon, in the next keystone position, is Silicon. This is the indispensable element of our computer age. What might now happen if we would cross modern carbon technology with modern silicon technology? Can we imagine a family of silicon-based fuels analogous to the carbon chain molecules?

Silicon does indeed have a series of chain molecules. They are called Silicon hydrides. Silicon hydrides are not as stable as the carbon chain molecules, but the smallest one among them, analogous to methane, is the most stable. It is called 'silane' or 'monosilane.' There also exists disilane, trisilane, tetrasilane, analogous to ethane, propane, or butane. The polysilicon hydride is the family name.

Do any of these polysilicon hydrides burn? The answer to that question may be moot, in as much as, in their instability, they liberate hydrogen gas, H_2, which most certainly will burn.

Given the story of HCB, what would the corresponding story of HSiB be? Methane CH_4 would become Silane, $Si\ H_4$, and the overall burn equation would become

$$2(SiH_4) + 4(O_2) \rightarrow 2(SiO_2) + 4(H_2O) \tag{11}$$

Observe that the product SiO_2 replaces CO_2. Unlike CO_2, SiO_2 cannot cause a greenhouse effect. It is not even gas. It is basically sand, and sand could help restore coastal areas flooded due to the greenhouse effect and resultant rise of sea levels.

Next, consider the individual combustion steps in the HCB story. Some of the other species involved change: formaldehyde H_2CO becomes H_2SiO, the formyl radical HCO becomes HSiO, and carbon monoxide CO becomes SiO. These are all less dangerous products.

Hydrogen gas H_2 does not change. It remains a potential explosion risk.

22.6 OPEN ISSUES

Some of what is discussed above are new, and awaits further research. One issue is weight—silicon-based fuels would be heavier than carbon-based fuels. So how could they be useful, for example, in vehicles? In response to that question, another possible exchange of elements comes to mind. Present day vehicles are made of steel; i.e., iron with a little cobalt and other additives. In the PA, Iron is adjacent to Cobalt, which is a keystone element. That keystone association triggers the thought of another candidate switch of elements: use carbon, not for the vehicle *fuel*, but rather for the vehicle *structure*. Nanotechnology gives us amazing new materials made of Carbon fibers. Some of these materials are strong enough for our *airplanes*, never mind our more ubiquitous surface vehicles.

KEYWORDS

- **algebraic chemistry**
- **ionic configurations**
- **ionization potentials**
- **new energy technology**

REFERENCE AND FURTHER READING

Whitney, C., (2013). *Algebraic Chemistry Applications and Origins.* Chapter 5, Nova Publishers: New York, USA.

CHAPTER 23

Information Entropy

FRANCISCO TORRENS[1] and GLORIA CASTELLANO[2]

[1]*Institute for Molecular Science, University of Valencia, PO Box 22085, E-46071 Valencia, Spain, Tel: +34-963-544-431, Fax: +34-963-543-274, E-mail: torrens@uv.es*

[2]*Department of Experimental Sciences and Mathematics, Faculty of Veterinary and Experimental Sciences, Valencia Catholic University Saint Vincent Martyr, Guillem de Castro-94, E-46001 Valencia, Spain*

23.1 DEFINITION

Shannon (1948a) proposed a quantitative measure of the amount of information or uncertainty supplied by a probabilistic experiment. The theory is explained in Eq. (2).

23.2 HISTORICAL ORIGIN

Clausius (1865) defined the concept of entropy. Based on the classical Boltzmann's work [H-($\eta\alpha$) theorem, 1872], Shannon (1948a) proposed a quantitative measure of the amount of information or uncertainty supplied by a probabilistic experiment. He developed an IE theory to solve some problems in message communication over primitive Morse code transmitters. The theory was applied to a varied range of situations that involve the transmission of messages, where term message is used in its widest sense.

23.3 NANO-SCIENTIFIC DEVELOPMENT

A significant connection exists between the entropy notion in thermodynamics and information/uncertainty ideas. Based on the classical Boltzmann's work (1872), Shannon (1948a,b) proposed a quantitative measure of the amount of information or uncertainty supplied by a probabilistic experiment. Consider an experiment in which event i results with probability p_i, which is described by the finite probability distribution:

$$p_i \geq 0, i = 1,\ldots,n; \sum_i p_i = 1 \qquad (1)$$

Shannon IE associated with this experiment is:

$$h_n(p_1, p_2, \ldots, p_n) = -\sum_i p_i \ln p_i \qquad (2)$$

The function has properties that give a reasonable measure of uncertainty in a probabilistic experiment, e.g.:

i. $h_n(p_1, p_2, \ldots, p_n) \geq 0$
ii. $h_n(p_1, p_2, \ldots, p_n) = 0$ if $p_i = 1$ for some i and $p_j = 0$ for $i \neq j$, which means that if only one result is possible, the uncertainty is null.
iii. $h_{n+1}(p_1, p_2, \ldots, p_n, 0) = h_n(p_1, p_2, \ldots, p_n)$, which means that adding the impossible event to the possible results of a given experiment does not change its uncertainty.
iv. $h_n(p_1, p_2, \ldots, p_n) \leq h_n(1/n, 1/n, \ldots, 1/n) = h_{max}$, which shows that the greatest uncertainty corresponds to equally likely outcomes.

Random events exist whose probabilities cannot be directly evaluated; e.g., the results of the measurements made on a microscale on microsystems are such mean values of some random variables. Many random distributions exist compatible with a given mean value. The problem is how to select the best one. The principle of maximum IE can be considered as such a criterion. According to this, systems choose the random distribution that maximizes IE or conditional IE subject to some set of restraints. The IE may be considered as the primary concept and use the probability distribution that maximizes IE subject to certain constraints for the statistical inference of the evolution (Jaynes, 1957). The principle appears to present a subjective character. As long as IE is accepted as being the most suitable measure of uncertainty, the system selects that particular random distribution, which contains the largest amount of uncertainty compatible with the given restraints. The success of the principle of maximum information in classical and quantum mechanics suggests extending its range of application. Making inferences on the basis of partial information, it is necessary to use that probability distribution that has maximum IE, as a measure of uncertainty, subject to whatever is known (Dewar, 2003). Boltzmann's H-theorem (1872) states the rise in IE as a measure of uncertainty. For a large class of stochastic evolutions of Markovian type, *H*-theorem holds. Denote by $p_{t,t+1}(w',w)$ the probability of transition from the state w, at the moment t to the state w', at the moment $t + 1$. Notice that $p_{t,t+1}(w',w) \geq 0$ and that:

$$\sum_{w'} p_{t,t+1}(w',w) = 1 \qquad (3)$$

Markovian evolution is described by:

$$p_{t+1}(w') = \sum_{w} p_t(w) p_{t,t+1}(w',w) \qquad (4)$$

At every moment, IE is given by:

$$h_t = -\sum_{w} p_t(w) \ln p_t(w) \qquad (5)$$

The *H*-theorem establishes that if the transition stochastic matrix $p_{t,t+1}(w',w)$ is bistochastic, i.e., if for any w and w':

$$\sum_{w} p_{t,t+1}(w',w) = \sum_{w'} p_{t,t+1}(w',w) = 1 \qquad (6)$$

then:

$$h_t \leq h_{t+1} \qquad (7)$$

The NA counterpart of the principle of maximum IE says that the systems choose the random distribution, which maximizes, in NA frame, IE or conditional IE subject to some restraints. However, maxima are not unique in NA frames. Consequently, a variety of acceptable distributions will result, and other choice criteria should be selected to ensure uniqueness. Specific real norms would be used to choose one of them (Iordache, 2011). The NA categorization of Boltzmann's *H*-theorem (1872) is discussed. Consider, e.g., two NA times $T = [n,t]$, $T' = [n,'t']$ and corresponding NA entropies: $H(T) = [h_0(n),h_1(n)]$, $H(T') = [h_0(n'),h_1(n')]$. The *H*-theorem NA counterpart would establish that NA IE $H(T)$ always rises in time, but the relation of order is a new one in NA frame; e.g., if $T < T'$ in NA order, by model categorization, NA valid *H*-theorem implies:

$$H(T) < H(T') \qquad (8)$$

which should be considered with the same NA order as for the time T. The NA inequality shows that it suffices that IE rise at the level $m = 0$, i.e., $h_0(n) < h_0(n')$ for $n < n,'$ to assure the rise in NA IE despite IE decay of other levels contribution. It is possible that $h_1(n) > h_1(n')$, i.e., on higher levels, the level-associated IE could decay. Restricting the analysis to a single level,

$m = 1$, may show results apparently contradicting the principle of rising IE but they are clarified in the multilevel frame. The multilevel IE rises, while allowing self-organization at the focused level $m = 1$.

23.4 NANO-CHEMICAL APPLICATION

Consider the use of a qualitative spot test to determine the presence of Fe in a water sample (Clegg & Massart, 1993). Without any sample history, the testing analyst must begin assuming that both outcomes are equiprobable with probabilities 1/2. When up to two metals may be present in the sample (e.g., Fe, Ni, both) four possible outcomes exist, ranging from neither (0,0) to both (1,1) being present with probabilities $1/4 = 1/2^2$. Which of these four possibilities turns up is determined *via* two tests, each having observable states. Similarly, with three elements, eight possibilities exist, everyone with a probability of $1/8 - 1/2^3$. Three tests are needed. The following pattern clearly relates the uncertainty and IE needed to resolve it. The number of possibilities is expressed to the power of 2. The power to which 2 must be raised to give the number of possibilities n is defined as the log to base 2 of that number. The IE and uncertainty are defined quantitatively in terms of the log to base 2 of the number of possible analytical outcomes:

$$I = H = \log_2 n \tag{9}$$

where I indicates the amount of IE, and H, the quantity of uncertainty. The initial uncertainty is defined in terms of the probability of the occurrence of every outcome; e.g., for the above-referred probabilities, the following definition results:

$$I = H = \log_2 n = \log_2 \frac{1}{p} = -\log_2 p \tag{10}$$

where I is IE contained in the answer, given that n possibilities existed, H, the initial uncertainty resulting from the need to consider the n possibilities, and p, the probability of every outcome if all n possibilities are equiprobable. The expression is generalized to the situation in which the probability of each outcome is not the same. If one knows from past experience that some elements be more likely to be present than others, Eq. (10) is adjusted so that the logarithms of individual probabilities, suitably weighted, be summed:

$$H = -\sum_i p_i \log_2 p_i \tag{11}$$

where:

$$\sum_i p_i = 1$$

Reconsider the original example, but now past experience showed that 90% of the samples contained no Fe. The degree of uncertainty is calculated *via* Eq. (11) as:

$$H = -(0.9 \log_2 0.9 + 0.1 \log_2 0.1) = 0.469 \text{ bit}$$

For a single event occurring with probability p, the degree of surprise is proportional to $-\ln p$. Generalizing result to a random variable X (taking N possible values $x_1, x_2, \ldots, x_n$, with probabilities $p_1, p_2, \ldots, p_n$) average surprise received on learning X value is: $-\Sigma p_i \ln p_i$.

23.5 MULTI-/TRANS-DISCIPLINARY CONNECTION

Use of IE in analytical chemistry originates from the idea: every analysis is a process of obtaining information about quali/quantitative sample composition or analyzed-substance structure (Cleij & Dijkstra, 1979; Eckschlager & Stepánek, 1978, 1979, 1981, 1982; Frank & Kowalski, 1982; Kowalski, 1980; Liteanu & Rica, 1979; Malissa, 1974). An analytical system possesses an input, output, and input-output relation (*cf.* Figure 23.1). System consists of two independent subsystems: (1) in the analytical device, information arises (Figure 23.1); (2) in the data processing, e.g., a computer, it is decoded. Subsystems are linked with a feedback channel. One judges either the analytical system as a whole or each subsystem separately. Content of IE provided by analytical procedures or devices enables one to compare their relative suitability for different applications. Optimization is performed *via* an information quantity as a response function. Prigogine (1980, 1989) reviewed basic aspects correlating entropy and evolution of complex systems. In this group, Barigye et al. (2013a,b,c) reported Shannon's, mutual, conditional and joint IE indices with global-indices generalization defined from local vertex invariants, relations frequency hypermatrices in mutual, conditional and joint IE-based information indices, and extended GT-STAF information indices based on Markov approximation models. Barigye et al. (2015) interpreted structural and physicochemically GT-STAF information theory-based indices. Pino Urias et al. (2015) published free software IMMAN for information theory-based chemometric analysis. Barigye & Marrero-Ponce (in press) analyzed digital communication and chemical-structure codification.

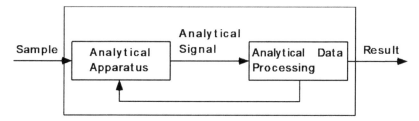

FIGURE 23.1 Analytical system.

23.5.1 OUT-OF-EQUILIBRIUM MICROSYSTEMS PHYSICS

Matter exploration in small regions and at short time intervals revealed the existence of reduced-size systems and new processes that happen at those scales (Rubí, 2012). Microsystems are not simply macrosystems miniatures, but they have their own structures and functions. Molecular motors, nanomotors, and active particles are autonomous microsystems that use energy-conversion mechanisms to carry out their work under the environmental influence and in out-of-equilibrium situations. Statistical physics offers a method able to describe the processes that such systems perform: complexity and nonadditive entropy S_q. A nonadditive entropy measure S_q was applied to certain complex natural, artificial and social systems (Tsallis & Plastino, 2012). The probability distributions that maximize S_q under restrictions showed ubiquity and robustness. The mechanisms that explain distributions origin imply a generalization (nonextensive statistical mechanics) of Boltzmann–Gibbs entropy theory.

23.5.2 STUDY OF EVOLVING-GENES IE

Shannon (1948a) and Larrea Lozano (2015) studied genes IE variation during evolution and its involvement in the theoretical explanation of evolutionary processes. Evolving-population organisms tend to self-organization and complex-structures creation. Phenomenon explanation from the thermodynamic viewpoint is not easy; however, as evolution is an irreversible process, one can assume that the organism, in an evolutionary context, behaves as a dissipative system far from thermodynamic equilibrium. After Prigogine (1955), it would explain structures-creation phenomena, in which conditions, a local IE decay is foreseeable, which is not observed at a macroscopic level as evolution acts as biodiversity source. He proposed that not the individual but the gene be considered as a fundamental evolutionary unit, from which viewpoint, a local IE decay could be found if one assumes genes IE; i.e., he foresaw that genes IE (relating to maximum possible IE) decayed as they evolved. In order to determine it, he calculated organisms diverse-genes IE *via* computing programs (see Chapters 28, 36–38).

23.6 OPEN ISSUES

The open issues are: (1) Attention to problems of hierarchy, stochastic-conditioning levels in the functioning of artificial-intelligence elements. The NA analysis and non-Markovian stochastic processes, polystochastic models are invocated; (2) Use of information criteria (e.g., IE and IE production); (3) Development based on models of neuromimetic elements. It is interesting to study the possible generalization of the ideas above to molecular classification, chemical (structural, quali/quantitative, trace) analyses, comparing/optimizing analytical methods/procedures, comparison of analytical systems, combining qualitative analysis systems, classification *via* expert systems and using information with inductive systems. Search for alternative methods in codifying information on chemical nets will continue to be of interest to chemists.

23.6.1 DATA PROCESSING

Signal processing is key in converting raw data to information. Analytical systems provide big data that cannot be processed manually. Obtaining samples and measuring them requires a huge amount of effort and obtained data matrices have more variables than measurements. The problem is called the curse of dimensionality and can induce overtraining or overfitting mathematical or pattern recognition models. Different preprocessing strategies overcame the problem and identified the best mathematical models to avoid overtraining so that a good generalization be obtained. A statistical analysis is key in order to test results generalization.

KEYWORDS

- **analytical chemistry**
- **chemometrics**
- **equipartition conjecture**
- **molecular classification**
- **uncertainty**

REFERENCES AND FURTHER READING

Alfonseca, M., (1996). *Diccionario Espasa: 1,000 Grandes Científicos*, Espasa Calpe: Madrid, Spain.

Barigye, S. J., & Marrero-Ponce, Y., (in press). Digital communication and chemical structure codification. In: Meyers, R. A., (ed.), *Encyclopedia of Complexity and Systems Science*. Springer: Heidelberg, Germany.

Barigye, S. J., Marrero-Ponce, Y., Alfonso-Reguera, V., & Pérez-Giménez, F., (2013a). Extended GT-STAF information indices based on Markov approximation models. *Chem. Phys. Lett.*, *570*, 147–152.

Barigye, S. J., Marrero-Ponce, Y., Martínez-López, Y., Torrens, F., Artiles-Martínez, L. M., Pino-Urias, R. W., & Martínez-Santiago, O., (2013b). Relations frequency hypermatrices in mutual, conditional and joint entropy-based information indices. *J. Comput. Chem.*, *34*, 259–274.

Barigye, S. J., Marrero-Ponce, Y., Martínez, S. O., Martínez, L. Y., Pérez-Giménez, F., & Torrens, F. (2013c). Shannon's, mutual, conditional and joint entropy information indices: Generalization of global indices defined from local vertex invariants. *Curr. Comput. -Aided Drug Des.*, *9*, 164–183.

Barigye, S. J., Marrero-Ponce, Y., Zupan, J., Pérez-Giménez, F., & Freitas, M. P., (2015). Structural and physicochemical interpretation of GT-STAF information theory-based indices. *Bull. Chem. Soc. Jpn.*, *88*, 97–109.

Clegg, D. E., & Massart, D. L., (1993). Information theory and its application to analytical chemistry. *J. Chem. Educ.*, *70*, 19–24.

Cleij, P., & Dijkstra, A., (1979). Information theory applied to the qualitative analysis. *Fresenius Z. Anal. Chem.*, *298*, 97–109.

Dewar, R. C., (2003). Information theory explanation of the fluctuation theorem, maximum entropy production, and self-organized criticality in nonequilibrium stationary states. *J. Phys. A.*, *36*, 631–641.

Eckschlager, K., & Stepánek, V., (1978). Information content of trace analysis results. *Microchim. Acta, 69*, 107–114.

Eckschlager, K., & Stepánek, V., (1979). *Information Theory as Applied to Chemical Analysis*. Wiley: New York, NY.

Eckschlager, K., & Stepánek, V., (1981). Information theory approach to trace analyses. *Microchim. Acta, 76,* 143–150.

Eckschlager, K., & Stepánek, V., (1982). Information theory in analytical chemistry. *Anal. Chem., 54,* 1115A–1127A.

Frank, I. E., & Kowalski, B. R., (1982). Chemometrics. *Anal. Chem., 54,* 232R–243R.

Iordache, O., (2011). *Modeling Multi-Level Systems.* Springer: Berlin, Germany.

Jaynes, E. T., (1957). Information theory and statistical mechanics. *Phys. Rev., 106,* 620–630.

Kowalski, B. R., (1980). Chemometrics. *Anal. Chem., 52,* 112R–122R.

Larrea, L. I., (2015). Estudio de la Entropía de la Información de Genes en Evolución. In: *Proceedings of the XXXVIII Congreso de la Sociedad Española de Bioquímica y Biología Molecular* (p. 71). València, Spain, SEBBM: València, Spain.

Liteanu, C., & Rica, J., (1979). Utilization of the amount of information in the evaluation of analytical methods. *Anal. Chem., 51,* 1986–1995.

Malissa, H., (1974). Automation in und mit der Analytischen Chemie. VI. Systemorientierte und informations theoretische gerechte Definition der analytischen Chemie. *Fresenius Z. Anal. Chem., 271,* 97–100.

Pino, U. R. W., Barigye, S. J., Marrero-Ponce, Y., García-Jacas, C. R., Valdes-Martiní, J. R., & Pérez-Giménez, F., (2015). IMMAN: Free software for information theory-based chemometric analysis. *Mol. Divers., 19,* 305–319.

Prigogine, I., (1955). *Introduction to the Thermodynamics of Irreversible Processes.* Thomas: Springfield.

Prigogine, I., (1980). *From Being to Becoming: Time and Complexity in the Physical Sciences.* Freeman: San Francisco.

Prigogine, I., (1989). What is entropy. *Naturwissenschaften, 76,* 1–8.

Rubí, J. M., (2012). La física de los microsistemas fuera del equilibrio. *Revista Española de Física, 26*(3), 64–69.

Shannon, C. E., (1948a). A mathematical theory of communication: Part I, discrete noiseless systems. *Bell Syst. Tech. J., 27,* 379–423.

Shannon, C. E., (1948b). A mathematical theory of communication: Part II, the discrete channel with noise. *Bell Syst. Tech. J., 27,* 623–656.

Tsallis, C., & Plastino, A. R., (2012). Complejidad y la entropía no-aditiva S_q. *Revista Española de Física, 26*(3), 70–75.

CHAPTER 24

Intermolecular Force Parameters

BOGDAN BUMBĂCILĂ[1] and MIHAI V. PUTZ[1,2]

[1]Laboratory of Computational and Structural Physical Chemistry for Nanosciences and QSAR, Biology-Chemistry Department, Faculty of Chemistry, Biology, Geography at West University of Timişoara, Pestalozzi Street No. 16, Timişoara, RO-300115, Romania

[2]Laboratory of Renewable Energies-Photovoltaics, R&D National Institute for Electrochemistry and Condensed Matter, Dr. A. Paunescu Podeanu Str. No. 144, RO-300569 Timişoara, Romania, Tel.: +40-256-592-638; Fax: +40-256-592-620; E-mail: mv_putz@yahoo.com or mihai.putz@e-uvt.ro

24.1 DEFINITION

Intermolecular forces are attraction or repulsion forces acting between neighboring particles (molecules, usually but also free atoms or ions). Compared to the intramolecular forces, which are keeping together the components of a molecule, they are much weaker forces. They can be "translated" as the "force fields," which are energy functions/interatomic potentials, functional relationships set to calculate the potential energy of a particle-composed system with given positions of the particles in space (Ponder et al., 2003).

The intermolecular forces are controlling the state of a substance but also other characteristics, like its boiling point, melting point, vapor pressure, viscosity, absorption data, and superficial tension. They are virtually responsible for all the physical properties of a material (Blaber, 1996).

24.2 HISTORICAL ORIGINS

The first scientist who wrote about "forces that keep the particles together" was Alexis Claude Clairaut (1713–1765), a French mathematician, geophysicist, and astronomer. But other pioneers of this theory also had notable contributions: Pierre-Simon, marquis de Laplace (1749–1827), Johann Carl Friedrich Gauss (1777–1855), James Clerk Maxwell (1831–1879), Ludwig Eduard Boltzmann (1844–1906).

24.3 NANO-CHEMICAL IMPLICATIONS

Intermolecular forces can be classified after several types (Margenau et al., 1969):

- ion-induced dipole forces
- ion-dipole forces
- hydrogen bonds
- halogen bonds
- van der Waals forces
- π-interactions
- hydrophobic effects

Ion-induced-dipole, ion-dipole interactions, hydrogen and halogen bonds are also called electrostatic interactions (Anslyn, 2004),

The state of matter (solid/liquid/gas) is determined by an equilibrium between the kinetic energy of the particles that are composing that substance (this energy tends to keep the particles moving) and the intermolecular forces of attraction (which tend to "group" the particles).

If the average kinetic energy of the particles is greater than the attraction forces, a substance will remain in the state of a gas. If the attraction forces are greater, a liquid or a solid will form, instead (Blaber, 1996).

An **ion-induced dipole** (Figure 24.1) force is a weak attraction force that results when the proximity of an ion induces a dipole in an atom or in a non-polar molecule, by rearranging the electrons in the non-polar species.

An **ion-dipole** (Figure 24.2) force is an attraction force between an ion and a neutral polar molecule. It is the most common interaction between particles in solutions of ionic compounds in polar liquids. Ion-dipole attractions become stronger as the charge of the ion increases or as the magnitude of the dipole increases (*see* https://www.chem.purdue.edu/gchelp/liquids/inddip2).

Hydrogen bond is a specific type of permanent dipole-dipole interaction, that involves the interaction between a partial positively charged hydrogen atom and a highly electronegative, partial negatively-charged oxygen, nitrogen, sulfur or fluorine atom, not covalently bound to the hydrogen atom (from another molecule–when we are speaking of intermolecular attraction forces, but intramolecular hydrogen bonds are partly responsible for the secondary and tertiary structures of proteins and nucleic acids) (Arunan et al., 2011; Sweetman et al., 2014).

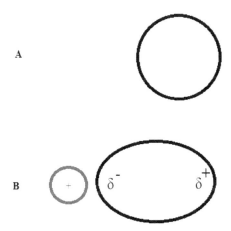

FIGURE 24.1   Ion-induced dipole force in a non-polar particle: A–non-polar particle, B–polarized particle with an ion in its proximity (https://www.chem.purdue.edu/gchelp/liquids/inddip2).

FIGURE 24.2   Ion-dipole force in a polar particle: the positively charged end of the dipole is oriented to anions and the negatively charged of a dipole to the cations.

Halogen bond is also a specific type of non-covalent dipole-dipole interaction which does not involve the formation or a breaking of existent bonds. A halogen atom displays an electrophile effect (acts like an *acceptor*) and settles a weak electrostatic interaction with a nucleophile species (a highly electronegative atom or an anion, a *donor*). It is important to mention that this bond is characteristic only between a halogen atom and monoatomic nucleophile species. If the donor is, for example, an aromatic cycle (which displays an electronic π cloud), we speak of halogen-aromatic interactions, different than these halogen bonds (Metrangolo et al., 2001; Politzer ct al., 2007).

Although both hydrogen and halogen bonding are forms of dipole-dipole interactions, they are not classified as van der Waals forces (Cockroft et al., 2007).

van der Waals forces are electrostatic interactions involving permanent dipoles and/or dipole-induced dipoles (or multipoles). A classification can be made, if:

- they take place between permanent dipoles–a.k.a. *Keesom forces*
- they take place between induced dipoles and dipoles–a.k.a. *Debye forces*
- they take place between non-permanent dipoles and induced dipoles–a.k.a. *London dispersion forces*

Dipole-dipole forces (Figure 24.3) are interactions between permanent dipoles. These interactions are responsible for reducing the potential energy of the particles and aligning the molecules over different rotational orientations (for increasing the attraction). Dipoles are associated especially with electronegative atoms: F, O, N, Cl, Br, S. One classical example is the one of formaldehyde. Because the Oxygen is more electronegative than the Carbon atom that is covalently bonded to it, the electrons associated with that bond will be closer to the Oxygen than to the Carbon, creating a partial negative charge ($\delta-$) on the oxygen, and a partial positive charge ($\delta+$) on the carbon. They are dependent by the temperature in the system (Brau, 2004).

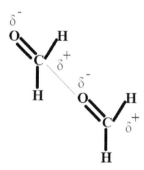

FIGURE 24.3   Dipole-dipole force between two permanent dipoles (formaldehyde molecules).

Forces between permanent dipoles are also called Keesom forces after Dutch physicist Willem Hendrik Keesom (1876–1956).

Interactions between permanent dipoles and induced dipoles are called Debye forces, after physicist Peter Joseph William Debye (1884–1966).

Because of the approach of a permanent dipole to a non-polar molecule with no permanent dipole, the electrons in the second one are polarized toward or away from the permanent dipole. These interactions are manifesting in the systems for short periods of time. In the induced dipoles, atoms with larger atomic radii are considered more polarizable, and they experience a greater attraction (higher partial electric charges and longer last in time). These forces are not as temperature-dependant as the dipole-dipole interactions because shifts for complementary charges orientations are permitted in various orientation degrees (Leite et al., 2012; Sapse et al., 1979).

Another type of van der Waals forces is represented by London interactions, named after physicist Fritz Wolfgang London (1900–1954), between fluctuating dipoles and induced dipoles. These forces appear because of the non-zero instantaneous dipole moments of all atoms and molecules. Polarization can be induced by a polar molecule but also by the repulsion of electronic clouds with the same electric charge, so the electronic density in an electronic cloud can fluctuate because of the electric conditions in the proximity (London, 1937). London forces between molecules increase with the atomic radius of the atoms in those molecules, because the larger the molecules are, the more dispersed the electron clouds are. Also, the larger the molecular surface areas are, the closer are the interactions between different molecules (Israelachvili, 1992).

The energy of the van der Waals interactions decreases rapidly in proportion to d^{-6} (where d is the distance between the two molecules. Although individual van der Waals forces have low energies they become significant when they sum up over a large molecular surface area, thus an important contact between particles (Selassie, 2003).

π-interactions/effects are non-covalent interactions where a negative charge region of a π-system interacts with a positive charge. Non-covalent interactions involving π-systems are mostly important to biological events such as receptor-ligand recognition (Meyer et al., 2003).

The **hydrophobic effect** is a manifestation of the non-polar molecules in aqueous solutions. They tend to form aggregates in order to separate from water. This aggregating phenomenon leads to a minimum exposure surface arca of the aggregate to the water molecule (Silverman, 2004).

Ionic and covalent interactions also may appear in the ligand-receptor interactions and/or in the process of receptor's folding. In the case of ionic interactions, the strength of these effects is directly related to the charge of the opposite ions and inversely dependent on the dielectric constant of the solvent and the distance between the charges (Selassie, 2003).

Parameters used in QSAR have a critical importance in determining the types of forces that are implied in drug-receptor interactions. These parameters can be divided in four major classes: electronic parameters, steric parameters, hydrophobic parameters, and biological parameters (Selassie, 2003). Table 24.1 summarizes a few of these parameters.

Hydrophobic effects of different substituents always represented points of high interest for the QSAR studies. For example, Corwin Herman Hansch introduced an equation to determine hydrophobic constants for substituents (Balaban et al., 1980):

$$\log\left(\frac{P_{C-X}}{P_{C-H}}\right) = \rho\pi_X \tag{1}$$

where P_{C-X} and P_{C-H} are the partition coefficients for the compound C-X (substituted) and C-H (unsubstituted). When the solvent pair is n-octanol/water, $\rho = 1$. π_X is the hydrophobic parameter for substituent X. In addition, if the compound is polysubstituted, the equation becomes (Balaban, et al., 1980):

$$\log\left(\frac{P_{C-X1...Xn}}{P_{C-H}}\right) = \sum_{i=1}^{n}\pi_{X_i} \tag{2}$$

Scientists Nyss and Rekker, introduced the concept of fragmental hydrophobic constant, defined as a "fragment" of the total hydrophobicity:

$$\log P_{C-X} = f_C + f_X \tag{3}$$

So, for a polysubstituted compound:

$$\log P_{C-X1......Xn} = f_C + \sum_{i=1}^{n}a_i f_i + \sum_{k}p_k \tag{4}$$

where a_i is the number of times fragment X_i is presented by the molecule, pk are corrections for some proximity effects which sometimes have to be introduced (Nys et al., 1973, 1974).

Because most of the studies show that electronic properties of the molecules are most important in establishing their chemical behavior and thus their biological activities, electronic parameters are most important to study.

Hydrophobic interactions are also important when studying biological systems. The estimation of the hydrophobic effects is based on the partition coefficients between the two-referral solvents: n-octanol and water (Balaban et al., 1980).

TABLE 24.1 Molecular Properties and Their Assessing Parameters in QSAR

Parameter types	Property described	Corresponding Parameter/Interaction	Detail
Hydrophobicity parameters	Lipophilicity	logP	n-octanol-water partition coefficient
		π	Hydrophobic parameter for a specific substituent
		R_m	$\log(1/R_f-1)$ where R_f retention factor in thin layer chromatography
		X	Contribution to molecular (total) logP of a particular substituent
		f	Hydrophobic fragmental constant
Electronic parameters	Electronic density	Ionic bonds	Measured by ionic strength, $\frac{1}{2}\sum c_i z_i^2$, where z_i is the valence and c_i the concentration of the ion
		Covalent interactions	Quantified by the dissociation energy (kcal/mol)
		Hydrogen bonds	Quantified by the dissociation energy (kcal/mol)
		σ level	The contour level or cut-off point in the intensity of any particular 3D representation of the electron density represented in standard deviation units
	Polarizability, van der Waals interactions	M_V	Molecular volume
		M_R	Molecular refractivity
		Parachor	$Pc = \frac{1}{\sqrt[4]{\gamma}}\frac{M}{d}$, where γ = surface tension, M = molar mass, d = density
Steric parameters	Steric hindrance, steric fit, steric misfit, molecular branching	E_S	Steric Taft parameter
		Distances: L, B_1, B_5	Sterimol parameters
		Volumes: MTD, MSD	Minimal topological difference, minimal steric difference
		Centric indices	BAC, C, C'
		Connectivity indices	R, J, χ
Biological parameters	Inhibitory effects, Affinities, Potencies, Bioconcentrations, In Vivo Reaction Rates, Bioavailability, Toxic effects, Mutagenicity effects	Inhibition constants	$\log(1/IC_{50})$
		Reaction rate constant	logk
		Bioconcentration factor	logBCF
		Total clearance	logT
		Lethal doses	$logLD_{50}$
		Mutagenic potency	$logTA_{98}$

24.4 TRANS-DISCIPLINARY CONNECTIONS

π-interactions are bringing an important contribution to the functionality of the biological systems because they provide a significant amount of binding enthalpy. For example, a cation π-interaction represents the basis of acetylcholine binding to acetylcholine esterase, the enzyme which breaks acetylcholine into choline and acetic acid, an important step for acetylcholine to fulfill its biological role. The primary structure of the enzyme includes 14 aromatic residues. The trimethylammonium group of acetylcholine binds through a cation π-interaction to the Trp residue because at this position, the partial negative charge is higher than at the Phe and Tyr spots (Kumpf et al., 1993; Dougherty, 1996).

van der Waals interactions, hydrogen bonds, electrostatic interactions also can be expressed at the intramolecular level. For example, a protein has to be "packed"/folded to superior structures in order to be biologically active. Usually, the nonpolar amino acid residues are exposed to the interior of its pseudoglobular structure (hydrophobic collapse), and the polar residues are oriented to the exterior for helping the solvation process of the protein by the water molecules. Hydrogen bonds are rapidly formed between the Hydrogen atoms in water and Nitrogen atoms in the amino groups. π-interactions and hydrophobic effects are expressed in the core till the most energetically minimized orientation of the protein is achieved for developing the three-dimensional structure. The folding process can also include covalent interactions, like the formation of disulfide bridges and sometimes, smaller polypeptides, named molecular chaperones assist the process of the protein's 3D structure formation and maintenance (Voetet al., 2010).

The intermolecular forces are particularly important to assess in order to design and discover new drugs. The biological activity of most of the present therapeutic agents is based on the process of ligand-receptor interaction. In order to bind to the receptor, the drug molecule has to fit its specific site on the receptor but also to establish some non-covalent interactions at the spot: hydrogen bonds, electrostatic interactions, van der Waals interactions, π-interactions. The parameters which express the potency of these interactions are useful in QSAR studies because they can show the best candidate in a series of molecules for the best fit on their biological receptor.

24.5 OPEN ISSUES

In the processes of molecular modeling and drug design, usually, scientists are trying to find a perfect match for a specific biological receptor which will trigger the perfect biological activity. Different requirements have to be fulfilled, especially the steric ones: the new drugs have to have structures as complementary as it is possible to the receptor, if the receptor's structure is known. Sometimes, if the receptor's structure is unknown, it can be "guessed" ("mapped") with the help of a standard ligand–usually the natural ligand. The binding of the molecule to the receptor is realized through very complex mechanisms, and many types of intermolecular interactions are settled between the two molecules. Because of the multitude of the implied factors, results are rarely attained only by human work and that is why software is needed. In the future, programs that make real-time estimations of intermolecular interactions during the docking process of the ligand in order to best describe the physical-chemical properties of binding site and the changes that maybe both the receptor and the ligand can suffer when being in proximity, will be developed and perfected.

The novel approaches in proteomics involve in vivo genetic mutations of some specific genes which are coding certain protein synthesis in order to "build up" proteins with perfected conformations or with higher abilities for binding ligands (multiplied binding sites or with higher steric compatibilities).

QSAR/QSPR researchers are trying to establish new equations which are correlating phyisico-chemical parameters of the molecules (induced by their intermolecular forces) with their properties/activities. For example, molar refractivity, refractive index, and electronic parameters have been used extensively, but the first study which correlated the surface tension with dissociation constants was Thakur's study. He showed that the surface tension can be successfully used to model the dissociation constant of some sulfonamide drugs. The dissociation constant pKa depends upon the polarity and the intermolecular forces. For a maximum activity, the sulfonamides should have a proper pKa, for penetrating in vivo membranes and best binding abilities to their target enzyme. These abilities depend upon their protonated/unprotonated form dissociation constants, expressed as pKa (Thakur, 2005).

KEYWORDS

- **ion-dipole forces**
- **ion-induced dipole forces**
- **van der Waals forces**

REFERENCES AND FURTHER READING

Anslyn, E. V., & Dougherty, D. A., (2005). *Modern Physical Organic Chemistry*. University Science Books, Sausalito, CA.

Arunan, E., Desiraju, G. R., Klein, R. A., Sadlej, J., Scheiner, S., Alkorta, I., et al., (2011). Definition of the hydrogen bond. *Pure and Applied Chemistry, 83*(8), 1637–1641.

Balaban, A., Chiriac, A., Moțoc, I., & Simon, Z., (1980). *Steric Fit in Quantitative Structure-Activity Relationships*. Springer Verlag, New York.

Blaber, M., (1996). *Intermolecular Forces*. www.mikeblaber.org (accessed on 18 February 2019).

Brau, C. A., (2004). *Modern Problems in Classical Electrodynamics*. Oxford University Press.

Cockroft, S. L., & Hunter, C. A., (2007). Chemical double-mutant cycles: Dissecting non-covalent interactions. *Chemical Society Reviews, 36*(2), 172.

Dougherty, D. A., (1996). Cation-pi interactions in chemistry and biology: A new view of benzene, Phe, Tyr, and Trp. *Science, 271*(5246), 163–168.

Israelachvili, J., (1992). *Intermolecular and Surface Forces* (2nd edn.). Academic Press.

Kumpf, R., & Dougherty, D., (1993). A mechanism for ion selectivity in potassium channels: Computational studies of cation-pi interactions. *Science, 261*(5129), 1708–1710.

Leite, F. L., Bueno, C. C., Da Róz, A. L., Ziemath, E. C., & Oliveira, O. N., (2012). Theoretical models for surface forces and adhesion and their measurement using atomic force microscopy. *International Journal of Molecular Sciences, 13*(12), 12773.

London, F., (1937). The general theory of molecular forces. *Transactions of the Faraday Society, 33,* 8–26.

Margenau, H., & Kestner, N., (1969). Theory of intermolecular forces. *International Series of Monographs in Natural Philosophy*. Pergamon Press.

Metrangolo, P., & Resnati, G., (2001). Halogen bonding: A paradigm in supramolecular chemistry. *Chemistry–A European Journal, 7*(12), 2511–2519.

Meyer, E. A., Castellano, R. K., & Diederich, F., (2003). Interactions with aromatic rings in chemical and biological recognition. *Angewandte Chemie (International Edition in English), 42*(11), 1210–1250.

Nys, G. G., & Rekker, R. F., (1973). Statistical analysis of a series of partition coefficients with special reference to the predictability of folding of drug molecules. The introduction of hydrophobic fragmental constants (f-values). *Chimie Thérapeutique, 8,* 521–535.

Nys, G. G., & Rekker, R. F., (1974). The concept of hydrophobic fragmental constants (f-values). *European Journal of Medicinal Chemistry (Chimica Terapeutica), 9,* 361–375.

Politzer, P., Lane, P., Concha, M. C., Ma, Y., & Murray, J. S., (2007). An overview of halogen bonding. *Journal of Molecular Modeling, 13*(2), 305–311.

Ponder, J. W., & Case, D. A., (2003). Interatomic potentials and their relative parameters for protein simulations. *Advances in Protein Chemistry and Structural Biology, 66,* 27–85.

Sapse, A. M., Rayez-Meaume, M. T., Rayez, J. C., & Massa, L. J., (1979). Ion-induced dipole H−n clusters. *Nature, 278*(5702), 332.

Selassie, C. D., (2003). In: Abraham, D. J., (ed.), *The History of Quantitative Structure-Activity Relationships in Burger's Medicinal Chemistry and Drug Discovery* (6th edn., Vol. 1). John Wiley and Sons Publishers, New York.

Silverman, R. B., (2004). *The Organic Chemistry of Drug Design and Drug Action* (2nd edn.). Amsterdam, Elsevier.

Sweetman, A. M., Jarvis, S. P., Sang, H., Lekkas, I., Rahe, P., Wang, Y., et al., (2014). Mapping the force field of a hydrogen-bonded assembly. *Nature Communications, 5*.

Thakur, A., (2005). QSAR study on benzenesulfonamide ionization constant: Physicochemical approach using surface tension. *Archive for Organic Chemistry, 14*, 49–58.

Voet, D., & Voet, J. G., (2010). *Biochemistry* (4th edn.). Hoboken, New Jersey, John Wiley & Sons Publisher. https://www.chem.purdue.edu/gchelp/liquids/inddip2.

CHAPTER 25

Ionic Configurations

CYNTHIA WHITNEY

Galilean Electrodynamics, 11660 239th Ave. NE, Redmond,
WA 98053-5613, USA, E-mail: Galilean_Electrodynamics@Comcast.net

25.1 DEFINITION

In algebraic chemistry (AC), the nature of chemical bonding between atoms is analyzed as instantaneously 'ionic.' The usual distinction word 'covalent,' for situations where, over time, electrons are shared among different atoms, does not come up because such sharing is fully accommodated by looking into all of the several ionic configurations that the molecule can have. The usual further distinction between 'polar' and 'non-polar' covalent bonding in molecules is also not needed. All distinctions become energy distinctions between the different ionic configurations. AC assigns an energy measure to every candidate ionic configuration. So instead of words, we have numbers; instead of three main categories of the chemical bond, named by three words, however, we have many distinct energy numbers and we may need to characterize a situation completely. With the energy numbers that AC provides, we gain insight into the statistics for the occurrence of competing for ionic configurations. This sort of analysis is illustrated below for the case of water. The typical bent shape of the water molecule, the dipole nature of that molecule, and the macrostructures it can form, all emerge from the analysis.

25.2 HISTORICAL ORIGIN(S)

The concept of an ion goes back to the beginning of modern Chemistry: an ion can be just an atom with an excess of, or a deficit of, electrons. Being charged, ions can attract each other, and bond to each other. The exact nature of chemical bonding is not yet completely understood. It is generally supposed that there must exist various different kinds of chemical bonds wherein electrons may have trajectories that involve two or more atomic nuclei. Or in the language of quantum mechanics, there may be 'orbitals' (electron wave functions) that honor two or more atomic nuclei.

The current way of characterizing the chemical bonding is inherently difficult to evaluate computationally. The problem lies in the continuity of quantum wave functions: continuous variables of any kind are just difficult to explore fully and fairly.

The concept of ionic configuration (IC) offers an alternative approach that reduces the problem of exploring so many continuous variables. Given the chemical formula for a molecule, it is possible to imagine a finite, discrete, set of possible ways the atoms could all be ionized, and so be made to stick together. All such possibilities can be evaluated using algebraic chemistry (AC) (*see* Chapter 1).

That AC evaluation exercise produces the energy required to be associated with each candidate IC. Some energy requirements may be positive, meaning those ICs are difficult to form, whereas other energy requirements may be negative, meaning those ICs will form spontaneously.

25.3   NANO-SCIENTIFIC DEVELOPMENT(S)

To illustrate the use of ICs, we can consider the case of what is probably the most important chemical on our planet; namely, water. For this analysis, the input AC data can be limited to the first ten elements. It consists of nuclear charge Z and mass M (with the naturally occurring proportions of isotopes), and model ionization potentials, which are rendered population-generic with M/Z scaling, and separated into the baseline $\Delta IP_{1,Z}$ for interaction between the nucleus and the electron population as a whole, and $\Delta IP_{1,Z}$ for interaction just between the electrons. We have:

Hydrogen: $Z = 1$, $M = 1.008$; model $IP_{1,1} = 14.250$ eV, $\Delta IP_{1,1} = 0$ eV.

Helium: $Z = 2$, $M = 4.003$; model $IP_{1,2} = 49.875$ eV, $\Delta IP_{1,2} = 35.625$ eV.

Lithium: $Z = 3$, $M = 6.941$; model $IP_{1,3} = 12.469$ eV, $\Delta IP_{1,3} = 1.781$ eV.

Beryllium: $Z = 4$, $M = 9.012$; model $IP_{1,4} = 23.327$ eV, $\Delta IP_{1,4} = 9.077$ eV.

Boron: $Z = 5$, $M = 10.811$; model $IP_{1,5} = 17.055$ eV, $\Delta IP_{1,5} = 2.805$ eV.

Carbon: $Z = 6$, $M = 12.011$; model $IP_{1,6} = 21.570$ eV, $\Delta IP_{1,6} = 7.320$ eV.

Nitrogen: $Z = 7$, $M = 14.007$; model $IP_{1,7} = 27.281$ eV, $\Delta IP_{1,7} = 13.031$ eV.

Oxygen: $Z = 8$, $M = 15.999$; model $IP_{1,8} = 27.281$ eV, $\Delta IP_{1,8} = 13.031$ eV.

Fluorine: $Z = 9$, $M = 18.998$; model $IP_{1,9} = 35.504$ eV, $\Delta IP_{1,9} = 20.254$ eV.

Neon: $Z = 10$, $M = 20.180$; model $IP_{1,10} = 43.641$ eV, $\Delta IP_{1,10} = 29.391$ eV.

Note that these data come from a semi-empirical model, and not from a fully developed new theory. This semi-empirical status is revealed, for example, in the fact that the *IP* and ΔIP numbers are the same for Nitrogen and Oxygen. These elements mark a transition where the spin quantum number being filled is switching sign. There is not a presently known reason to guide the model to different numbers before and after this switch, so they are the same.

The current semi-empirical status of these data means the four or five digits given are not necessarily all significant; really, extra digits are given just help the reader follow the calculations.

25.4 NANO-CHEMICAL APPLICATION

Now consider the water molecule, H_2O. If asked to propose an ionic configuration for it, most people will propose $2H^+ + O^{2-}$. This proposal is basically a 'Scientific Convention,' i.e., a statement that is accepted for purposes of communication, but not actually tested. But we can, in fact, test it with AC. Creation of an H^+ ion takes

$$IP_{1,1} \times Z / M = 14.250 \times 1 / 1.008 = 14.1369 \text{ eV} \tag{1}$$

(Observe that this especially simple example just returns the original model ionization potential, i.e., stripped of its M/Z scaling.)

Clearly, from (1) the creation of *two* H^+ ions takes $2 \times 14.1369 = 28.2738$ eV. Next, the creation of an O^{2-} ion takes:

$$-IP_{1,1}\left[\sqrt{Z(Z+1)} + \sqrt{Z(Z+2)}\right]/M_Z + \left[\Delta IP_{1,Z}Z - \Delta IP_{1,Z+2}(Z+2)\right]/M_Z =$$
$$-IP_{1,1}\left[\sqrt{8 \times 9} + \sqrt{8 \times 10}\right]/15.999 + \left[\Delta IP_{1,8} \times 8 - \Delta IP_{1,10} \times 10\right]/15.999 = \tag{2}$$
$$\left[-14.250 \times (8.4853 + 8.9443) + 13.031 \times 8 - 29.391 \times 10\right]/15.999 = -27.3788 \text{eV}$$

So altogether, the creation of the conventional ionic configuration $2H^+ + O^{2-}$ takes

$$28.2738 - 27.3788 = 0.8950 \ \text{eV} \tag{3}$$

Although small, this energy is positive. This fact makes the conventional ionic configuration $2H^+ + O^{2-}$ somewhat suspect.

Therefore, let us consider an alternative ionic configuration: $2H^- + O^{2+}$. Creation of an H^- ion takes

$$-IP_{1,1}\sqrt{Z(Z+1)}\Big/M_Z - \Big[\Delta IP_{1,Z+1} \times (Z+1) - \Delta IP_{1,Z} \times Z)\Big]\Big/M_Z =$$
$$-IP_{1,1}\sqrt{1\times 2}\Big/1.008 - \Big[\Delta IP_{1,2} \times 2 + \Delta IP_{1,1} \times 1\Big]\Big/1.008 \tag{4}$$
$$-\Big[14.250\times 1.4142 + 35.652\times 2 + 0\Big]\Big/1.008 = -19.9924 - 70.6845 = -90.6769\text{eV}$$

So the creation of two H^- ions takes $2 \times (-90.6769) = -181.3538$ eV. Creation of an O^{2+} ion takes

$$IP_{1,1}\Big[Z + \sqrt{Z\times(Z-1)}\Big]\Big/M_Z + \Big[\Delta IP_{1,Z} \times Z - \Delta IP_{1,Z-2} \times (Z-2)\Big]\Big/M_Z =$$
$$IP_{1,1}\Big[8 + \sqrt{8\times 7}\Big]\Big/15.999 + \Big[\Delta IP_{1,8} - \Delta IP_{1,6}\Big]\Big/15.999 = \tag{5}$$
$$14.250\times\Big[8 + 7.320\Big]\Big/15.999 + \Big[104.248 - 43.920\Big]\Big/15.999 =$$
$$14.250\times 15.4833\,/\,15.999 + 60.328\,/\,15.999 = 13.7907 + 3.7707 = 17.5614\text{eV}$$

So altogether, the creation of the ionic configuration $2H^- + O^{2+}$ takes

$$-181.3538 + 17.5614 = -163.7924\text{eV} \tag{6}$$

This number is solidly negative, so this ionic configuration $2H^- + O^{2+}$ for water is very believable. The author believe $2H^- + O^{2+}$ is what we all know as ordinary water.

In Nature, however, both ionic configurations can exist, although only in proportion to their so-called 'Boltzmann factors.' A Boltzmann factor has the form $\exp(- E/kT)$, where E is energy, and T is temperature, and k is Boltzmann's constant. Therefore, $2H^+ + O^{2-}$ does exist too. So what is it? The author believe the conventional ionic configuration $2H^+ + O^{2-}$ is what the rather exploratory part of the scientific literature calls 'Bown's gas.' Its slightly positive energy means that, upon impact, it can deliver energy, comparable to the energy that burning a methane molecule delivers–but without burning (oxidizing) anything at all! Just the relaxation from the ionic configuration $2H^+ + O^{2-}$ to the ionic configuration $2H^- + O^{2+}$ delivers the energy. Because of this feature, Brown's gas finds a practical application in fine welding.

Note the two candidate ionic configurations for water imply completely different molecule shapes and properties. Brown's gas $2H^+ + O^{2-}$ has nothing about it that could make it be anything other than linear. It just looks like $H^+ \cdot O^{2-} \cdot H^+$; i.e., two naked protons in line with an O^{2-} ion. Apart from its rarity, and its practical utility, it is a boring molecule.

But ordinary water $2H^- + O^{2+}$ is more interesting. Each of the hydrogen atoms has an electron pair. Such pairs are strong subsystems unto themselves. This is obvious from the high ionization potential of Helium–the highest ionization potential of all the elements. And it is further supported by the low ionization potential of Lithium–a physical realization of the old social adage: "Two is a company, but three is a crowd."

The ionic configuration $2H^- + O^{2+}$ invites the mental image of a water molecule with one O^{2+} at the center, and with four extremities; namely, two protons, and two pairs of electrons, tightly bound to each other. This image makes a tetrahedron. It accounts for the polarization of the water molecule, and for the bond angles that water exhibits. The tetrahedron 'arms' are at $109°$ from each other. This is not far from $120°$, so with only a slight ruffling out of a plane, the $109°$ angle can account for the hexagonal shape of a macroscopic snowflake.

Water also makes some interesting ions; namely, the so-called hydroxyl and hydronium ions, and the so-called 'exclusion zone' water ion (Pollack, 2013). Here we have space enough to look at the hydroxyl ion $(OH)^-$. It could have an ionic configuration $O^{2-} + H^+$. We already know that O^{2-} requires -27.3788 eV, and H^+ requires 14.1369 eV, so $O^{2-} + H^+$ takes $-27.3788 + 14.1369 = -132419$ eV. So this ionic configuration is feasible.

But there is also another possible ionic configuration: $O^+ + H^{2-}$. The ion O^+ takes:

$$IP_{1,1} \times 8 / M_8 + \Delta IP_{1,8} \times 8 / M_8 - \Delta IP_{1,7} \times 7 / M_8 \qquad (7)$$

Compare this example to that of Eq. (1). There, the one-and-only term returned the original model ionization potential before its M/Z scaling. But here, there are three terms. The first two of them return the original model ionization potential before its M/Z scaling, but the third term is something new. It says that the removal of one electron leads to a resettling of the remaining electrons. Since $\Delta IP_{1,7}$ is a positive number, this resettling is exothermic.

Evaluating (7) for O^+,

$$IP_{1,1} \times 8 / M_8 + \Delta IP_{1,8} \times 8 / M_8 - \Delta IP_{1,7} \times 7 / M_8 =$$
$$14.250 \times 8 / 15.999 + 13.031 \times 8 / 15.999 - 13.031 \times 7 / 15.999 = \qquad (8)$$
$$7.1254 + 6.5159 - 5.7014 = 7.9399 \, eV$$

which is positive. But then H^{2-} takes

$$-IP_{1,1} \times \left(\sqrt{1 \times 2} + \sqrt{1 \times 3} \right) \Big/ M_1 + \Delta IP_{1,1} \times 1 / M_1 - \Delta IP_{1,3} \times 3 / M_1 =$$
$$-14.250 \times (1.4142 + 1.7321)/1.008 + 0 - (-1.781) \times 3/1.008 = \qquad (9)$$
$$-14.250 \times 3.1463 / 1.008 + 5.3006 / 1.008 = -44.4789 + 5.2585 = -39.2204\,eV$$

This result is solidly negative. Altogether, the ionic configuration $O^+ + H^{2-}$ takes $-39.2204 + 7.9399 = -31.2805\,eV$, which is still solidly negative. Indeed, it is more solidly negative than the ionic configuration $O^{2-} + H^+$ at $-13.2419\,eV$. So it will certainly dominate.

For similar analyses of ICs for the hydronium ions and the exclusion-zone water ions, see Whitney (2014).

25.5   MULTI-/TRANS-DISCIPLINARY CONNECTION(S)

Since ICs suggest something about the shapes of molecules and ions, they provide a handy tool for helping to bring about desired engineering results, such as the new energy technology (NET) of cold fusion (CF).

KEYWORDS

- **algebraic chemistry**
- **computational chemistry**
- **ionization potentials**

REFERENCES AND FURTHER READING

Pollack, G., (2013). *The Fourth Phase of Water, Beyond Solid Liquid Vapor*. Chapter 4, Ebner & Sons, Seattle, USA.

Whitney, C., (2013). *Algebraic Chemistry Applications and Origins*. Chapter 5, Nova Publishers: New York, USA.

Whitney, C., (2014). EZ water: A new quantitative approach applied–many numerical results obtained. *Water, 5*, 105–120.

CHAPTER 26

Ionization Potentials

CYNTHIA WHITNEY

Galilean Electrodynamics, 11660 239th Ave. NE, Redmond, WA 98053-5613, USA, E-mail: Galilean_Electrodynamics@Comcast.net

26.1 DEFINITION

The term ionization potential (IP) refers to the energy required to remove one electron (a first-order IP), or several electrons (a higher-order IP), from an atom of a particular chemical element. A great deal of data about IPs exists. Within these data, clear patterns exist, and they are made even clearer when the appropriate scaling is applied. For the element with nuclear charge Z and nuclear mass M, the appropriate scale factor is M/Z. This scaling produces a new numerical variable, a scaled ionization potential, here called IP, that has the same rise overall Periods in the periodic table. For first-order IPs, the universal rise factor on each period is a simple ratio of integers: 7/2. In addition, the rises on the sub-periods within each period follow a simple pattern detailed herein. Finally, higher-order IPs all follow similar definite patterns, also detailed herein. The regularity of the IPs invites an algebraic approach to Chemistry overall (*see* Chapter 1).

26.2 HISTORICAL ORIGINS

The raw experimental data about IPs is readily available – a copious amount of it has long been published in standard handbooks on chemistry and/or Physics. The data set looks very idiosyncratic. That circumstance has inhibited many searches for meaning in it.

But some patterns do begin to emerge when we apply a scale factor intended to convert element-specific information into population-generic information. The scale factor is just M/Z, where M is average nuclear mass (or the individual isotope mass), and Z is the nuclear charge.

The scale factor itself emerges from a new analysis of the hydrogen atom (*see* Chapter 19). That analysis accounts for the ground state of hydrogen, not the conventional way, by postulate, but rather by a balance between three mechanisms: (a) energy loss caused by radiation from circular orbiting; (b) energy gain caused by finite speed of electromagnetic signals between the nucleus and the electron; (c) further energy loss, caused by further radiation, due to inevitable Thomas rotation [an effect first discovered in the context of special relativity theory (SRT) (*see* Thomas, 1927), but not really dependent on SRT].

Consideration of all three effects mentioned accounts for the Hydrogen ground state without injecting Planck's constant as a Postulate.

$$P_{tourque} = \left(e^4 / m_p c\right) / \left(r_e + r_p\right)^3 \tag{1}$$

$$P_{total\ radiation} = 32e^6 / 3c^3 m_e^2 r_e^4 \tag{2}$$

$$r_e + r_p \approx r_e = 32 m_p e^2 / 3 m_e^2 c^2 \approx 5.5 \times 10^{-9}\ \text{cm} \tag{3}$$

That makes the energy of the hydrogen ground state

$$-e^2 / r_e \propto 1/m_p \tag{4}$$

All atoms are to some degree similar to the prototypical atom, Hydrogen. For another element, we can plausibly expect the ground state energy to scale as:

$$Ze^2 \text{ and } 1/M \tag{5}$$

That is, scaling the measured ionization potential of an arbitrary element by its *M/Z* factor should create a new numerical variable, here called *IP*, that is, less disparate to hydrogen. That enables a search for new meaning in the *IP* data set.

26.3   NANO-SCIENTIFIC DEVELOPMENT(S)

There exists a very reliable pattern for the *M/Z* scaled values of measured ionization potentials of all orders and all elements (*see* Figure 26.1). It shows the *IP* values of ionization order *IO* for all elements that have $Z \geq IO$. The points represent the scaled data points, and the lines represent a mathematical model that fits the data points.

The very first data point, located at the lower left, is $IP_{1, Z=1}$. This is the *M/Z*-scaled ionization potential of hydrogen. The remaining data points from the $IO = 1$ line are $IP_{1,Z} = IP_{1,1} + \Delta IP_{1,Z}$, where $\Delta IP_{1,Z}$ is the M/Z-scaled increment in energy for first ionization that is required due to interactions between the multiple electrons in element Z.

The bottom line, for $IO = 1$, correlates with the periods in the Periodic Table (PT). There are seven periods (Figure 26.1). The rise in every full period is 7/2. The length L of every period can be characterized in terms of an integer $N = 1$ to 4: $L = 2N^2$. There are six declines D from the start of one period to the start of the next. Empirically, the first three decline ratios are 7/8, and the second three decline ratios are unity, meaning no change.

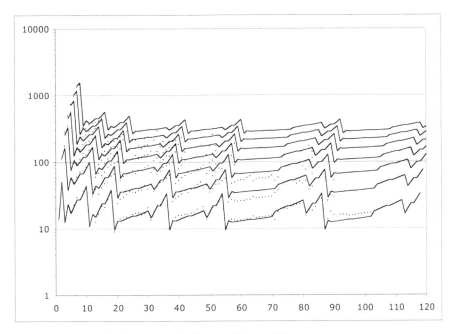

FIGURE 26.1 Ionization potentials of orders 1 through 7.

The periods generally have sub-periods, over which there are rises that are fractions of the total rise, 7/2. These can be expressed in terms of N and a parameter from quantum mechanics: the orbital angular momentum of the quantum state being filled, l. For $N = 1$, $l = 0$ and the rise is the full period rise, 7/2, so the rise fraction is 1. For $N > 1$ and $l = 0$, the rise fractions are presently just empirical. For $N > 1$ and $l > 0$, the rise fraction is:

$$\text{rise fraction} = \left[(2l+1)/N^2 \right]\left[(N-l)/l\right] \tag{6}$$

All the values of the decline and the rise fractions are laid out in the following table:

period	N	L	decline	l	fraction	l	fraction	l	fraction	l	fraction
1	1	2		0	1						
2	2	8	$7/8$	0	$1/2$	1	$3/4$				
3	2	8	$7/8$	0	$1/3$	1	$3/4$				
4	3	18	$7/8$	0	$1/4$	2	$5/18$	1	$2/3$		
5	3	18	1	0	$1/4$	2	$2/18$	1	$2/3$		
6	4	32	1	0	$1/4$	3	$7/48$	2	$5/16$	1	$9/16$
7	4	32	1	0	$1/4$	3	$7/48$	2	$5/16$	1	$9/16$

Values of $IP_{IO,Z}$ for $IO > 1$ are interesting, because they rise so abruptly. For $IO = 2$, $IP_{2,2} = IP_{1,1} \times 8$, where one might have expected $IP_{2,2} = ? = IP_{1,1} \times 2^2 = IP_{1,1} \times 4$. This fact means that pulling two electrons off a helium nucleus takes equal amounts of energy for separating opposite charges, and for separating like charges. This fact speaks to the durability of electron pairs, evident throughout all of chemistry.

For all $IO > 1$, the whole pattern can be expressed as follows:

$$IP_{IO,Z=IO} = IP_{1,Z=1} \times \left(2 \times IO^2\right)$$

$$IP_{IO,Z=IO+1} = IP_{1,Z=1} \times \left(2 \times IO^2 + \frac{1}{2} \times IO\right) + \Delta IP_{1,Z=2} \times \left(\frac{1}{2} \times IO\right)$$

$$IP_{IO,Z=IO+2} = IP_{1,Z=1} \times \left(\frac{1}{2} \times IO^2 + \frac{1}{2} \times IO\right) + \Delta IP_{1,Z=3} \times \left(\frac{1}{2} \times IO^2 + \frac{1}{2} \times IO\right)$$

$$IP_{IO,Z=IO+3} = IP_{1,Z=1} \times \left(\frac{1}{2} \times IO^2 + \frac{1}{2} \times IO\right) + \Delta IP_{1,Z=3} \times \left(\frac{1}{2} \times IO^2 - \frac{1}{2} \times IO\right) + \Delta IP_{1,Z=4} \times IO$$

etc.

$$IP_{IO,Z=IO+10} = IP_{1,Z=1} \times \left(\frac{1}{2} \times IO^2 + \frac{1}{2} \times IO\right) + \Delta IP_{1,Z=11} \times \left(\frac{1}{2} \times IO^2 + \frac{1}{2} \times IO\right)$$

$$IP_{IO,Z=IO+11} = IP_{1,Z=1} \times \left(\frac{1}{2} \times IO^2 + \frac{1}{2} \times IO\right) + \Delta IP_{1,Z=11} \times \left(\frac{1}{2} \times IO^2 - \frac{1}{2} \times IO\right) + \Delta IP_{1,Z=13} \times IO$$

etc., etc.

Observe that the pattern for any IO is similar to the pattern for $IO = 1$, but shifted right to element IO, increased in absolute amplitude for all elements, but muted in relative amplitude among elements. The 'shift' behavior suggests that each successive element reveals information about all subsequent elements. The 'increase' behavior says that removing more electrons always takes more energy. The muting behavior suggests that a distinction exists between ionization events occurring 'all-at-once' *vs.* events occurring 'one-at-a-time.'

For ordinary lab bench chemistry, 'one-at-a-time' is the main subject of interest; 'all-at-once' represents the kind of outdoor field event that occurs in the testing of high explosives. The 'one-at-a-time' scenario, being of greater practical utility, is discussed at length in Chapter 1. For either kind of situation, the pattern can serve as an important tool for predicting what happens.

KEYWORDS

- **first-order ionization potentials**
- **high-order ionization potentials**
- **scaling laws for chemistry**
- **visual displays in chemistry**

REFERENCES AND FURTHER READING

Thomas, L. H., (1927). The kinematics of an electron with an axis. *Phil. Mag. Series, 7, 3*(13), 1–22.

Whitney, C. K., (2013). *Algebraic Chemistry–Applications and Origins*. Nova Science Publishers, New York, USA, Chapter 9.

CHAPTER 27

Mass Transfer in Electrochemistry

MIRELA I. IORGA

National Institute for Research and Development in Electrochemistry and Condensed Matter, Department of Applied Electrochemistry, 300569 Timisoara, Romania

27.1 DEFINITION

Mass transfer is one of the main stages of an overall electrochemical reaction where the cationic transport can occur in three main modes. Migration represents the transport under the influence of the electric field, diffusion–under the influence of a concentration gradient or by natural or forced convection, from solution mass to the electrode surface, to replace particles consumed during the electrochemical reaction.

27.2 HISTORICAL ORIGIN(S)

Among the pioneers in the study of mass transfer could be mentioned H. Darcy (1803–1858), who discovered an empirical equation which predicts permeability – water flow through a porous bed (sand or gravel), named Darcy's law (Craig, 2008; Rolle, 2014). Darcy and Weisbach also modified the Prony equation (Bradley, 1998); this predicts pressure or head loss in a pipe or tube due to viscous friction. These two equations are used today in hydraulic systems design.

Another one was A.E. Fick (1829–1901), whose studies conduct him to predict the flow of gas into a fluid or solid membrane, called Fick's diffusion law. Nowadays, this is mentioned as the fundamental equation to predict the diffusion or flow into another different substance (Fick, 1855, 1995; Bird et al., 1976; Crank, 1980; Bockshtein et al., 2005).

But the theoretical basis of the main mass transfer phenomena, like diffusion, gas evaporation into another one, permeability, absorption, adsorption, are ascribed to H. von Helmholtz (1821–1894), (Bowler & Morus, 2005), and L.E. Boltzmann (1844–1906, 1994).

Meanwhile, in America, J.W. Gibbs (1839–1903) was the earliest theoretical scientist who developed physical chemistry, specifically thermodynamics (Hastings, 1909); Gibbs' phase rule is critical in mass transfer, phase change phenomena and metallurgy. He was also a mathematician and developed much of the vector notation – dot product and cross product – used nowadays in electrodynamics, fluid mechanics, field theories, and three-dimensional phenomena (Cropper, 2001).

In electrochemistry, the overpotential term was for the first time used by Caspari in 1899 (Agar, 1953). The introduction of the overvoltage concept, η, was necessary to prevent the difficulties arising because the overall electrode reaction is usually composed of several partial reactions (Paunovic, 1998). If an electrode is connected to an electrochemical cell passed by the current, its potential will differ from equilibrium potential, and the difference between those two potentials is called overpotential.

27.3   NANO-SCIENTIFIC DEVELOPMENT(S)

An experimental investigation of cathodic metal deposition kinetics is a complicated operation, due to the process characteristic features. During electrolysis cathode surface does not remain intact; it is continuously changing while metal deposition occurs. Deposit growth depends essentially on the metal nature and the electrolysis conditions (Antropov, 2001).

For an electrochemical reaction to occur, an essential prerequisite must be met: the charge transfer must take place. It may be preceded or succeeded by other phenomena: particles moving to the position where they can receive/give electrons; substances dissociation (i.e., ions occurrence); atom/ion resulting from electron transfer can take part in some chemical reactions or the crystal growth, etc.

The main stages of an electrochemical reaction are:

- *mass transfer* (may occur by migration–under an electric field, by diffusion–under concentration gradient or by natural or forced convection) from the solution to the electrode surface, to replace the consumed particles during the electrochemical reaction;
- *particles adsorption* to the external Helmholtz plan, where the electrochemical reaction takes place;

- *desolvation* of chemical species adsorbed before or after the charge transfer reaction;
- *transformations* of chemical species adsorbed before or after the charge transfer reaction;
- *charge transfer reaction* (release or acceptance of electrons through the electrode), which is the real electrochemical reaction and whose presence in an overall reaction development gives the electrochemical character;
- *adsorption* of primary products formed electrochemically;
- *desorption* of these products;
- *transfer* of created products from the electrode surface into the solution.

It is not mandatory that in an overall electrochemical reaction, all the above steps to be present. Thus, in the case of electrode processes, some of these reactions may be missing (except the charge transfer reaction).

As are represented in Figure 27.1, in metal ion discharging, the main four stages are (Facsko, 1969):

- cation transfer by diffusion, migration or convection from solution mass to electrode surface – *mass transfer*;
- solvated cation desolvation or complex ion dissolution in simple ion – *chemical reaction*;
- cation neutralization on cathode surface – *charge transfer*;
- atom transition to the stable form of deposited matter (hydrogen atoms unification in molecules or metal atoms penetration in the crystal lattice) – *crystallization*.

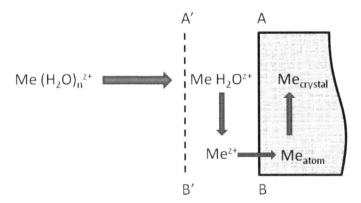

FIGURE 27.1 Scheme of metal ions discharging on cathode surface: AB – electrode surface; A'B' – reaction zone limit (Antropov, 2001; Paunovic, 1998).

The significations are as follows: AB is the electrode surface; A'B' – the layer from the immediate electrode vicinity (also called reaction zone); $Me(H_2O)_n{}^{z+}$ – solvated ion; M^{z+} – metal ion.

For the discharge process to take place in reversible conditions, at a potential given by Nernst relation (1), all four stages should proceed at high speed.

$$\varepsilon = \varepsilon^0 + \frac{RT}{zF} \ln \frac{a_{ox}}{a_{red}} \tag{1}$$

where ε is the electrode potential; ε^0 is the standard electrode potential; R is the perfect gas universal constant; T is the temperature (°K); F is the Faraday number; a_{ox} is the oxidized species activity; and a_{red} is the reduced species activity.

If the speed of one stage is much lower than other, that comparison may be deemed to arise under reversible conditions, the additional electric field created in electrochemical double layer serves to accelerate this slow stage. Ensuring the necessary conditions for this is required extra energy beside equilibrium reaction. A measure of this power is overvoltage (Oniciu & Constantinescu, 1982).

The four possible partial reaction types determine the existence of four stages types that can control the overall process speed: mass transfer, chemical reaction, charge transfer and crystallization (Facsko, 1969).

In general it was too complicated to determine the "participation" of each of the stages of electrochemical reactions at the overvoltage, but theoretically, limit cases can be treated in which the overvoltage is due to a single stage. Considering these boundary cases, one can assume that the rate-determining stage (which gives the overvoltage type) is irreversible, and all the other steps were being carried on at equilibrium.

27.3.1 OVERVOLTAGE

When an electrode is connected to an electrochemical cell, passed by a current, its voltage will be different from the equilibrium potential. If the electrode equilibrium potential (potential in the absence of an external current) is E and the same electrode potential passed by the current is $E(I)$, then the difference between these two potentials, η, is called overvoltage and is given by Eq. (2):

$$\eta = E(I) - E \tag{2}$$

Overvoltage shows how far of its equilibrium state an electrode is. It shows, however, what additional electrode voltage to be applied to the electrode to provide a specific current density and the desired reaction rate. From overvoltage definition, one can observe that for an anodic process the overvoltage has a positive value, while for a cathodic process, it has a negative value (Facsko, 1979).

The four possible partial reactions types mentioned above determine the existence of four stage types which can control the global process rate: mass transfer, chemical reaction, charge transfer and crystallization (Facsko, 1969; Oniciu & Constantinescu, 1982; Vaszilcsin, 1995).

Depending on the nature of the slowest stage (Vaszilcsin, 1995), there are the following overvoltage types:

- *Mass transfer (or transport) overvoltage, η_d*

It is a consequence of reactants concentration decreasing on the electrode surface or of reaction products accumulation, due to lower overall transport rate of electroactive species from solution to the reaction surface and respectively inversely. In the mass transfer processes, the substances consumed or formed during the electrode reaction are transported from solution mass to the interface (electrode surface) and from the interface into the solution. When diffusion determines concentration variation, it is called diffusion overvoltage.

- *Reaction overvoltage, η_r*

The overvoltage occurs when a chemical reaction takes place at a rate lower than all other stages of global electrode process, preceding or following the charge transfer stage and modifies the electroactive species concentration (reactants or reaction products). Reactions can be homogenous – in the solution–and heterogeneous – on the surface. The chemical reactions rate constant is independent of potential. Vetter introduced this term.

- *Charge transfer (or activation) overvoltage, η_t*

It is due to slow electrons transfer from the electrode to the electroactive species or reverse. Charge transfer involved double layer crossing of electrically charged particles (ions or electrons). This transfer occurs between the electrode and an ion or a molecule. The charge transfer reaction is the only partial reaction directly affected by the electrode potential. Thus, the charge transfer reaction rate is determined by the electrode potential.

- *Crystallization overvoltage, η_k*

Occurs in the case of metal electrodeposition or electro-dissolution and since that the electrode overall reaction product remains on the cathode surface, crystallizing on it, or if crystal lattice decomposition initiates the overall electrode reaction in anodic dissolution. This means that one of the stages of the overall electrode process will be deposit crystallization (the final stage) or de-crystallization (the initial stage) in the case of electro-dissolution. It occurs when the reaction products are embedded in the electrode crystal lattice with at a lower rate than any other stage of electrode overall process.

Thus, from the above, there are four different overvoltage types, and one can consider that global overvoltage, η, according to Eq. (3), consists of four components:

$$\eta = \eta_d + \eta_r + \eta_t + \eta_k \tag{3}$$

where η, η_d, η_r, η_t, η_k have the meanings mentioned above.

Sometimes the transfer and the reaction overvoltage act simultaneously. Since both are determined by reactant or reaction product concentration changes in the reaction zone, it is usually to be considered together, as *concentration overvoltage, η_c*. Regardless of the reason that determines the potential moving from its equilibrium value, occurs the possibility to achieve and even exceed the equilibrium potential of a particular reaction electrode process.

In addition to the overvoltages mentioned above, sometimes *resistance overvoltage* occurs, is due to the occurrence of electric resistance on the electrode surface, by an oxide or salt film forming, covering the interface. Obviously, as the electrical resistance of the film is higher, than the overvoltage will be higher. Since this overvoltage is not the result of a low-speed progress of an electrode process stage but is an ohmic drop, it is called ohmic pseudo-overvoltage. It can have very high values (tens, even hundred volts) and therefore should be excluded from polarization measurements.

The overvoltage size depends on the electrode process nature. As in other electrode reactions, in the case of metal electrodeposition, the polarization depends on the current density and increases with it.

Other researchers (Fleig & Jamnik, 2005) present quantitative models for calculating overvoltage influence on current density for different species and kinetic situations. Even for the same metal deposition, the results of polarization measurement (Antropov, 2001), depending on the current densities, the solution composition, and temperature, may take the form of straight lines in some of the coordinates systems: $\eta - i$, $\eta - \log i$, $1/\eta - \log i$, $1/\eta^2 - \log i$.

27.3.2 MASS TRANSFER OVERVOLTAGE

Mass transfer overvoltage is determined by concentration changes near the electrode, as a result of a too slow transport of reactant species from or into the solution (Facsko, 1979). The cations were discharging leading to a decrease in their concentration near the cathode. If this loss is not entirely and immediately compensate by diffusion, migration or convection, the cations concentration near the cathode will be smaller than into the solution volume, and the electrode potential, according to Nernst relation (1), more negative than in the absence of current.

In the case of diluted solutions activities are equal with concentrations.

Normally, in practice, there are three models of mass transfer (Bond, 1990; Facsko, 1979; Vaszilcsin, 1995) and therefore three phenomena that mass transfer overvoltage depends on: migration, diffusion, and convection.

27.3.2.1 MIGRATION – THE EXISTENCE OF A VOLTAGE GRADIENT AND ELECTRIC CHARGE OF REACTIVE SPECIES

Mass transfer by migration is a result of the force exerted on charged particles by an electric field. Thus, the positive particles will move toward the negative pole of the field, and the negative ones to the positive pole. Still, if a large excess of indifferent electrolyte exists, the migration process is taken up almost entirely by this electrolyte ions, and thus electroactive species migration can be neglected. Due to the necessity of using electrolytes support (indifferent) in excess, the mass transfer equations can be established only for dilute solutions.

27.3.2.2 DIFFUSION – THE EXISTENCE OF A CONCENTRATION GRADIENT

Mass transport by diffusion is the result of the emergence of a concentration gradient between the external Helmholtz plan and solution mass, which occurs under the influence of the electroactive species (metal ions) movement.

Due to the electrode reaction, the electroactive species concentration decreases in the electrode vicinity determines a concentration gradient in the layer near the electroactive surface ($<10^{-2}$ cm thick).

As the electrolyte concentration at electrode surface changes immediately after the start of the electrolysis, it is evident that the diffusion phenomenon plays a major role in any overall reaction where the slow stage is mass transfer.

27.3.2.3 CONVECTION – THE EXISTENCE OF PRESSURE, TEMPERATURE OR DENSITY GRADIENT IN THE VICINITY OF THE ELECTRODE

Two types of convection exist: natural or forced. Mass transfer by natural convection occurs as a result of the solution density gradient, which is the product of thermal disturbances or other characteristics of the electrolyte solution. Mass transfer by forced convection occurs as a result of electrolyte solution stirring (by various methods) and represents the species movement due to a mechanical force (constraints).

In practice, convection is usually performed by electrolyte solution mixing or stirring, or by its flowing through the reactor. Another possibility is the electrode movement. If one form of forced convection occurs in the reactor, it imposes the dominant mass transfer mode.

Even if not intended to impose such a mass transport, in a period less than 10 s from electrical circuit closing (start), natural convection occurs. The factors that determine this phenomenon are the small differences in solution densities, the chemical changes occurring at the electrode surface and/or, more pronounced, the formation and release of the gaseous products.

Figure 27.2 shows the three mass transfer models of the electroactive species to the electrode surface.

The concentration of the species consumed at the electrode is maintained by migration from the electrolyte mass, where the concentration is maximum. The generation of a species on the electrode causes a concentration gradient, but this time, the highest concentration is on the electrode's surface and the lower into the electrolyte. Diffusion occurs in reverse, from the electrode to the electrolyte.

In the case of convection (Bagotzky, 1993), in a very thin layer of the solution near the electrode, motionless, the transport occurs by diffusion. Also always one has to do with the combined transfer–by convection and diffusion–which means a *convective diffusion* (Koryta et al., 1993).

In Figure 27.3, the dependence of the mass transfer, charge transfer and heat transfer (Tobias, 1994; Newman & Tobias, 1962; Selman & Tobias, 1978) is shown schematically.

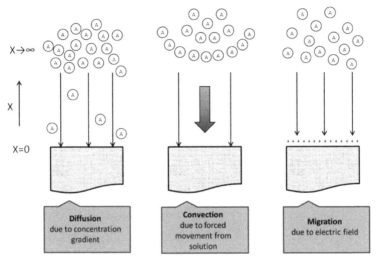

FIGURE 27.2 Mass transfer models of the electroactive species to the electrode surface (Bond, 1990).

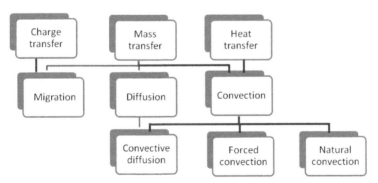

FIGURE 27.3 The dependence of the mass transfer, charge transfer and heat transfer (Tobias, 1994; Newman & Tobias, 1962; Selman & Tobias, 1978).

When current passes and cations are discharging, the space near the immediate vicinity of the cathode becomes poor in ions; than concentration gradient occurs and begins the cation diffusion from the solution mass to the electrode surface (Facsko, 1979).

In Figure 27.4, the concentration profile of metal ions is represented in the cathode vicinity, where the slow stage is the metal ions diffusion (Schlesinger & Paunovic, 2000).

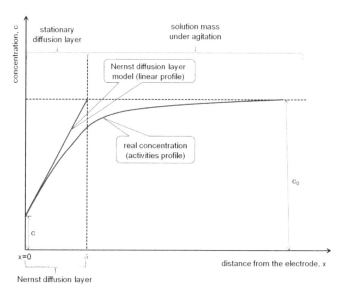

FIGURE 27.4 The concentration profile of metal ions (slow stage – metal ions diffusion; c_o – metal ions concentration in the solution mass; c – metal ions concentration near the immediate vicinity of the cathode) (Schlesinger & Paunovic, 2000).

According to the Nernst equation, the equilibrium potential of the cathode will be given by Eq. (4):

$$\varepsilon_{rev} = \varepsilon^o + \frac{RT}{zF}\ln c_o \tag{4}$$

The Eq. (5) will give the modified potential at current passing:

$$\varepsilon = \varepsilon^o + \frac{RT}{zF}\ln c \tag{5}$$

The Eq. (6) will provide the transport overvoltage:

$$\eta = \varepsilon - \varepsilon_{rev} = \frac{RT}{zF}\ln\frac{c}{c_o} \tag{6}$$

The Eq. (7) will give the cations discharging rate:

$$\frac{dN}{dt} = -\frac{i_c}{zF} \tag{7}$$

$$i_c = -\frac{zFD}{\delta}(c_o - c) \tag{8}$$

$$i_c = -K(c_o - c) \tag{9}$$

$$\eta = \frac{RT}{zF}\ln\frac{Kc_o + i_c}{Kc_o} \tag{10}$$

From Eq. (10) one can observe that at low current densities, overvoltage is small and increases with current density.

When the cathodic current, i_c, approaching $-Kc_o$ value, this increase becomes sharp. At $i_c = -Kc_o$ the transport overvoltage theoretically becomes infinitely high. In reality, this cannot occur, because the cathode potential only will increase until it reaches the discharging potential of another cation, hydrogen in this case.

The value of cathodic current density $i_c = -Kc_o$ is called *limit current*; in this case, because the slow stage is mainly the diffusion – it is called diffusion limit current density, $i_{c,d}$, and is set according to the relation (11):

$$i_{c,d} = -Kc_o = -\frac{zFD}{\delta}c_o \tag{11}$$

Overvoltage, according to the limit current will have the value given by Eq. (12):

$$\eta = \frac{RT}{zF}\ln\left(1 - \frac{i_c}{i_{c,d}}\right) \tag{12}$$

where i_c and $i_{c,d}$ are cathodic currents, and they will have negative values.

If in addition to diffusion, ions are transported through migration, the transport overvoltage is changing.

By definition, the transport number of an ion is the ratio of the transported current intensity (density) and the total current intensity (density), according to Eq. (13).

$$t_c = -\frac{i_c}{i} \tag{13}$$

where t_c is the transport number of an ion.

The sum of all the solution's species transport numbers is 1. In this case, the transport number of the other ions from solution will be given by the Eq. (14):

$$1 - t_c = t \tag{14}$$

Equation (8) becomes:

$$i_c = -\frac{zFD}{t\delta}(c_o - c) \tag{15}$$

and the maximum value corresponding to limit current density (when $c = 0$) will be:

$$i_{c,l} = -\frac{zFD}{t\delta}c_o \tag{16}$$

In practical and economic terms it is desirable the extension of diffusion limit current values and the limitation of hydrogen discharge – an undesirable process that decreases the electrolysis process yield (Facsko, 1979).

The cathodic transport overvoltage is even lower as the cathodic current density is lower and the higher is the ion concentration discharging within the solution. For a given current density and concentration, the transport overvoltage is even lower as higher is the value of K constant, expressed as Eq. (17):

$$K = \frac{zFD}{\delta} \tag{17}$$

To intensify the cathodic process is recommended the use of solutions with a higher concentration of reducing particles. The optimal level is determined in this case by the fact that the solution in the anode vicinity (in the metal electrowinning processes and applications of various electroplating) becomes saturated as soon as the solution concentration is higher (e.g., at a lower current density).

According to Eq. (17), the constant K value can be increased by increasing of diffusion coefficient (D) or decreasing of diffusion layer thickness (δ). Increasing of diffusion coefficient D can be done by electrolyte temperature rising.

The diffusion layer thickness decreasing is achieved by increasing the flow rate or electrolyte mixing degree by solution stirring, as in the case of a turbulent flowing mass transfer which is much more efficient than the case of laminar flowing (Wendt & Kreisa, 1999).

In conclusion, for a given current density and concentration, both the solution heating and stirring will determine the transport overvoltage decreasing.

27.4 NANO-CHEMICAL APPLICATION(S)

If it is necessary to increase the productivity of electrodeposition processes, it is imperative that both limit current density and electroactive surface to be as higher as, this fact favoring high deposition rates, therefore higher productivities.

According to the Eq. (18):

$$I = j \cdot A \tag{18}$$

the current working intensity is directly proportional to the current density, j, and with the electrochemically active electrode surface.

The increasing of the current limit density can be achieved by:

- metal ions high concentrations;
- increasing the temperature;
- mass transfer intensification through electrode-electrolyte relative movement.

The practice limits the first two options, became apparently the importance of studying the mass transport phenomenon, namely transport overvoltage in the metal ions electrodeposition process.

27.5   METHODS FOR MASS TRANSFER IMPROVEMENT

27.5.1   *INCREASING OF ELECTRODES SURFACE*

The profitable operation of an electrochemical reactor depends on the reactor's productivity. If all other process parameters (current efficiency, electrolyte composition, temperature, etc.) are constant, the reactor productivity is proportional to the amount of current, Q, used per unit of time.

$$Q = I \cdot t = j \cdot A \cdot t \tag{19}$$

where: I is the current intensity, [A]; t is the time, [sec]; j is the current density, [A/m²]; and A is the electroactive surface of the working electrode, [m²].

Current density is given by Eq. (20):

$$j = -zFI\frac{C_b - C_0}{\delta} \tag{20}$$

where: z is the charge of the useful ion; F is the Faraday constant, [C/mol]; D is the diffusion coefficient, [m²/s]; C_0 is the active species concentration, [mol/m³]; C_b is the species concentration in solution, [mol/m³]; and δ is the Nernst diffusion layer thickness, [m].

From Eqs. (19) and (20) result:

$$Q = -zFD\frac{C_b - C_0}{\delta} \cdot A \cdot t \tag{21}$$

Under constant conditions, the increase of Q can be achieved by increasing the active surface of the working electrode (A) and/or by reducing the Nernst diffusion layer thickness (δ).

The increase of working electrode active surface is possible if three-dimensional electrodes are used, and the decrease of diffusion layer thickness—by electrode-electrolyte relative movement. To increase the electrodes surface are build reactors with a vast number of laminated metal electrodes or with filler electrodes in the form of small pieces, which during cell operation may be in fixed or fluidized bed. The increasing of surface and mass transfer allows the use of continuous processes and more compact systems. Fixed or fluidized bed electrodes are particularly suitable for producing a very high specific active electrode surface (5÷150 cm^{-1}).

The primary applications of these systems are as follows: metal extraction from very poor ores, leaching waters and conventional spent electrolytes, and metal recovery from effluents; in all these cases the metal concentration is low.

The problem of achieving the optimum conditions for mass transfer in diluted solutions electrolysis (e.g., galvanic wash waters) can be solved by using a cathode of electrically conductive particles filler through the liquid to be treated flows. Thus, a large cathodic surface is created, available for the metal electrodeposition, realizing a multiple mass transfer coefficient besides cells with plane electrodes. The fluidized bed electrodes may be inert or active.

A substantial improvement in mass transfer to the electrode surface can be realized by the use of a fluidized bed of inert particles (e.g., glass beads), fluid around the electrode to reduce the diffusion layer. The particles forming the fluidized bed are inert and have the diameter range from hundredths of a millimeter to 1 centimeter.

As applications of fluidized bed electrodes in metal ions recovery could be mentioned: cobalt recovery from 200 ÷ 0.2 ppm (Berzins & Delahay, 1953), copper recovery from concentrated wastewaters of copper nitrate and cadmium recovery to 1 ÷ 5 ppm (Agar, 1947).

27.5.2 *ELECTRODE-ELECTROLYTE RELATIVE MOVEMENT*

The vast majority of the electrodeposition processes are carried out using a DC power source to provide a constant current to the cathode. The optimum current density range for adequate deposits is much lower than the range of limit current density but related to it.

Thus, for a given electrolyte, the deposition degree increasing is favored by I_L increasing through:

- high concentration of dissolved metal;
- high temperatures;
- a relative movement between cathode and electrolyte.

Thus, in practice, is often more advantageous the intensification of electrode-electrolyte movement. In addition to allowing a higher current density was leading to increased production rate, an improvement in the flow regime can help to remove entrapped air or gaseous hydrogen generated on the work-piece or to provide a more stable pH and temperature in the cathode zone.

In electrochemistry, one of the main ways of improving transfer processes is the relative movement of the electrode-electrolyte.

The most critical processes to create a relative rate of the electrolyte to the electrodes are represented by the electrolyte mechanical stirring, recirculation, sonication, and electrode movement by rotation or vibration. The phenomenon is crucial both from the theoretical point of view and by analytical and industrial applications that generate.

27.5.2.1 THE USE OF TURBULENCE PROMOTERS

In the case of electrolytic processes, a method for mass transfer improvement is the use of turbulence promoters, which by gas bubbles releasing, results in turbulence formation in the vicinity of the electrode surface.

This alternative for mass transfer improvement was applied to a system in which the electrochemical reaction at the electrode surface is controlled by the diffusion and gas bubbles release (Vilar et al., 2011). It was noticed that the mechanism which leads to mass transfer improvement, varies depending on the metal electrode type and orientation, and electrolyte flowing direction. Thus, mass transfer is improved if electrode geometry does not prevent the release of gas bubbles generated during the electrochemical process. Also, the authors investigated the effect of using of triangle/semi-circular/cylindrical promoters, placed at the base of a rectangular electrolytic cell. The obtained results have shown that the mass transfer coefficient increases with the increase of height/radius ratio of turbulence promoters (Venkateswarlu et al., 2010).

27.5.2.2   ELECTROLYTE RECIRCULATION

The electrolysis cells with pumping electrolyte circulation are characterized by the absence of electrodes or other moving parts, which leads to a lower energy consumption and mechanical wear. Dynamic flowing regime allows the use of relatively high current densities, even in the case of low concentrations, achieving a well-advanced exhaustion degree of the electrolyzed solution (Pletcher & Walsh, 2000). The cathodes wherein the recovered metal is deposed are periodically removed from the electrolytic bath and replaced with new ones.

27.5.2.3   MECHANICAL STIRRING OF THE ELECTROLYTE

The electrolysis cells with electrolyte mechanical stirring are characterized by an intensive movement of the electrolyte solution, due by mechanical stirrers.

27.5.2.4   ELECTRODES ROTATION

The electrolysis cells with rotated electrodes carried out the relative movement of the electrolyte to the electrode by one of the electrodes rotation, usual by anode rotation.

The literature mentioned experiments studying the rotation effect on the mass transfer in the case of rotated electrode (Eisenberg et al., 1954), and also referred to turned electrodes used to study the dissolution rate of magnesium and its alloys in HCl solution.

For mass transfer study in turbulent flow regime, Recendiz and collaborators have used rotating cylindrical electrodes. Mass transfer control is imposed by the rotation speed of the internal cylinder and by the applied limit current density. Studies regarding mass transfer were realized in the case of rotating cylindrical electrodes (for copper electrodeposition) in turbulent regime (Re > 100). The experimental measurements of the mass transport coefficients were carried out by electrolysis at a given potential (Recendiz et al., 2011).

27.5.2.5   THE VIBRATION OF ELECTRODE/ELECTROLYTE

One of the most efficient methods for mass transfer intensification in electrolysis cells is electrode vibration.

The vibration technique is a method for electrode-electrolyte relative movement intensification defined as initiating and sustaining of a periodic motion with well-defined mechanical parameters (amplitude – A, frequency – ω) between electrode's active surface and electrolyte.

This technique was introduced in electrochemistry by De Bretevile, Harris and Lindsey, around the 1950 years, and was later applied by other authors. By the strong agitation effect on the solution layer near the electrode, the diffusion layer thickness is much diminished, and the limit current density may be increased 10–15 times. As such, the current efficiencies are noticeably enhanced, electrolysis voltage decreases, the specific consumption of electricity is reduced accordingly, and the electrolysis cell productivity considerably increases (Radovan et al., 1982).

The vibrated electrode offers particularly favorable conditions for electrochemical processes carrying out, controlled by mass transfer by ensuring a substantial increase in ionic species transport to and from the electrode surface (mass transfer coefficient k increases as d decreases, which happens in forced convection conditions by h_c and i_{lim} decreasing).

The vibration technique may be applied to remove the metal ions from diluted solutions (at low concentrations). It is known that transport overvoltage occurs only at low concentrations and/or high current densities.

It is given by Eq. (22):

$$\eta_t = \frac{RT}{zF} \ln\left(1 - \frac{i}{i_{lim}}\right) \tag{22}$$

where i_{lim} is the limit current density.

The transport overvoltage can be diminished by solution heating, but mostly by agitation, because in the case of a plane electrode, the cathodic limit current density is given by Eq. (23):

$$i_{lim,c} = \frac{zFD}{t\delta} c \tag{23}$$

where: D is the ion diffusion coefficient that increases with temperature increasing by about 2%/°C; t is the transport number of all ions present in the solution minus transport number of discharging cations; d is the diffusion layer thickness, which appreciably decreases with solution agitation.

The literature records a series of researches on mass transfer intensification by vibrating. There are many equations for mass transfer and some applications with electrodes with various shapes (Fakeeha et al., 1995; Venkata Rao & Venkateswarlu, 2010). In the requests mentioned above, the generating of oscillating field at electrode-electrolyte interface is achieved either by vibrating the electrode surface or the electrolyte volume. Although as relative movement, both processes perform the same, the specific energy consumption is clearly for above process, because the energy associated with the vibration occurs in the limit layer adjacent to the electrode surface. It does not take place in the solution volume, as in the second alternative (Gomaa et al., 2004). Regarding the energy consumption compared to the mechanical vibration parameters (A, ω) was observed (Takahashi et al., 1993) that both amplitude and frequency are increasing, have similar effects on mass transfer enhancing.

Regarding the electrodes vibration, there are two main modes of vibration:

- longitudinal (the movement is parallel to the active surface);
- transverse (the movement is perpendicular to the active surface).

As an example, some devices for the intensification of the electrode-electrolyte movement were patented. They are a vertically vibrated electrode (Mirica et al., 2014) or a vibrating system with multiple possibilities for the electrode movement: longitudinal (horizontally or vertically) or transverse according to the process necessities (Buzatu et al., 2011). Subsequently, some studies were performed on these devices (Iorga et al., 2012).

27.5.2.6   SONICATION OF ELECTRODE/ELECTROLYTE

In recent years began to talk more and more about sono-electrochemistry, the science of ultrasonic field influence on electrochemical processes. In a fluid, the propagation of ultrasonic pressure waves produces the cavitations effect, which consists in bubbles formation, which contains gas and liquid vapors. These bubbles, at low intensity of ultrasonic field, oscillate in the electrolyte volume, contributing mostly to mass transfer and modification of diffusion layer near the electrodes.

From the physical point of view, sounds and ultrasounds as they are perceived represent the result of pressure wave propagation through an elastic medium, where the mechanical energy is converted and transmitted to the propagation medium as sound or ultrasound wave. The spread of ultrasounds in liquids lead to the occurrence of cavitations phenomenon-formation and disintegration of bubbles filled with gas and vapors. Because inside of these bubbles temperature and pressure are very high, their breaking leads to unusual reactions, compared to the situation when the ultrasonic activity is missing.

The "ultrasound" term is used as the name for mechanical particles oscillations of a fluid or solid elastic medium having the upper audible frequency limit considered at 20 kHz. The top boundary of the ultrasonic frequencies is found 100 MHz (Kirk Othmer, 1994).

The application of ultrasounds in electrochemical processes increases the electroactive species transport to/from the electrode, both in the diffusion layer and inside of the double layer, by electrolyte agitation due to pressure wave and cavitations.

In electrochemistry are two main modes for ultrasounds applying:

- working electrode sonication by using the ultrasonic transducer with concentrator;
- electrolyte sonication.

As an example, based on previous works (Iorga et al., 2011), some authors investigated different procedures to improve the mass transfer in the case of Cu electrodeposition from $CuSO_4/H_2SO_4$ solutions. Since it was necessary to observe the influence of the electrode-electrolyte relative movement, the electrodeposition processes were investigated in several operating modes, as follows: electrolyte mechanical stirring, electrode vibration, and electrolyte ultrasonication. Subsequently, comparisons between the obtained results were made, and the conclusion is that the best performance was obtained for the ultrasound-assisted electrodeposition, followed by the vibrating electrode method and by the mechanical stirring of the electrolyte, respectively (Iorga et al., 2012).

27.6 MULTI-/TRANS-DISCIPLINARY CONNECTION(S)

The study of mass transfer in heterogeneous systems and also of hydrodynamic processes that determine it is the primary key of electrochemical reactors engineering and not only. Electrochemical reactors design took full advantage of chemical engineering accumulated knowledge since transport mechanisms and criteria equations properly defined are independent of heterogeneous reaction or by material flow sets on the phase limit.

From an electrochemical point of view, the mentioned effects are studied regarding their influence on electroplating, electrodeposition, and electrosynthesis, increasing the reaction rates and changing the deposits quality

(Doche et al., 2001; Hardcastle et al., 2000; Mason & Lorimer, 2002; Walton and Phull, 1996).

The electrodeposition of various metals and alloys with simultaneous application of ultrasounds is frequently mentioned in the literature. This process provides a considerable improvement in mass transfer, as compared with the conventional electrodeposition methods (Grunwald, 1995). It has been found that the ultrasonic agitation during electrodeposition produces smoother and more compact deposits with high corrosion resistance while decreasing the coating time.

In the case of electrodeposition, the influence of ultrasound intensity and frequency on deposits hardness, particle size, and surfaces morphology is studied.

The use of ultrasonic in electrochemistry presents many advantages, of which one can mention limiting of gas bubbles accumulation on the electrode and a more efficient ions transport through the double layer. Moreover, in these circumstances occurs a permanent cleaning and activation of the electrode surface.

All these improvements determine beneficial effects on diffusion processes, electrodeposition yields, due to the cavitation effect applied to the electrolyte (Malik & Ray, 2009). Collision with the electrode surface causes bubbles breaking and leads to the formation of a liquid jet which is directed at high speed onto the surface. Thus, the double layer is broken, and consequently, mass transfer is improved.

27.7   OPEN ISSUES

In the case of chemical engineering, mass transfer finds broad applications. The main domains where it is used are as follows: engineering of reaction, separations, heat transfer, and more sub-disciplines of chemical engineering.

In electrochemistry, identifying the influence factors on the electrode kinetics and the possibilities of controlling the reaction rate opens up new opportunities for practical electrochemical applications in situations inaccessible by now (treatment of diluted effluents containing metal ions, obtaining of higher yields than those currently accessible in electrochemical reactors, etc.).

Another critical domain where mass transfer is implied is astrophysics. In this field, the mass transfer is the process in which the matter is bound to a body (typically a star) by gravitation. Further on it is filling its Roche lobe and becomes bound by gravity to another body (generally a compact object, which could be a black hole, a neutron, or a white dwarf), and in the end, maybe accreted onto it. In binary systems, it is a general phenomenon and may play a significant role in some types of supernovae and pulsars.

ACKNOWLEDGMENT

This contribution is part of the Nucleus-Programme under the project "Deca-Nano-Graphenic Semiconductor: From Photoactive Structure to The Integrated Quantum Information Transport," PN-18-36-02-01/2018, funded by the Romanian National Authority for Scientific Research and Innovation (ANCSI).

KEYWORDS

- **electrochemistry**
- **mass transfer**
- **mass transport**
- **metal deposition**

REFERENCES AND FURTHER READING

Agar, J. N., (1947). Diffusion and convection at electrodes. *Discussions Faraday Soc., 1*, 26–37.

Agar, J. N., (1953). Electrochemistry and corrosion. *Trans. Faraday Soc., 49*, 533–539.

Antropov, L. I., (2001). *Theoretical Electrochemistry*. University Press of the Pacific, Honolulu, Hawaii.

Bagotzky, V. S., (1993). *Fundamentals of Electrochemistry*. Plenum Press, New York.

Berzins, T., & Delahay, P., (1953). Oscillographic polarographic waves for the reversible deposition of metals on solid electrodes. *J. Am. Chem. Soc., 75*(3), 555–559.

Bird, R. B., Stewart, W. E., & Lightfoot, E. N., (1976). *Transport Phenomena*. John Wiley & Sons.

Blackmore, J., (1995). *Ludwig Boltzmann–His Later Life and Philosophy, 1900–1906, Book One: A Documentary History*. Kluwer.

Bokshtein, B. S., Mendelev, M. I., & Srolovitz, D. J., (2005). *Thermodynamics and Kinetics in Materials Science: A Short Course* (pp. 167–171). Oxford University Press: Oxford.

Bond, A. M., (1990). *Broadening Electrochemical Horizons: Principles and Illustration of Voltammetric and Related Techniques*. John Wiley, New York.

Bowler, P. J., & Morus, I. R., (2005). *Making Modern Science: A Historical Survey*. University of Chicago Press.

Bradley, M., (1998). *A Career Biography of Gaspard Clair Francois Marie Riche De Prony.* Bridge-Builder, Educator, and scientist, Mellen Press.

Buzatu, D., Iorga, M., Mirica, M. C., Urmosi, Z., Pop, R., Balcu, I., & Mirica, N., (2011). *Sistem de Vibrare a Elementelor statice/Electrozilor cu Aplicații în procese chimice și electrochimice (In English: Vibration System for Static Elements / Electrodes with Applications in Chemical and Electrochemical Processes)*, Patent Application A/00478/2011, OSIM Romania.

Crank, J., (1980). *The Mathematics of Diffusion.* Oxford University Press.

Cropper, W. H., (2001). The greatest simplicity. In: Willard Gibbs, (ed.), *Great Physicists* (pp. 106–123). Oxford University Press.

Doche, M. L., Hihn, J. Y., Touyeras, F., Lorimer, J. P., Mason, T. J., & Plattes, M., (2001). Electrochemical behavior of zinc in 20 kHz sonicated NaOH electrolytes. *Ultrasonics Sonochemistry, 8*(3), 291–298.

Eisenberg, M., Tobias, C. W., & Wilke, C. R., (1954). Ionic mass transfer and concentration polarization at rotating electrodes. *Journal of the Electrochemical Society, 101,* 306–320.

Facsko, G., (1969). *Tehnologie electrochimică* (In English: Electrochemical Technology). Tehnică, București, Romania.

Facsko, G., (1979). *Electrochimie și coroziune–Curs* (in English: Electrochemistry and Corrosion-Lectures). Inst. Politehnic"TraianVuia"Timișoara, Romania.

Fakeeha, A. H., Abdel, A. F. A., & Inamah, I. A., (1995). Mass transfer at vibrating electrodes in continuous flowing electrolyte systems. *The 4th Saudi Engineering Conference,* 81–88.

Fick, A., (1855). On liquid diffusion, *Poggendorffs Annalen* 94, 59–(1995). *Reprinted in Journal of Membrane Science, 100,* pp. 33–38.

Fleig, J., & Jamnik, J., (2005). Work function changes of polarized electrodes on solid electrolytes. *Journal of the Electrochemical Society, 152*(4), E138–E145.

Gomaa, H., Al Taweel, A. M., & Landau, J., (2004). Mass transfer enhancement at vibrating electrodes. *Chemical Engineering Journal, 97,* 141–149.

Grünwald, E., (1995). *Tehnologii moderne de galvanizare în industria electronică și electrotehnică* (Engl: Modern Electroplating Technologies in the Industry of Electronics and Electrotechnics). Casa Cărții de Știință, Cluj-Napoca, Romania.

Hardcastle, J. L., Ball, J. C., Hong, Q., Marken, F., Compton, R. G., Bull, S. D., & Davies, S. G., (2000). Sonoelectrochemical and sonochemical effects of cavitation: Correlation with interfacial cavitation induced by 20 kHz ultrasound. *Ultrasonics Sonochemistry, 7*(1), 7–14.

Hastings, C. S., (1909). Josiah Willard Gibbs, *Biographical Memoirs of the National Academy of Sciences, 6,* 373–393.

Iorga, M. I., Mirica, N., Pop, O. R., Buzatu, D., & Mirica, M. C., (2012). Mass transfer improvement in electrochemical copper recovery using an electrode vibrating system. In: *Proceedings of 20th International Congress of Chemical and Process Engineering CHISA 2012* (pp. 1451). Prague, Czech Republic.

Iorga, M. I., Pop, R., Mirica, M. C., & Buzatu, D., (2012). Methods for the enhancement of mass transport at the recovery of metal ions from hydroelectrometallurgical processes. In: *Dr Mohammad, N.,* (ed.), *Recent Researches in Metallurgical Engineering: From Extraction to Forming.* ISBN: 978–953–51–0356–1, InTech, doi: 10,5772/38717.

Iorga, M., Mirica, M. C., Balcu, I., Mirica, N., & Rosu, D., (2011). Influence of relative electrode-electrolyte movement over productivity for silver recovery from diluted solutions. *Journal of Chemistry and Chemical Engineering–JCCE*, *5*(4), 296–304.

Kirk-Othmer, (1994). *Encyclopedia of Chemical Technology* (IV edn., Vol. 9). Wiley & Sons.

Koryta, J., Dvorak, J., & Kavan, L., (1993). *Principles of Electrochemistry*. John Wiley & Sons Ltd., New York.

Mallik, A., & Ray, B. C., (2009). Morphological study of electrodeposited copper under the influence of ultrasound and low temperature. *Thin Solid Films*, *517*, 6612–6616.

Mason, T. J., & Lorimer, J. P., (2002). *Applied Sonochemistry. The Uses of Power Ultrasound in Chemistry and Processing*, Wiley VCH, Weinheim.

Mirica, M. C., Iorga, M., Balcu, I., & Mirica, N., (2014). *Dispozitiv pentru realizarea mişcării electrod-electrolit prin vibrarea electrodului* (Engl: *Device for Electrode-Electrolyte Movement by Electrode Vibration*), Patent no. 127414, OSIM Romania.

Newman, J. S., & Tobias, C. W., (1962). Theoretical analysis of current distribution in porous electrodes. *J. Electrochem. Soc., 109*(12), 1183–1191.

Oniciu, L., & Constantinescu, E., (1982). *Electrochimie şi coroziune* (Engl: Electrochemistry and Corrosion), Did. Ped., Bucureşti, Romania.

Paunovic, M., (1998). *Fundamentals of Electrochemical Deposition*. John Wiley & Sons Inc., New York.

Pletcher, D., & Walsh, F. C., (2000). *Industrial Electrochemistry* (2nd edn., p. 228).

Radovan, C., Mirica, N., & Angheluş, A., (1982). *Scientific and Technical Bulletin of IPTV Timisoara. Chemistry Series*, *27*, 41.

Recendiz, A., Leon, S., Nava, J. L., & Rivera, F. F., (2011). Mass transport studies at rotating cylinder electrode during zinc removal from dilute solutions. *Electrochimica Acta*, *56*, 1455–1459.

Rolle, K. C., (2014). *Heat and Mass Transfer*. Cengage Learning, USA.

Schlesinger, M., & Paunovic, M., (2000). *Modern Electroplating*. John Wiley & Sons Inc., New York.

Selman, J. R., & Tobias, C. W., (1978). In: Drew, T., (ed.), *Advances in Chemical Engineering* (Vol. 10). Academic Press, New York.

Simmons, C. T., (2008). Henry Darcy (1803–1858), immortalized by his scientific legacy. Hydrogeology Journal, 16, 1023–1038.

Takahashi, K., Kawabata, S., Mori, S. K. S., & Tanimoto, A., (1993). Some features of combined convection mass transfer for pulsating fluid flow in a pipe. *Kagaku KogakuRombunshu*, *19*(1), 127–130.

Tobias, C. W., (1994). *The beginnings of electrochemical engineering at Berkeley. Interface*, *3*(3), 17.

Vaszilcsin, N., (1995). *Electrochimie–Curs* (Engl: Electrochemistry-Lectures). Univ. Politehnica, Timişoara, Romania.

Venkata R. P., & Venkateswarlu, P., (2010). Effect of vibrating disc on ionic mass transfer in an electrolytic cell. *International Communications in Heat and Mass Transfer*, *37*, 1261–1265.

Venkateswarlu, P., Chittibabu, N., & Subba, R. D., (2010). Effect of turbulence promoters on ionic mass transfer in a rectangular electrolytic cell. *Indian Journal of Chemical Technology*, *17*, 260–266.

Vilar, E. O., Cavalcanti, E. B., & Albuquerque, I. L. T., (2011). A mass transfer study with electrolytic gas production. In: Mohamed El-Amin, (ed.), *Advanced Topics in Mass Transfer*. ISBN: 978-953-307-333-0, InTech.

Walton, D. J., & Phull, S. S., (1996). Sonoelectrochemistry. *Advances in Sonochemistry, 4,* 205–284.

Wendt, H., & Kreisa, G., (1999). Electrochemical engineering. *Science and Technology in Chemical and Other Industries*, Springer Verlag, Berlin.

Willard, G. J., (*2012*). *Physics History. American Physic*al Society. Retrieved 16 Jun 2012.

CHAPTER 28

Molecular Classification

FRANCISCO TORRENS[1] and GLORIA CASTELLANO[2]

[1]*Institute for Molecular Science, University of Valencia, PO Box 22085, E-46071 Valencia, Spain, Tel: +34-963-544-431, Fax: +34-963-543-274, E-mail: torrens@uv.es*

[2]*Department of Experimental Sciences and Mathematics, Faculty of Veterinary and Experimental Sciences, Valencia Catholic University Saint Vincent Martyr, Guillem de Castro-94, E-46001 Valencia, Spain*

28.1 DEFINITION

A molecular classification is defined based on a similarity matrix $\mathbf{R}$, in which IE is calculated by Eq. (4). The distance D between two similarity matrices R and $\mathbf{W}$ is computed by Eq. (5).

28.2 HISTORICAL ORIGIN

The further development of Clausius phenomenological thermodynamics and its extension to heterogeneous systems in the 1880s is primarily because of Helmholtz and Gibbs. In the 20th century, the focus of interest moved to irreversible processes, with particular emphasis on biophenomena (Prigogine, Onsager, and Gyarmati). Glansdorff & Prigogine (1971) devised a theorem, according to which the stationary states of open homogeneous systems are cases of maximum EP. Many works were devoted to molecular classification (Hohne & Pierce, 1989; Pierce & Hohne, 1986; Varmuza, 1980). Because of the strongly combinatory nature of the problem, in practice, it is impossible to study all the possible classification trees. Heuristic and qualitative rules are used to generate or select quasi-optimum classification trees (Benzécri, 1980).

28.3 NANO-SCIENTIFIC DEVELOPMENT

The initial step in quantifying the concept of similarity for chemical species is to list their most important structural elements or properties (Iordache, 2011, 2012, 2014). A vector is associated with every species. The vector components take the value either "1" or "0," where "1" means the presence of a given structural element or property, whereas "0" means its absence; e.g., "1" may correspond to a high value of the property, e.g., hydrophilicity, whereas "0" corresponds to a low value. Binary bit string representations of molecular structure and properties (fingerprints) are standard tools to analyze chemical similarity (Willett, 1998). The vectors associated with the chemicals are denoted: $i = <i_1, i_2, \ldots, i_k, \ldots>$, where i_k are either "1" or "0." The binary characterization according to the presence ("1") or absence ("0") of a given property was initially used. The application of multi-valued digits or real properties instead of Boolean ones was tested. A hierarchy of the structural elements or properties is required; e.g., assume that the property indexed by i_1 is more significant than that indexed by i_2, this more significant than i_3, etc. in the order of the coordinates in the associated vectors. A similarity matrix is associated with any set of chemicals, and an IE to the former. A similarity index r_{ij} between two different species $i = <i_1, i_2, \ldots, i_k, \ldots>$ and $j = <j_1, j_2, \ldots, j_k, \ldots>$ is defined as:

$$r_{ij} = \sum_k t_k (a_k)^k \, ; \, k = 1, 2 \ldots \tag{1}$$

where $0 \leq a_k \leq 1$ and $t_k = 1$ if $i_k = j_k$, $t_k = 0$ if $i_k \neq j_k$ for all k. The entire system is characterized by matrix $\mathbf{R} = [r_{ij}]$. The similarity index should have the natural properties of reflexivity ($r_{ii} = 1$) and symmetry ($r_{ij} = r_{ji}$). The definition assigns a coefficient of weight a_k to any property involved in the description of the species i and j, provided that Boolean values i_k and j_k be the same for both chemicals. The fact that the relationship r_{ij} be reflexive and symmetric allows a partition of the set into not-necessarily-disjoint classes. A class consists of a number of similar chemical species gathered together. In order to limit the study to partition into disjoint classes, the defined similarity must be transitive, i.e., $\min_k(r_{ik}, r_{kj}) \leq r_{ij}$. The procedure to ensure transitivity is that the classification algorithm starts from the stable similarity matrix. In order to obtain a stable matrix, the sequence $\mathbf{R}, \mathbf{R}(2), \ldots, \mathbf{R}(k), \ldots$ with $\mathbf{R}(2) = \mathbf{R}o\mathbf{R}$ and $\mathbf{R}(k) = \mathbf{R}(k-1)o\mathbf{R}$ is calculated. The composition rule o is given by:

$$(\mathbf{R}o\mathbf{W})_{ij} = \max_k \left[\min(r_{ik}, w_{kj}) \right] \tag{2}$$

where $\mathbf{R} = [r_{ij}]$, $\mathbf{W} = [w_{ij}]$ are two arbitrary matrices of the same type. The composition equation calculates the (i,j)-th element of the matrix $\mathbf{R}o\mathbf{W}$, which consists in taking the smallest of the two elements r_{ik} and w_{kj}, for a given row i of R and a column j of $\mathbf{W}$, then repeat the procedure for all k and select the largest of all resulting elements. An integer n exists such that from $n$on, the matrix is stable to the composition rule o, so that $\mathbf{R}(n) = \mathbf{R}(n+1)$, etc. The elements of the stable similarity matrix $\mathbf{R}(n)$ verify symmetry, reflexivity, and transitivity. Denote by $r_{ij}(n)$ the elements of the stable matrix $\mathbf{R}(n)$. The partition is based on the degree of classification T with $0 \leq T \leq 1$. The classification rule follows: two species i and j are assigned to the same class if $r_{ij}(n) \geq T$. Applying the rule, the set of classes at the degree of classification T is obtained. For $T = 0$, a unique class result including all species, whereas for $T = 1$, every class includes one species. When T varies from 0 to 1, different sets of classes arise. Actually, a new set of classes arises every time T crosses the value of a similarity index r_{ij} of matrix $\mathbf{R}$. A general tree of classes is built, which is the expected flow sheet. The class of i, noted $\hat{i}$, is the set of species j that satisfy the rule: $r_{ij} \geq T$. The similarity matrix of classes $\hat{\mathbf{R}}$ is built as:

$$\hat{\mathbf{R}}_{\hat{i}\hat{j}} = \max(r_{wt}); w \in \hat{i}, t \in \hat{j} \tag{3}$$

where w designates any index of species belonging to the class of i and, similarly, t, any index referring to the class of j. To any similarity matrix $\mathbf{R}$, IE $H(\mathbf{R})$ is associated:

$$H(\mathbf{R}) = -\sum r_{ij} \ln r_{ij} - \sum (1 - r_{ij}) \ln(1 - r_{ij}) \tag{4}$$

which expresses the quantity of IE associated with the matrix of design. The defined IE is a measure of the imprecision in classifying experiments. In order to compare two similarity matrices $\mathbf{R} = [r_{ij}]$ and $\mathbf{W} = [w_{ij}]$, a distance D is introduced:

$$D(\mathbf{R}, \mathbf{W}) = -\sum r_{ij} \ln(r_{ij}/w_{ij}) - \sum (1 - r_{ij}) \ln\left[(1 - r_{ij})/(1 - w_{ij})\right] \tag{5}$$

which measures the discrepancy between two similarity matrices and associated classification. In the classification algorithm, every hierarchical tree corresponds to an IE dependence *vs.* the grouping level b and an H–b diagram is obtained Tondeur & Kvaalen (1987). EC is proposed as a selection criterion between different variants, resulting from classification between hierarchical trees (Corriou et al., 1991; Iordache et al., 1988, 1990, 1991, 1993a,b). According to EC, for a given task, the best configuration of a flow sheet is that in which EP is the most uniformly distributed. The EC implies an EP linear dependence *vs.* grouping level, so that EL is:

$$H_{eqp} = H_{max} b \tag{6}$$

As the classification is discrete, EC is a regular staircase function. The best variant is chosen to be that minimizing the sum of the squares of the deviations:

$$S = \sum_{bi} \left(H - H_{eqp} \right)^2 \tag{7}$$

28.4   NANO-CHEMICAL APPLICATION

The situation above allows organizing and systematizing the synthesis of classification trees with the help of EP notion, which is interesting in questions of organization, system structure with the help of a concept coming from the thermodynamics of irreversible processes. In the longer view, other analogous situations are identified in physics, mathematics, and thermodynamics. In the last, the approach was made between EC and Glansdorff & Prigogine's theorem (1971), according to which the stationary states of open homogeneous systems are cases of maximum EP. In the restrictive framework of the linear thermodynamics of irreversible processes, EC applied to the time variable is equivalent to define the stationary state as a case with minimum EP and includes Glansdorff & Prigogine's theorem (1971). The approach comprises a supplementary side, which is that of the hierarchical, multiscale systems organized at different complexity levels. The non-Archimedean analysis is the mathematical tool adapted to the study of the hierarchical and multi-scale systems. The complexity levels correspond to the successive branching levels in the splitting schemes (classification trees), and the optimality criterion is the most uniform possible distribution of EP on the different complexity levels. In other situations, the complexity levels correspond to different scales of time or space. It was proposed the conjecture that the natural evolution of a system out of equilibrium, in a state of minimum EP, remains variable in the case of evolution on different time scales. In the systems far from equilibrium and that organize themselves, i.e., a morphology appears, it seems possible to have an EP nonuniform distribution, on the different organization levels, whereas the higher-order derivatives of entropy with respect to time have the property of stationarity. The properties are linked to the nonlinear character of the flow-motive power relationships. In morphogenesis, the formulation of a GEC is related to the selection laws among different used morphologies, e.g., in the study of phase transitions (crystallization). Every morphology corresponds to a complexity level. The forms have different dimension scales and, taking into account the crossing velocities, different time scales. Hill conjectured that the transitions from a crystalline form always take place to those corresponding to the greatest EP. In agreement with the formulation of the presented GEC, the systems pass from one complexity level to another via all possible intermediate levels. On the contrary, if one passes from one level to the following, avoiding the intermediate levels without occupying them, the curve of entropy vs. level goes more away from EL and a greater value of the global criterion, Eq. (7), results. Another class of relationships of the theory with the preceding ones lies in the observation that, in the neighborhood of some changes of the splitting schemes or classification trees (structure), the global criterion presents a relative extreme (minimum) followed by a discontinuity. It is possible to relate the minimum to the trend to the self-organized criticality in percolation. A conjecture states that certain systems automatically tend to function in the neighborhood of their transition threshold. It is considered that regulating mechanisms exist, which allow the operation of a complex system at the limit where it can just work, at which place a phase transition is a possible fact, but it is not produced. The search of global variational criteria, ruling the spontaneous evolution of complex systems, morphogenesis or processes optimization is an avenue of research. It is about finding one (or some) function having the properties of extreme (Lyapunov type) in the neighborhood of the studied state, regime or path. It does not seem a priori possible but starting from general principles of the type minimum EP or even principle of minimum action. Nicolis & Prigogine (1977) reviewed self-organization in nonequilibrium systems.

28.5 MULTI-/TRANS-DISCIPLINARY CONNECTION

The theory was applied to problems of PR in chemistry: (1) structural elucidation of spectra; (2) chromatography; (3) materials classification and analysis; (4) prediction of bioactivity; (5) medicinal chemistry; (6) pharmacological chemistry, drug design and development; (7) environmental chemistry; (8) synthetic organic/inorganic chemistry; (9) codification of chemical structures and taxonomy (*see* Chapter 23).

28.6 OPEN ISSUES

The value of the methods of PR in chemistry is not sure. Their applications are limited to situations in which one has not detailed theories to the interpretation of the results. The domain raises problems of cooperation between different specialists: experts in the particular domain, connoisseur in PR and master in information technology. A necessity of collaboration exists as chemists are increasingly confronted with a great amount of data to interpret. It is interesting to study a possible generalization of ideas to IE.

KEYWORDS

- **classification rule**
- **entropy production**
- **equipartition conjecture**
- **information entropy**
- **the degree of classification**

REFERENCES AND FURTHER READING

Alfonseca, M., (1996). *Diccionario Espasa: 1,000 Grandes Científicos*. Espasa Calpe: Madrid, Spain.

Benzécri, J. P., (1980). L'Analyse des Données I. La Taxonomie. Dunod : Paris, France.

Corriou, J. P., Iordache, O., & Tondeur, D., (1991). Classification of biomolecules by information entropy. *J. Chim. Phys. Phys. Chim. Biol.*, *88*, 2645–2652.

Glansdorff, P., & Prigogine, I., (1971). *Thermodynamic Theory of Structure, Stability, and Fluctuations*. Wiley: London, UK.

Hohne, B. A., & Pierce, T. H., (1989). *Expert Systems: Applications in Chemistry*. ACS Symp. Ser. No. 408, ACS: Washington, DC.

Iordache, O. M., Maria, G. C., & Pop, G. L., (1988). Lumping analysis for the methanol conversion to olefins kinetic model. *Ind. Eng. Chem. Res.*, *27*, 2218–2224.

Iordache, O., (2011). *Modeling Multi-Level Systems*. Springer: Berlin, Germany.

Iordache, O., (2012). *Self-Evolvable Systems: Machine Learning in Social Media*. Springer: Berlin, Germany.

Iordache, O., (2014). *Polytope Projects*. CRC: Boca Raton, FL.

Iordache, O., Bucurescu, I., & Pascu, A., (1990). Lumpability in compartmental models. *J. Math. Anal. Appl.*, *146*, 306–317.

Iordache, O., Corriou, J. P., & Tondeur, D., (1991). Neural network for systemic classification and process fault detection. *Hung. J. Ind. Chem. Veszprém*, *199*, 265–274.

Iordache, O., Corriou, J. P., & Tondeur, D., (1993b). Separation sequencing: Use of information distance. *Can. J. Chem. Eng.*, *71*, 955–966.

Iordache, O., Corriou, J. P., Garrido-Sanchez, L., Fonteix, C., & Tondeur, D., (1993a). Neural network frames: Application to biochemical kinetic diagnosis. *Computers Chem. Eng., 17,* 1101–1113.

Nicolis, G., & Prigogine, I., (1977). *Self-Organization in Nonequilibrium Systems: From Dissipative Structures to Order through Fluctuations.* Wiley: New York, NY.

Pierce, T. H., & Hohne, B. A., (1986). *Artificial Intelligence: Applications in Chemistry.* ACS Symp. Ser. No. 306, ACS: Washington, DC.

Tondeur, D., & Kvaalen, E., (1987). Equipartition of entropy production: An optimality criterion for transfer and separation processes. *Ind. Eng. Chem. Res., 26,* 50–56.

Varmuza, K., (1980). *Pattern Recognition in Chemistry.* Springer: New York, NY.

Willett, P., (1998). Chemical similarity searching. *J. Chem. Inf. Comput. Sci., 38,* 983–996.

CHAPTER 29

Molecular Devices and Machines

MARGHERITA VENTURI

Department of Chemistry "Giacomo Ciamician," University of Bologna, Italy

29.1 DEFINITION

The concept of a macroscopic device and machine can be extended to the molecular level (Scheme 29.1). A molecular-level device can be defined as an assembly of a discrete number of molecular components designed to achieve a specific function. A molecular-level machine is a particular type of molecular-level device in which the component parts can display changes in their relative position as a result of some external stimuli.

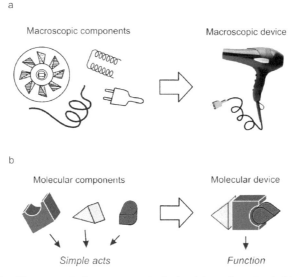

a

Macroscopic components   Macroscopic device

b

Molecular components   Molecular device

Simple acts   Function

SCHEME 29.1   The concept of a macroscopic device (a) can be extended to the molecular level (b).

29.2   FROM THE MACROSCOPIC WORLD TO THE NANOWORLD

In everyday life, we make extensive use of macroscopic devices. A macroscopic device is an assembly of components designed to achieve a specific function. Each component of the device performs a simple act, while the entire device performs a more complex function, characteristic of the assembly. For example, the function performed by a hairdryer (production of hot wind) is the result of acts performed by a switch, a heater, and a fan, suitably connected by electric wires and assembled in an appropriate framework (Scheme 29.1a).

The new trend is to reduce the dimensions and weight of the component parts as much as possible. The most important and common examples are in the information technology domain, but there are some other fields that will be the beneficiary of this trend, such as medicine, energy, materials, and environment. Miniaturization is, therefore, a must for the actual progress and the scientific achievements of the last few years show that the concept of device and machine can be extended at the molecular level, i.e., at the nanometer scale, that at present seems to be the ultimate frontier to miniaturization.

In agreement with the definition of devices and machines of the macroscopic world, a molecular device is an assembly of a discrete number of molecular components designed to achieve a specific function. Each molecular component performs a single operation; meanwhile, the assembly as a whole performs a complex function, as a result of the component cooperation (Scheme 29.1b). A molecular machine is, instead, a particular type of device in which the constituent molecular parts can display changes in their relative positions upon the influence of an external stimulus (Balzani et al., 2008).

Molecular-level devices operate via electronic and/or nuclear rearrangements and, like the macroscopic ones, are characterized by (i) the kind of energy input supplied to make them work, (ii) the way in which their operation can be monitored, (iii) the possibility to repeat the operation at will (cyclic process), (iv) the time scale needed to complete a cycle, and (v) the performed function.

As far as point (i) is concerned, a chemical reaction can be used, at least in principle, as an energy input. In such a case, however, if the device has to work cyclically [point (iii)], it will need the addition of reactants at any step of the working cycle, and the accumulation of by-products resulting from the repeated addition of matter can compromise the operation of the device. On the basis of this consideration, the best energy inputs to make a molecular device work are photons and electrons. It is indeed possible to design very interesting molecular devices based on appropriately chosen photochemically (Credi et al., 2012) and electrochemically (Credi et al., 2011) driven reactions (Balzani et al., 2008).

In order to control and monitor the device operation [point (ii)], the electronic and/or nuclear rearrangements of the component parts should cause readable changes in some chemical or physical property of the system. In this regard, photochemical and electrochemical techniques are very useful since both photons and electrons can play the dual role of "writing" (i.e., causing a change in the system) and "reading" (i.e., reporting the state of the system).

The operation time scale of molecular devices [point (iv)] can range from less than picoseconds to seconds, depending on the type of rearrangement (electronic or nuclear) and the nature of the components involved.

Finally, as far as point (v) is concerned, molecular-level devices performing various kinds of functions can be imagined that open the way to possible and interesting applications of these systems.

29.3   THE WAY TO CONSTRUCT MOLECULAR DEVICES

The miniaturization of device and machine components is currently pursued by a "top-down" approach. This approach, which is in the hands of physicists and engineers, consists of manipulating progressively smaller pieces of matter by photolithography and related techniques. However, it is becoming clear that the top-down approach is subject to drastic limitations, including a severe cost escalation when the components approach the nanometer dimension. But "there is plenty of room at the bottom" for further miniaturization, as Richard P. Feynman (1960) stated in his famous address to the American Physical Society on December 29, 1959. Then, to proceed toward miniaturization at the nanometer scale, new avenues have to be found.

A promising strategy to exploit science and technology at the nanometer scale is the "bottom-up" approach, which starts from nano- or subnanoscale objects (namely atoms or molecules) to build up nanostructures.

29.3.1   *BOTTOM-UP ATOM-BY-ATOM*

The idea that atoms could be used to construct nanoscale devices and machines was first raised by Feynman during his address already mentioned ("The principles of physics do not speak against the possibility of maneuvering things atom by atom"), and depicted in a visionary way in the mid-1980s by K. Eric Drexler (Drexler 1986, 1992). He claimed that it would be possible to build a general purpose "nanorobot," nicknamed an assembler. Such a device could, in principle, build atom-by-atom almost anything, including copies of itself. The fascinating but, admittedly, abstract atom-by-atom construction of nanoscale devices and machines, however, which was so much appealing to some physicists, did not convince chemists, who are well aware of the high reactivity of most atomic species and of the subtle aspects of chemical bond. Chemists, indeed, know that atoms are not simple spheres that can be moved from one place to another place at will. Atoms do not stay isolated; they bond strongly to their neighbors, and it is difficult to imagine that the atoms can be taken from a starting material and transferred to another material.

29.3.2 BOTTOM-UP MOLECULE-BY-MOLECULE

In the late 1970s, a new branch of chemistry, called *supramolecular chemistry*, emerged and expanded very rapidly, consecrated by the award of the Nobel Prize in Chemistry to C.J. Pedersen, D.J. Cram, and J.-M. Lehn in 1987.

In the frame of research on supramolecular chemistry, the idea began to arise that molecules are much more convenient building blocks than atoms to construct nanoscale devices and machines. The main reasons at the basis of this idea are as follows: (i) molecules are stable species, whereas atoms are difficult to handle; (ii) Nature starts from molecules, not from atoms, to construct the great number and variety of nanodevices and nanomachines that sustain life; (iii) most of the laboratory chemical processes are dealing with molecules, not with atoms; (iv) molecules are objects that already exhibit distinct shapes and carry device-related properties; and (v) molecules can self-assemble or can be connected to make larger structures.

In the following years, supramolecular chemistry grew very rapidly, and it became clear that the "bottom-up" approach based on molecules opens virtually unlimited possibilities concerning the design and construction of artificial molecular-level devices and machines.

It must be added that quite recently new and very powerful techniques have confirmed the concept of molecules as nanoscale objects exhibiting their own shape, size, and properties (Giessibl 2003; Otero et al., 2006; Rigler et al., 2001). Single-molecule fluorescence spectroscopy and the various types of probe microscopes are, indeed, capable of "seeing" (Gimzewski et al., 1999)

(Figure 29.1, top) or "manipulating" (Hla et al., 2001) (Figure 29.1, bottom) single molecules, and can even investigate bimolecular chemical reactions at the single-molecule level.

Much of the inspiration to construct molecular-level devices and machines comes from the outstanding progress of molecular biology that has begun to reveal the secrets of the natural molecular-level devices and machines (Cramer, 1993; Goodsell, 1996, 2004; Jones, 2004) which constitute the material base of life (Figure 29.2).

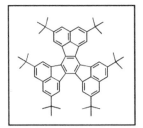 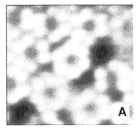 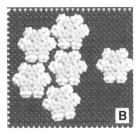

FIGURE 29.1 Top: model of the molecular structure of hexa-*tert*-buthyldecacylene (HBD); (A) STM image of HBD molecules on an atomically clean Cu(100) surface; (B) snapshot of the molecular mechanical simulations based on the molecular coordinates taken from the STM data. Bottom: the millennium year number 2000 (16.3 nm length) that has been written by using 47 CO single molecules; each protrusion represent an individual CO molecule, and the background vertical lines denote the intrinsic Cu surface step edges.

Bottom-up construction of devices and machines as complex as those present in Nature is, of course, an impossible task. Therefore, chemists have tried to construct much simpler systems, without mimicking the complexity of the biological structures.

In the last 20 years, synthetic talent that has always been the most distinctive feature of chemists, combined with a device-driven ingenuity evolved from chemists' attention to functions and reactivity, have led to outstanding achievements in this field (Balzani et al., 2008).

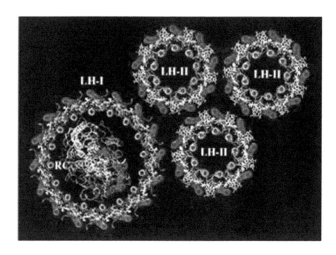

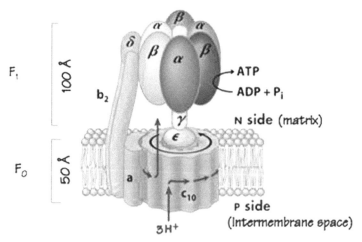

FIGURE 29.2 Top: architecture of the light-harvesting device of purple bacteria. Bottom: the protein motor that accomplishes the ATP synthesis.

It should be pointed out, however, that nanoscale devices and machines cannot be considered merely as 'shrunk' versions of macroscopic counterparts because physics is different at the nanoscale. Several phenomena at the nanoscale are, indeed, governed by the laws of quantum mechanics and, most importantly, some intrinsic properties of molecular level entities are quite different from those of macroscopic objects. For example, molecules are in a state of constant random motion and are subjected to continual collisions (Brownian motion); furthermore, in the nanoworld, things are somewhat floppy and stick strongly to each other because of electromagnetic interactions.

29.4 ARTIFICIAL MOLECULAR DEVICES

Some paradigmatic examples (developed in the research group where the author of this contribution works) of three families of molecular-level devices are illustrated in the following: (1) devices for the transfer of electrons or electronic energy; (2) devices capable of performing extensive nuclear motion, also called molecular machines; and (3) devices whose function implies the occurrence of both electronic and nuclear rearrangements.

29.4.1 DEVICES FOR THE TRANSFER OF ELECTRONS OR ELECTRONIC ENERGY

Apart from futuristic applications, related, e.g., to the construction of a chemical computer, the design and realization of a molecular-level electronic set (i.e., a set of molecular-level systems capable to play functions that mimic those performed by macroscopic components in electronic devices) is of great scientific interest. It introduces, indeed, new concepts in the field of chemistry and stimulates the ingenuity of research workers engaged in the emerging field of nanotechnology. For space reasons, only two examples are illustrated.

29.4.1.1 MOLECULAR WIRES

An important function at the molecular level is photoinduced energy- and electron-transfer over long distances and/or along predetermined directions. This function can be obtained by linking donor and acceptor components by a rigid spacer. An example (Schlicke et al., 1999) is given by the $[Ru(bpy)_3]^{2+} - (ph)_n - [Os(bpy)_3]^{2+}$ compounds (ph = 1,4-phenylene; n = 3, 5, 7) in which excitation of the $[Ru(bpy)_3]^{2+}$ unit is followed by electronic energy transfer to the ground state of the $[Os(bpy)_3]^{2+}$ unit, as shown by the sensitized emission of the latter. For the system with n = 7 shown in the top of Figure 29.3, the rate constant of energy transfer over the 4.2 nm metal-to-metal distance is 1.3×10^6 s^{-1}.

In the $[Ru(bpy)_3]^{2+} - (ph)_n - [Os(bpy)_3]^{3+}$ compounds, obtained by chemical oxidation of the Os-based moiety of the previously described systems, photoexcitation of the $[Ru(bpy)_3]^{2+}$ unit causes the transfer of an electron to the Os-based one with a rate constant of 3.4×10^7 s^{-1} when $n = 7$ (Figure 29.3, bottom). Unless the electron added to the $[Os(bpy)_3]^{3+}$ unit is rapidly removed, a back electron transfer (rate constant 2.7×10^5 s^{-1} for $n = 7$) takes place from the $[Os(bpy)_3]^{2+}$ unit to the $[Ru(bpy)_3]^{3+}$ one.

Spacers with energy levels or redox states in between those of the donor and acceptor units may help energy- or electron-transfer (hopping mechanism). Spacers whose energy or redox levels can be manipulated by an external stimulus can play the role of switches for the energy- or electron-transfer processes.

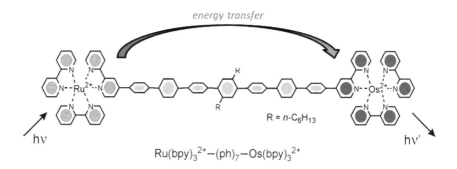

$$Ru(bpy)_3^{2+}-(ph)_7-Os(bpy)_3^{2+}$$

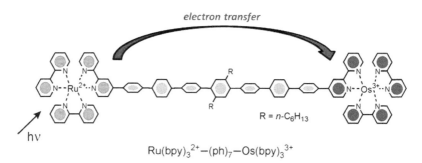

$$Ru(bpy)_3^{2+}-(ph)_7-Os(bpy)_3^{3+}$$

FIGURE 29.3 Photoinduced energy-transfer (top) and electron-transfer (bottom) processes occurring in $[Ru(bpy)_3]^{2+}-(ph)_n-[Os(bpy)_3]^{2+}$ and $Ru(bpy)_3^{2+}-(ph)_n-[Os(bpy)_3]^{3+}$ systems, respectively.

29.4.1.2   ANTENNA SYSTEMS

Dendrimers (Campagna et al., 2012) are well-defined macromolecules exhibiting a tree-like structure (Figure 29.4). Like trees, dendrimers usually exhibit aesthetically pleasant structures. However, as in the case of a tree, the interest in a specific dendrimer does not depend on its beauty, but on the 'fruit' (i.e., the specific function) that it is able to produce.

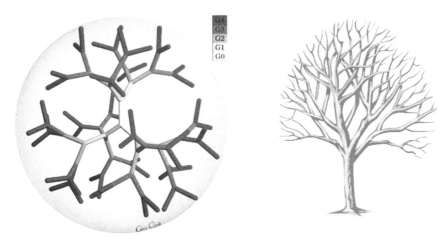

FIGURE 29.4 Schematic representation of a dendrimer, a macromolecule exhibiting a tree-like structure.

In suitably designed dendrimers, electronic energy transfer can be channeled towards a specific position of the array. Compounds of this kind play the role of antenna systems for light harvesting. A number of tree-like multi-center transition-metal complexes based on Ru(II) and Os(II) as metals, 2,3- and 2,5-bis(2-pyridyl)pyrazine (2,3- and 2,5-dpp) as bridging ligands, and 2,2'-bipyridine (bpy) and 2,2'-biquinoline (biq) as terminal ligands (Figure 29.5, top) have been prepared with this aim (Balzani et al., 1998).

Dendrimers containing up to 22 metal ions, 21 bridging ligands (2,3-dpp), and 24 terminal ligands (bpy) have been obtained and carefully characterized. They comprise 1090 atoms, with a molecular weight of 10,890 Dalton (for the docosanuclear Ru-based complex, Figure 29.5, bottom), and an estimated size of 5 nm.

Since the properties of the modular components are known, and different modules can be located in the desired positions of the dendrimer structure, synthetic control of the properties can be pursued. It is, therefore, possible, as schematically shown in Figure 29.6, to construct arrays where the electronic energy-migration pattern can be predetermined (Balzani et al., 1998), so as to channel the energy created by light absorption on the various components towards a selected module (antenna effect).

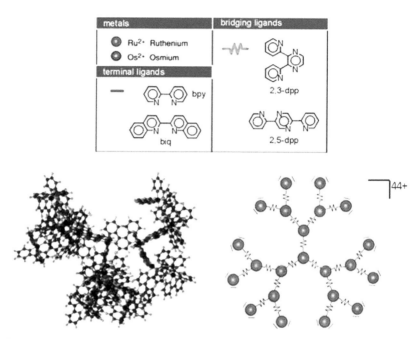

FIGURE 29.5 Top: formulae of the 2,3- and 2,5-dpp bridging ligands, and bpy and biq terminal ligands used to obtain tree-like multicenter transition-metal complexes, and their schematic representation. Bottom: schematic representation and spatial structure of a dendrimer containing 22 Ru²⁺ ions placed in each branching center of the dendritic array.

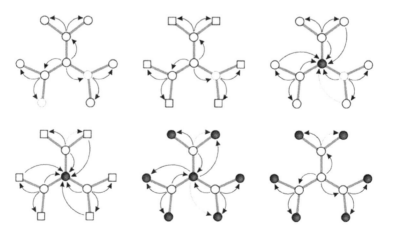

FIGURE 29.6 Schematic representation of the different energy-transfer patterns that can be obtained in 20+ electric charged decanuclear dendrimers on choosing different metals and ligands. White and black circles indicate Ru(II) and Os(II), respectively; in the peripheral positions, circles and squares indicate $M(bpy)_2$ and $M(biq)_2$ moieties, respectively.

29.4.2 DEVICES CAPABLE OF PERFORMING EXTENSIVE NUCLEAR MOTION: MOLECULAR-LEVEL MACHINES

As already mentioned a *molecular-level machine* is a particular type of molecular-level devices in which the component parts can display changes in their relative position as a result of some external stimuli.

In principle, molecular machines can be designed starting from several kinds of molecular and supramolecular systems, including DNA (Balzani et al., 2008; Credi et al., 2014). However, most of the artificial systems constructed so far are based on interlocked molecular species such as rotaxanes and catenanes. The names of these compounds derive from the Latin words *rota* and *axis* for ring and axle, and *catena* for chain. Rotaxanes are composed of a dumbbell-shaped molecule surrounded by (at least) a macrocyclic compound (the ring) and terminated by bulky groups (stoppers) that prevent disassembly (Figure 29.7, top left). Catenanes are made of (at least) two interlocked rings (Figure 29.7, top right). Important features of these systems derive from noncovalent interactions between components that contain complementary recognition sites. Such interactions that are also responsible for the efficient template-directed syntheses of rotaxanes and catenanes include charge-transfer (CT) ability, hydrogen bonding, hydrophobic-hydrophilic character, π-π stacking, electrostatic forces and, on the side of the strong interaction limit, metal-ligand bonding (Balzani et al., 2008).

Rotaxanes and catenanes are appealing systems for the construction of molecular machines because (i) the mechanical bond allows a large variety of mutual arrangements of the molecular components, while conferring stability to the system; (ii) the interlocked architecture limits the amplitude of the intercomponent motion in the three directions; (iii) the stability of a specific arrangement (co-conformation) is determined by the strength of the intercomponent interactions; and (iv) such interactions can be modulated by external stimulation. The large-amplitude motions that can be achieved with rotaxanes and catenanes are represented schematically in Figure 29.7a–c. Particularly, two interesting molecular motions can be envisaged in rotaxanes, namely (a) translation, i.e., shuttling, of the ring along the axle (Figure 29.7a), and (b) rotation of the ring around the axle (Figure 29.7b). Hence, rotaxanes are good prototypes for the construction of both linear and rotary molecular motors. Systems of type (a), termed molecular shuttles, constitute the most common implementation of the molecular machine concept with rotaxanes (Balzani et al., 2008).

For space reasons, only two examples of such kind of rotaxanes and two examples of ring rotation in catenanes are illustrated.

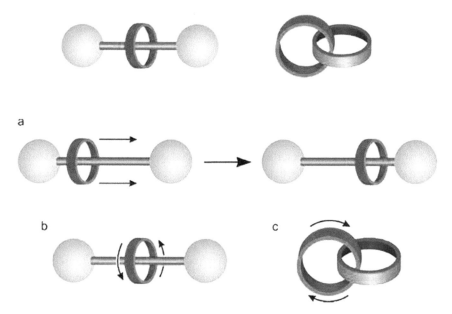

FIGURE 29.7   Top: schematic representation of a rotaxane (left) and a catenane (right) structure. Bottom: intercomponent motions that can be obtained with these simple interlocked molecular architectures; they are ring shuttling in rotaxanes (a), and ring rotation in rotaxanes (b) and catenanes (c).

29.4.2.1   CHEMICALLY CONTROLLABLE MOLECULAR SHUTTLES AND ELEVATORS

An example (Ashton et al., 1998) of chemically driven molecular machines is represented by a rotaxane that behaves as a reversible molecular shuttle controlled by acid-base stimulation (Figure 29.8). It is made of an axle-like component containing an ammonium ion and an electron acceptor bipyridinium unit that can establish hydrogen-bonding and charge-transfer interactions, respectively, with the ring component, namely a dibenzo[24] crown-8 ether. An anthracene moiety is used as a stopper because its absorption, luminescence, and redox properties are useful to monitor the state of the system (Figure 29.8, top).

Since the N^+–H···O hydrogen-bonding interactions between the macrocyclic ring and the ammonium center are much stronger than the charge-transfer interactions of the ring with the bipyridinium unit, the rotaxane exists as only one of the two possible translational isomers, i.e., that in which the ring surrounds the ammonium ion (state 0, Figure 29.8a, bottom). However, deprotonation of the ammonium center with a base (Figure 29.8b, bottom) causes 100% displacement of the ring by Brownian motion to the bipyridinium unit (state 1, Figure 29.8c, bottom); reprotonation of the rotaxane with an acid (Figure 29.8d, bottom) directs the ring back on the ammonium center. Such a switching process can be investigated in solution by ¹H NMR spectroscopy and by electrochemical and photophysical measurements.

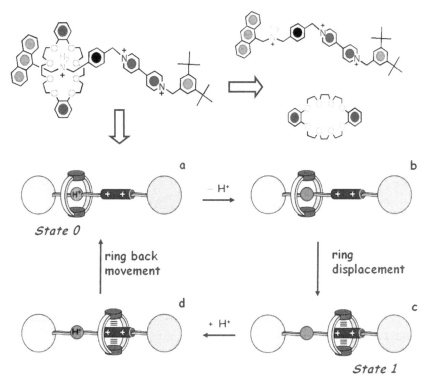

FIGURE 29.8 Chemical structure and molecular components (top), and operation scheme of an acid-base controllable rotaxane (bottom).

The full chemical reversibility of the energy supplying acid-base reactions guarantees the reversibility of the mechanical movement, in spite of the formation of waste products. The kinetics of ring shuttling was also studied in detail by stopped-flow spectroscopic experiments (Garaudée et al., 2005) finding that the base-induced shuttling is slower (seconds) than the acid-induced backward shuttling (tens of milliseconds).

It should be added that, in the deprotonated rotaxane, it is possible to displace the ring from the bipyridinium station by destroying the charge transfer interactions through reduction of this station. Therefore, in the system, mechanical movements can be induced by two different types of stimuli (acid-base and reduction-oxidation). Furthermore, this rotaxane could be useful for information processing since it is a bistable system and can exhibit a binary logic behavior.

By using an incrementally staged strategy, the architectural features of the acid-base controllable molecular shuttle above described were implemented to design and construct a two-component molecular device that behaves as a nanoscale elevator (Badjic et al., 2004). This interlocked compound (Figure 29.9a), which is ca. 2.5 nm in height and has a diameter of ca. 3.5 nm, consists of a tripod component containing two different notches – one ammonium center and one bipyridinium unit – at different levels in each of the three legs. This component is interlocked by a tritopic host, which plays the role of a platform that can be made to stop at the two different levels. The three legs of the tripod carry bulky feet that prevent the loss of the platform.

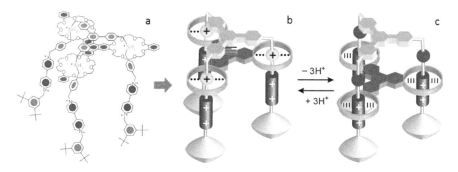

FIGURE 29.9   Chemical formula and operation scheme of a molecular elevator.

As it happens for the rotaxane previously described, initially, the platform resides exclusively on the 'upper' level, i.e., with the three rings surrounding the ammonium centers (Figure 29.9b). Upon deprotonation of the ammonium centers, the platform moves to the 'lower' level, that is, with the three macrocyclic rings surrounding the bipyridinium units (Figure 29.9c), and subsequent addition of acid restores the ammonium centers with the concomitant back-movement of the platform to the upper level. The 'up and down' elevator-like motion, which can be repeated many times, can be monitored by NMR spectroscopy, electrochemistry, and absorption and fluorescence spectroscopy. It can also be noted that the acid-base controlled mechanical motion is associated to remarkable structural modifications, such as the opening and closing of a large cavity and the control of the positions and properties of the bipyridinium legs. This behavior can in principle be used to control the uptake and release of a guest molecule, a function of interest for the development of drug delivery systems.

29.4.2.2 MOLECULAR SHUTTLE POWERED BY SUNLIGHT

The chemically powered artificial nanomachines, like those described above, are *not autonomous* since, after the mechanical movement induced by a chemical input, they need another, opposite chemical input to reset, which also implies generation of waste products. However, the addition of a reactant (fuel) is not the only means by which energy can be supplied to a chemical system. In fact, Nature shows that, in green plants, the energy needed to sustain the machinery of life is ultimately provided by sunlight.

Energy inputs in the form of photons can, indeed, cause mechanical movements by reversible chemical reactions without the formation of waste products (Ballardini et al., 2001; Balzani et al., 2008, 2009). The design and construction of molecular shuttles powered exclusively by light energy is, therefore, a fascinating yet challenging subject.

An example of such kind of systems is represented by the thoroughly designed rotaxane shown in the top of Figure 29.10 (Balzani et al., 2006). It is made of a bis-*p*-phenylene-34-crown-10 electron donor ring (**R**) and an axle-like component which contains as many as five molecular units suitably chosen and assembled in order to achieve the devised function. It, indeed, comprises two electron acceptor recognition sites for the ring, namely a bipyridinium (A_1) and a dimethyl substituted bipyridinium (A_2) units, that can play the role of "stations" for the ring **R**, a $[Ru(bpy)_3]^{2+}$-type electron-transfer photosensitizer **P**, which is able to operate with visible light and also plays the role of a stopper, a *p*-terphenyl-type rigid spacer **S**, which has the task of keeping the photosensitizer far from the electron acceptor units, and finally a tetraarylmethane group **T** as the second stopper.

The stable structure of the rotaxane, characterized by mass spectrometry and NMR spectroscopy, is the one in which the R component encircles the A_1 unit, in keeping with the fact that this station is a better electron acceptor than the other one.

The photoinduced shuttling movement of the ring between the two stations, triggered by visible light excitation of the **P** unit, can be obtained in solution at room temperature according to a purely intramolecular photochemical mechanism. It is based on the following four operations (Figure 29.10, bottom):

a) *Destabilization of the stable translational isomer*: light excitation of the photoactive unit **P** (Step 1) is followed by the transfer of an electron from the excited-state of **P** to the A_1 station, which is encircled by the ring R (Step 2), with the consequent "deactivation" of this station.

b) *Ring displacement*: the ring moves from the reduced A_1^- station to A_2 (Step 3).

c) *Electronic reset*: a back electron-transfer process from the "free" A_1^- station to P^+ (Step 4) reactivates the electron-acceptor power of the A_1 station.

d) *Nuclear reset*: as a consequence of the electronic reset, back movement of the ring from A_2 to A_1 takes place (Step 5), which restores the initial structure of the rotaxane.

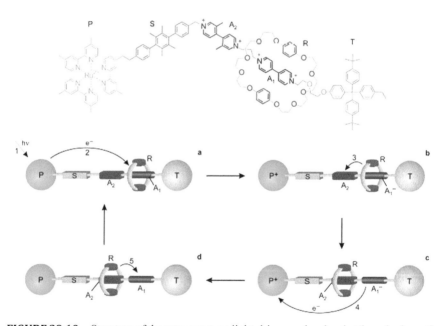

FIGURE 29.10   Structure of the rotaxane-type light-driven molecular shuttle and schematic representation of the intramolecular mechanism for the photoinduced ring-shuttling movements.

According to this mechanism, each light energy input (i.e., each absorbed photon) causes the occurrence of a forward and backward movement (i.e., a complete cycle) of ring R between the two stations. The rotaxane, therefore, operates as an autonomous "four-stroke" linear motor, powered by visible light; indeed each phase corresponds, in kind, to the (a) fuel injection and combustion, (b) piston displacement, (c) exhaust removal, and (d) piston replacement of a four-stroke motor. However, the timing of the strokes is intrinsic to the molecular motor and cannot be tuned up externally, such as in modern fuel-injected car engines.

At 303 K the quantum yield for the shuttling process is about 2%. The low efficiency of this motor may seem disappointing, particularly if compared with the performance of natural motors. It should be considered, however, that the operation of the system is based on the challenge that a complex nuclear motion can compete with an electron-transfer process. Furthermore, the low efficiency is compensated by several advantages: (i) the operation of the system relies exclusively on intramolecular processes; (ii) this artificial molecular motor does not need the assistance of external species; (iii) the fuel, sunlight, is totally free; and (iv) in principle, it can work at the single-molecule level.

29.4.2.3 *RING ROTATION IN CATENANES*

In a catenane, structural changes caused by the rotation of one ring with respect to the other (Figure 29.7c) can be evidenced only when one of the two rings contains two non-equivalent units.

In such a case the system can exist as two different equilibrating conformations, the populations of which reflect their relative free energies as determined primarily by the strengths of the two different sets of non-covalent bonding interactions (Figure 29.11a→b).

The two possible conformational isomers of the catenane can be interchanged by appropriate stimuli. In the schematic representation shown in Figure 29.11, it has been assumed that the symmetric ring resides preferentially around station A (state 0), until a stimulus, S_1, is applied that switches off this recognition site (A→A') (Figure 29.11c). As a consequence, the system is brought into a non-equilibrium state that subsequently relaxes (equilibrates) according to the new potential energy landscape. Such a process implies the rotation of the non-symmetric molecular ring to bring the second recognition site (B) inside the symmetric ring, until a new equilibrium is reached (Figure 29.11d, state 1). If station A is switched on again by an opposite stimulus, S_2, (Figure 29.11b), the original potential energy landscape is restored, and another conformational equilibration occurs through the rotation of the non-symmetric ring (Figure 29.11b→a). Hence, the relative populations of the two species can be controlled reversibly by switching off and on again the recognition properties of one of the two recognition sites of the non-symmetric ring. It should be pointed out, however, that repeated switching between the two states does not need to occur through a complete rotation (Balzani et al., 2008).

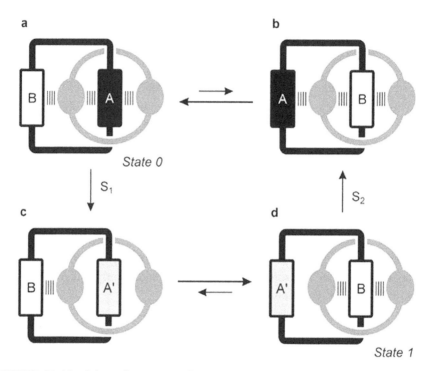

FIGURE 29.11 Schematic representation of the equilibrium between the two conforma-tions available for a catenane containing two different sites in one of its molecular rings (a and b), and ring motion induced by external stimuli (c and d).

The catenane shown in Figure 29.12 (Balzani et al., 2000) is an interesting example of a system in which the ring rotation can be chemically controlled. It incorporates a dioxybenzene-based macrocycle and a tetracationic ring comprising a bipyridinium (BPY^{2+}) and a diazapyrenium (DAP^{2+}) unit. Its major isomer is that in which the DAP^{2+} unit is located inside the cavity of the dioxybenzene-based macrocycle while the BPY^{2+} unit is positioned alongside, as evidenced by electrochemical techniques.

The ability of *n*-hexylamine to form adducts with diazapyrenium has been exploited to displace the equilibrium in favor of the isomer having the diaza-pyrenium unit alongside the cavity of the macrocycle that occurs through the rotation of the non-symmetric ring (Figure 29.12). The electrochemical analysis is once again very useful to evidence this conformational change. Upon addition of triflic acid, protonation of *n*-hexylamine takes place and adduct formed between *n*-hexylamine and the DAP^{2+} unit of the catenane is disrupted. Consequently, the original conformation is restored with another rotation of the non-symmetric ring (Figure 29.12), as revealed by the fact that the electrochemical behavior recorded after the addition of triflic acid is identical to that obtained before the addition of *n*-hexylamine.

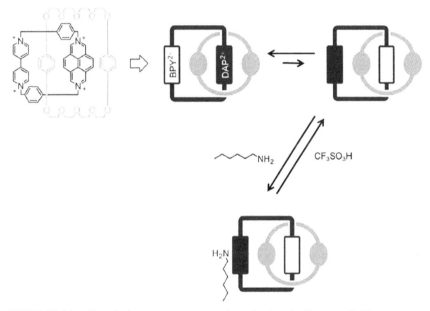

FIGURE 29.12 Chemical structure and operation of a chemically controllable catenane.

The controlled ring rotation discussed in Figure 29.11 can also be obtained by electrochemical stimulation, if at least one of the two sites contained in the non-symmetric ring is a redox switchable unit. Catenanes that exhibit such a behavior can be seen as electrochemically driven molecular machines. An example is offered by the catenane shown in Figure 29.13 (Asakawa et al., 1998), which is made of a symmetric electron-acceptor tetracationic ring and a non-symmetric macrocycle comprising two different electron donor units, namely a tetrathiafulvalene (TTF) group, and a dioxynaphthalene (DON) unit. Because the TTF unit is a better electron donor than the DON one, as witnessed by the potential values for their oxidation, the stable conformation of the catenane is that in which the symmetric ring encircles the TTF unit of the non-symmetric one (Figure 29.13a).

On electrochemical oxidation, the TTF unit is oxidized first (Figure 29.13b), thereby losing its electron-donating properties. Because of the disruption of the CT interaction and the electrostatic repulsion between TTF$^+$ and the tetracationic ring, a rotation of the non-symmetric ring occurs: TTF$^+$ is expelled from the cavity of the symmetric ring, where it is replaced by the neutral DON unit (Figure 29.13c). Upon subsequent reduction of TTF$^+$ (Figure 29.13d), another rotation of the non-symmetric ring takes place: the neutral TTF unit is repositioned inside the cavity of the tetracationic ring with the restore of the original conformation (Figure 29.13a).

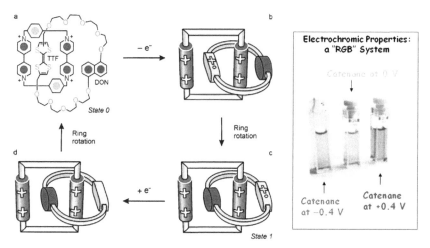

FIGURE 29.13 Left: chemical structure and operation of an electrochemically controllable catenane. Right: color changes observed upon electrochemical oxidation (+0.4 V *vs.* SCE) or reduction (−0.4 V *vs.* SCE) of the catenane.

The oxidation/reduction cycle can be followed by ^{1}H NMR spectroscopy, UV/Vis spectroscopy, and cyclic voltammetry. Interestingly, the TTF oxidation causes a clearly detectable color change from green (starting catenane) to red, and another clear color change from green to blue can be induced upon reduction of the bipyridinium unit of the electron-acceptor ring (Figure 29.13, right). Therefore, because of these electrochromic properties, the catenane could be used as a RGB system.

29.4.3 DEVICES BASED ON ELECTRONIC AND NUCLEAR MOTION

In the two preceding families of devices, the functions performed by the molecular-level systems are based on either electronic or nuclear movements.

However, there are devices working on the basis of both electronic and nuclear rearrangements that take place in distinct steps. Three selected examples are described that, because of the functions performed, could be useful for the realization of a molecular-level electronic set like the systems illustrated in point (1).

29.4.3.1 MOLECULAR PLUG/SOCKET SYSTEMS

Supramolecular species whose components are connected by means of non-covalent forces can be disassembled and re-assembled by modulating the interactions that keep the components together, thereby allowing switching of electron- or energy-transfer processes. Two-component systems of this type are reminiscent of plug/socket electrical devices and, like their macroscopic counterparts, must be characterized by (i) the possibility of connecting/disconnecting the two components in a reversible way, and (ii) the occurrence of an electron or electronic energy flow from the socket to the plug when the two components are connected (Balzani et al., 2008). Hydrogen bonding interactions between ammonium ions and crown ethers are particularly convenient for constructing molecular-level plug/socket devices since they can be switched on and off quickly and reversibly by means of acid-base inputs.

In the system illustrated in Figure 29.14 (Ishow et al., 1999), the plug-in function is related to the threading, driven by the formation of strong N^+–H...O hydrogen bonds in non-polar solvents, of a luminescent (±)-binaphthocrown ether by an ammonium ion carrying a methyl group on one side and a potentially luminescent anthracene moiety on the other side, obtained by protonation of the corresponding amine. In the plugged-in state, the quenching of the binaphthyl-type fluorescence is accompanied by the sensitization of the fluorescence of the anthracene unit of the ammonium ion. This result clearly shows that in the plug-in state energy transfer from the binaphthyl unit of the crown (the molecular socket) to the anthracene moiety of the ammonium ion (the molecular plug) has taken place. Addition of a stoichiometric amount of base, which deprotonates the ammonium ion, causes the revival of the binaphthyl fluorescence and the disappearance of the antracene-based fluorescence, demonstrating that plug out has occurred.

Interestingly and exactly as it happens for the macroscopic plug/socket device, the plug-in process does not occur when a plug component incompatible with the size of the socket, like the benzyl- instead of the methyl-substituted amine (Figure 29.14), is employed.

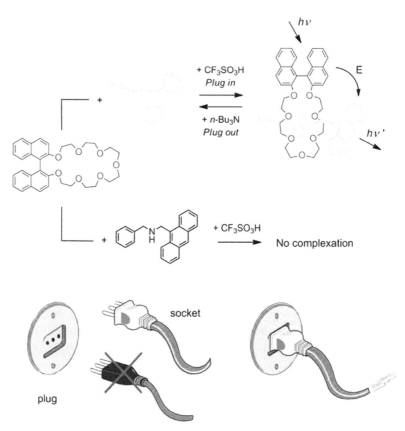

FIGURE 29.14 Acid/base controlled plug-in/plug-out of an ammonium ion carrying a methyl group on one side and a potentially luminescent anthracene moiety on the other side with a luminescent (±)-binaphthocrown ether. The occurrence of photoinduced energy transfer in the plug-in state is also schematized.

29.4.3.2 MOLECULAR EXTENSION-CABLE SYSTEMS

The plug-socket concept has been used to design a molecular system which mimics the function of a macroscopic electrical extension cable (Ferrer et al., 2006). It comprises three components (Figure 29.15a). (a) The first one consists of two moieties, namely a $[Ru(bpy)_3]^{2+}$ unit, which behaves as an electron donor under light excitation, and a dibenzo[24]crown-8 macrocycle, capable of playing the role of a hydrogen-bonding socket. (b) The second component, which plays the role of the extension cable, is made up of a dialkylammonium ion, that can insert itself as a plug into the dibenzo[24] crown-8 socket of the first component by virtue of hydrogen-bonding interactions, a biphenyl spacer, and a benzonaphtho[36]crown-10 moiety, which fulfills the role of a π-electron rich socket. (c) Finally, the third component is the 1,1'-dioctyl-4,4'-bipyridinium dication that can play the role of an electron drain when plugged into the benzonaphtho[36]crown-10 moiety of the second component.

In such system, reversible connection–disconnection of the two plug and socket junctions can be controlled independently by acid–base and red-ox stimulation, respectively. In the fully assembled system (Figure 29.15b) light excitation of the Ru-based unit of the first component is followed by electron transfer to the third component, with the second component playing the role of an extension cable. Interestingly, the photoinduced electron-transfer process can be powered by sunlight capable of exciting the $[Ru(bpy)_3]^{2+}$ moiety which, in such a way, behaves as a photovoltaic power station.

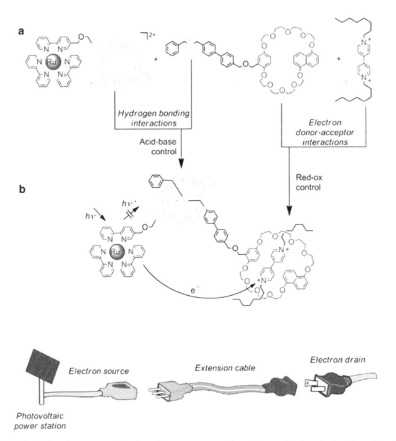

FIGURE 29.15 A supramolecular three-component system which mimics the function played by a macroscopic extension cable.

29.4.3.3 A THREE-POLE SWITCH

In host-guest systems based on electron donor-acceptor interactions, association/dissociation can be driven by redox processes so that it is possible to design electrochemical switches that can be used to control energy- and electron-transfer processes (Balzani et al., 2008).

An example of a system that can be switched reversibly in three different states through electrochemical control of the guest properties of one component is illustrated in Figure 29.16 (Ashton et al., 1999). It consists of three components: TTF (top center of Figure 29.16) and two rings, namely an electron-donor macrocycle and an electron-acceptor tetracationic ring, shown respectively in the top left and right of Figure 29.16. Because it is well known that tetrathiafulvalene is stable in three different oxidation states, TTF(0), TTF$^+$, and TTF^{2+}, and since oxidation decreases the electron-donor properties of a molecule with a concomitant increase in its electron-acceptor properties, it is expected that (i) TTF(0) gives a 1:1 charge-transfer complex with the electron-acceptor tetracationic ring, playing the role of electron donor, (ii) TTF$^+$ does not exhibit any electron donor/acceptor character, while (iii) TTF^{2+} can play the role of an electron-acceptor guest, giving a 1:1 charge-transfer complex with the electron-donor macrocycle. These expectations open the way to the use of TTF as a switchable electron-donor/electron-acceptor guest and to control its association with electron-acceptor/electron-donor hosts by electrochemical techniques. Indeed, electrochemical experiments carried out on acetonitrile solutions containing TTF and the two previously mentioned rings evidence that, depending on the potential range, TTF is: (i) free in the TTF$^+$ state, (ii) complexed with the electron-acceptor host in the TTF(0) state, or (iii) complexed with the electron-donor host in the TTF^{2+} state (Figure 29.16, bottom).

The reversibility of the electrochemical processes shows that the complexation-decomplexation reactions and, as a consequence, the exchange of the guest between the two hosts, are fast processes compared to the time scale of the electrochemical experiments. In such a three-state system, the switching ("writing") can be performed electrochemically while the monitoring of the state of the system ("reading") can be done by absorption, emission, and NMR spectroscopy, and electrochemically.

The mechanical movements taking place in the described supramolecular system, where a "free" molecule can be electrochemically driven to associate with either of two different receptors, open the way to futuristic applications in the field of molecular-level signal processors. One can, indeed, conceive second generation systems where the electrochemically driven movements

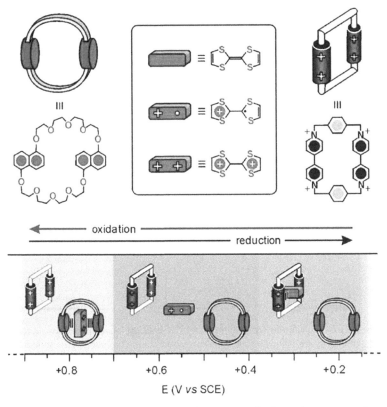

FIGURE 29.16 Components of the three–pole system and schematic representation of the ranges of electrochemical stability of the three states available to the system.

can control the occurrence, and the selection of the partner in energy- or electron-transfer processes. Consider, e.g., a system (Figure 29.17) where a chromophoric group **A** is appended to the potential guest and chromophoric groups **B** and **C**, whose lowest excited state is lower than that of **A**, are appended to the potential hosts, respectively. In such a system, light excitation of **A** could lead to no energy transfer (OFF), energy transfer to **B** (ON 1) or energy transfer to **C** (ON 2), depending on the potential value selected by the operator. Admittedly, the donor/acceptor-based molecular-level connection poses severe limitations to the choice of the chromophoric groups to be used if energy transfer has to proceed through the charge-transfer connections. However, it does not seem unlikely that systems can be designed where molecular association (i) plays only the role of bringing the two chromophoric groups at a suitable distance for through–space energy transfer, or (ii) relies on a different kind of interaction (e.g., hydrogen bonding). Similar switching of electron-transfer processes could also be performed, and more complex energy- and/or electron-transfer patterns are conceivable. The strategy described could also be used, at least in principle, to catalyze chemical reactions.

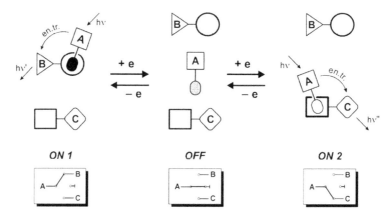

FIGURE 29.17   Schematic representation of the concept of a three-pole supramolecular switching of energy-transfer. The cartoons correspond to the molecular components shown in Figure 29.16, while **A**, **B**, and **C** are suitably chosen chromophoric groups.

29.5 OUTLOOK

The construction of simple prototypes of molecular devices and machines has been achieved by using careful, incremental design strategies, the tools of modern synthetic chemistry, and the paradigms of supramolecular chemistry

– together with some inspiration from Nature. The results obtained so far show that the extension of the concept of a device or machine to the nanoscale is a fascinating topic; looking at molecular and supramolecular species from the viewpoint of functions with reference to devices in the macroscopic world is, indeed, a very interesting exercise that introduces novel concepts into chemistry as a scientific discipline.

At present, the studies performed mainly concern basic research, however, it is expected, because of the fast growth and the staggering potential of this field, that artificial molecular devices and machines of practical use will see the light in a not too distant future. This optimistic view is unequivocally supported by the fact that the Nobel Prize in Chemistry 2016 has been assigned to three chemists, Jean-Pierre Sauvage, J. Fraser Stoddart, and Bernard L. Feringa, to recognize their work in the design and synthesis of molecular machines (Leigh, 2016).

Once again, however, the Swedish Academy of Sciences neglected Italy: Vincenzo Balzani is, indeed, another and important pioneer in the field of molecular machines. He collaborated with two of the awarded scientists, and most of their molecular machines were designed and put in operation in the Balzani laboratory.

KEYWORDS

- **antenna systems**
- **Balzani**
- **catenanes**
- **Feringa**
- **Feynman**
- **molecular switches**
- **molecular wires**
- **rotaxanes**
- **Sauvage**
- **Stoddart**
- **supramolecular chemistry**

REFERENCES AND FURTHER READING

Asakawa, M., Ashton, P. R., Balzani, V., Credi, A., Hamers, C., Mattersteig, G., et al., (1998). A chemically and electrochemically switchable [2] catenane incorporating a tetrathiafulvalene unit. *Angew. Chem. Int. Ed., 37,* 333–337.

Ashton, P. R., Ballardini, R., Balzani, V., Baxter, I., Credi, A., Fyfe, M. C. T., et al., (1998). Acid-base controllable molecular shuttles. *J. Am. Chem. Soc., 120,* 11932–11942.

Ashton, P. R., Balzani, V., Becher, J., Credi, A., Fyfe, M. C. T., Mattersteig, G., et al., (1999). A three-pole supramolecular switch. *J. Am. Chem. Soc., 121*, 3951–3957.

Badjic, J. D., Balzani, V., Credi, A., Silvi, S., & Stoddart, J. F., (2004). A molecular elevator. *Science, 303*, 1845–1849.

Ballardini, R., Balzani, V., Credi, A., Gandolfi, M. T., & Venturi, M., (2001). Artificial molecular-level machines: Which energy to make them work? *Acc. Chem. Res., 34*, 445–455.

Balzani, V., Campagna, S., Denti, G., Juris, A., Serroni, S., & Venturi, M., (1998). Designing dendrimers based on transition-metal complexes. Light-harvesting properties and predetermined redox patterns. *Acc. Chem. Res., 31*, 26–34.

Balzani, V., Clemente-Leon, M., Credi, A., Ferrer, R. B., Venturi, M., Flood, A. H., & Stoddart, J. F., (2006). Autonomous artificial nanomotor powered by sunlight. *Proc. Natl. Acad. Sci. USA, 103*, 1178–1183.

Balzani, V., Credi, A., & Venturi, M., (2008). *Molecular Devices and Machines – Concepts and Perspectives for the Nano World*. Wiley-VCH: Weinheim, Germany.

Balzani, V., Credi, A., & Venturi, M., (2009). Light powered molecular machines. *Chem. Soc. Rev., 38*, 1542–1550.

Balzani, V., Credi, A., Langford, S. J., Raymo, F. M., Stoddart, J. F., & Venturi, M., (2000). Constructing molecular machinery: A chemically-switchable [2] catenane. *J. Am. Chem. Soc., 122*, 3542–3543.

Campagna, S., Ceroni, P., & Puntoriero, F., (2012). *Designing Dendrimers*. John Wiley & Sons: Hoboken, USA.

Cramer, F., (1993). *Chaos and Order. The Complex Structure of Living Systems*. VCH: Weinheim, Germany.

Credi, A., Semeraro, M., Silvi, S., & Venturi, M., (2011). Redox control of molecular motion in switchable artificial nanoscale devices. *Antioxid. Redox Signal, 14*, 1119–1165.

Credi, A., Silvi, S., & Venturi, M., (2012). Photochemically driven molecular devices and machines. In: Gale, P. A., & Steed, J. W., (eds.), *Supramolecular Chemistry: From Molecules to Nanomaterials* (Vol. 8, 3719–3750). John Wiley & Sons: Hoboken, USA.

Credi, A., Silvi, S., & Venturi, M., (2014). Molecular machines and motors. Recent advances and perspective. *Top. Cur. Chem., 354*, 1–342, Springer-Verlag: Berlin, Germany.

Drexler, K. E., (1986). *Engines of Creation, The Coming Era of Nanotechnology*. Anchor Books: New York, USA.

Drexler, K. E., (1992). *Nanosystems: Molecular Machinery, Manufacturing, and Computation*. Wiley: New York, USA.

Ferrer, B., Rogez, G., Credi, A., Ballardini, R., Gandolfi, M. T., Balzani, V., et al., (2006). Photo-induced electron flow in a self-assembling supramolecular extension cable. *Proc. Natl. Acad. Sci. USA, 103*, 18411–18416.

Feynman, R. P., (1960). There's plenty of room at the bottom. *Caltech's Engineering and Science Magazine, 23*, 22–36.

Garaudée, S., Silvi, S., Venturi, M., Credi, A., Flood, A. H., & Stoddart, J. F., (2005). Shuttling dynamics in an acid-base switchable [2]rotaxane. *Chem. Phys. Chem., 6*, 2145–2152.

Giessibl, F. J., (2003). Advances in atomic force microscopy. *Rev. Mod. Phys., 75*, 949–983.

Gimzewski, J. K., & Joachim, C., (1999). Nanoscale science of single molecules using local probes. *Science, 283*, 1683–1688.

Goodsell, D. S., (1996). *Our Molecular Nature: The Body's Motors, Machines, and Messages*. Copernicus: New York, USA.

Goodsell, D. S., (2004). *Bionanotechnology*, Wiley-Liss: Hoboken, USA.

Hla, S. W., Meyer, G., & Rieder, K. H., (2001). Inducing single-molecule chemical reactions with a UHV-STM: A new dimension for nanoscience and technology. *Chem. Phys. Chem., 2,* 361–366.

Ishow, E., Credi, A., Balzani, V., Spadola, E., & Mandolini, L., (1999). A molecular-level plug/socket system. Electronic energy transfer from a binaphthyl unit incorporated into a crown ether to an anthracenyl unit linked to an ammonium ion. *Chem. Eur. J., 5,* 984–989.

Jones, R. A. L., (2004). *Soft Machines, Nanotechnology and Life*. Oxford University Press: Oxford, UK.

Leigh, D. A., (2016). Genesis of the nanomachines: The 2016 Noble Prize in Chemistry. *Angew. Chem. Int. Ed., 55,* 14506–14508.

Otero, R., Rosei, F., & Besenbacher, F., (2006). Scanning tunneling microscopy manipulation of complex organic molecules on solid surfaces. *Annu. Rev. Phys. Chem., 57,* 497–525.

Rigler, R., Orrit, M., & Basche, T., (2001). *Single Molecule Spectroscopy*. Springer-Verlag: Berlin, Germany.

Schlicke, B., Belser, P., De Cola, L., Sabbioni, E., & Balzani, V., (1999). Photonic wires of nanometric dimensions. Electronic energy transfer in rigid rod-like $Ru(bpy)_3^{2+}(ph)_nOs(bpy)_3^{2+}$ compounds (ph = 1, 4-phenylene, n = 3, 5, 7). *J. Am. Chem. Soc., 121,* 4207–4212.

CHAPTER 30

Molecular Dynamics Calculation

ISTVÁN LÁSZLÓ

Department of Theoretical Physics, Budapest University of Technology and Economics, H-1521 Budapest, Hungary, E-mail: laszlo@eik.bme.hu

30.1 DEFINITION

The molecular dynamics method is a time-dependent computer simulation of the physical, chemical or thermodynamical properties of a system. Usually, this system has a large number of particles which can be atoms, molecules or other compounds. At each time step, the forces acting on the particles are calculated with the help of a potential function or of a quantum chemical calculation. During a given time step the motion of the particles is described by classical Newton equations. In the next time step the forces are calculated once more and in this way the trajectories of the particles are determined by numerical solving Newton's equations of motion. The method can be used in theoretical physics, solid-state physics, chemical physics material science and modeling biomolecules. In ergodic systems, the thermodynamical quantities are calculated as time averages. Such kind of calculations can be seen as computer experiments. Analyzing and managing the data has many similarities with realistic experiments.

30.2 HISTORICAL ORIGIN(S)

The basis of modern computer simulation was laid down by Metropolis et al. (1953) when they described "a general method suitable for fast electronic computer machines, of calculating the properties of any substance which maybe considered as composed of interacting individual molecules." In our day we call this algorithm metropolis algorithm. Fermi et al. (1955) performed numerical studies of the dynamics of an an-harmonic, one-dimensional crystal. As the Metropolis et al., calculation was a Monte Carlo calculation and the Fermi et al. (1955) calculation was a numerical solution of a vibrating string by replacing it with a finite number of points, these calculation were not yet veritable molecular dynamics calculations. The first proper molecular dynamics simulation was published in 1956 by Alder and Wainwright (1956). In this work, they studied the dynamics of hard spheres where the spheres were moving at constant velocity in straight lines between elastic collisions. Gibson et al. (1960) simulated radiation damage in crystalline copper. Their model contained about 500 atoms on their lattice sites. One atom was initially endowed with arbitrary kinetic energy and direction of motion, as though it had just been struck by a bombarding particle. According to the classical equation of motion, numerical solutions were used to follow the motion of the atoms and to study the energy transfer and damages in the lattice. Rahman (1964) put 864 argon atoms in a box interacting with a Lennard-Jones potential and obeying a classical equation of motion. In this molecular dynamics calculation, the liquid argon pair correlation function and constant of self-diffusion was calculated. Using Rahman's work as a starting point, Verlet (1967) obtained good agreements with the experimental data of the thermodynamical properties of the argon. Molecular dynamics calculation of the pressure and configuration energy of a Lennard-Jones fluid were reported for 108 state conditions by Nicolas et al. (1979). Harp and Berne (1968) were the first who studied liquid of diatomic molecules in molecular dynamics simulation, and this method was first used by Rahman and Stillinger (1971) for the liquid water. Step by step the application of the molecular dynamics simulations were extended to large molecules such as proteins (McCammon, Gelin and Karplus, 1977).

In molecular dynamics calculations, a crucial point is the calculation of the forces acting on the atoms, molecules or other particles. In most of the simulations, these forces are calculated from classical potential functions. These potential functions are continuously developed and in extreme cases can be used in calculations with several ten or hundred thousand particles. There is a recent publication of huge-scale molecular dynamics simulation (Watanabe, Suzuki, &Ito, 2013) where 1,449,776,020 particles were used with Lennard-Jones interaction during 178,000-time steps. There are problems where the forces calculated from potential functions are not precise enough to describe the interactions. In these case, some kind of quantum mechanical methods are used. There are molecular dynamics calculations where the interactions are determined with the help of tight binding methods (Xu et al., 1992; Porezag et al., 1995; László, 1998; Zheng et al., 2004; László, Zsoldos, 2012), or ab initio methods (Car and Parrinello, 1985).

There are several books about the molecular dynamics method (Allen and Tildesley, 1996; Frenkel and Smit, 1996; Billing and Mikkelsen, 1996; Gubbins and Quirke, 1996).

30.3 NANO-SCIENTIFIC DEVELOPMENT(S)

The molecular dynamics calculations have great similarity with realistic experiments. First, we prepare the system we want to study just like in an experiment. If we want to measure the temperature, pressure, or other quantity we make contact a measuring instrument with the sample. In molecular dynamics calculation, we simulate the system, and from the calculated time-dependent positions, and momentums of the particles, we calculate the desired quantity.

Let the system have N particles, and the mass of the i-th particle be M_i. We put them, for example, in a box of sizes $L_x \times L_y \times L_z$. The position of the i-th particle is described by the position vector.

$$R_i(t) = (X_i(t), Y_i(t), Z_i(t)) \tag{1}$$

Depending on the particle positions the total energy of the system can be given with a potential analytic function:

$$E(\mathbf{R}) = E(\mathbf{R_1}, \mathbf{R_2}, ..., \mathbf{R_i}, ... \mathbf{R_N}) \tag{2}$$

and the forces acting on the particles are calculated as:

$$F_{X_i} = -\frac{\partial E(\mathbf{R})}{\partial X_i}, \quad F_{Y_i} = -\frac{\partial E(\mathbf{R})}{\partial Y_i}, \quad F_{Z_i} = -\frac{\partial E(\mathbf{R})}{\partial Z_i} \tag{3}$$

In more sophisticated calculations the forces are calculated from tight-binding or *ab-initio* calculations. In most of the cases, the Newton equations are solved numerically using a time step of Δt. Applying the Verlet algorithm (Verlet, 1967) the velocities $\mathbf{V_i}(t)$, momentums $\mathbf{P_i}(t)$ and the positions $\mathbf{R_i}(t + \Delta t)$ are obtained as:

$$\mathbf{V_i}(t) = \frac{\mathbf{R_i}(t + \Delta t) - \mathbf{R_i}(t - \Delta t)}{2\Delta t} \tag{4}$$

$$\mathbf{P_i}(t) = M_i \mathbf{V_i}(t) \tag{5}$$

$$\mathbf{R_i}(t + \Delta t) = 2\mathbf{R_i}(t) - \mathbf{R_i}(t - \Delta t) + \frac{\mathbf{F_i}(t)}{M_i}\Delta t^2 \tag{6}$$

In this way knowing the position of the N particles (atoms or molecules or others) of masses M_i at the time step t and $(t-\Delta t)$ we can calculate their positions at the time $(t + \Delta t)$.

With the help of the equipartition theorem, we attribute a temperature T to the system as:

$$T(t) = \frac{1}{3Nk_B} \sum_{i=1}^{N} \frac{P_i^2(t)}{2M_i}. \tag{7}$$

Here k_B is the Boltzmann constant.

The constant energy (E), constant volume (V) and constant particle number (N) simulation that is a microcanonical (constant – NVE) ensemble simulation has the following steps:

1. Start with an initial set of positions $R_i(0)$ and momenta $P_i(0)$.
2. Integrate the equation of motions according to the relations (4–MD-6).

By periodic repeating the unit cell of sizes $L_x \times L_y \times L_z$, the periodic boundary condition can be used.

In the canonical ensemble (constant – NVT) there are various possibilities for the simulation. In the simpler case, there is a scaling of the velocities by the factor $\lambda = \sqrt{\dfrac{T}{T(t)}}$, where $T(t)$ is the temperature by Eq. (7) and T is the desired temperature of the thermostat. This scaling can be done in each time step or randomly with a given probability. Berendsen et al. (1984) scaled the temperature in such a way that the relation

$$\frac{\Delta T}{\Delta t} = \frac{1}{\tau}(T - T(t)) \tag{8}$$

be valid at each time step, where the optimal value for the constant τ is $\dfrac{\Delta t}{\tau} \approx 0.0025$. The corresponding λ is

$$\lambda = \sqrt{\frac{T(t) + \frac{\Delta t}{\tau}(T - T(t))}{T(t)}} \tag{9}$$

In the Andersen thermostat (Andersen, 1980), the velocity of a randomly chosen particle is replaced by one drawn from a Maxwell-Boltzmann distribution corresponding to the desired temperature T. Thus, the constant temperature is achieved by stochastic collisions with a heat bath. Nosé (1984) defined an extended Lagrangian that contains additional, artificial coordinates and velocities. In this way, he could describe the influence of the heat bath with the help of extra force acting on the particle. Using Hoover (1985) formulation, we call this method Nosé-Hoover thermostat.

There are similar constant pressure simulations in the references (Allen and Tildesley, 1996; Frenkel and Smit, 1996; Billing and Mikkelsen, 1996; Gubbins and Quirke, 1996).

30.4   NANO-CHEMICAL APPLICATION(S)

In molecular dynamics calculations, we can follow the motion of the atoms and can have on nanometer sizes a deeper insight into the changing and development of the physical and chemical properties. An interesting case is the study of the C_{60} fullerene formation. Many authors tried to produce the C_{60} fullerene in molecular dynamics calculations, but mostly they could reach a cage-like structure with many defects. They also were using some kind of artificial conditions. Ballone and Milani (1990) kept the carbon atoms on the surface of a sphere for $T > 4000K$, Chelikowsky (1991) removed and randomly replaced the energetically unfavorable atoms and Wang et al., (1992) confined the carbon atoms into a sphere. László (1998) studied C_{60} fullerene formation in molecular dynamics simulation at helium atmosphere. Irle et al., (2006) developed the "shrinking hot giant" road that leads to the formation of buckminsterfullerene C_{60}, C_{70}, and larger fullerenes. László and Zsoldos (2012) presented molecular dynamics calculations where the final buckminsterfullerene C_{60} and C_{70} fullerenes were coded in the initial structures cut out from the graphene. Thus, the final structure was predetermined by the initial one.

KEYWORDS

- **electronic structure**
- **molecular dynamics**
- **optical properties**
- **thermodynamics**

REFERENCES AND FURTHER READING

Alder, B. J., & Wainwright, T. E., (1956). Molecular dynamics by electronic computers. In: Prigogine, I., (ed.), *Proc. of the Int. Symp. on Statistical Mechanical Theory of Transport Processes (Brüssels)* (pp. 97–131). Interscience, Wiley New York.

Allen, M. P., & Tildesley, D. J., (1996). *Computer Simulation of Liquids.* Clarendon Press: Oxford.

Andersen, H. C., (1980). Molecular dynamics at constant pressure and/or temperature. *J. Chem. Phys., 72,* 2384–2393.

Ballone, P., & Milani, P., (1992). Simulated annealing of carbon clusters. *Phys. Rev., B42,* 3201–3204.

Berendsen, H. J. C., Postma, J. P. M., Van Gunsteren, W. F., DiNola, A., & Haak, J. R., (1984). Molecular dynamics with coupling to an external bath. *J. Chem. Phys., 81,* 3684–3690.

Billing, G. D., & Mikkelsen, K. V., (1996). *Molecular Dynamics and Chemical Kinetics.* A Wiley-Interscience Publication John Wiley and Sons Inc.: New York, Chichester, Brisbane, Toronto, Singapore.

Car, R., & Parrinello, M., (1985). Unified approach for molecular dynamics and density-functional theory. *Phys. Rev. Lett., 55,* 2471–2474.

Chelikowsky, J. R., (1991). Nucleation of C60 clusters. *Phys. Rev. Let., 67,* 2970–2973.

Fermi, E., Pasta, J. G., & Ulam, S. M., (1955). Studies of non-linear problems. *ELASL Report LA–1940.*

Frenkel, D., & Smit, S., (1996). *Understanding Molecular Simulation, From Algorithms to Applications.* Academic Press, Inc.: San Diego, London, Boston, New York, Sydney Tokyo, Toronto.

Gibson, J. B., Goland, A. N., Milgram, M., & Vineyard, G. H., (1960). Dynamics of radiation damage. *Phys. Rev., 120,* 1229–1253.

Gubbins, K. E., & Quirke, N., (1996). *Molecular Simulation and Industrial Applications. Methods, Examples and Prospects.* Gordon and Breach Science Publishers: Austria, Canada, China, France, Germany, India, Japan, Luxemburg, Malaysia, The Netherland, Russia, Singapore, Switzerland, Thailand, United Kingdom.

Harp, G. D., & Berne, B. J., (1968). Linear and angular momentum autocorrelation functions in diatomic liquids. *J. Chem. Phys., 49,* 1249–1254.

Hoover, W. G., (1985). Canonical dynamics: Equilibrium phase-space distributions. *Phys. Rev. A., 31,* 1695–1697.

Irle, S., Zheng, G., Whang, Z., & Morokuma, K., (2006). The C60 formation puzzle "Solved": QM/MD simulation reveal the shrinking hot giant road of the dynamic fullerene self-assembly mechanism. *J. Phys. Chem. B., 110,* 14531–14545.

László, I., (1998). Formation of cage-like C60 clusters in molecular dynamics simulations. *Europhys. Lett., 44,* 741–746.

László, I., & Zsoldos, I., (2012). Graphene-based molecular dynamics nanolithography of fullerenes, nanotubes and other carbon structures. *Europhys. Lett., 99,* 63001, p5.

McCammon, J. A., Gelin, B. R., & Karplus, M., (1977). Dynamics of folded proteins. *Nature, 267,* 585–590.

Metropolis, N., Rosenbluth, A. W., Rosenbluth, M. N., & Teller, A. H., (1953). Equation of state calculations by fast computing machines. *J. Chem. Phys., 21,* 1087–1092.

Nicolas, J. J., Gubbins, K. E., Streett, W. B., & Tildesley, D. J., (1979). Equation of State for the Lennard-Jones fluid. *Mol. Phys., 37,* 1429–1454.

Nosé, S., (1984). EA unified formulation of the constant temperature molecular dynamics method. *J. Chem. Phys., 81*, 511–519.

Porezag, D., Frauenheim, T., Köhler, T., Seifert, G., & Kaschner, R., (1995). Construction of tight-binding-like potentials on the basis of density-functional theory: Application to carbon. *Phys. Rev. B., 51*, 12947–12957.

Rahman, A., (1964). Correlation of the motion of atoms in liquid argon. *Phys. Rev., 136*, A405–A411.

Rahman, A., & Stillinger, F. H., (1971). Molecular dynamics study of liquid water. *J. Chem. Phys., 455*, 3336–3359.

Verlet, L., (1967). Computer "experiment" on classical fluids. I. Thermodynamical properties of Lennard-Jones molecules. *Phys. Rev., 159*, 98–103.

Wang, C. Z., Xu, C. H., Chan, C. T., & Ho, K. M., (1992). Disintegration and formation of C60. *J. Phys. Chem., 96*, 3563–3565.

Watanabe, H., Suzuki, M., & Ito, N., (2013). Huge-scale molecular dynamics simulation of multi-bubble nuclei. *Comput. Phys. Commun., 184*, 2775–2784.

Xu, C. H., Whang, C. Z., Chan, C. T., & Ho, K. M., (1992). A transferable tight-binding potential for carbon tight-binding study of grain boundaries in silicone. *J. Phys.: Condens. Matter, 4*, 6047–6054.

Zheng, G., Irle, S., Elstner, M., & Morokuma, K., (2004). Quantum chemical molecular dynamics model study of fullerene formation from open-ended carbon nanotubes. *J. Phys. Chem. A., 108*, 3182–3194.

CHAPTER 31

Monte Carlo Molecular Simulations

BOGDAN BUMBĂCILĂ[1] and MIHAI V. PUTZ[1,2]

[1]*Laboratory of Computational and Structural Physical Chemistry for Nanosciences and QSAR, Biology-Chemistry Department, Faculty of Chemistry, Biology, Geography at West University of Timişoara, Pestalozzi Street No. 16, Timişoara, RO-300115, Romania*

[2]*Laboratory of Renewable Energies-Photovoltaics, R&D National Institute for Electrochemistry and Condensed Matter, Dr. A. Paunescu Podeanu Str. No. 144, RO-300569 Timişoara, Romania, Tel.: +40-256-592-638; Fax: +40-256-592-620; E-mail: mv_putz@yahoo.com, mihai.putz@e-uvt.ro*

31.1 DEFINITION

Monte Carlo molecular modeling is an application of Monte Carlo methods to molecular issues. These problems can also be modeled by the molecular dynamics method. The difference is that this approach relies on equilibrium statistical mechanics rather than molecular dynamics. Instead of trying to reproduce the dynamics of a system, it generates states according to appropriate Boltzmann probabilities. It is, therefore, also a particular subset of the more general Monte Carlo method in statistical physics (Binder et al., 2002; Frenkel et al., 2001).

31.2 HISTORICAL ORIGINS

Previously Monte Carlo method, statistical sampling was used to estimate uncertainties in the simulations. A similar early method was developed and described by Georges-Louis Leclerc, Count of Buffon, in the Buffon's needle experiment, where the π value was estimated by dropping needles on a set made of parallel and equidistant strips. The value of π number was the probability that the needle will lie across a line between two strips, using integral geometry (Schroeder, 1974).

In the 1930s, Enrico Fermi, an Italian physicist, who created the world's first nuclear reactor, experimented with the Monte Carlo method while studying neutron diffusion, but nothing was published. The modern version of the method was invented in the late 1940s by Stanislaw Ulam, while working on nuclear weapons projects. Another scientist, Margittai Neumann János understood its importance and programmed the Electronic Numerical Integrator and Computer (the first electronic general purpose computer) to perform Monte Carlo calculations. Stanislaw Ulam proposed the algorithm for determining the average distance a neutron would travel in a substance before a collision with an atomic nucleus, and for appreciating how much energy the neutron is likely to give off following a collision. Being secret, because it was developed during the Second World War for defense purposes, the work of Neumann and Ulam was given a code name. A colleague of Neumann and Ulam, Nicholas Metropolis, suggested using the name Monte Carlo, which refers to the Monte Carlo Casino in Monaco where Ulam's uncle would borrow money from relatives to gamble.

31.3   NANO-CHEMICAL IMPLICATIONS

Monte Carlo method is a broad class of computational algorithms that rely on repeated random sampling to obtain numerical results.

The Monte Carlo method in the field of molecular modeling is a technique for performing computer simulations of a molecular system. A Monte Carlo simulation is generating configurations of a system by making random changes to the positions of the species (atoms) present and their orientations and spatial conformations. A Monte Carlo method uses a technique called importance sampling. This means that the method is able to generate low-energy states and so, the measurements can be accurately calculated. The potential energy of each configuration of the system can be appreciated from the positions of the atoms. The Monte Carlo method samples thus a 3D space of the positions of the particles in that space (Leach, 2001).

Monte Carlo method for molecular modeling is the application of the Monte Carlo algorithm to molecular systems. These systems can also be modeled by the molecular dynamics method. The difference is that this approach relies on equilibrium statistical mechanics rather than molecular dynamics. Instead of trying to reproduce the dynamics of a system, it generates states according to appropriate Boltzmann probabilities. Thus, it is the application of the Metropolis Monte Carlo simulation to molecular systems. It is therefore also a particular subset of the more general Monte Carlo method in statistical physics (Allen et al., 1987).

Monte Carlo method involves a Markov chain procedure in order to determine a new state for a system from a previous one. This new state is accepted at random. Each trial usually counts as a move. The avoidance of dynamics restricts the method to studies of static quantities only, but the freedom to choose moves makes this method a flexible one. These moves must only satisfy a basic condition of balance for the equilibrium to be properly described, but detailed balance is usually imposed when designing new algorithms. An additional advantage is that some systems, such as the Ising model, lack a dynamical description and are only defined by an energy prescription; for these the Monte Carlo approach is the only one feasible (Binder et al., 2002).

A Markov chain (discrete-time Markov chain or DTMC), named after Andrey Markov (1856–1922, a Russian mathematician knew for his work in the stochastic statistics field), is a random process that undergoes transitions from one state to another on a state space (a set of values which a process can adopt). It must possess a property that is usually characterized as "memorylessness": the probability distribution of the next state depends only on the current state and not on the sequence of events that preceded it, or the properties of random variables related to the future depend only on relevant information about the current time, not on information from further in the past. This specific kind of "memorylessness" is called the Markov property. In a Markov process, the evolution is determined only from what happened at the preceding time step. So at each time step, the system loses the memory of its previous evolution (Everitt, 200; Serfozo, 2009; Durett, 2012).

The great success of this method in statistical mechanics has led to various generalizations such as the method of simulated annealing for optimization, in which a fictitious temperature is introduced and then gradually lowered (Allen et al., 1987; Frenkel et al., 2001).

Some basic steps to be followed in a Monte Carlo simulation are presented:

1) A particle is selected randomly. It has the position "r."
2) An elementary move is made. The selected particle is displaced to a new position:
   $x' = x + \Delta x$
   The displacement–Δx, can be chosen as a randomly orientated vector within a sphere with a radius R or a parallelepiped with a linear dimension of $2R$.

3) The potential energy difference (ΔV) due to the move is calculated.
4) A number, α, is randomly generated. $A \in [0;1]$.
5) Accept the move if $\Delta V = V(x')-V(x) < 0$, $\alpha < e^{-\beta \Delta V}$. If the condition is not fulfilled, reject the move and let the system stay in its original configuration. The rejection also counts as a move but with 0 displacements.
6) The 1–5 steps are repeated.

In a molecular dynamics simulation, the successive configurations of a system of particles are connected in time. In a Monte Carlo simulation, each configuration depends only upon its very previous one and not upon any other configuration. The method is generating random configurations and uses a special set of criteria to decide whether the new configuration can be or can't be accepted. This set of criteria ensures that the probability of obtaining a certain configuration is equal to the Boltzmann factor (Leach, 2001):

$$\exp\left\{\mathcal{9}\left(r^{N}\right)\big|k_{B}T\right\} \tag{1}$$

where $\mathcal{9}(r^{N})$ is calculated using the potential energy function.

Configurations with low energy are thus generated with a higher probability than configurations with higher energy. For each configuration that is accepted, the values of the desired properties are calculated and at the end of the calculation the average of these properties is obtained by simply averaging over the number of values calculated, M (Leach, 2001):

$$\langle A \rangle = \frac{1}{M}\sum_{i=1}^{M} A\left(r^{N}\right) \tag{2}$$

Most Monte Carlo simulations of molecular systems are more properly referred to as Metropolis Monte Carlo simulations, after the team who performed the first such calculation. The distinction can still be important because there are other methods in which an ensemble of states can be generated. The Metropolis variant is though, the most popular one.

In a Monte Carlo simulation each new configuration of the system may be generated by randomly moving a single atom or molecule. In some cases, new configurations may also be obtained by moving several atoms or molecules or by rotating about one or more bonds. The energy of the new configuration is then calculated using the potential energy function. Two situations can exist:

a. If the energy of the new configuration is lower than the energy of its predecessor, then the new configuration is accepted.
b. If the energy of the new configuration is higher than the energy of its predecessor, then the Boltzmann factor of the energy difference is calculated with the formula:

$$\exp\left\{-\frac{\vartheta_{new}\left(r^N\right)-\vartheta_{old}\left(r^N\right)}{k_B T}\right\} \tag{3}$$

A random number between 0 and 1 is then generated that is compared with the Boltzmann factor. If the number is higher than the factor, the move is rejected, and the original configuration is retained for next analysis. If the random number is lower than the factor, the move is accepted, and the new configuration becomes the next state of the system, with higher energy. The smaller the value of the move, expressed as $\left[\vartheta_{new}\left(r^N\right)-\vartheta_{old}\left(r^N\right)\right]$, the greater is the probability that the move will be accepted (Leach, 2001).

31.4 TRANS-DISCIPLINARY CONNECTIONS

Minimal topological difference (MTD) method in QSAR is a topological method to assess receptor-ligand interactions which can be used for a series of compounds with wide differences in their chemical structures. Both minimal steric difference (MSD) and MTD methods are using the concept of a standard molecule, considered as a steric match of the receptor's cavity and a "copy" of the hypermolecule, which sets the 3D spatial coordinates of the perfect ligand when studying a series of analog compounds (Balaban et al., 1980).

In order to superpose a molecule on a standard one, it is necessary to use the rule of obtaining minimal steric differences – minimal unsuperposable volumes. But this rule is not always a sufficient condition, because the problem of binding to the receptor is more complex: the orientation of the atoms has to have a certain conformation with internal atomic bonds having a correct position and the intermolecular bond-forming groups being exposed for a perfect steric fit (Balaban et al., 1980).

The Monte Carlo version (MCD) of MSD is an improvement for the computation of non-overlapping volumes superpositions with the standard molecule by introducing the metric measurement system instead of the topological formalism (Balaban et al., 1980).

Random numbers $\xi \in (0;1)$ with random distribution within this interval are used. Numbers $\xi \in (a;b)$ with $a, b \in R$, are obtained following the relation:

$$\xi' = a + (b-a)\xi \qquad (4)$$

The numbers ξ' are randomly distributed within the (a,b) interval. The precision of Monte Carlo calculations is conditioned by the quality of the random number sequence used in calculations. Random number generators existing in mathematical electronic libraries are fulfilling the precision requirements (Balaban et al., 1980).

In order to determine the length L of the sequence of random numbers ξ' which are assuring the calculation of Monte Carlo distribution figures with an error ε, of course, a function can be used:

$$\varepsilon = f(a, L) \qquad (5)$$

The function can be calibrated to implement itself into the computer program, and the length L_0 of the sequence which assures the required precision ε_0 is automatically determined (Balaban et al., 1980).

So the Monte Carlo techniques offer more accurate computations of Minimal Steric Differences, overlapping volumes and non-overlapping volumes as volumes of intersecting or non-intersecting molecular van der Waals envelopes. The two molecules, M_0 and M_i for which such parameters have to be estimated in the proper superposition, are included in a parallelepiped. A sufficient number of points is randomly dispersed within the parallelepiped, and the number of points within the intersecting molecular envelopes has to be counted. MCD can offer a relatively simple calculation of overlapping volumes thus this method avoids the problem of multiple intersecting atomic van der Waals volumes required in geometric calculations (Bruns et al., 1981; Voiculetz et al., 1993).

For determining the MCD, the molecules have to be characterized by Cartesian coordinates and van der Waals radii of their atoms. For calculations of the coordinates, standard values of geometric parameters like bond angles and lengths and van der Waals radii have to be used. All molecules of an analogue series are represented in the same Cartesian coordinates system (Balaban et al., 1980).

For example, for two analogue molecules M_a and M_b we can consider: N_a atoms of coordinates A_{aI}, B_{aI}, C_{aI} and van der Waals radii R_{aI} with $I = 1,2,...$ N_a, respectively N_b atoms of coordinates A_{bJ}, B_{bJ}, C_{bJ} and van der Waals radii R_{bI} with $J = 1,2...N_b$.

The van der Waals envelopes of the molecules M_a and M_b are thus described with the following equations:

$$\left(\xi_x' - A_{aI}\right)^2 + \left(\xi_y' - B_{aI}\right)^2 + \left(\xi_z' - C_{aI}\right)^2 < R_{aI}^2, \; I = 1,2...N \tag{6}$$

$$\left(\xi_x' - A_{aJ}\right)^2 + \left(\xi_y' - B_{aJ}\right)^2 + \left(\xi_z' - C_{aJ}\right)^2 < R_{aJ}^2, \; J = 1,2...N \tag{7}$$

MCD represents the total volume of the unsuperposable van der Waals envelopes which do not overlap at the standard molecule superposition (Balaban et al., 1980).

Point $P_{\xi'}\left(\xi_x', \xi_y', \xi_z'\right) \in R^3$ belongs to a non-overlapping zone of van der Waals envelopes of molecule M_a or M_b if it satisfies a single inequality in the system formed by MC-MS-3 and MC-MS-4 equations.

ξ_x', ξ_y', ξ_z' the coordinates of the random point $P_{\xi'}$ are generated with a random number generator. If point $P_{\xi'}$ satisfies at least one "a" type and one "b" type equation of the system, it falls into a volume in which the two vans der Waals envelopes of the molecules are overlapping. If $P_{\xi'}$ satisfies neither of the equations of the system, it falls outside the envelopes of both molecules (Balaban et al., 1980).

The manifold of points $P_{\xi'}$ satisfying a single equation is noted as N_I and the manifold of points satisfying at least one "a" and one "b" equation or neither of the equations is noted as N_A. The manifold N_T is the reunion of N_I and N_A, $N_I \cup N_A = N_T$.

If a is the volume of the parallelepiped which is containing the superposed van der Waals envelopes of the molecules M_a and M_b, MCD is calculated by the following relation:

$$MCD = a\frac{|N_I|}{|N_T|} \tag{8}$$

where $|N|$ is the cardinal of manifold N_I, respectively N_T (Balaban et al., 1980)

In Figure 31.1, the red color respectively the blue color is delimiting the van der Waals envelopes of the two superposed molecules. The white zone represents the superposable volume and the grey zone the unsuperposable volume. The orange circles are representing points in these volumes or outside them.

31.5   OPEN ISSUES

The Monte Carlo simulation is involving random trial steps, like in gambling. It has the advantage that does not require a continuous energy function as in molecular dynamics, and the number of points/particles can easily vary, unlike in the molecular dynamics method. Also unlike molecular dynamics simulations, Monte Carlo simulations are free from the restrictions of solving Newton's equations of motion. This freedom allows for cleverness in the proposal of moves that generate trial configurations within the statistical mechanics ensemble of choice (Earl et al., 2008).

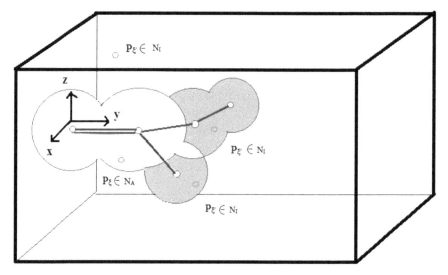

FIGURE 31.1   Spatial representation of MCD (Balaban et al., 1980).

Still, highly correlated movements are hard to simulate, and this is leading to a poor sampling of changes on a large scale. For some molecules, Monte Carlo simulations are inaccurate. Because Newton's equations of motion aren't solved, no dynamical information can be gathered from a traditional Monte Carlo simulation. In the case of proteins, one of the main difficulties of Monte Carlo simulations is represented by the conducting of large scale moves of the protein in a particular solvent. Any move that significantly alters the internal coordinates of the protein without moving the solvent particles, also and will likely result in a large overlap of atoms and, thus, a rejection of the trial configuration (Earl et al., 2008).

A 2014 study proved that the Monte Carlo method is an efficient approach to build up a robust model for estimating HIV-1 integrase inhibition of coumarin compounds. The aim of this study was the building of QSAR models for coumarin derivatives as HIV-1integrase inhibitors with the application of the Monte Carlo method. A dataset of 26 coumarin derivatives with known HIV-1 integrase inhibition activity was selected for the study. Canonical simplified molecular input-line entry system (SMILES) for all the compounds were generated with the ACD/Chem Sketch program (ACD/Chem Sketch v.11.0) in order to preserve consistency because different software may generate different SMILES notations. SMILES is a representation of the molecular structure by a sequence of symbols. Some symbols represent molecular fragments, such as atoms or bonds (e.g., 'C,' 'N,' '=,' '#,' etc.). Some of these fragments are represented by two symbols (e.g., 'Br,' 'Cl,' '@@,' etc.) which cannot be separated. Optimal SMILES-based descriptors, determined by descriptor correlation weight ($DCW(T,N_{epoch})$), were calculated with CORAL software (Veselinović et al., 2014).

In 2001, Jorgensen and Duffy, have carried out Monte Carlo statistical mechanics simulations for more than 250 organic solutes in water. Physically significant descriptors such as number of hydrogen bonds, the solvent-accessible surface area and indices for cohesive interactions in solids are correlated with pharmacologically important properties including the octanol/water partition coefficient ($\log P$), aqueous solubility ($\log S$), and brain/blood concentration ratio ($\log BB$), important factors for assessing the bioavailability of a drug. They found that $\log P$, $\log S$, and $\log BB$ can be predicted well-using regression equations with only 4–8 descriptors. Monte Carlo simulation was helpful for example, for finding that $\log P$ is well predicted by an equation where the dominant terms are the total surface area and the number of hydrogen bonds accepted by the solute (Jorgensen et al., 2001).

In some situations, it can be shown a particular interest in the behavior of just one part of the system. For example, when simulating a solvent-solute system that is containing a single solute molecule surrounded by a large number of solvent molecules, the behavior of the soluted molecule and its interactions with the solvent are the most interested in characteristics. Solvent molecules that are in a far distance from the solute molecule can behave like a bulk agent. The Monte Carlo method can be used in these cases, for exploring the solvent space. One simple procedure is the preferential sampling, when the molecules in the proximity of the solvated molecule are moved more frequently than the others, situated further away. This can be possible by defining a cut-off region around the solute. Molecules from outside the cut-off region are moved less frequently than those inside the region, as determined by a "p" probability parameter. At each Monte Carlo "move," a molecule is randomly chosen. If the molecule is inside the cut-off region than it is moved. If it is situated outside this region, a random number between 0 and 1 is generated, and this number is compared to the "p" probability. If "p" is greater than the number, a move is made. If "p" is smaller, no move is made, no averages are calculated, and a new molecule is randomly selected. The closer the "p" number is to 0, the more often closer molecules are moved from further molecules (Leach, 2001).

In preferential sampling, it is necessary to ensure that the correct procedures are followed when deciding whether to accept or reject a move in a manner that is in concordance with the principle of the microscopic reversibility. If a molecule situated inside the cut-off region is moved outside of it, in the preferential sampling mode, the chances of the molecule now being selected for a "from outside to inside" move are less than for the original "from inside to outside" move and this is important when establishing the acceptance of the move criteria (Leach, 2001).

KEYWORDS

- **Boltzmann probabilities**
- **importance sampling**
- **Markov chain**

REFERENCES AND FURTHER READING

Allen, M. P., & Tildesley, D. J., (1987). *Computer Simulation of Liquids*. Oxford University Press.

Balaban, A., Chiriac, A., Moţoc, I., & Simon, Z., (1980). *Steric Fit in Quantitative Structure-Activity Relationships*. Springer Verlag, New York.

Binder, K., & Heermann, D. W., (2002). *Monte Carlo Simulation in Statistical Physics, an Introduction* (4th edn). Springer Verlag.

Bruns, W., Motoc, I., & O'Driscoll, K. F., (1981). *Monte Carlo Applications in Polymer Science*. Springer Verlag.

Durrett, R., (2012). *Essentials of Stochastic Processes* (p. 37). Springer Science & Business Media.

Earl, D. J., & Deem, M. W., (2008). Chapter 2. Monte Carlo simulations. In: *Molecular Modeling of Proteins, Molecular Modeling of Proteins* (Vol. 443), Humana Press.

Everitt, B. S., (2002). *The Cambridge Dictionary of Statistics*. 2nd edition. Cambridge University Press, Cambridge, UK.

Frenkel, D., & Smit, B., (2001). *Understanding Molecular Simulation*. Academic Press.

Jorgensen, W. L., & Duffy, E. M., (2001). *Prediction of Pharmaceutically Important Properties from Monte Carlo Simulations, Beilstein-Institut, Chemical Data Analysis in the Large*. May 22nd–26th 2000, Bozen, Italy.

Leach, R. A., (2001). *Molecular Modeling*: *Principles and Applications* (2nd edn.). Pearson Education Limited.

Schroeder, L., (1974). Buffon's needle problem: An exciting application of many mathematical concepts. *Mathematics Teacher, 67*(2), 183–186.

Serfozo, R., (2009). *"Basics of Applied Stochastic Processes," Probability and its Applications* (p. 35). (Berlin: Springer-Verlag).

Veselinović, J., Veselinović, A., Toropov, A., Toropova, A., Damnjanović, I., & Nikolić, G., (2014). Monte Carlo method based QSAR modeling of coumarin derivates as potent HIV-1 integrase inhibitors and molecular docking studies of selected 4-phenyl hydroxycoumarins. *Scientific Journal of the Faculty of Medicine in Niš, 31*(2), 95–103.

Voiculetz, N., Moţoc, I., & Simon, Z., (1993). *Specifically Interactions and Biological Recognition Processes*. CRC Press, Inc.

CHAPTER 32

Nanotubes

ISTVÁN LÁSZLÓ

Department of Theoretical Physics, Budapest University of Technology and Economics, H-1521 Budapest, Hungary, E-mail: laszlo@eik.bme.hu

32.1 DEFINITION

The most known versions of nanotubes are the carbon nanotubes. Carbon nanotubes are cylindrical carbon structures of few nanometer in diameter and several micrometers in length. From a geometrical point of view they can be constructed by rolling up a hexagonal or honeycomb graphite sheet. Depending on the rolled up unit cell of the graphite sheet we speak about zig-zag, arm-chair or chiral nanotubes. One or both ends of these tubes can be closed by half of a fullerene. In these cases, we speak closed end nanotubes. If both ends are closed in some cases, we can call them fullerenes as well. Nanotubes are also categorized as single-walled nanotubes (SWNT) and multi-walled nanotubes (MWNT). Individual nanotubes can be held together by van der Waals forces in nanotube ropes as well. Generally, we say that carbon nanotubes are allotropes of carbon with a cylindrical nanostructure.

Depending on the material of the rolled up unit cell we can construct other nanotubes as well as the silicone nanotube and the boron nitride nanotube. There are also other tubular nanostructures as the inorganic nanotube, the DNA nanotube and the membrane nanotube.

32.2 HISTORICAL ORIGIN(S)

The fullerenes, nanotubes and the graphene are three allotrope families of carbon, and their production in the last twenty-seven years have triggered intensive researches in the field of carbon structures. Each of them marks a breakthrough in the history of science. The fullerenes, the multi-walled carbon nanotubes, the single-walled carbon nanotubes, and the graphene jumped into the center of interest in 1985, 1991, 1993, and 2004 (in order) (Kroto et al., 1985; Iijima, 1991; Iijima and Ichihashi, 1993; Bethune, 1993; Geim et al., 2004). All of these breakthroughs are good examples; for the fact that a breakthrough can be realized only if the science has a certain level of maturity concerning the combination of a set of favorable conditions; for example, having the right materials available, as well as the related theory, investigation tools and scientific minds (Montihioux and Kuznetsov, 2006). The first authors who presented electron transmission images of multi-walled carbon nanotubes were perhaps Radushkevich and Lukyanovich in a paper written in the Russian language in 1952 (Radushkevich and Lukyanovich, 1952). But as we have mentioned, it was a publication without a breakthrough. Our present-day study of nanotubes was triggered by the papers of Iijima, Ichihashi, and Bethune (Iijima, 1991; Iijima and Ichihashi, 1993; Bethune, 1993). Carbon nanotubes can be produced by arc discharge, laser ablation, plasma torch or chemical vapor deposition using various metallic catalysts (Dresselhaus et al., 1996).

32.3   NANO-SCIENTIFIC DEVELOPMENT(S)

The electronic properties of carbon nanotubes are usually obtained with the help of the zone folding method, which is based on the graphene electronic structure (Dresselhaus et al., 1996). During the rolling up procedure, both the angles and the inter-atomic distances are changing in the hexagonal carbon network. The two-dimensional hexagonal honeycomb structure is described by the unit vectors $\mathbf{a_1} = a\left(\sqrt{3},-1\right)/2$ and $\mathbf{a_2} = a\left(\sqrt{3},1\right)/2$. These vectors are given in a two-dimensional orthonormal basis. Each unit cell contains $r = 2$ atoms at relative positions of $(\mathbf{a_1} + \mathbf{a_2})/3$ and $2(\mathbf{a_1} + \mathbf{a_2})/3$. Using the lattice translation vector $\mathbf{t} = t_1\mathbf{a_1} + t_2\mathbf{a_2}$ we can describe the hexagonal lattice with the integers t_1 and t_2. The (m,n) single wall carbon nanotube is specified by the chiral vector $\mathbf{C_h} = m\mathbf{a_1} + n\mathbf{a_2}$, where the perimeter of the nanotube equals to the length of $\mathbf{C_h}$. The length of the nanotube is the multiple of the nanotube translation vector $\mathbf{T} = -\mathbf{a_1}\left(m+2n\right)/d_R + \mathbf{a_2}\left(2m+n\right)/d_R$. The vector $\mathbf{T}$ is perpendicular to the chiral vector $\mathbf{C_h}$ and its length is the shortest translation period of the nanotube. Here d_R is the greatest common divisor of (m + 2n) and (2m + n). The chiral vector and the integer multiple of the nanotube translation vector $\mathbf{T}$ defines a rectangle which is the supercell in the rolling up construction of the nanotube. It is cut out from the hexagonal lattice and rolled up in order to have a tube.

The reciprocal unit cell vectors $\mathbf{b}_1$, $\mathbf{b}_2$ come from the vectors $\mathbf{a}_1$, $\mathbf{a}_2$ by the relations (László and Zsoldos, 2009).

$$\mathbf{b}_1 = 2\pi \left(\mathbf{a}_2 \times \mathbf{z}\right) / \left(\mathbf{a}_1 \cdot \mathbf{a}_2 \times \mathbf{z}\right) \tag{1}$$

$$\mathbf{b}_2 = 2\pi \left(\mathbf{z} \times \mathbf{a}_1\right) / \left(\mathbf{a}_1 \cdot \mathbf{a}_2 \times \mathbf{z}\right) \tag{2}$$

The unit vector $\mathbf{z}$ has the same direction as the product $\mathbf{a}_1 \times \mathbf{a}_2$. As the chiral vector $\mathbf{C}_h$ and the translation vector $\mathbf{T}$ are unit cell vectors of the nanotube, they also determine the unit cell vectors $\mathbf{Cb}_h = \pi \left(\sqrt{3}(m+n), n-m\right) / \left(a\left(m^2+mn+n^2\right)\right)$ and $\mathbf{Tb} = \pi d_R \left(m-n, \sqrt{3}(m+n)\right) / \left(a\sqrt{3}\left(m^2+mn+n^2\right)\right)$ in the reciprocal lattice, using similar relations like Eqs. (1) and (2).

The electronic wave function of the wave vector $\mathbf{k} = k_1 \mathbf{b}_1 + k_2 \mathbf{b}_2$ has the form (Laszlo, 2004)

$$\left|\mu\mathbf{k}\right\rangle = \sum_{v=1}^{r} c_{\mu\mathbf{k}}^{v} \left|\mathbf{k}\right\rangle_v, \tag{3}$$

where

$$\left|\mathbf{k}\right\rangle_v = \frac{1}{\sqrt{N}} \sum_{\mathbf{t}} e^{i\mathbf{k}\cdot\mathbf{t}} \left|\mathbf{t}\right\rangle_v. \tag{4}$$

The number of unit cells is N, and $\left|\mathbf{t}\right\rangle_v$ is the atomic basis function centered at the site v of the unit cell $\mathbf{t}$. The $c_{\mu\mathbf{k}}^{v}$ coefficients are the μ-th eigenvectors of the $\mathbf{H}(\mathbf{k})$ Fourier transform of the Hamiltonian and $E_\mu(\mathbf{k})$ is the corresponding eigenvalue, where

$$\mathbf{H}(\mathbf{k}) = \begin{pmatrix} 0 & -\gamma_1 - \gamma_2 e^{-i2\pi k_1} - \gamma_3 e^{-i2\pi k_2} \\ -\gamma_1 - \gamma_2 e^{i2\pi k_1} - \gamma_3 e^{i2\pi k_2} & 0 \end{pmatrix} \tag{5}$$

and

$$E_\mu(\mathbf{k}) = \pm \left\{ \gamma_1^2 + \gamma_2^2 + \gamma_3^2 + 2\gamma_1\gamma_2 \cos 2\pi k_1 + 2\gamma_1\gamma_3 \cos 2\pi k_2 \right.$$
$$\left. + 2\gamma_2\gamma_3 \cos 2\pi \left(k_2 - k_1\right) \right\}^{1/2} \tag{6}$$

The γ_i parameters are the first neighbor hopping parameters (László & Zsoldos, 2009).

The special points of the honeycomb Brillouin zone are $\mathbf{0}$ for the Γ point; $\mathbf{b}_1/2$, $\mathbf{b}_2/2$ and $(\mathbf{b}_1 + \mathbf{b}_2)/2$ for the M points; $(\mathbf{b}_1 + 2\mathbf{b}_2)/3$ and $(2\mathbf{b}_1 + \mathbf{b}_2)/3$ for the K points. In the nanotube (m, n), only those $\mathbf{k}$ points are selected for which holds the following relation:

$$\mathbf{C}_h \mathbf{k} = 2\pi \left(mk_1 + nk_2\right) = 2\pi\sigma \tag{7}$$

This condition comes from the periodic boundary condition for the period of the vector $\mathbf{C}_h$. Here $\sigma = 1, 2\ldots\ m + n$ is an integer.

32.4   NANO-CHEMICAL APPLICATION(S)

Usually, nanotubes with the value $n = m$ are metals. The nanotubes for them $n–m = 3k$ and k is an integer are semiconductors or semimetals with a band-gap of less than 0.1 eV. Other nanotubes are semiconductors that have band-gap of about 1 eV. Its size is inversely proportional to the tube radius.

Carbon nanotubes are the strongest and stiffest materials yet discovered. Its tensile strength and elastic modulus result from the covalent sp^2 bond and it is stronger than the sp^3 bond in the diamond. The carbon nanotube young's modulus is more than 1 TPa and the corresponding value for steel is about 0.2 TPa.

The present application of nanotubes is mostly limited as a composite material in order to improve mechanical, electrical or thermal properties of a bulk material. There are however several possible applications as nano-electric circuits, transistors and diodes (Postma et al., 2001), applications in biomedicine as microfluidic channels (Sharei et al., 2013), production of materials with extraordinary toughness (Zhang et al., 2005), applications in solar cells (Guldi et al., 2005), and so on.

KEYWORDS

- **electrical properties**
- **fullerenes**
- **mechanical properties**
- **nanotubes**
- **optical properties**

REFERENCES AND FURTHER READING

Bethune, D. S., Kiang, C. H., De Vries, M. S., Gorman, G., Savoy, R., Vazquez, J., & Beyers, R., (1993). Cobalt-catalyzed growth of carbon nanotubes with single-atomic-layer walls. *Nature, 363*, 605–607.

Dresselhaus, M. S., Dresselhaus, G., & Eklund, P. C., (1996). *Science of Fullerenes and Carbon Nanotubes* (p. 801). Academic Press: San Diego, Boston, New York, London, Sydney, Tokyo, Toronto.

Guldi, D. M., Rahman, G. M. A., Prato, M., Jux, N., Qin, S., & Ford, W., (2005). Single-wall carbon nanotubes as integrative building blocks for solar-energy conversion. *Angewandte Chemie, 117*, 2051–2054.

Iijima, S., (1991). Helical microtubules of graphite carbon. *Nature, 354*, 56–58.

Iijima, S., & Ichihashi, T., (1993). Single-shell carbon nanotubes of 1nm diameter. *Nature, 363*, 603–605.

Kroto, H. K., Heath, J. R., O'Brien, S. C., Curl, R. F., & Smalley, R. E., (1985). C_{60}: Buckminsterfullerene. *Nature, 318*, 162–163.

László, I., (2004). The electronic structure of nonpolyhex carbon nanotubes. *J. Chem. Inf. Comput. Sci., 244*, 315–322.

László, I., & Zsoldos, I., (2009). Calculation of curvature effects and hybridization in zigzag carbon nanotubes. *Phys. Status Solidi. B., 246*, 2610–2613.

Monthioux, M., & Kuznetsov, V. L., (2006). Who should be given credit for the discovery of carbon nanotubes? *Carbon, 44*, 1631–1623.

Novoselov, K. S., Geim, A. K., Morozov, S. V., Jiang, D., Zhang, Y., Dubonos, S. V., Grigorieva, I. V., & Firsov, A. A., (2004). Electric field effect in atomically thin carbon films. *Science, 306*, 666–669.

Postma, H. W. C., Teepen, T., Yao, Z., Grifoni, M., & Dekker, C., (2001). Carbon nanotube single-electron transistors at room temperature. *Science, 293*, 76–79.

Radushkevich, L. V., & Lukyanovich, V. M., (1952). About the structure of carbon formed by thermal decomposition of carbon monoxide on iron substrate *Zurn. Fisic. Chim., 26*, 88–95 (in Russian).

Sharei, A., Zoldan, J., Adamo, A., Sim, W. Y., Cho, N., Jackson, E., et al., (2013). *A Vector-Free Microfluidic Platform for Intracellular Delivery P.N.A.S., 110*, 2082–2087.

Zhang, M., Fang, S., Zakhidov, A. A., Lee, S. B., Aliev, A. E., Williams, C. D., Atkinson, K. R., & Baughman, R. H., (2005). Strong, transparent, multifunctional carbon nanotube sheets. *Science, 309*, 1215–1219.

CHAPTER 33

New Energy Technology

CYNTHIA WHITNEY

Galilean Electrodynamics, 11660 239th Ave. NE, Redmond,
WA 98053-5613, USA, E-mail: Galilean_Electrodynamics@Comcast.net

33.1 DEFINITION

New energy technology (NET) is a short name appropriate for any power-generation concept that is not yet in widespread use. One of these NET concepts is called cold fusion (CF). Within the traditional narrative of Physics, only so-called hot fusion (HF), i.e., thermonuclear fusion, is understood, attempted, and somewhat developed. By contrast, CF is much doubted, since any kind of fusion requires that nuclei come into close proximity, despite the fact that nuclei are typically shielded behind electrons, and electrons repel each other. It, therefore, seems that putting nuclei into close proximity requires crashing atoms together with brute force; i.e., at a high temperature. But Chemistry is a wily science, and it may yet offer new tools useful for engineering CF. In particular, the new approach to chemistry called algebraic chemistry (AC) (*see* Chapter 1) has something important to offer. AC does not deal in the usual computer models of atoms in terms of electron wave functions associated with atoms; instead, it deals in what is called ionic configurations (ICs) (*see* Chapter 25). An IC is the list of atoms involved in a molecule or an ion, along with the charge attributed to each atom. Every molecule or ion therein has some fairly short list of plausible ICs, and each IC has its own characteristic energy, which can be estimated using AC. Furthermore, this energy-estimation exercise can be conducted for all the different isotopes that the elements involved can present. This aspect is especially important for analyzing CF, inasmuch as CF involves heavy isotopes of Hydrogen, which have one or more neutrons along with the single proton that is the nucleus of the every-day light Hydrogen that dominates our Universe.

33.2   HISTORICAL ORIGIN

New energy technology (NET) is the hope of a world that seems to be exhausting its present energy resources, such as fossil fuels, and overloading its environment with combustion products, such as CO_2. Of course, we are looking for alternative energy technologies. We presently have a lot of experience with nuclear fission for power production, but some of that experience is discouraging, or even frightening. There have been accidents, there can be toxic waste problems, there could be military diversions, and there is much political resistance.

We also have nuclear fusion power as a concept, but it isn't commercially viable yet. We can do thermonuclear fusion with magnetic confinement – hot fusion (HF). But with HF, we can so far hardly even break even.

Cold fusion (CF) is an even newer idea. It is presently surrounded with secrecy, hype, and doubt. One often-cited reason for doubt is the lack of neutrons produced in any claimed CF reaction. People expect neutrons from all nuclear reactions just because fission reactions produce neutrons. But in the case of *fusion* reactions, the expectation is wrong-headed: fusion reactions do *not* produce neutrons; they *consume* them; that is why they all start with *heavy* isotopes.

Another often-cited reason for doubt about CF is its variability of results. But that is to be expected when we do not yet understand what we are doing! The third reason for doubt about CF is the lack of a credible theory about CF. The theory is one area where near-term progress now seems possible: AC can help here.

33.3   NANO-SCIENTIFIC DEVELOPMENT

Let us begin by comparing naturally occurring Hydrogen, which has average nuclear mass 1.008, to the individual isotopes in naturally occurring Hydrogen: pure Hydrogen with nuclear mass 1, Deuterium with nuclear mass 2, and Tritium with nuclear mass 3.

CF requires, first of all, that neutral atoms be stripped of electrons. Stripping an average Hydrogen atom of its one electron to make an average H^+ is estimated to take:

$$IP_{1,1} / 1.008 = 14.2500 / 1.008 = 14.1369 \text{ eV} \tag{1}$$

But we need to distinguish isotopes. Stripping the lightest isotope to make H^+ is estimated to take:

$$IP_{1,1}/1 = 14.2500 \text{ eV} \tag{2}$$

But stripping the deuterium isotope to make D⁺ is estimated to take:

$$IP_{1,1}/3 = 14.2500/3 = 4.7500 \text{ eV} \tag{3}$$

And stripping the tritium isotope to make T⁺ is estimated to take:

$$IP_{1,1}/3 = 14.2500/3 = 4.7500 \text{ eV} \tag{4}$$

The message is clear: heavier positive ions are easier to make. This is good news for CF, because CF needs those heavy positive ions as reactants.

CF also requires that heavy isotopes not be diverted into receiving additional electrons. Stripped electrons have to go somewhere, so let us study where they go. First, in the case of naturally occurring hydrogen, receiving an electron to make natural H⁻ involves energy

$$\begin{aligned} &-IP_{1,1} \times \sqrt{1 \times 2}/1.008 - \Delta IP_{1,2} \times 2/1.008 + \Delta IP_{1,3} \times 1/1.008 = \\ &-14.2500 \times 1.4142/1.008 - 35.6250 \times 2/1.008 + 0 = \\ &-19.9924 - 70.6845 = -90.6769 \text{ eV} \end{aligned} \tag{5}$$

This big negative number speaks to the much-favored status of electron pairs!

Next, consider just the lightest isotope of hydrogen. The energy to make this H⁻ is

$$\begin{aligned} &-IP_{1,1} \times \sqrt{1 \times 2}/1 - \Delta IP_{1,2} \times 2/1 + \Delta IP_{1,3} \times 1/1 = \\ &-14.2500 \times 1.4142/1 - 35.6250 \times 2/1 + 0 = \\ &-20.1524 - 71.250 = -91.4024 \text{ eV} \end{aligned} \tag{6}$$

This is a huge energy, but for CF, we are really interested in the heavier isotopes. For deuterium, the energy to make the negative ion D⁻ is:

$$\begin{aligned} &-IP_{1,1} \times \sqrt{1 \times 2}/2 - \Delta IP_{1,2} \times 2/2 + \Delta IP_{1,3} \times 1/2 = \\ &-14.2500 \times 1.4142/2 - 35.6250 \times 2/2 + 0 = \\ &-10.0762 - 35.6250 = -45.7012 \text{ eV} \end{aligned} \tag{7}$$

For tritium, the energy to make T⁻ is:

$$\begin{aligned} &-IP_{1,1} \times \sqrt{1 \times 2}/3 - \Delta IP_{1,2} \times 2/3 + \Delta IP_{1,3} \times 1/3 = \\ &-14.2500 \times 1.4142/3 - 35.6250 \times 2/3 + 0 = \\ &-6.7175 - 23.7500 = -30.4675 \text{ eV} \end{aligned} \tag{8}$$

The message is: making negative ions is rewarding, but at least making heavier negative ions is *less* rewarding than making lighter ones. So more of the heavier nuclei can remain available for CF.

Now consider heavy water molecules. Recall the entry in this volume on ICs: the conventional IC for ordinary water, $2H^+ + O^{2-}$ had positive energy, and so could not occur very frequently. But what do we know about the heavy-water molecule $D^+ + H^+ + O^{2-}$? We know D^+ takes 7.1250 eV, half what H^+ takes, 14.250 eV, and we know O^{2-} takes –27.3788 eV, so the IC $D^+ + H^+ + O^{2-}$ takes:

$$7.1250 + 14.2500 - 27.3788 = -6.0038 \text{ eV} \tag{9}$$

This energy is negative, so this IC can readily occur. That is good for CF, which needs the D^+. For the IC similarly involving tritium, $T^+ + H^+ + O^{2-}$, the energy requirement is

$$4.7500 + 14.2500 - 27.3788 = -8.3788 \text{ eV} \tag{10}$$

This energy is even more negative than that for the IC involving deuterium, $D^+ + H^+ + O^{2-}$, and so it favors this IC involving Tritium even more. That fact means the T^+ ion is preferentially exposed for subsequent CF.

33.4 NANO-CHEMICAL APPLICATION

Exposing D^+ and/or T^+ ions is just the first requirement for CF. Next comes the job of getting them together. There is presently more folklore than a physical theory about how to do this job.

One thing that is known to be helpful is leaving heavy water to rest undisturbed for a long time before inserting electrodes and attempting to provoke CF events. So what is it about still water that can help the CF scenario?

The author believe the significance of still water for CF is related to the phenomenon of exclusion zone water, or EZ water, discussed in Pollack (2013). The basic phenomenon is this: still, water in a gravitational field forms a surface layer that is tough enough to permit a boat to float.

This surface layer of water is slightly charged negatively. That means it expels positive ions (Hydrogen nuclei). That is, it makes an exclusion zone. This behavior accounts for the name 'EZ water.'

This EZ water layer is highly structured into a hexagonal pattern of macroscopic extent. The basic unit in this pattern has three Hydrogen atoms and two Oxygen atoms, and the net charge of this unit is –1. One can write it as $(H_3O_2)^-$.

We can begin to understand EZ water by looking at its possible ionic configurations (ICs). One possibility is $3H^+ + 2O^{2-}$. We already know that naturally occurring H^+ takes $14.250/1.008 = 14.1369$ eV, so $3H^+$ takes $3 \times 14.1369 = 42.4107$ eV. And we already know that O^{2-} takes -27.3788 eV, so $2O^{2-}$ takes $2 \times (-27.3788) = -54.7576$ eV. So $3H^+ + 2O^{2-}$ takes $42.4107 - 54.7576 = -12.3469$ eV.

Another possible IC is $3H^- + 2O^+$. We already know that naturally occurring H^- takes $3 \times (-90.6769) = -272.0307$ eV. And O^+ takes

$$(IP_{1,1} \times 8 + \Delta IP_{1,8} \times 8 - \Delta IP_{1,7} \times 7) / M_8 =$$

$$(14.250 \times 8 + 13.031 \times 8 - 13.031 \times 7) / 15.999 = \qquad (11)$$

$$(114.000 + 104.248 - 91.217) / 15.999 = 7.9399 \text{ eV}$$

So $2O^+$ takes $2 \times 7.9399 = 15.8798$ eV, and $3H^- + 2O^+$ takes $-272.0307 + 15.8798 = -256.1509$ eV. This energy is much more negative than that of the first IC, so this IC is very much preferred.

What does it mean? For one thing, an H^- ion is very much larger than an H^+ ion, which is, after all, just a proton! As for O^+, it is somewhat smaller than O^{2-}, but not by much. Overall, $3H^- + 2O^+$ fills space better than $3H^+ + 2O^{2-}$ does. That is why it can provide the exclusion property.

Having the exclusion property means having something that can contribute to pushing those heavy nuclei closer together.

33.5   OPEN ISSUES

It seems clear that CF is an engineering problem, rather than a science problem *per se*, but it does reflect some light back onto Physics. Any eventual success with CF engineering will surely stimulate some kind of review and renewal in Physics.

KEYWORDS

- **algebraic chemistry**
- **ionic configurations**
- **ionization potentials**

REFERENCES AND FURTHER READING

Pollack, G., (2013). *The Fourth Phase of Water, Beyond Solid Liquid Vapor*. Chapter 4, Ebner & Sons, Seattle, USA.

Whitney, C., (2013). *Algebraic Chemistry Applications and Origins*. Chapter 5, Nova Publishers: New York, USA.

Whitney, C., (2014). EZ water: A new quantitative approach applied–many numerical results obtained. *Water*, *5*, 105–120.

CHAPTER 34

The Orbital Exponent Efficacy to Study the Periodic Parameters

NAZMUL ISLAM

Department of Chemistry and Theoretical and Computational Chemistry Research Laboratory, Techno Global-Balurghat, Balurghat, D. Dinajpur, 733103, India, E-mail: nazmul.islam786@gmail.com, nazmulislam@tgbtechno.org

We know that the effective 'nucleus-electron' attraction is proportional to the effective nuclear charge (Z_{eff}) and inversely proportional to the effective principal quantum number (n^*). The effective nuclear charge increases gradually while we are going across a period where the effective quantum number remains constant. Thus, along a period, the effective 'nucleus-electron' attraction depends mainly on the effective nuclear charge as other factors (n^*) remains constant. But in the next period, they both jump in number. The second factor, the effective principal quantum number, is the dominating one while we are going across a group. Thus, the consideration of only effective nuclear charge to explain the periodicity of periods and groups is erroneous. Alternatively, the effective attraction power of the nucleus upon the electron can be described by the orbital exponent (ξ) – the ratio of the effective nuclear charge and the effective principal quantum numbers, i.e., Z_{eff}: n^*. The orbital exponent, ξ, is simply defined as:

$$\xi = Z_{eff}/n^*$$

Reed (1999) modified Slater's method (1930) by simply refining the Slater's (1930) grouping of the orbital. He (1999) also proposed two sets of rules for the computation of the screening constants. Reed (1999) also considered the pairing energy for p and d orbitals. Now relying upon the periodic law and following Reed, we have evaluated the effective nuclear charge and the orbital exponents for the atoms of the 118 elements of the Periodic Table with some modifications. We propose a simple approach for the calculation of the ionization potential, electronegativity, atomic radius of atoms in terms of their orbital exponents. The prescription is simple and utilizes only the simple Bohr equation with some modifications. We have pointed out that the atomic first ionization energy does not only depend on the principal quantum number (n) but also on the azimuthal quantum number (l) of the orbital (n, l) on which the electron of interest is present. The formalism is tested through the calculation of the atomic ionization potentials of 118 elements of the Periodic Table. The orbital exponent values for 118 elements of the Periodic Table are computed following the suggestions of Reed. The calculated numerical results for a number of atoms are shown to agree quite well with their experimental counterparts. To perform the validity tests of the present scale of ionization potential, various physicochemical properties of the atoms are also correlated on the basis of the computed ionization potential data. It is found that the stability of the half-filled configuration depends on the orbital (n, l) on which the electron is present. The express periodic behavior and correlation of the most important physicochemical properties of elements suggest that present method of evaluation of the periodic parameters in terms of their orbital exponents' ionization potential of the atoms is a quite a successful venture.

Slater's rule (1930) is based on experiments and Clementi and Raimondi's method (1963) is based on HF calculation. There is no any analytic concept why and how the strength of nuclear charge does lose? Moreover, the effective nuclear charge (Z_{eff}) and the radius evaluated using the Z_{eff} by Slater's empirical method is not periodic always. Again Ghosh and Biswas (2002) pointed out that the radius of lanthanides did not show the periodic character if one evaluates the atomic radius of Lanthanides using the Clementi and Raimondi effective nuclear charge. In order to overcome the difficulties of Slater's (1930) method and Clementi and Raimondi's (1963) method, Reed (1999) modified Slater empirical rule for the evaluation of screening constants. In this work, we have considered Reed's suggestions for the evaluation of screening constants of the s, p, and d-block elements and extended Reed's rule for the evaluation of screening constants of f-block elements. When electron entire in the 5f, 6p and higher we have used the contribution of 4f as 1. In the same shell, f electrons shield each other by a factor 0.3228. It is here important to mention that in order to evaluate the orbital exponent, we have used the Eq. (4) and the value of n* proposed by Slater (1930) for $n = 1$ to $n = 6$ and for $n = 7$, we have used the value of $n* = 4.3$ proposed by Ghosh and Biswas (2002) (Table 34.1 and Figure 34.1).

TABLE 34.1 Computed Orbital Exponent for Atoms of 118 Elements of the Periodic Table

Atom	ξ	Atom	ξ	Atom	ξ
H	1.00000	Nb	0.81038	Tl	0.92557
He	1.67720	Mo	0.83180	Pb	1.08681
Li	0.66340	Tc	0.85323	Bi	1.24805
Be	1.00200	Ru	0.87465	Po	1.40929
B	1.34060	Rh	0.89608	At	1.56945
C	1.67920	Pd	0.91750	Rn	1.73176
N	2.01780	Ag	0.93892	Fr	0.53656
O	2.35640	Cd	0.96035	Ra	0.69405
F	2.69500	In	1.12965	Ac	0.71398
Ne	3.03360	Sn	1.29895	Th	0.73391
Na	0.76907	Sb	1.46825	Pa	1.02895
Mg	0.99480	Te	1.63755	U	1.18644
Al	1.22053	I	1.80685	Np	1.34393
Si	1.44627	Xe	1.97615	Pu	1.63898
P	1.67200	Cs	0.54933	Am	1.79647
S	1.89773	Ba	0.67714	Cm	1.95395
Cl	2.12347	La	0.73098	Bk	1.97388
Ar	2.34920	Ce	1.03305	Cf	2.26893
K	0.62357	Pr	1.19429	Es	2.42642
Ca	0.80659	Nd	1.35552	Fm	2.58391
Sc	0.82976	Pm	1.51676	Md	2.74140
Ti	0.85292	Sm	1.67800	No	2.89888
V	0.87608	Eu	1.83924	Lr	2.91881
Cr	0.89924	Gd	2.00024	Rf	2.93874
Mn	0.92241	Tb	2.16171	Db	2.95867
Fe	0.94557	Dy	2.32295	Sg	2.97860
Co	0.96873	Ho	2.48419	Bh	2.99853
Ni	0.99189	Er	2.64543	Hs	3.01847
Cu	1.01505	Tm	2.80667	Mt	3.03840
Zn	1.03822	Yb	2.96790	Uun	3.05833
Ga	1.22124	Lu	2.98831	Uuu	3.07826
Ge	1.40427	Hf	3.00871	Uub	3.09819
As	1.58730	Ta	3.02912	Uut	3.10888
Se	1.77032	W	3.04952	Uuq	3.26637
Br	1.95335	Re	3.06993	Uup	3.42386
Kr	2.13638	Os	3.09033	Uuh	3.58135
Rb	0.57680	Ir	3.11074	Uus	3.73884
Sr	0.74610	Pt	3.13114	Uuo	3.89633
Y	0.75178	Au	3.15155		
Zr	0.78895	Hg	3.17195		

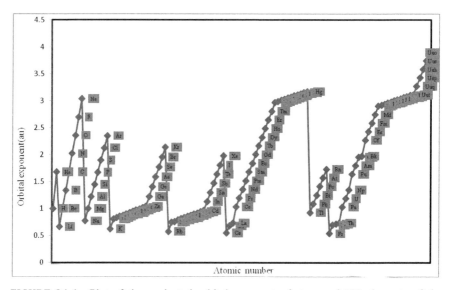

FIGURE 34.1 Plot of the evaluated orbital exponent of atoms of 118 elements of the periodic table as a function of their atomic number.

KEYWORDS

- atomic radii
- Gordy atomic electronegativity scale
- ionization energy
- Mulliken atomic electronegativity scale
- orbital exponent

REFERENCES AND FURTHER READING

Clementi, E., & Raimondi, D. L., (1963). *J. Chem. Phys.*, *38*, 2686–2689.
Clementi, E., Raimondi, D. L., & Reinhardt, W. P., (1967). *Ibid*, *47*, 1300–1307.
Ghosh, D. C., & Biswas, R., (2002). *Int. J. Mol. Sci.*, *3*, 87–113.
Reed, J., (1999). *Chem. Edu.*, *76*, 802–804.
Slater J. C., (1930). *Phys, Rev.*, *36*, 57–64.

CHAPTER 35

Periodic Arch

CYNTHIA WHITNEY

Galilean Electrodynamics, 11660 239th Ave. NE, Redmond,
WA 98053-5613, USA, E-mail: Galilean_Electrodynamics@Comcast.net

35.1 DEFINITION

The Periodic Arch (PA) is one of many modern visual displays that have been developed to augment the traditional Periodic Table (PT) of the chemical elements. The motivation for all these developments has been that the traditional PT, which is all rows and columns, seems to defy nature, which seems to want more flexibility than that. For example, a traditional PT usually identifies functional categories, such as light metals, brittle metals, ductile metals, low-melting metals, non-metals, and noble gases. But on the traditional PT, some of these category boundaries are more diagonal than vertical or horizontal. This fact means that rows and columns are not a very function-oriented format for a PT. For another example, a traditional PT usually has two long rows of elements completely excised, and displayed separately, below the main body of the PT. This fact means that the rectangular format of the PT is not very space-efficient. The PA answers to these difficulties by inverting the PT and making it into nested arches around and over its founding element, hydrogen. The essay that follows displays the PA, notes some of the elements that the PA brings into new visual proximity, discusses some relationships they actually have, and points to further issues for future investigation.

35.2 HISTORICAL ORIGIN(S)

The idea of a PT goes back to Mendeleyev, who studied the behaviors of chemical elements, and realized that as atomic number increases, behaviors repeat. He put the elements into a table, in which the elements aligned in a column behave similarly.

From its inception, the idea of the PT has intrigued the minds of new researchers, many of whom have been inspired to produce all sorts of new, but related, visual displays. This activity has continued into modern times. The reader can now find a wonderfully rich collection of modern displays of the PT online, through: http://www.meta-synthesis.com/webbook/35_pt/pt_database.php?PT_id = 277.

One of these modern displays is the PA. A picture of this particular type of display appears in Figure 35.1.

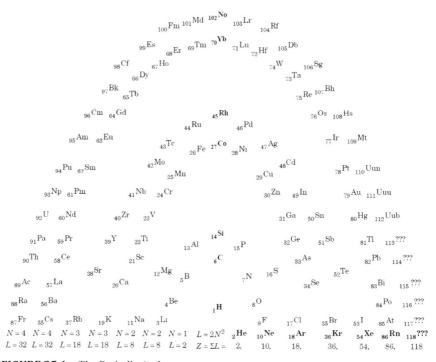

FIGURE 35.1 The Periodic Arch.

The PA is designed to focus the mind differently than the traditional PT does. For example, a traditional PT calls attention to the lengths of the periods, whereas the PA has this information at the bottom, where the variable L for length is displayed. More important than L, and so displayed higher, is a new variable called N. Observe that $L = 2N^2$. So N is the root from which L arises. N itself goes by units from 1 to 4, with repeats for all N except $N = 1$.

The PA draws attention to different aspects of behavior than the PT does. Look at the elements stacked straight up the middle of the PA, in 'keystone' positions: Hydrogen ($_1$H), carbon ($_6$C), silicon ($_{14}$Si), iron ($_{26}$Fe), rubidium ($_{44}$Ru), ytterbium ($_{70}$Yb), and nobelium ($_{102}$No). These elements are important to chemistry as facilitators of complex molecule formation, because they can both give, or take, electrons, whichever may be needed.

The terminology 'keystone element' is appropriate for many other reasons as well. Hydrogen $_1$H is certainly the 'keystone' for all of the present-day physical analysis of atoms, from deep quantum mechanics to practical chemical engineering. Carbon $_6$C is certainly the 'keystone' for all of the Organic Chemistry and hence all of biological life. Silicon $_{14}$Si is certainly the 'keystone' for present-day technological life. Cobalt $_{27}$Co is not so famous, but it lies between Iron $_{26}$Fe and Nickel $_{28}$Ni, and is functionally better than either of them: harder, more corrosion resistant, and more heat resistant. And, like iron, cobalt is much strengthened by the addition of a trace of carbon – that other keystone element. We, humans, are currently more than three millennia into our 'Iron Age,' which, with the help of carbon and other trace additives, has morphed into our 'Steel Age.' Had cobalt been more plentiful on this planet, this might have been our 'Cobalt/Steel Age.' Rhodium ($_{45}$Rb) is also not so famous, mainly because it is not so plentiful, but it is good for plating and alloying. The remaining keystone elements, ytterbium ($_{70}$Yb), and nobelium ($_{102}$No) remain to be significantly exploited.

35.3 NANO-SCIENTIFIC DEVELOPMENT(S)

The periodic arches above and around hydrogen can be divided into local neighborhoods corresponding to traditional categories: noble gasses, non-metals, and several functionally distinct families of metals that together constitute the great preponderance of all elements. The traditionally named metal categories are light metals, brittle metals, ductile metals, and low-melting metals.

To explore all these named neighborhoods, start from hydrogen ($_1$H) in Figure 35.1. Hydrogen itself has always been recognized as unique, and has not been assigned to any named neighborhood. On the PA, hydrogen ($_1$H) is in the central hub position, with all the keystone elements above it. The first two of the keystone elements are carbon ($_6$C) and silicon ($_{14}$Si). These are non-metals, and like them, hydrogen could rationally be grouped with the non-metals.

Then to the lower right of hydrogen come the seven 'noble gasses': helium ($_2$He), neon ($_{10}$Ne), argon ($_{18}$Ar), krypton ($_{36}$Kr), xenon ($_{54}$Xe), radon ($_{86}$Rn), and one more not yet discovered or synthesized ($_{118}$???). They are all 'out of the way, down in the basement,' so to speak. This positioning befits elements that participate in chemistry only very reluctantly!

Just above hydrogen and the noble gasses, there are 15 elements traditionally identified as 'non-metals': boron ($_5$B), carbon ($_6$C), nitrogen ($_7$N), oxygen ($_8$O), fluorine ($_9$F), silicon ($_{14}$Si), phosphorus ($_{15}$P), sulfur ($_{16}$S), chlorine ($_{17}$Cl), arsenic ($_{33}$As), selenium ($_{34}$Se), bromine ($_{35}$Br), tellurium ($_{52}$Te), iodine ($_{53}$I), astatine ($_{85}$At). There is also one more element not yet discovered or synthesized ($_{117}$???) that will probably be a non-metal. If we include both $_1$H and $_{117}$???, we have 17 non-metals.

Everything else on the PA consists of metals. The whole left side, the top, and most of the right side of the PA consists of metals.

The metals start to the left of hydrogen ($_1$H), with 12 elements traditionally identified as 'light metals': lithium ($_3$Li), beryllium ($_4$Be), sodium ($_{11}$Na), magnesium ($_{12}$Mg), potassium ($_{19}$K), calcium ($_{20}$Ca), rubidium ($_{37}$Rb), strontium ($_{38}$Sr), cesium ($_{55}$Cs), barium ($_{56}$Ba), francium ($_{87}$Fr), and radium ($_{88}$Ra), are to the left of hydrogen. Viewed from the PA perspective, aluminum $_{13}$Al also seems to belong with the light metals. It has traditionally been grouped with low-melting metals instead. If we include $_{13}$Al with the light metals, then we have 13 light metals.

The combination of hydrogen, non-metals, and light metals constitute the entire base of the chemically active part of the PA. The rest of the metals then make the superstructure of the PA.

Starting on the upper left, we have the 'actinide series' ($_{89}$Ac through $_{103}$Lr) and the 'lanthanide series' ($_{57}$La through $_{71}$Lu). Observe that they have their natural pride of place in the PA. They are *not* any kind of awkward footnote, as they have been in traditional PT's.

Toward the lower middle of the PA superstructure come the 'brittle metals': scandium ($_{21}$Sc), titanium ($_{22}$Ti), vanadium ($_{23}$V), chromium ($_{24}$Cr), magnesium ($_{25}$Mn), yttrium ($_{39}$Y), zirconium ($_{40}$Zr), and niobium ($_{41}$Nb).

From the lower middle to the right come the 'low-melting metals.' Traditionally, aluminum ($_{13}$Al) was included here, but the PA suggests re-categorizing aluminum as a light metal. Then we have zinc ($_{30}$Zn), gallium ($_{31}$Ga), germanium ($_{32}$G3), cadmium ($_{48}$Cd), indium ($_{49}$In), tin ($_{50}$Sn), antimony ($_{51}$Sb), mercury ($_{80}$Hg), thallium ($_{81}$Tl), lead ($_{82}$Pb), bismuth ($_{83}$Bi), polonium ($_{84}$Po), the oddly named $_{112}$uub, and the not yet discovered or synthesized: $_{113}$??? through $_{116}$???. That makes 17 low-melting metals.

Then on the upper right of the PA come the 'ductile metals': from iron ($_{26}$Fe) through copper ($_{29}$Cu), and from molybdenum ($_{42}$Mo) through silver ($_{47}$Ag), and from hafnium ($_{72}$Hf) through gold ($_{79}$Au), and from rutherfordium

($_{104}$Rf) through the oddly named $_{111}$uuu. The combination of the actinide and lanthanide series and the ductile metals make up about half of the PA.

Let us now investigate the physical significance of the numerical parameter N that is the root variable giving rise to the lengths of element lists in each arch of the PA (or each row of the traditional PT). We have the parameter values

$$N = 1, 2, 3, 4 \tag{1}$$

The value $N = 1$ occurs once, and the values $N = 2$, 3, and 4 each occur twice.

From N we then have the element list lengths given by

$$L = 2N^2 \tag{2}$$

The values of L that occur are then

$$L = 2 \times 1 = 2 \tag{3}$$
$$L = 2 \times (1 + 3) = 2 \times 4 = 8 \tag{4}$$
$$L = 2 \times (1 + 3 + 5) = 2 \times 9 = 18 \tag{5}$$
$$L = 2 \times (1 + 3 + 5 + 7) = 2 \times 16 = 32 \tag{6}$$

In the PA (or the PT), $L = 2$ occurs once, $L = 8$ occurs twice, $L = 18$ occurs twice, $L = 32$ occurs twice.

All these facts are interpreted traditionally in terms of the quantum numbers for orbit radius, orbital angular momentum, electron spin, and electron total angular momentum, as they occur in solutions for Schrödinger's wave equation. This approach was initially developed for the Hydrogen atom, and later adapted for atoms of all elements, and conceptually extended to describe molecules.

The facts about the PA (or the PT) can also be interpreted in a simple pictorial way. Electrons apparently like to form rings. The simplest and most famous electron ring corresponds to the Cooper pair. This pair consists of one electron, said to have spin 'up,' and another electron, said to have spin 'down.' The concept of 'spin' permits situations with electron pairs to be reconciled with the idea that electrons within an atom all have to occupy different quantum states.

Less famous, and possibly more contestable, is the idea of rings of three or more electrons, all with the same spin. As we look at the successive elements in the PA (or PT), we can work out how successive electron states are filled. The empirical picture is not perfect, but over most of the sub-periods corresponding to given orbital angular momentum, it appears that usually all of the spin states of one direction fill first, and only then do all of the spin states of the other direction fill.

Why would such a filling order even be allowed, much less be the norm? We are tempted to invoke another quantum number, this one for the direction of the total orbital angular momentum, including both orbit angular momentum and spin angular momentum. In the presence of a magnetic field, the direction of the total angular momentum would cause an energy distinction. So we imagine that, even without any magnetic field, the latent ability to create an energy distinction can permit a succession of same-spin electron states to be filled.

But then, why do there exist occasional exceptions to the same-spin filling order?

The present author believes that an answer to the problem of electron state filling order requires new ideas about how electrons experience each other; i.e., how electromagnetic signals travel back and forth between electrons. Assumptions about that issue were seemingly finalized early in the 20th century, with the advent of special relativity theory (SRT). But the whole science of information theory (IT) was not even available until the middle of the 20th century. So we are now very much overdue for a post-action review on the issue of electromagnetic signals and SRT (*see* Chapter 11). With SRT appropriately updated, we see how same-charge rings can form at superluminal speeds, and electron rings of three or more electrons having the same spin are not impossible.

35.4   NANO-CHEMICAL APPLICATION(S)

The PA is an illustrative technique that supports an approach to chemistry that focuses on the visually obvious, including trends in available chemical data that enable one to make very good estimates for data that is *not* directly available, either because it has not to be measured, or it *cannot* be measured. The overall approach is called algebraic chemistry (AC) (*see* Chapter 1). The data exploited consists of ionization potentials (IPs).

35.5   MULTI-/TRANS-DISCIPLINARY CONNECTION(S)

The PA represents the kind of mindset that is often shared by engineers and artists.

35.6   OPEN ISSUES

Electromagnetic signals are the root issue that needs to be revisited. Until this revisit is fully accomplished, and widely disseminated, nanotechnology and other technologically important areas will be unnecessarily hampered.

KEYWORDS

- **algebraic chemistry**
- **electron state filling order**
- **visual displays in chemistry**

REFERENCE AND FURTHER READING

Whitney, C., (2013). *Algebraic Chemistry Applications and Origins* (pp. 155–177). Chapter 3, Nova Publishers: New York, USA.

CHAPTER 36

Periodic Law

FRANCISCO TORRENS[1] and GLORIA CASTELLANO[2]

[1]*Institute for Molecular Science, University of Valencia, PO Box 22085, E-46071 Valencia, Spain, Tel: +34-963-544-431, Fax: +34-963-543-274, E-mail: torrens@uv.es*

[2]*Department of Experimental Sciences and Mathematics, Faculty of Veterinary and Experimental Sciences, Valencia Catholic University Saint Vincent Martyr, Guillem de Castro-94, E-46001 Valencia, Spain*

36.1   DEFINITION

The Periodic Law (PL) establishes that the chemical and physical properties of the elements and their combinations are a periodic function of the atomic number of the element.

36.2   HISTORICAL ORIGIN

The evolution of the discovery of elements with time followed several methods: blowlamp, Lavoisier's theory of combustion, electrolysis, spectroscopy, Mendeleev's Periodic Law (PL), fractional distillation, radioactivity, particle accelerators, etc. Atomic theory (Dalton, 1803) is the most important of all chemical theories. Döbereiner (since 1817) observed five *triads*: Li–Na–K, Ca–Sr–Ba, Cl–Br–I, P–As–Sb, and S–Se–Te. Dumas et al., (1827–1858) expanded them with new elements: F, Mg, O, N, and Bi. Dalton (1803), Avogadro (1811) and Gerhardt (1850) were the base of the discussions at the First International Congress of Chemistry (Karlsruhe, 1860), which adopted the atomic weights of Gerhardt (Cannizzaro's system, 1858 after Dulong & Petit (1818), based on Avogadro's law, 1811). Chancourtois (1862) built the spiral periodic system tellurian snail (*vis tellurique*). Meyer (1864) adopted Cannizzaro's system (1858) in his monograph. Odling (1865) published a

PTE with 13 rows and seven columns. Newlands (1865) reported PTE law of the octaves. Lenssen (1867) ordered the elements in 20 triads. Hinrichs (1867) arranged the elements in a periodic graphic with the chemically similar elements in radial lines. Mendeleev (1869) applied his PL of the atomic weights to PTE. When the elements were represented by order of their atomic weights, their properties were repeated in a series of periodic intervals. Meyer (1870) discovered a close relationship between the elements and their atomic weights. Without PL, people would have no reason to predict the properties of the unknown elements, even could not talk about the lack of one of them. The discovery of the elements was a fact of observation: chance, perspicacity, and spirit of scrutiny. From PTE structure, it was seen spaces, which corresponded to new elements (Trifonov & Trifonov, 1990). Based on his PL, Mendeleev (1871) predicted the existence and properties of some new elements: Sc, Ga, Ge, etc. For physicists, PTE served as a challenge to discover the underlying physical structures that would account for the recognized regularities. Mendeleev's PL difficulties are: (1) the inversion of three pairs of elements (Ar–K, Co–Ni, Te–I); (2) discovery of the *inert* elements (Ramsay, 1894–1910); (3) accommodation of lanthanoids; (4) he classified the nonexistent elements ether and coronium; (5) Moseley (1913) discovered the atomic number Z, reformulating PL on atomic number; (6) actinoids accommodation (Seaborg's hypothesis, 1944; PTE, 1945). From Moseley's measurements (1913), it was possible to discern the patterns of the enigmatic periodic lengths in PTE. However, a periodicity interpretation came about only with Bohr's theory (1913).

36.3 NANO-SCIENTIFIC DEVELOPMENT

Meyer (1870) considered the volume occupied by determined fixed weights of the elements. In such conditions, every weight contained the same number of atoms of its element, which meant that the ratio of the volumes of the elements was equivalent to the ratio of the volumes of the atoms that composed the elements. One could talk of atomic volumes. On representing the atomic volumes of the elements *vs.* atomic weights (*cf.* Figure 36.1), a series of waves was obtained that reached maximum values in the alkalines (Na, K, Rb, Cs). Every decay and rise to a maximum or minimum would correspond to a period in PTE. In every period, other physical properties decayed and rose, together with the atomic volume. The H, the first of PTE, is a special case and constituted period-1. Periods-2 and -3 comprised seven elements each, and they repeated Newlands' law of octaves (1865). However, both following waves included more than seven elements, which clearly showed that Newlands made an error. It was not possible to force that the law of octaves is fulfilled throughout all PTE, with seven elements in every row. The last periods should be longer than the first ones. The important PL fact is that this dependence implies a certain periodicity, which shows in the periodic repetition of the characteristic properties: the elements with atomic numbers $Z = 2, 10, 18, 36, 54, 86$ and 118 are all chemically inert elements. The chemical and physical properties of the elements ranged between the inert elements mark after the first very short period H–He, two short periods with eight each, two long periods with 18 and two very long periods with 32 elements.

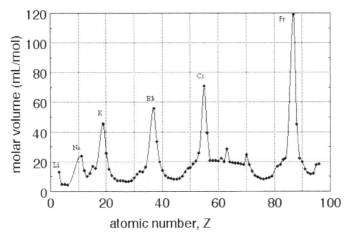

FIGURE 36.1   Variation of atomic volumes with atomic number.

Mendeleev, aware of PTE heuristic value, indicated research directions. He pointed out PTE holes to be filled: Sc, Ga, Ge, etc. ($Z = 21, 31, 32$; OSs 3, 3, 4). From the position of an unknown element in PTE, it was possible to predict its physical and chemical properties, so that one gave indications to where such a missing element might be sought and with what methods. Mendeleev predicted Ge properties (*cf.* Table 36.1), etc.

36.4   NANO-CHEMICAL APPLICATION

The most characteristic representative points of atomic volume (V_a)-atomic number (Z)-diagram are placed on the maxima and correspond to alkalines (Garzón-Ruipérez, 1988). The values of their atomic volumes are not repeated as PL states, but they rise regularly. An empirical function $P(Z)$ exists, which reproduces with precision the different V_a values for those elements. On referring to those representative points that are placed on the maxima, notice that a maximum value of a function $P(Z)$, $P_m(Z)$, for a given value of Z, has only meaning if it is compared with those of the preceding $P(Z-1)$ and later $P(Z+1)$ points, fulfilling the following conditions:

$$P_m(Z) > P(Z-1) \tag{1}$$

$$P_m(Z) > P(Z+1)$$

TABLE 36.1　Mendeleev Predictions (*eka*-Si, Es, 1871) and Values (Ge, 1886 & ff.)

Mendeleev predictions	Values found for Ge
Atomic weight 72 g·mol^{-1}	Atomic weight 72.6 g·mol^{-1}
Specific gravity 5.5 g·cm^{-3}	Specific gravity 5.47 g·cm^{-3}
Atomic volume 13 cm^3·mol^{-1}	Atomic volume 13.4 cm^3·mol^{-1}
Color: dark grey	Color: greyish white
Valency: 4	Valency: 4
Formula of the upper oxide EsO$_2$	Formula of the upper oxide GeO$_2$
Specific gravity of the oxide 4.7 g·cm^{-3}	Specific gravity of the oxide 4.7 g·cm^{-3}
Fusible metal that will volatilize at a high-temperature	Melting point: 960°C, and volatilizes at higher temperatures
By the action of air, it will give a white powder of EsO$_2$	By the action of air, it gives a white powder of GeO$_2$
The element will separate water into its elements with difficulty	The element does not separate water into its elements
Acids will have a rather light action	It is not attacked by HCl but aqua regia
Alkalis will not attack it	Solutions of KOH do not attack it, but it is oxidized by molten alkalis
The oxide will be fireproof and reduce easily to metal	The oxide is fireproof and reduces to metal with C
The hydroxide will be weakly basic	The basic hydroxide properties are weak
The chloride will have formula EsCl$_4$; it will be liquid with boiling point ≈90°C and specific gravity of 1.9 g·cm^{-3}	The GeCl$_4$ is liquid; boiling point 86°C and specific gravity 1.89 g·cm^{-3}
It will form an unstable gaseous compound EsH$_4$ that will be, however, more stable than SnH$_4$	The GeH$_4$ is an unstable gas, but more stable than SnH$_4$
The fluoride EsF$_4$ will be gaseous	The fluoride GeF$_4$ is a gas
It will form an organometallic compound Es(C$_2$H$_5$)$_4$ that will boil at 160°C and have a specific gravity of 0.96 g·cm^{-3}	The organometallic compound Ge(C$_2$H$_5$)$_4$ melts at 90°C and boils at 163.5°C. Its specific gravity is 0.99 g·cm^{-3}

Conditions (1) order relationships, being precisely these relationships which, in the case of PL being rigorously fulfilled, should be repeated at determined intervals equal to the size of the periods. Relationships (1) are equivalent to the following:

$$P_m(Z)-P(Z-1) > 0 \tag{2}$$
$$P(Z+1)-P_m(Z) < 0$$

Relationships (2) are valid only for the maxima of $P(Z)$, and others more general are desired for all Z values, so that the differences $P(Z+1) - P(Z)$ were calculated and every value is assigned to element Z. Calling $D(Z)$ to this value, one obtains:

$$D(Z) = P(Z+1)-P(Z) \tag{3}$$

Instead of $D(Z)$, one can take the values $R(Z) = P(Z+1)/P(Z)$, assigning them to element Z. If PL were general, the elements belonging to the same group, having to occupy analogous positions in the different waves, would satisfy, without exception, the same relationship, i.e., they would fulfill:

$$\text{Either } D(Z) > 0 \text{ or } D(Z) < 0 \tag{4}$$
$$\text{Either } R(Z) > 1 \text{ or } R(Z) < 1 \tag{5}$$

However, the results show that this is not the case, so that PL is not general, existing some anomalies; e.g., the number of irregularities is not negligible. What considerations can one make faced with these results? If the properties are not repeated in the sense of the traditional statement, which on the other hand would not be a serious inconvenience, and neither the relationships do it rigorously, what is PL reduced to? Reconsidering the atomic volume, it is observed that a rough periodicity exists and this is more evident for periods 4–6 (if one does without lanthanoids in the last) than 2 and 3. However, in detail, anomalies exist that are manifested in the corresponding diagrams, in which the representative points of the different elements for the successive periods are not always in phase (they do not occupy analogous positions in the waves). The information could be attributed to little precise values of the corresponding magnitudes, and to the fact that the properties of the elements are not strictly those of the atom but the ones of the corresponding net, which typically does not repeat for the elements belonging to the same period. However, the ionization energy and OSs are atomic properties, which values are precise, and their periodicity shows exceptions in a number similar to that of other nonatomic properties. Among all properties, OSs are just those that present the greatest precision, because they are essentially integer values. In agreement with this fact, it would be convenient, in order to eliminate the possible influence over $D(Z)$ and $R(Z)$ of the lack of precision in P values, to contrast PL for the mentioned property. Anomalies exist in a number similar to other properties.

36.5 MULTI-/TRANS-DISCIPLINARY CONNECTION

Historically, the difficulties that found Mendeleev (1869), which were solved with the discovery of the atomic number Z (Moseley, 1913), made thinking that his work was perfect and his PL (changing the atomic weight for Z) was an exact natural law. After the study of the optical spectra (Wollaston, 1802; Fraunhofer, 1814; Kirchhoff and Bunsen, 1859; Balmer, 1885; Zeeman, 1897; Rydberg, 1899; Ritz, 1908), Bohr's theory (1913) and quantum mechanics (Heisenberg, 1925; Schrödinger, 1926; Dirac, 1928), PL became interpreted in the terms of electronic configurations (Aufbau principle, after Pauli exclusion principle, 1925), admitting that PL is a finished and perfect work. The arrangement in s, p, d-blocks, etc., is as much as allowed in organizing PTE, contributed to overestimate the mentioned theoretical results.

36.6 OPEN ISSUES

Lanthanoids only could correspond, in the sense of the phase concordance, to their homologs, whether they exist, actinoids; but, on the one hand, it is to be decided the accommodation of the latter in period-7 and, on the other, assuming Seaborg's hypothesis (1944) and PTE (1945), one would find 14 new groups with two elements each, which would not authorize one to establish group relationships. The PL, despite its importance in chemistry systematization, has not the status of the laws of physics. In sum: (1) the properties of the elements are not repeated; only, and in any case, their chemical character; (2) order relationships are repeated, although with some exceptions; (3) electron configurations are repeated, with exceptions. Periodicity is not general. However, if one accepts a natural order in the elements, its underlying law should be of phenomenic character. The problem is to find the underlying structure to the order. It is interesting to study the generalization of the ideas to periodic properties, PTE, etc.

KEYWORDS

- **periodic system**
- **group of the element**
- **congeneric**
- **transition element**
- **successive period**

REFERENCES AND FURTHER READING

Alfonseca, M., (1996). *Espasa Dictionary: 1,000 Great Scientists*. Espasa Calpe: Madrid, Spain.

Garzón-Ruipérez, L., (1988). *De Mendeléiev a los Superelementos*. Universidad de Oviedo: Gijón, Spain.

Mendeleev, D., (1905). *The Principles of Chemistry*. Longmans, Green & Co.: London, UK.

Meyer, L., (1864). *Die Modernen Theorien der Chemie und ihre Bedeutung für die chemische Statik*, Maruschke & Berendt: Breslau, Germany.

Trifonov, D. N., & Trifonov, V. D., (1990). *How Were Discovered the Chemical Elements* [in Russian], Mir: Moscow, Russia.

CHAPTER 37

Periodic Property

FRANCISCO TORRENS[1] and GLORIA CASTELLANO[2]

[1]*Institute for Molecular Science, University of Valencia, PO Box 22085, E-46071 Valencia, Spain, Tel: +34-963-544-431, Fax: +34-963-543-274, E-mail: torrens@uv.es*

[2]*Department of Experimental Sciences and Mathematics, Faculty of Veterinary and Experimental Sciences, Valencia Catholic University Saint Vincent Martyr, Guillem de Castro-94, E-46001 Valencia, Spain*

37.1 DEFINITION

If the chemical elements are ordered by increasing atomic weight, then their chemical properties reveal a characteristic periodicity. Since the discovery of the atomic number (Moseley, 1913), the true independent variable of PL, PP is redefined: If the chemical elements are ordered by increasing atomic number, then their chemical properties reveal a characteristic periodicity.

37.2 HISTORICAL ORIGIN

In chemistry, the first systematic correlations come from the conservation laws of mass (van Helmont, 1648; Lavoisier, 1774) and energy (Mayer, 1842), followed by Dalton conception of structural matter (1803). Cannizzaro revived Avogadro's law (1811) in the First International Congress of Chemistry (Karlsruhe, 1860), which adopted the atomic weights of Gerhardt (1850) in Cannizzaro's system (1858) after Dulong & Petit (1818), based on Avogadro's law. Mendeleev (1869) and Meyer (1870) were the first ones to place the quantitative structure-activity relationships/QSPRs in the center of chemistry, with their vision of the empirical PTE. In Mendeleev PTE (1869), elements that occupy the same column present similar properties, which greatest discontinuity is between alkalines and halogens. Atomic volume was the first PTE property (cf. Figure 37.1), measured by Meyer (1870) in an attempt to systematize chemical-elements ordering: again alkaline volume results in the greatest from the elements in the corresponding period. However, with the advent of QT, QSPRs among elements of periods and down groups of PTE (Mendeleev, 1871) acquired in-depth quantitative meaning, relating the elementary electronic structure to the observed atomic reactivity, which in every aspect results a certain manifestation of QSPR pair, which is quantified since the derivation of the associated equation. When two atoms come into contact with each other, they perturb the electron distribution within one another, which interplay between the electrostatic potentials of the two atoms in many instances is rather subtle but, in some cases, it leads to unexpected effects when the two atoms are allowed to bind to each other. Problems (e.g., reactivity, mutual polarizability) that occur when two atoms come into intimate contact are, however, able to be tackled via QT. Optical-spectra study (Wollaston, 1802; Fraunhofer, 1814; Kirchhoff and Bunsen, 1859; Balmer, 1885; Zeeman, 1897; Rydberg, 1899; Ritz, 1908), Bohr's theory (1913) and quantum mechanics (Heisenberg, 1925; Schrödinger, 1926; Dirac, 1928) consolidated PTE. Kirchhoff and Bunsen (1859) stated that the atoms of a gas or vapor emit characteristic line spectra when supplied with energy. The same atoms are also capable of absorbing light of the same wavelength as the emitted radiation. Absorption lines appear at these frequencies in the continuous spectrum emitted by glowing solids and liquids. The PTE gave chemists a road map with checkpoints; in other words, it soon became a practical tool. Once more, in modern PTE form, the greatest discontinuity in the chemical (after Mendeleev, 1869) and physical (after Meyer, 1870) properties continues to be between alkalines and halogens.

37.3 NANO-SCIENTIFIC DEVELOPMENT

37.3.1 ATOMIC WEIGHT

The atomic mass of an element is the average mass of an atom in a conventional sample of the element and is expressed in atomic mass units (amu) (Gray, 2009). The amu is defined as the 1/12 part of the mass of a ^{12}C-atom. One amu is the mass of a proton or a neutron, so the atomic mass of an element is approximately equal to the total number of protons and neutrons in the nucleus. The relationship between the atomic mass and 1 amu is called atomic weight. However, the atomic weights of some elements are not integers. In the case that a conventional sample of the element includes two or more natural isotopes (Soddy, 1912), the weighted average of the isotopes masses justifies the fractional value of the atomic mass.

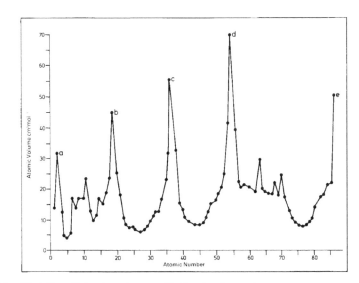

FIGURE 37.1   Variation of atomic volumes with the atomic number.

37.3.2   DENSITY

The density of an element is that corresponding to an ideal crystal without defect of the pure element. It is calculated from the atomic mass and crystallographic distances. It is expressed in grams by a cubic centimeter ($g \cdot cm^{-3}$).

37.3.3   ATOMIC RADIUS

The density of a substance depends on the mass of the atoms that compose it and the space that they occupy. The atomic radius is the average distance between the nucleus and the most peripheral electrons in the atom, expressed in picometres (pm) or angstroms ($1 \text{Å} = 10^2 \, pm = 10^{-10} \, m$).

37.3.4   CRYSTAL STRUCTURE

The crystal structure (arrangement able to give a diffraction pattern) indicates the atoms arrangement in the unit cell (the component that is repeated till building all the crystal) when the element is found in the most pure crystal form. In the case of the elements that are gases or liquids at RT, obtaining the crystal form requires that the temperature be equal to or lesser than the melting point.

37.3.5 ORDER OF INCORPORATION OF THE ELECTRONS

The order of incorporation of the electrons (Aufbau principle, after the Pauli exclusion principle, 1925) is the progression according to which the electrons occupy the available atomic orbitals.

37.3.6 ATOMIC EMISSION SPECTRUM

The atoms of the elements, when they are heated to high temperatures, emit light of determined wavelengths (colors), which are related with the energy differences between the orbitals. Every spectral line corresponds to a determined energy gap, arranged from the red to the violet section.

37.3.7 MATTER AGGREGATION STATES

A temperature scale, in Celsius degrees, shows the temperature range in which an element is solid, liquid, or gas. The border between solid and liquid corresponds to the melting point, while the limit between liquid and gas accords with the boiling point.

37.4 NANOCHEMICAL APPLICATION

A number of PPs are of interest in nanochemical applications, e.g., melting point, boiling point, melting heat, vaporization heat, density, atomic volume, ionization energy, atomic radius, covalent radius, electronegativity, binding energy, etc.

Halogens make the utility of PTE clear. The four elementary substances form diatomic molecules X_2; all their hydrogenated compounds present HX form, and their Na salts NaX formula. The free elements are oxidant agents and their oxidant power regularly decays in the order F_2, Cl_2, Br_2, I_2; in fact, F_2 displaces from their salts any of the three remaining elements:

$$F_2 + 2\,KCl \rightarrow 2\,KF + Cl_2$$
$$F_2 + 2\,KBr \rightarrow 2\,KF + Br_2$$
$$F_2 + 2\,KI \rightarrow 2\,KF + I_2$$

The Cl_2 displaces both Br and I:

$$Cl_2 + 2\,KBr \rightarrow 2\,KCl + Br_2$$
$$Cl_2 + 2\,KI \rightarrow 2\,KCl + I_2$$

while Br_2 only does it with I:

$$Br_2 + 2\,KI \rightarrow 2\,KBr + I_2$$

The color of the free elements becomes more noticeable in rising the atomic number, from pale yellow to almost black. The salts also show a color gradation, from AgF and AgCl, colorless, to AgBr, pale yellow, and AgI, yellow. Generally, the weak intermolecular (van der Waals) forces, which keep the molecules joined in liquids and crystals, rapidly rise in magnitude with the increasing atomic number of the atoms in the molecules, which is shown, e.g., by the sequence of the melting and boiling points of the inert elements, which means in the variation of the matter aggregation state of the free halogens, from a hardly condensable gas (F_2), passing *via* an easily condensable gas and a liquid, to a solid (I_2). The melting points show an almost regular rise of 100°C from a period to the immediate one, and boiling points, analogous rises. The complete regularity above is not shown by the hydrogen halides (melting, boiling points of HF are rather higher than those that should be expected from the sequence shown by the three other substances). The anomaly indicates that HF presents certain structural peculiarity, which is also shown by water and a few simple substances.

37.5 MULTI-/TRANS-DISCIPLINARY CONNECTION

The known elements include some familiar and many rare ones. Some elemental substances are metals, some other ones, nonmetals; at RT some are gases, some others, liquids, and others, solids. They show a variety in their chemical properties and in the nature of the compounds that they form.

Torrens & Castellano (2015) reported reflections on the nature of PTE with implications in chemical education (*see* Chapters 36 and 38).

37.6 OPEN ISSUES

The study of chemistry is neither simple nor easy; to obtain a wide knowledge of general (introduction to) chemistry, it is necessary to learn a great set of facts. Chemical facts cannot be completely coordinated by a unifying theory. However, the development of chemical theories was rapid to help the student, who can simplify his study, on substances properties and reactions relating the empirical information with theories (e.g., theory of atomic structure, PL). It is interesting to study the possible generalization of the ideas above to PL, PTE, etc.

KEYWORDS

- **empirical chemometrics**
- **Periodic Law**
- **periodic system**
- **Periodic Table**
- **periodicity**

REFERENCES AND FURTHER READING

Alfonseca, M., (1996). *Espasa Dictionary: 1,000 Great Scientists*, Espasa Calpe: Madrid, Spain.

Gray, T. W., (2009). *The Elements: A Visual Exploration of Every Known Atom in the Universe*. Black Dog–Leventhal: New York, NY.

Torrens, F., & Castellano, G., (2015). Reflections on the nature of the Periodic Table of the elements: Implications in chemical education. In: Seijas, J. A., Vázquez, T. M. P., & Lin, S. K., (eds.), *Synthetic Organic Chemistry* (Vol. 18, pp. 1–15). MDPI: Basel, Switzerland.

CHAPTER 38

Periodic Table

FRANCISCO TORRENS[1] and GLORIA CASTELLANO[2]

[1]*Institute for Molecular Science, University of Valencia, PO Box 22085, E-46071 Valencia, Spain, Tel: +34-963-544-431, Fax: +34-963-543-274, E-mail: torrens@uv.es*

[2]*Department of Experimental Sciences and Mathematics, Faculty of Veterinary and Experimental Sciences, Valencia Catholic University Saint Vincent Martyr, Guillem de Castro-94, E-46001 Valencia, Spain*

38.1 DEFINITION

When the elements were represented by the order of their atomic weights, their properties were repeated in a series of periodic intervals, PTE. Since the discovery of the atomic number (Moseley, 1913), the true independent variable of PL, PTE is redefined. When the elements were represented by the order of their atomic numbers, their properties were repeated in a series of periodic intervals, PTE. However, the complete description of PTE foundation was much later and came with the atomic theory. The elements are a reality (inasmuch as an elemental grouping of matter exists with particular, differentiated and permanent properties) that, although the diversity that is esteemed in the universe seems infinite, a limited number of constituting elements exists, which can be ordered vs. PPs (atomic weight, ionization energy, electron affinity, electronegativity, atomic volume, etc.). Nowadays, the most striking PTE points continue to be: (1) the elements that occupy the same column present similar properties; (2) the opposed behavior between both extreme groups: alkalines and halogens.

38.2 HISTORICAL ORIGIN

Scientists work over the years revealed much about the atoms, which are matter building blocks, which make up the elements. Scientists of the

19th century discovered more than 50 different atoms, every one with its own mysterious properties. In Prout's *protil* hypothesis (1815), every element was an H-atoms composite. First International Chemical Congress (Karlsruhe, 1860) raised questions. (1) Is it necessary to establish a difference between the terms atom and molecule, in such a way that it be called molecules the smallest parts of the substances that take part in a chemical transformation or that result from it, and that are comparable among them with regard to their physical properties; whereas the atoms are the smallest particles of the substances that are contained in the molecules? (2) Can the expression composed atom be substituted by the expression radical or rest? The KC adopted Gerhardt's atomic weights (1850) in Cannizzaro's system (1858) after Dulong & Petit (1818), based on Avogadro's law (1811). Periodic systems of the elements appeared: cylindrical tellurian snail *vis* tellurique (Chancourtois, 1862), periodic graphic (Hinrichs, 1867), etc. The main fault of the earlier PTEs was to assume that their periods might always have the same length (Newlands, 1965). Mendeleev (1869) proposed PTE based on chemical properties (*cf.* Figure 38.1) and Meyer (1870), on physical ones (Román Polo, 2002). New questions raised. (1) Why some atomic weights are integers but others not? (2) What is the relationship between elements spectra and their PTE place? (3) How to incorporate the materials that resulted from the transmutation processes? (4) Should they be considered chemically uniform when they behave uniformly in all chemical tests, yet they have different radioactive behaviors? Thomson (1913) discovered nonradioactive isotopes (Ne, atomic number $Z = 10$, mass number $A = 20$, 22). Chemical properties of all isotopes of an element are substantially the same, so that they are essentially determined by Z, not A.

			Ti = 50	Zr = 90	? = 180
			V = 51	Nb = 94	Ta = 182
			Cr = 52	Mo = 96	W = 186
			Mn = 55	Rh = 104,4	Pt = 197,4
			Fe = 56	Ru = 104,4	Ir = 198
		Ni = Co = 59	Pd = 106,6	Os = 199
H = 1			Cu = 63,4	Ag = 108	Hg = 200
	Be = 9,4	Mg = 24	Zn = 65,2	Cd = 112	
	B = 11	Al = 27,4	? = 68	Ur = 116	Au = 197?
	C = 12	Si = 28	? = 70	Sn = 118	
	N = 14	P = 31	As = 75	Sb = 122	Bi = 210?
	O = 16	S = 32	Se = 79,4	Te = 128?	
	F = 19	Cl = 35,5	Br = 80	J = 127	
Li = 7	Na = 23	K = 39	Rb = 85,4	Cs = 133	Tl = 204
		Ca = 40	Sr = 87,6	Ba = 137	Pb = 207
		? = 45	Ce = 92		
		?Er = 56	La = 94		
		?Yt = 60	Di = 95		
		?In = 75,6	Th = 118?		

FIGURE 38.1   First version of Mendeleev's periodic table of the elements (1869).

The PTE is empirical: more important than nomenclature is chemical behavior, depending on the outer-core shells and frontier orbitals. It (*cf.* Figure 38.2) illustrates chemists' ability to summarize a lot of information in a 2D representation, e.g., alkalines, with the greatest volume in the corresponding period, lie in group 1A.

```
  1A   2A                                               3A   4A   5A   6A   7A   8A
  -----                                                 -----
1 | H |                                                                     |He |
  |---+----                                             --------------------+---|
2 |Li |Be |                                           | B | C | N | O | F |Ne |
  |---+---|                                             |---+---+---+---+---+---|
3 |Na |Mg |3B   4B   5B   6B   7B |      8B      |1B   2B |Al |Si | P | S |Cl |Ar |
  |---+---+---+---+---+---+---+---+---+---+---+---+---+---+---+---+---|
4 | K |Ca |Sc |Ti | V |Cr |Mn |Fe |Co |Ni |Cu |Zn |Ga |Ge |As |Se |Br |Kr |
  |---+---+---+---+---+---+---+---+---+---+---+---+---+---+---+---+---+---|
5 |Rb |Sr | Y |Zr |Nb |Mo |Tc |Ru |Rh |Pd |Ag |Cd |In |Sn |Sb |Te | I |Xe |
  |---+---+---+---+---+---+---+---+---+---+---+---+---+---+---+---+---+---|
6 |Cs |Ba |LAN|Hf |Ta | W |Re |Os |Ir |Pt |Au |Hg |Tl |Pb |Bi |Po |At |Rn |
  |---+---+---+---+---+---+---+---+---+---+---+---+---+---+---+---------
7 |Fr |Ra |ACT|
  --------------

          Lanthanide |La |Ce |Pr |Nd |Pm |Sm |Eu |Gd |Tb |Dy |Ho |Er |Tm |Yb |Lu |
                     |---+---+---+---+---+---+---+---+---+---+---+---+---+---+---|
          Actinide   |Ac |Th |Pa | U |Np |Pu |Am |Cm |Bk |Cf |Es |Fm |Md |No |Lw |
                     ------------------------------------------------------------
```

FIGURE 38.2   American Standard Code for Information Interchange (ASCII)-art PTE.

Moseley (1912) discovered a law defining the atomic number, a concept that replaced the atomic weight in Mendeleev's PTE (1869), rearranging pairs (Ar–K, Co–Ni, Te–I). The study of the optical spectra (Wollaston, 1802; Fraunhofer, 1814; Kirchhoff and Bunsen, 1859; Balmer, 1885; Zeeman, 1897; Rydberg, 1899; Ritz, 1908), Bohr's theory (1913) and quantum mechanics (Heisenberg, 1925; Schrödinger, 1926; Dirac, 1928) consolidated PTE, which was related to electron configurations (Bohr, 1913). Bohr (especially since 1920) elaborated several proposes to explain PTE *via* a series of quantum numbers that characterize the different electron shells. Quantum physics development entwined in this way with a fundamental chemistry section: the order of the elements. The journal *Boletín das Ciencias* devoted a special issue to PTE (*cf.* Figure 38.3).

Interpretation of PTE via the Pauli exclusion principle. Elements series in PTE manifests the subjacent symmetry in atomic physics (*cf.* Figure 38.4). Pauli exclusion principle (1925) made it possible to explain PTE, before the introduction of electron spin (Goudsmit–Uhlenbeck, 1925). All electrons must differ within the atom by at least one of the four quantum numbers that characterize their state, meaning that in the first orbit at most two electrons

FIGURE 38.3 *Boletín das Ciencias* special issue No. 67 (Mar., 2009) devoted to PTE.

exist, in the second $\leq 8 = 2 \times 2^2$ and in the third $\leq 18 = 2 \times 3^2$. From the rules, all the particular qualities of the elements that are reflected in PTE follow. The model does not offer a picture of the symmetry relationships in the arrangement of electrons. The elliptical model (Sommerfeld, Wilson and Ishiwara, 1915) is a better approximation. In general, for every quadruplet (n, l, m, s), one possible state exists. For every triple (n, l, m), two associated states exist corresponding to the two possible spin alignments. For every pair (n, l), one has also the $(2l + 1)$ states that belong to the possible values of m (from $-l$ to l). For every n, $2\Sigma_{l=0}^{n-1}(2l + 1) = 2n^2$ different states exist.

FIGURE 38.4   MS Spanish flag (warship cannons battery); not MS Yin-Yang symbol.

Burbidge et al. (1957) explained elements emergence (nucleosynthesis) in physics (*cf.* Figure 38.5). The stabilization hypothesis may be of help for polystochastic-model developments (Baez & Dolan, 1995; Leinster, 2004). A braided category is a monoidal category with additional structure; a sylleptic category is a braided category with additional structure, etc. (Crans, 2000).

38.3   NANO-SCIENTIFIC DEVELOPMENT

The basic structure of PTE emerges from the universal and fundamental laws of quantum mechanics (Gray, 2009). Every chemical element is defined by its atomic number, which is an integer. The atomic number of an element is the number of protons that are found in the nucleus of every atom of the element, which determines the number of electrons that orbit around this nucleus. The electrons, in particular, those that occupy the outer shells, determine element chemical properties. Elements in PTE are ordered by rising atomic number. Elements sequential arrangement leaves some spaces, which allow placing in the same column elements with the same number of electrons in the outer shell, which explains most important PTE trait: elements in the same column present similar chemical properties. Mendeleev predicted six elements (Sc, Ga, Ge, Tc; Z = 21, 31, 32, 43; OSs 3, 3, 4, 7; *cf.* Figure 38.6; Re, Po; Z = 75, 84; OSs 7, 4). Bohr forecasted Hf (Z = 72; OS 4).

FIGURE 38.5 From Escher's landscape without atoms to other landscape with atoms (image modified from J. J. Gómez Cadenas).

FIGURE 38.6 PTE-commemorative Spanish postage stamp (February 2, 2007): 4 holes.

After columns ordering, PTE identifies groups of elements. Those in the same group are congeneric; they have related chemical and physical properties. In *s*-block (columns 1–2), column 1, the first element H is somewhat anomalous. Conventionally, it shows in column 1, as it shares some chemical properties with them, especially the trend to form compounds losing one electron H^+ like Na^+. However, H_2 is a gas while the other column-1 elements are soft metals named alkali metals, react with water and release $H_{2(g)}$, which is explosively inflammable. Column-2 elements are named alkaline-earth metals. Like alkalines, they are soft but slowly release $H_{2(g)}$ on reacting with water, which allows, e.g., using Ca in portable $H_{2(g)}$ generators. The wide central *d*-block in PTE corresponds to the transition metals. They are used in the industry (e.g., Fe, W, *cf.* Figure 38.7). All but Hg are hard and have a typical metallic appearance, even radioactive Tc. They are stable under air, but someone slowly rusts; e.g., Fe rusting is the most destructive undesired chemical reaction. Other transition metals (e.g., Au, Pt) are valued by their high resistance to rush. Both boxes at the bottom left in *d*-block are reserved for lanthanoid and actinoid *f*-block. Following PTE logic, between columns 2 and 3 a space of two rows and 14 columns of elements should be to include *f*-block. However, to avoid an excessively wide PTE, the convention is that the spaces are saved and *f*-block is shown in two rows at PTE bottom.

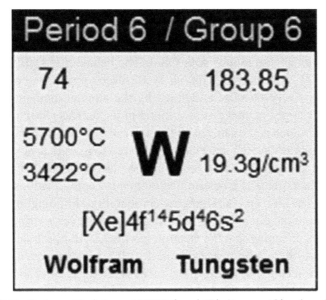

FIGURE 38.7 De Luyart & de Luyart (1783) found W in Bergara (Gipuzkoa, Spain).

p-block (the last six columns) bottom-left triangle includes the ordinary metals, although common metals are the transition ones. The top-right triangle comprehends the nonmetals. They are electric insulators, unlike metals, which are electric conductors. The C originates organic chemistry (*cf.* Figure 38.8). Between metals and nonmetals, a diagonal exists where the border elements (semimetals) are found. They have something of metal and some other things of a nonmetal. They hardly pass the electric current. They include the semi-conductors, which became important for modern life. The diagonal breaks the general rule of placing the elements with common traits in the same column. Column 2 from the right include the halogens, which in the pure state are disagreeable. All are reactive and give off a repulsive smell. $F_{2(g)}$ attacks almost all and $Cl_{2(g)}$ was used as toxic in World War I. However, as chemical compounds (e.g., fluoride toothpaste, kitchen salt NaCl) they were adapted to quotidian life. The F is unique in highest electronegativity with moderate size, and modulates physicochemical properties, e.g., lipophilicity, acidity, etc., and pharmacological activities. The last column comprehends the inert elements (inert means the lowest reactivity). In general, they form compounds with neither themselves nor other elements. Their chemical inertia allows them to be used as reactive-elements protective, e.g., Na under Ar.

The *f*-block at the bottom is formed by two rows. The upper row, which begins with La, includes the lanthanoids, and the lower one, which starts with Ac, comprehends the actinoids. Lanthanoids stand out by their similar chem-ical behavior: trend to form ionic compounds with OS 3. All actinoids (e.g., U, Pu) are radioactive. Transuranic actinoids show an increasing tendency to develop ionic compounds with OS 3. The behavior is similar to lantha-noids. In PTE, transuranic actinoids were placed under the corresponding lanthanoids. Every element is defined by the atomic number: the number of positive charges named protons, found in its nucleus (Scerri, 2007). The number of protons is the same as the number of electrons (negative charges) found in orbits around the nucleus. The electron exists as a probability cloud, with more likelihood of being found in one place than another, but without being in a fixed place at a determined moment. Representations correspond to orbitals. The *s*-orbital is symmetric (probability of finding the electron is independent of the direction from the nucleus). However, *p*-orbital is formed by two lobes, meaning that the electron has the greatest probability of being found in two regions in space opposed to the nucleus, and it is the least probable that it be found in any direction between the two sections. Although only one *s*-orbital exists, three *p*-orbitals are, which lobes point following the three orthogonal directions in space (*x,y,z*). Five different types of *d*-orbital and seven of *f*-orbital exist, with a rising number of lobes. Every orbital

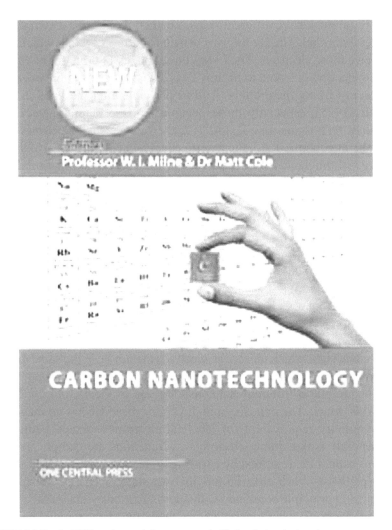

FIGURE 38.8 In PTE, carbon originates organic (living beings) chemistry.

form has different sizes: e.g., 1*s* orbital is a small sphere, 2*s* orbital is a greater one, 3*s* orbital is even larger, etc. The energy required in order that an electron occupies a given orbital rise with orbital size. In the same conditions, the electrons always occupy the orbital having the least energy. It is not possible that two particles be in the same quantum state (Pauli exclusion principle, 1925). As electrons have an internal state called spin, which takes the value of either 1/2 or –1/2, both electrons that, at maximum, occupy the same orbital must have different spins. The H only has one electron, which occupies 1s orbital. The He has two, which occupy 1s orbital, which cannot yet receive more electrons. The Li has three and, as 1s orbital is already full, the third electron is forced to occupy 2s orbital, which has more energy, etc. The order of filling the orbitals (Aufbau principle after the Pauli exclusion principle, 1925), especially those occupied by the outer (valence) shell, determines PTE shape. In columns 1–2, the valence-shell electrons occupy s-orbitals; in columns 3–12, the five d-orbitals; in columns 13–18, the three p-orbitals; throughout the 14 elements of f-block, the seven f-orbitals are filling – identical electrons. No way exists to tell one electron from another, and the ambiguity must be reflected in the wavefunctions that describe more than one electron (Kimball, 2015). The result is *quantum rule*, where a multielectron wavefunction must change sign whenever two electrons are interchanged, and *Pauli exclusion principle*, where no two electrons can be in the same quantum state. The exclusion principle is a Quantum-Rule application. The simplest example is two-electron He. Assume each electron is described by its wavefunction. If the electrons were explained by the same wavefunction and if both had the same spin direction, an interchange would leave the wavefunction unchanged, which is not good. The wavefunction must change sign, which means the electrons cannot be in the same *quantum state* (means wavefunction and spin direction). For ground-state He, the electrons are put in different states making one of them '*spin up*' and the other '*spin down.*' The opposite spin directions mean He has no magnetism. The asymmetry requirement for identical electrons is exact. Pauli exclusion principle is not. Many-electron wavefunctions are complicated. Two electrons cannot be described by two single-electron wavefunctions, so one cannot really say each electron has its own state or energy. Despite this, the exclusion principle combined with approximations yields results. The Li has three electrons, which means two of them must have parallel spins (up/down are the only spin choices). The presence of three electrons and Pauli principle require that a wavefunction of a different shape describe one of the electrons. An electron explained by the second wave is typically further from the nucleus, which means relatively small energy is needed to remove an electron from Li. In an approximate shell model, the third electron must be placed in an outer shell because the first two electrons fill the inner shell. Extended applications of ideas related to the shell model allowed developing approximations that describe atoms, molecules, and more complicated systems. Despite complications associated with Pauli-principle approximations, PTE can be explained.

38.4 NANO-CHEMICAL APPLICATION

Some PPs are atomic weight, density, atomic radius, crystal structure, order of incorporation of the electrons, atomic emission spectrum, matter aggregation states, etc. The International Union of Pure and Applied Chemistry (IUPAC, 2015) updates PTE systematic names, synonyms, symbols, properties, etc. Winter (2015) programmed the interactive PTE database WebElements™ (*cf.* Figure 38.9).

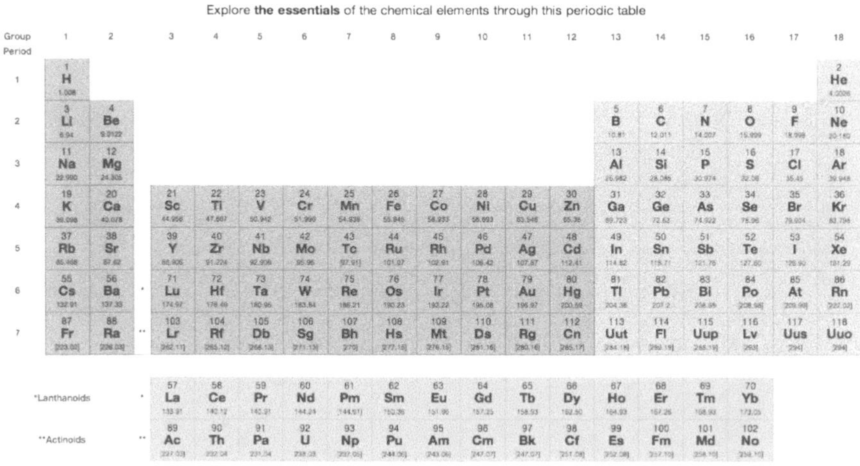

Explore **the essentials** of the chemical elements through this periodic table

Group	1	2	3	4	5	6	7	8	9	10	11	12	13	14	15	16	17	18
Period																		
1	H																	He
2	Li	Be											B	C	N	O	F	Ne
3	Na	Mg											Al	Si	P	S	Cl	Ar
4	K	Ca	Sc	Ti	V	Cr	Mn	Fe	Co	Ni	Cu	Zn	Ga	Ge	As	Se	Br	Kr
5	Rb	Sr	Y	Zr	Nb	Mo	Tc	Ru	Rh	Pd	Ag	Cd	In	Sn	Sb	Te	I	Xe
6	Cs	Ba	Lu	Hf	Ta	W	Re	Os	Ir	Pt	Au	Hg	Tl	Pb	Bi	Po	At	Rn
7	Fr	Ra	Lr	Rf	Db	Sg	Bh	Hs	Mt	Ds	Rg	Cn	Uut	Fl	Uup	Lv	Uus	Uuo

*Lanthanoids	La	Ce	Pr	Nd	Pm	Sm	Eu	Gd	Tb	Dy	Ho	Er	Tm	Yb
**Actinoids	Ac	Th	Pa	U	Np	Pu	Am	Cm	Bk	Cf	Es	Fm	Md	No

MERCK (2015) coded another interactive PTE database (*cf.* Fig. 10).

FIGURE 38.9 Interactive WebElements™ Periodic Table.

Proposition. Atoms of the same class exist. It is possible to define classes of atoms so that inside a certain group ones could be replaced by others. In quantum mechanics, the proposition has consequences. It underlies behind the special phenomenon of superfluidity of liquid-He, which flows *via* pipes without resistance. It is the origin of all PTE and force that keeps one from slipping him away *via* the ground. *d*/*f*-blocks in PTE match magnetic elements (*cf.* Figure 38.11) with colored compounds and OS 3, 2, 4, 5, 6, 7, 1, etc.

There are 29 life-essential elements: 11 massive (H, C–O, Na, Mg, P–Cl, K, Ca, *cf.* Figure 38.12, black), 14 trace (B, F, Si, V–Zn, Se, Mo, I, red) and four possible trace (As, Br, Sn, W, green), with OSs 2, –1, 3, etc.; e.g., K^+ controls heart, Ca^{2+} manages proteins, etc.

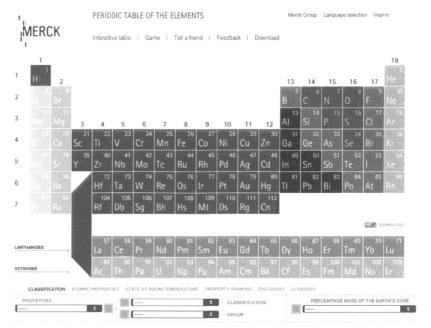

FIGURE 38.10 Interactive MERCK Periodic Table of the Elements.

```
    1A    2A                                              3A   4A   5A   6A   7A   8A
   -----                                                                        -----
1 |   |   |                                                               |   |   |
  |---+----                                         ---------------------------+---|
2 |   |   |                                               |   |   |   |   |   |   |
  |---+---|                                               |---+---+---+---+---+---|
3 |   |   |3B  4B  5B  6B  7B |    8B      |1B  2B |       |   |   |   |   |   |   |
  |---+---+----                           ----+---+---+---+---+---+---+---+---+---|
4 |   |   |Sc |Ti | V |Cr |Mn |Fe |Co |Ni |Cu |Zn |       |   |   |   |   |   |   |
  |---+---+---+---+---+---+---+---+---+---+---+---+---+---+---+---+---+---+---+---|
5 |   |   | Y |Zr |Nb |Mo |Tc |Ru |Rh |Pd |Ag |Cd |       |   |   |   |   |   |   |
  |---+---+---+---+---+---+---+---+---+---+---+---+---+---+---+---+---+---+---+---|
6 |   |   |LAN|Hf |Ta | W |Re |Os |Ir |Pt |Au |Hg |       |   |   |   |   |   |   |
  |---+---+---+---+                                                               |
7 |   |   |ACT|                                                                   |
   -------------
```
```
  Lanthanide |La |Ce |Pr |Nd |Pm |Sm |Eu |Gd |Tb |Dy |Ho |Er |Tm |Yb |Lu |
             |---+---+---+---+---+---+---+---+---+---+---+---+---+---+---|
    Actinide |Ac |Th |Pa | U |Np |Pu |Am |Cm |Bk |Cf |Es |Fm |Md |No |Lw |
```

FIGURE 38.11 Periodic table of magnetic elements.

Metal oxides for gas sensors are SnO_2, ZnO, In_2O_3, TiO_2, WO_3, Fe_2O_3, Ga_2O_3, CuO, NiO, and V_2O_5; their PTE (*cf.* Figure 38.13) shows diverse groups, periods and OSs 2, 3, 4, etc.

```
   1A   2A                                            3A   4A   5A   6A   7A   8A
   -----                                              -----            -----
1 | H |                                                                     |   |
  |---+----                                           ----------------------+---|
2 |   |   |                                          | B | C | N | O | F |   |   |
  |---+---|                                          |---+---+---+---+---+---+---|
3 |Na |Mg |3B  4B  5B  6B  7B |    8B    |1B  2B |    |Si | P | S |Cl |   |   |
  |---+---+---+---+---+---+---+---+---+---+---+---+---+---+---+---+---+---+---|
4 | K |Ca |   |   | V |Cr |Mn |Fe |Co |Ni |Cu |Zn |   |   |As |Se |Br |   |   |
  |---+---+---+---+---+---+---+---+---+---+---+---+---+---+---+---+---+---+---|
5 |   |   |   |   |   |Mo |   |   |   |   |   |   |   |Sn |   |   | I |   |   |
  |---+---+---+---+---+---+---+---+---+---+---+---+---+---+---+---+---+---+---|
6 |   |   |LAN|   |   | W |   |   |   |   |   |   |   |   |   |   |   |   |   |
  |---+---+---+---+---+---+---+---+---+---+---+---+---+---+---+---+---+---+---|
7 |   |   |ACT|
  ---------------

 Lanthanide |   |   |   |   |   |   |   |   |   |   |   |   |   |   |   |
            |---+---+---+---+---+---+---+---+---+---+---+---+---+---+---|
 Actinide   |   |   |   |   |   |   |   |   |   |   |   |   |   |   |   |
            -----------------------------------------------------------
```

FIGURE 38.12 Life essential elements: massive (*black*); trace (*red*); possible (*green*).

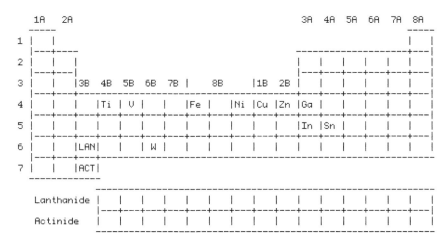

FIGURE 38.13 PTE of metals oxides most used in gas sensors.

Mendeleev (1871) predicted with great accuracy almost all the properties of the element Ge, unknown in his time, which later was used in the first transfer variator (transistor; Shockley, Bardeen and Brattain, 1948). From a two-dimensional to a three-dimensional PTE (Cerdà, 2012). Before nanoscience, people played chemistry as to sink the fleet: D9, cobalt. It was because the elements were placed and their PPs determined by their position in PTE: melting point, boiling point, density, etc. Now it is all over; e.g., C has not a unique melting point because its properties are different if it is the case of a GR, NT or fullerene. The GR is a C-atoms single layer in the form of a honeycomb lattice. Its properties suggest applications. Many GR layers placed on top of each other become ordinary graphite. A C-NT is a GR lattice curled up to make a tube. Fullerenes are roughly a spherical form of the structure. The simplest fullerene is made of 60 atoms that lie at points described by the geometry of a black-and-white-television soccer ball. Now one has a heap of different materials, which share the same nature and atomic composition but completely different properties. Torrens & Castellano (2015b,c) reviewed the history of nanominiaturization (Feynman, 1959, 1960, 1982). Nanotechnology adds a third dimension to PTE; e.g., graphite-like monolayer MoS_2 is the most known and stable 2DL TMD. MoS_2 has exceptional charge-carrier mobility and is a contender for the next electronics generation, as Si chip reaches its fundamental limits. It is an *n*-type semiconductor, useful for PVCs. Chhowalla et al. (2013) revised the chemistry of 2DL TMD nanosheets, which are used as solid lubricants. In PVCs, a need exists to collect all EMR spectrum (e.g., IR, VIS). Some NPs of certain size and shape collect IR; some other NPs of different size/shape, VIS. Intercalating both types of NPs, PVCs collect a greater portion of the EMR spectrum. Monolayers proved to be stable under ambient conditions (room temperature in air, GR; hexagonal BN, *white GR*; BCN; fluorographene; GR oxide; MoS_2, WS_2, $MoSe_2$, WSe_2; micas, Bi/Sr/Ca/Cu oxide; H, B–F, Si, S, Ca, Cu, Se, Sr, Mo, W, Bi, *cf.* Figure 38.14, *black*), those probably stable in air (semiconducting dichalcogenides: $MoTe_2$, WTe_2, ZrS_2, $ZrSe_2$, etc.; MoO_3, WO_3; layered Cu oxides; Zr, Te, *red*) and those unstable in air but that may be stable in inert atmosphere (metallic dichalcogenides: $NbSe_2$, NbS_2, TaS_2, TiS_2, $NiSe_2$, etc.; layered semiconductors: GaSe, GaTe, InSe, Bi_2Se_3, etc.; Ti, Ni, Ga, Nb, In, Ta, *green*); 3D compounds that were successfully exfoliated to monolayers [TiO_2, MnO_2, V_2O_5, TaO_3, RuO_2, etc.; perovskite-type: $LaNb_2O_7$, $(Ca,Sr)_2Nb_3O_{10}$, $Bi_4Ti_3O_{12}$, $Ca_2Ta_2TiO_{10}$, etc.; hydroxides: $Ni(OH)_2$, $Eu(OH)_2$, etc.; V, Mn, Ru, La, Eu, *blue*]; OSs 3, 4, 2, 5, 6, etc.

Materials obtained by the incorporation of metals, e.g., Ti–Co, Cu, Zr, Mo, Sn, and W, in the frame or at the surface of nanoporous oxides or hydroxides (*cf.* Figure 38.15) with OSs 2, 4, 3, 6, etc. are able to catalyze selective reactions between organic compounds and hydroperoxides or H_2O_2 (Kholdeeva, 2014).

Research Project Nanotechnology and Labor Risks Prevention (Díaz Soler, 2015) surveys practices in preventive health matters, with questions on the nanomaterials to which workers are exposed: $CaCO_3$, dendrimers, SiO_2, TiO_2, fullerene, GR, nanoclays, single/multiple-wall C-NTs, ferrous, organic, C-black, Au, Al_2O_3, Ce_2O_3, ZnO, Ag, polystyrene, polymers, quantum dots and BaTiO$_3$. Matching PTE (*cf.* Figure 38.16) shows different groups, periods and OSs 2, 0, 3, 4, etc.

```
   1A   2A                                        3A   4A   5A   6A   7A   8A
   -----                                          -----------------------------
1 | H |                                                                  |    |
  |---+----                                       ------------------------+---|
2 |   |   |                                       | B | C | N | O | F |   |    |
  |---+---|                                       |---+---+---+---+---+---+---|
3 |   |   |3B  4B  5B  6B  7B |  8B   |1B  2B |   |Si |   | S |   |   |    |
  |---+---+---+---+---+---+---+---+---+---+---+---+---+---+---+---+---+---+---|
4 |   |Ca |   |   | U |   |Mn |   |   |   |Cu |   |   |   |   |Se |   |    |
  |---+---+---+---+---+---+---+---+---+---+---+---+---+---+---+---+---+---+---|
5 |   |Sr |   |Zr |   |Mo |   |Ru |   |   |   |   |   |   |   |Te |   |    |
  |---+---+---+---+---+---+---+---+---+---+---+---+---+---+---+---+---+---+---|
6 |   |   |LAN|   |   | W |   |   |   |   |   |   |   |   |Bi |   |   |    |
  |---+---+---+                                                            |
7 |   |   |ACT|
  -------------

  Lanthanide |La |   |   |   |   |   |Eu |   |   |   |   |   |   |   |    |
             |---+---+---+---+---+---+---+---+---+---+---+---+---+---+---|
  Actinide   |   |   |   |   |   |   |   |   |   |   |   |   |   |   |    |
             -----------------------------------------------------------
```

FIGURE 38.14 Monolayers PTE: stable (*black*); probably (*red*); unstable (*green*); 3D compounds exfoliated to monolayers (*blue*).

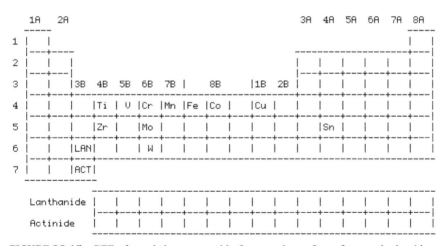

```
   1A   2A                                        3A   4A   5A   6A   7A   8A
   -----                                          -----------------------------
1 |   |                                                                   |    |
  |---+----                                       ------------------------+---|
2 |   |   |                                       |   |   |   |   |   |   |    |
  |---+---|                                       |---+---+---+---+---+---+---|
3 |   |   |3B  4B  5B  6B  7B |  8B   |1B  2B |   |   |   |   |   |   |    |
  |---+---+---+---+---+---+---+---+---+---+---+---+---+---+---+---+---+---+---|
4 |   |   |   |Ti | U |Cr |Mn |Fe |Co |   |Cu |   |   |   |   |   |   |    |
  |---+---+---+---+---+---+---+---+---+---+---+---+---+---+---+---+---+---+---|
5 |   |   |   |Zr |   |Mo |   |   |   |   |   |   |Sn |   |   |   |   |    |
  |---+---+---+---+---+---+---+---+---+---+---+---+---+---+---+---+---+---+---|
6 |   |   |LAN|   |   | W |   |   |   |   |   |   |   |   |   |   |   |    |
  |---+---+---+
7 |   |   |ACT|
  -------------

  Lanthanide |   |   |   |   |   |   |   |   |   |   |   |   |   |   |    |
             |---+---+---+---+---+---+---+---+---+---+---+---+---+---+---|
  Actinide   |   |   |   |   |   |   |   |   |   |   |   |   |   |   |    |
             -----------------------------------------------------------
```

FIGURE 38.15 PTE of metals incorporated in frame at the surface of porous hydroxides.

38.5 MULTI-/TRANS-DISCIPLINARY CONNECTION

Chemistry presents an important idea ever elucidated in 400 years of experimental science: existence and organization of PTE (the other is bioevolution). Chemical-elements classification constitutes a solid, systematic and deductive base for chemistry study, so that its elaboration was a fundamental step for learning the experimental science. It is a non-trivial classification: a system that reflects something beyond the criterion on which it was arranged. When an order is in place, one can make predictions about if something presents properties of a gas, metal, salt, etc. The periodicity of chemical elements, valence, physical nature of chemical bond and reactions, all are determined by spins. It is essential to understand human nature and place in the whole of beings, food, drugs, materials, daily life, etc. (Valero Molina, 2013). The PTE is a source of culture, art, geography, history, philology, grammar, languages, mythology, etc. It is the guiding star in chemistry, physics, mineralogy, technique, nanotechnology, etc.

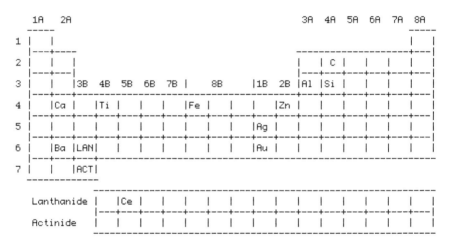

FIGURE 38.16   PTE of nanomaterials in a survey of practices in preventive matters.

Synthetic biology. Synthetic biologists developed a focus on which they split productive processes into the constituting elements to identify the unitary operations (A. D. Little) of the synthetic process (Furter, 1980). Synthetic biologists as Craig Venter talked about using the techniques to design living organisms.

PTE in education. Education abilities and skills are mnemonic rules, handling program packages (e.g., Microsoft Office, OpenOffice), navigation *via* the Internet, etc. Román Polo (2007) raised PTE questions. Schwarz & Rich (2010) provided a PTE background. Torrens & Castellano (2015a) reflected on PTE nature in education. It is still to be explored the method-ological application of the approach, enriching present teaching techniques: implementing new methods in learning situations is technically simple, but checking their usefulness in the practice of teaching is extraordinarily complex and requires a research that hardly began.

Elements, atomism, and religion. The Au and Ag, Fe and Cu appear many times in the Bible because of their monetary or utilitarian value. The Pb and Sn are mentioned in passing. Another additional element has a symbolic value of a completely different type: S. From the three great monotheistic religions arisen from Abraham's stock, Christianity was not the first in accepting atomism but Islam, because (1) the atom is eternal and uncreated; (2) atomism is opposed to eucharistic transubstantiation (*cf.* Figure 38.17); e.g., atomist Newton never said atom but corpuscle.

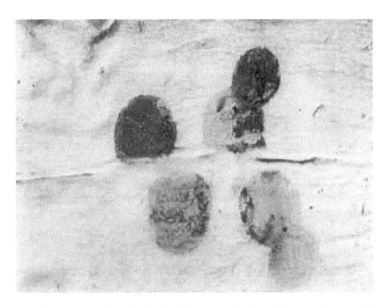

FIGURE 38.17 Corpus Christi Miracle of Corporal Clothes of Daroca, Aragon, 1238.

Atomism and philosophy. Democritus proposed that the material would be formed by tiny indivisible particles: atoms (indivisible). Plato objected that if atoms could occupy some space, then they could be divided. 19th-century empiricism rejected physical objects that could not be observed (e.g., atoms, molecules), which attitude made it difficult for Maxwell and Boltzmann to gain acceptance for their pioneering ideas on the atomic theory of statistical physics, which contributed to Boltzmann's suicide. Einstein Thesis (1905) and article (1906) were devoted to atomic-dimensions estimation.

Chemistry–physics philosophical conflict. Confrontation, expressed in terms of elements *vs.* atoms, leads one to the consideration of Mendeleev's element concept, which is a philosophical and historical concept in chemical history (Bensuade-Vincent & Simon, 2012). Mendeleev's viewpoint allows raising the reductionism subject, which became important after quantum mechanics. The PTE reflects a delicate abstract philosophical understanding of the element, which was questioned after isotopes and atomic physics. Chemists were misunderstood as naïf positivists when they refused to accept atoms existence in the 19th century. After Compte and Mach, chemistry is positivist and not positivist at once in its approximation. Ostwald and Duhem showed positivism limit in chemistry. Atomism discussion generated by positivism allows exploring atomisms variety that existed and exists. The PTE represents chemist's distinctive atomism, which focuses on the atom as a chemical-relationships node. Does nanotechnology mark chemistry end as a subject? A strong continuity exists in a science-society relationship that will survive the chemistry→nanotechnology transition. A further examination of the natural entities and new interest in these at nanoscale produced a revival of Faustian ambitions associated with chemistry. Nanotechnology does not only search for imitating nature but also surpassing it, with an increasing number of scientific visionaries that announce artificial life and self-propagating nanomachines as precursors of life control by humanity. A general philosophical guide could help to construct an ethics adequate for contemporary research in the nanorevolution context.

Chemistry and ethics. Colborn et al. (1996) alerted to endocrine disruptors (León et al., 2009a,b; Torrens & Castellano, 2010, 2013; Torrens et al., 2008a,b; Torrens Zaragozá, 2011). The American Chemical Society reported The Chemical Professional's Code of Conduct (ACS, 2012); however, it does not explain what to do if the science policy is in conflict with demands related to the obligation with the boss or other employees. Two designers (McDonough & Braungart, 2002) promoted using synthetics as raw materials.

PTE and literature. The PTE (*cf.* Figure 38.18) is related to literature (Baldridge & Abendroth, 2008) (*see* Chapters 36 and 37).

38.6   OPEN ISSUES

Some problems remain in PTE. (1) The H lies in group 1 (it presents a unique electron), but it has nothing to do with alkalines. Which group of PTE should one place H in? Chemically, H is more similar to halogens than to alkalines. (2) The placing of He, Lu and Lr does not follow the electron configuration criterion after the Pauli exclusion principle. (3) *p*-Block diagonal breaks the general rule of placing the elements with common traits in the same

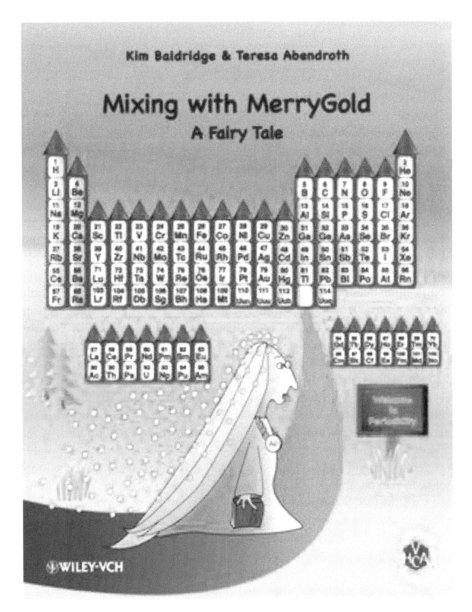

FIGURE 38.18 PTE like a castle in the fairy tale, *Mixing with MerryGold.*

column. (4) In different elements, the chemical species is dissimilar: atom, diatomic molecule, etc.; e.g., when PTE indicates that H and Ar are gases, H_2 is diatomic, but Ar is an atom: The sentence Ar is the most abundant element in air, is both true (atomic element) and false (molecular element). (5) Some properties of the solid elements are not strictly those of the atom but the corresponding net, which typically does not repeat for the elements belonging to the same period. (6) In general, inert elements do form compounds with neither other elements nor themselves. (7) The natural elements are mixtures of stricto sensu elements (isotopes, Soddy, 1912), which atomic weights are (approximately) integers measured in the unit: proton weight, which is close to the natural H-atom, element which is, in turn, a mixture of isotopes (H, D, radioactive T, $Z = 1$, $A = 1, 2, 3$) that fulfill the general rule of the integers referred to the proton. (8) The standard atomic weights are not a constant of nature (Bustelo Lutzardo et al., 2012). (9) The PTE elements, divisible atoms, are they constituted by other common elementary entities that determine their characteristics and properties? What way in? What number with? What are these elementary entities? How are they? (10) In nanomaterials, one has a heap of different materials, which share the same nature and atomic composition but different properties (e.g., C-NT, GR, fullerene). (11) Should one add a zero element to PTE: the neutron ($Z = 0$)? (12). The models used in chemistry must be revised according to representations of the electron based on absorbed and re-emitted energy (Giertz, 2015). (13) Is it possible to create a spatially periodic model of H-formation from the oversaturated state of elementary particles in the evolving universe (Gladyshev, 2015)? (14) While geologists and mining engineers measure the abundance of an element in the Earth's crust in terms of mass as a proportion of total mass, does it make more sense for chemists to quote abundance in terms of the number of atoms per n atoms, i.e., molar abundance? This may be suitable for geologists and mining engineers but, for chemists, it would make more sense to reckon abundance in terms of the number of atoms per n atoms. Val Castillo (2015) foresaw how PTE future will be: (1) Will researchers continue creating elements with a half-life greater than 10^{-1}s? (2) How will the new elements accommodate in Seaborg PTE (1945)? (3) Will another period be filled in the next 50 years with the elements of $5g$ configuration?

Towards a periodic table of nucleons. A man could deliberately reorder the atomic structure of molecules in ordinary chemical reactions, why not to rearrange the protons and neutrons in the atomic nucleus in nuclear reactions?

Fundamental particles. All simple particles are existing in Nature experiment reactions by which they become other particles or radiations, or are obtained from them. There are not immutable particles, which is truly fundamental.

Towards a periodic table of molecules. The PTE is complete (Gray, 2014). However, no catalog exists of all molecules of the universe, and it cannot be.

Chemistry assumes that molecules are formed by the static binding energy created by static Coulomb force (Heitler & London, 1927). Aufbau principle is applicable to atoms, ions and molecules, which, whether natural or synthetic, are made from atoms, which are put together, in turn, from three components: protons, neutrons, and electrons. Protons, which are positively charged, and neutrons, which are electrically neutral, form a dense nucleus within the heart of every atom, which is surrounded by a swarm of negatively charged electrons. However, when one puts atoms together to build molecules, certain atoms attract more electron density to themselves than others, which leads to typical-molecules surface being electron rich (electrostatically negative) in some areas and electron deficient (electrostatically positive) in others, which have implications in rational drug design, quantitative structure-activity/property relationships, pharmacophore models, etc.

Chemistry and culture. Chemical science is embedded in cultural values, which matter for the public acceptance of scientific and technological innovations. Ethics of chemistry needed to negotiate priorities *via* a democratic collective deliberation. Comparing physics and chemistry, is chemistry the future? New trends (nanotechnology, etc.) show decaying differences between physics and chemistry, science and technology, economy and capitalism, etc. It is interesting to study the possible generalization of the ideas above to PPs, PL, etc.

KEYWORDS

- **periodic law**
- **periodic property**
- **periodic system**
- **periodicity**
- **valence**

REFERENCES AND FURTHER READING

ACS, (2012). URL: http://www.acs.org/content/acs/en/careers/career-services/ethics/the-chemical-professionals-code-of-conduct. html (accessed on 27 11 2015).

Alfonseca, M., (1996). *Espasa Dictionary: 1,000 Great Scientists*, Espasa Calpe: Madrid, Spain.

Baez, J., & Dolan, J., (1995). Higher-dimensional algebra and topological quantum field theory. *J. Math. Phys., 36,* 6073–6105.

Baldridge, K., & Abendroth, T., (2008). *Mixing with Merry Gold: A Fairy Tale*. Wiley–VCH: Weinheim, Germany.

Bensuade-Vincent, B., & Simon, J., (2012). *Chemistry: The Impure Science*, Imperial College Press: London, UK.

Burbidge, E. M., Burbidge, G. R., Fowler, W. A., & Hoyle, F., (1957). Synthesis of the elements in stars. *Rev. Mod. Phys.*, *29*, 547–654.

Bustelo, L. J. A., García, M. J., & Román, P. P., (2012). Los elementos perdidos de la tabla periódica: Sus nombres y otras curiosidades. *Anal. Quim.*, *108*, 57–64.

Cerdà, H., (2012). Entrevista: Javier garcía martínez. *Técnica Industrial*, *300*, 82–85.

Chhowalla, M., Shin, H. S., Eda, G., Li, L. J., Loh, K. P., & Zhang, H., (2013). The chemistry of two-dimensional layered transition metal dichalcogenide nanosheets. *Nat. Chem.*, *5*, 263–275.

Colborn, T., Dumonski, D., & Myers, J. P., (1996). *Our Stolen Future: Are We Threatening Our Fertility, Intelligence, and Survival–A Scientific Detective Story*. Penguin: New York, NY.

Crans, S., (2000). On braidings, syllepses, and symmetries. *Cahiers Topol. Géom. Diff. Categ.*, *41*, 2–74.

De Luyart, J. J., & De Luyart, F., (1783). Análisis químico del volfram, y examen de un nuevo metal, que entra en su composición. In: *Extractos de las Juntas Generales Celebradas por la Real Sociedad Bascongada de los Amigos del País* (pp. 46–88). Vitoria, Spain.

Díaz, S. B. M., (2015). *Nanotecnología y Prevención de Riesgos Laborales*, Universidad de Granada: Granada, Spain.

Einstein, A., (1906). Eine neue Bestimmung der Molüküldimensionen. *Ann. Phys.*, *324*, 289–306.

Feynman, R. P., (1960). There is plenty of room at the bottom. *Caltech Eng. Sci.*, *23*, 22–36.

Furter, W. F., (1980). *History of Chemical Engineering*. American Chemical Society: Washington, DC.

Geim, A. K., & Grigorieva, I. V., (2013). Van der Waals heterostructures. *Nature (London)*, *499*, 419–425.

Giertz, H. W., (2015). A novel model of the electron based on absorbed and re-emitted energy. *J. Basic Appl. Res. Int.*, *9*, 166–173.

Gladyshev, G. P., (2015). Physicochemical stages of evolution: Ring-like structures in the Universe. *Nat. Sci.*, *7*, 266–269.

Gray, T. W., (2009). *The Elements: A Visual Exploration of Every Known Atom in the Universe*. Black Dog & Leventhal: New York, NY.

Gray, T., (2014). *Molecules: The Elements and the Architecture of Everything*. Black Dog & Leventhal: New York, NY.

Heitler, W., & London, F., (1927). Wechselwirkung neutraler Atome und homöopolare Bindung nach der Quantenmechanik. *Z. Phys.*, *44*, 455–472.

IUPAC, URL: http://www.iupac.org (accessed on 27 11 2015).

Kholdeeva, O. A., (2014). Recent developments in liquid-phase selective oxidation using environmentally benign oxidants and mesoporous metal silicates. *Catal. Sci. Technol.*, *4*, 1869–1889.

Kimball, J., (2015). *Physics: Curiosities, Oddities, and Novelties*. CRC: Boca Raton, FL.

Leinster, T., (2004). *Higher Operands, Higher Categories*. Cambridge University Press: Cambridge, UK.

León, L. M., Laza, M., Puchol, V., Torrens, F., Abad, C., & Campos, A., (2009a). Dynamic mechanical measurements of epoxy matrix-silica nanocomposites II. *Polym. Polym. Compos.*, *17*, 313–324.

León, L. M., Laza, M., Torrens, F., Puchol, V., Abad, C., & Campos, A., (2009b). Incorporation of silica nanospherical particles in epoxy–amine crosslinked materials II.

Dynamic mechanical measurements of epoxy matrix-silica nanocomposites. *Polym. Polym. Compos.*, *17*, 457–465.

McDonough, W., & Braungart, M., (2002). *Cradle to Cradle: Remaking the Way We Make Things*. North Point: New York, NY.

MERCK Periodic Table of the Elements. URL: http://pse.merck.de/merck.php?lang = EN (accessed on 27 11 2015).

Román, P. P., (2002). *El Profeta del Orden Químico: Mendeléiev*, Científicos para la Historia No. 9, Nivola: Tres Cantos (Madrid), Spain.

Román, P. P., (2007). Jugando con la tabla periódica de los elementos químicos. In: Pinto, C. G., (ed.), *Aprendizaje Activo de la Física y la Química* (pp. 35–42). Equipo Sirius: Madrid, Spain.

Scerri, E. R., (2007). *The Periodic Table: Its Story and Its Significance*. Oxford University Press: New York, NY.

Schwarz, W. H. E., & Rich, R. L., (2010). Theoretical basis and correct explanation of the periodic system: Review and update. *J. Chem. Educ.*, *87*, 435–443.

Torrens, F., & Castellano, G., (2010). Incorporation of silica nanospheres into epoxy-amine materials: Polymer nanocomposites. In: Lechkov, M., & Prandzheva, S., (eds.), *Encyclopedia of Polymer Composites: Properties, Performance, and Applications* (pp. 823–844). Polymer Science and Technology No. 5, Nova: New York, NY.

Torrens, F., & Castellano, G., (2013). Bisphenol, diethylstilbestrol, polycarbonate and the thermomechanical properties of epoxy–silica nanostructured composites. *J. Res. Updates Polym. Sci.*, *2*, 183–193.

Torrens, F., & Castellano, G., (2015a). Reflections on the nature of the periodic table of the elements: Implications in chemical education. In: Seijas, J. A., Vázquez, T. M. P., & Lin, S. K., (eds.), *Synthetic Organic Chemistry* (Vol. 18, pp. 1–15). MDPI: Basel, Switzerland.

Torrens, F., & Castellano, G., (2015b). Reflections on the cultural history of nanominiaturization and quantum simulators (computers). In: *Sensors and Molecular Recognition* (Vol. 9, pp. 1–7). Universidad Politécnica de Valencia: València, Spain.

Torrens, F., & Castellano, G., (2015c). Ideas in the history of nano/miniaturization and (quantum) simulators: Feynman, education and research reorientation in translational science. In: Seijas, J. A., Vázquez, T. M. P., & Lin, S. K., (eds.), *Synthetic Organic Chemistry* (Vol. 19, pp. 14–16). MDPI: Basel, Switzerland.

Torrens, F., Monzó, I. S., Gómez-Clarí, C. M., Abad, C., & Campos, A., (2008a). Study and comparison of interaction parameters and phase behavior of epoxy/polystyrene and epoxies copolymer polystyrene–*b*–poly(methyl methacrylate) blends. *Polym. Compos.*, *29*, 1337–1345.

Torrens, F., Solar, L., Puchol, V., Latorre, J., Abad, C., & Campos, A., (2008b). Incorporation of silica nanospherical particles into epoxy-amine cross-linked materials. *Polym. Polym. Compos.*, *16*, 139–152.

Torrens, Z. F., (2011). Polymer bisphenol-A, the incorporation of silica nanospheres into epoxy-amine materials and polymer nanocomposites. *Nereis, 3*, 17–23.

Val, C. O., (2015). Historia de la evolución de la tabla periódica de los elementos químicos: Un ejemplo más de la aplicación del método científico. *An. Quím*, *111*, 109–117.

Valero, M. R., (2013). El sistema periódico y su relación con la vida cotidiana. Parte I. *An. Quím.*, *109*, 301–307.

Winter, M., (2015). URL: http://www.webelements.com (accessed on 20 February 2019).

CHAPTER 39

π-π Interaction in Fullerenes

LEMI TURKER[1] and CAGLAR CELIK BAYAR[2]

[1]*Middle East Technical University, Department of Chemistry, 06800 Ankara, Turkey, Tel: +90 312 210 3244, Fax: +90 312 210 3200, E-mail: lturker@metu.edu.tr, lturker@gmail.com*

[2]*Bulent Ecevit University, Department of Metallurgical and Materials Engineering, 67100 Zonguldak, Turkey*

39.1 DEFINITION

Of the carbon clusters, above about C_{30}, the lowest energy structures up to several hundred-carbon atoms appear to be the fullerenes (Curl, 1993). Fullerenes are carbon cage molecules of definite compositions shaped as spheroidal shells which possess 12 pentagons and an arbitrary number of hexagons, such that the number of carbon atoms (*n*) is given by $n = 20 + 2 h$, where *h* is the number of hexagons in each structure (Albert et al., 2010).

Fullerenes, as π conjugated spherical guest molecules, are attractive nano-sized structures with about 1 nm in diameter (Yuan et al., 2014). The electron accepting the ability of fullerenes – C_{60}, C_{70} and their derivatives (Figure 39.1) – is probably their most characteristic chemical property (Sun et al., 2002).

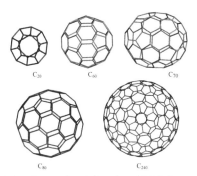

FIGURE 39.1 Chemical structures of certain spheroidal fullerenes.

This may happen via electrostatic interactions. Electrostatic interactions include charge transfer (CT) reactions which require a donor and an acceptor molecule. However, non-covalent and weaker π-π interactions may be considered as additional forces that have an important role in describing the stability of host-guest supramolecular systems. Generally, the host molecule contains a conjugated π system while the guest molecule acts as an electron poor fullerene derivative.

39.2 HISTORICAL ORIGINS

The pioneering work by Kroto et al., (1985) and Krätschmer et al., (1990) regarding the initial discovery of fullerenes C_{60} and C_{70}, and the development of a large-scale method for their preparation, have prompted a lively interest to employ these 3-dimensional carbon allotropes to undergo supramolecular interaction with various host molecules (Mukherjee & Bhattacharya, 2012). Since the discovery of C_{60}, additional cages as small as C_{20} and as large as C_{240} were discussed (Churilov et al., 2007). Additionally, some other fullerene structures consist of non-carbon atoms were identified or proposed (Oku, 2007). It seems that other structures might have comparable stability and in particular structures with just a few of their hexagons enlarged to heptagons might be worth theoretical investigation. Generation and motion of heptagonal defects on the array were studied (Saito et al., 1992). Large, negatively curved assemblies of carbon atoms were proposed as building blocks for giant *hyperfullerenes* (Scuseria, 1992). Curvature in conjugated organic molecules leads the σ-bonds at the carbon atom to deviate from planarity, and there is a change in hybridization, namely rehybridization of the carbon atoms so that π-orbital is no longer of purely *p*-orbital character. Thus, fullerenes are of intermediate hybridization (Haddon et al., 1986). The electronic structure of spheroidal fullerenes including giant ones such as C240 and C540 have been discussed theoretically based on topological considerations (Pincak & Pudlak, 2007). Fullerenes form supramolecular complexes with varieties of π-conjugated molecules, including charge transfer complexes with strong electron donors, co-crystals with curved π-conjugated molecules through convex-concave π-π interactions and complexes with planar conjugated molecules such as porphyrins and metal-loporphyrins through intermolecular π-π interactions (Zhang et al., 2013). It was shown that C_{60} and C_{70} are among the very few molecules which can bind two electrons in the gas phase (Limbach et al., 1991).

The π-π interactions have great importance and should be understood clearly since they give extra strength to supramolecular complexes that they belong to in addition to other intermolecular forces such as hydrogen bonding, van der Waals forces, and electrostatic interactions. π-conjugated molecules with tunable electronic properties may be employed as suitable building blocks for the construction of functional materials with exceptional electrochemical and photophysical properties (Mukherjee & Bhattacharya, 2012).

39.3 NANO-SCIENTIFIC DEVELOPMENTS

π-stacking interactions are sometimes possible in which the planar π-systems of aromatic molecules or aromatic macrocycles lie one on top of the other with nearly parallel orientation (Figure 39.2). Such interactions are responsible for the stacking of hydrogen-bonded base pairs in DNA (Atkins & de Paula, 2002).

Although the term, aromatic-aromatic interactions, imply stacking of two π-systems on top of each other, it is not always the case. For instance, two benzene molecules orient themselves perpendicular to each other. To directly stack two benzenes on top of one another leads to an adverse repulsion of electrostatic origin (Anslyn & Dougherty 2006). Nevertheless, simple aromatics do have favorable interactions with each other. Systems like benzene prefer T-shaped or edge-to-face interaction better than stacking. In some more complicated structures, it is best to form a displaced or slipped stack (Figure 39.2).

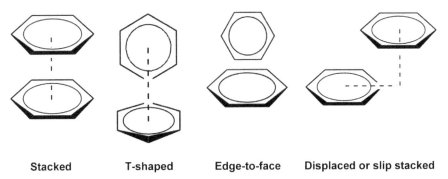

| Stacked | T-shaped | Edge-to-face | Displaced or slip stacked |

FIGURE 39.2   Types of π-π stacking interactions.

This still aligns regions of positive electrostatic potential with regions of negative electrostatic potential which leads to the energetically favorable system (π-staking). π-donor-acceptor interaction occurs between any two molecules or regions of a molecule where one has a low energy empty orbital (acceptor) and the other high energy filled orbital (donor). When these two orbitals have a proper orientation in space, some extent of charge transfer can occur from the donor to the acceptor molecule (Anslyn & Dougherty, 2006).

39.3.1   π-π INTERACTION IN FULLERENE-PORPHYRIN SUPRAMOLECULAR STRUCTURES

As porphyrins have distinctive electronic and photophysical properties, they are particularly attractive building blocks for the construction of supramolecular electron donor-acceptor composites using π-π interaction (Zhang et al., 2013).

Porphyrin and metalloporphyrin dimers (bis-porphyrins) having jaw-like clefts were reacted with C_{60} in order to investigate the host-guest supramolecular recognition in detail (Sun et al., 2002). Jaw-like clefts in these bis-porphyrins were effective hosts to fullerene guests. The electron-rich 6:6 ring-juncture bonds of C_{60} showed an unusually close approach to the porphyrin or metalloporphyrin plane. Binding constants in toluene solution increased in the order of Fe(II) < Pd(II) < Zn(II) < Mn(II) < Co(II) < Cu(II) < 2H and detected in the range of 490–5200 M^{-1}. Unexpectedly, the free-base porphyrin (2H) bound C_{60} stronger than the metallated porphyrins. This was ascribed to electrostatic forces between the electron rich 6:6 ring-juncture bonds of the fullerene to the electropositive N-H center of the porphyrin. This highlighted the importance of an electrostatic component enhancing the van der Waals forces of the π-π interaction between bis-porphyrins and C_{60} fullerene.

Two points (rigidly) bound supramolecular porphyrin-C_{60} fullerene conjugates were synthesized (D'Souza et al., 2005). To achieve this *meso*-tetraphenylporphyrin was functionalized to possess one or four [18]crown-6 moieties at different locations on the porphyrin macrocycle while fullerene was functionalized to possess an alkylammonium cation, and a pyridine or phenyl entity. Both Zn of porphyrin–pyridine axial coordination and alkylammonium cation–crown ether complexation were observed in the case of using C_{60} fullerene having a pyridine entity. Besides alkylammonium cation–crown ether complexation, π-π interaction between porphyrin and C_{60}π-systems were observed differently in the case of using C_{60} fullerene having a phenyl entity. The former type complexes were found to be more stable than the latter because of two point rigid bindings. The existence of π-π interaction in the latter was supported by DFT B3LYP/3-21G(d) calculation, and edge-to-edge porphyrin-fullerene distances were found in the range of 2.2 to 2.3 Å. Spectral and photochemical studies were performed in a polar solvent, benzonitrile. This was unlike previous studies on one-point bound self-assembled porphyrin-fullerene dyads, where the spectral and photochemical studies could be performed only in nonpolar solvents such as toluene and dichlorobenzene, due to the limited stability of these types of conjugates in such solvents.

New tripyridyl tripodal ligand appended with C_{60} fullerene was synthesized and introduced into a porphyrin macro ring, N-(trisporphyrinatozinc(II))$_3$ (Kuramochi et al., 2008, 2011). From UV-vis absorption and fluorescence titration data, the binding constant of 3×10^8 M^{-1} was estimated in benzonitrile. The larger binding constant and the almost complete fluorescence quenching indicated that the interactions were via two pyridyl groups of C_{60}-tripod to the porphyrinatozinc(II) coordination and a π-π interaction of the fullerene to the porphyrin(s).

A glycyl-substituted porphyrin was used as a starting compound for the synthesis of a π-π stacked porphyrin-C_{60} fullerene dyad with a frozen geometry (Lembo et al., 2009). By this way, a dyad with a completely frozen geometry similar to that found in co-crystallized porphyrin-fullerene systems was obtained. The close proximity between the two chromophores induced a ground state electronic interaction as shown by the ^{1}H-NMR, ^{13}C-NMR, and UV-vis studies.

A supramolecular complex of a novel porphyrin tripod, TPZn$_3$, was formed with a fullerene derivative containing a pyridine moiety, PyC$_{60}$ (Takai et al., 2010). TPZn$_3$ acted as a donor while PyC$_{60}$ acted as an acceptor in the complex. The photodynamics of the supramolecular complex (TPZn$_3$-PyC$_{60}$) was investigated by femtosecond and nanosecond laser flash photolysis measurements in nonpolar media, toluene, and dichlorobenzene. It was concluded that PyC$_{60}$ was captured inside the cavity of TPZn$_3$ not only by an axial coordination bond formed between Zn^{2+} and the pyridine moiety but also π-π interactions. According to the DFT B3LYP/3-21G(d) calculation, the distance of an axial coordination bond (Zn-N) was found to be 2.00 Å. The minimum distances between the other porphyrin planes and fullerene were found to be 2.58 and 2.63 Å, respectively. These values of spatial separation suggested possible π-π interactions.

The UV-vis and fluorescence spectroscopic studies on spontaneous non-covalent interaction between a designed octaethyl monoporphyrin; 2,3,7,8,12,13,17,18-octaethyl-21H,23H-porphyrin; and C_{60} and C_{70} fullerenes in toluene medium were highlighted (Mukherjee & Bhattacharya, 2012). While UV-vis studies revealed a considerable amount of porphyrin-fullerene interaction in the ground state, steady-state fluorescence study established remarkable quenching of fluorescence intensity of considered porphyrin in the presence of fullerenes C_{60} and C_{70}. The porphyrin-C_{70} complex was found to be more stable than that of C_{60} in the experiments. Theoretical calculations at MM level in vacuum also supported this by illustrating the presence of intermolecular interaction between the graphitic 6:6 plane of C_{70} and the porphyrin. The attractive interactions between C_{70} and the porphyrin were driven by the presence of dispersive forces associated with π-π interactions rather than electrostatic mechanism associated with charge transfer. The most concrete evidence of this statement was illustrated by the side-on rather than end-on binding of C_{70} with the plane of porphyrin. The equatorial face of C_{70} was centered over the electropositive center of the porphyrin plane.

39.3.2   *π-π INTERACTION IN FULLERENE CONTAINING DIELS-ALDER PRODUCT*

The reaction of 6,13-diphenylpentacene and an excess amount of C_{60} fullerene yielded a complete *syn*-diastereoselective Diels-Alder product, C_{2v} symmetric *cis*-bis[60]fullerene-diphenylpentacene (Briggs & Miller, 2006). The driving force of the reaction was the π-π stacking (van der Waals) interactions between the adjacent C_{60} fullerene moieties. The X-ray crystal structure of the product illustrated that the distance between the centers of the abutting pentagons on adjacent C_{60} fullerene moieties was 3.284 Å.

39.3.3   *π-π INTERACTION IN FULLERENE ENCAPSULATED DIMERIC CAPSULES*

The formation of C_{60} fullerene encapsulated superstructures (dimeric capsules) was performed by Park et al. (2008). It was done by the addition of an appropriate halo-acetic acid to a toluene solution of the resorcin[4] arene derivatives (hosts), which have four pyridine units, and C_{60} fullerenes (guests). The formation of the superstructures was demonstrated by ^{1}H-NMR, ^{13}C-NMR, UV-vis and fluorescence spectroscopies. The C_{60} fullerene encapsulation process could be controlled by changing the temperature, adding an acid/base and even through varying the solvent polarity. These complexes were self-assembled through pyridinium−anion (the conjugate base of the acid used), pyridinium–hydrogen bonding interactions and by host-guest π-π and van der Waals interactions. This study proposed a new strategy by which the simple addition of acids to appropriately designed macromolecules with basic pendent moieties led to supramolecular nanocavities capable of encapsulating large molecules, such as C_{60} fullerene.

39.3.4 *π-π INTERACTION IN ALKYL SUBSTITUTED FULLERENE ASSEMBLIES*

Polymorphism of alkyl substituted fullerene assemblies was investigated (Asanuma et al., 2010). Alkyl chains were connected to the C_{60} with one amide and three ester moieties. Two different kinds of intermolecular forces, the π-πinteractions of C_{60} moieties and the van der Waals interactions of aliphatic chains, took place between assembled molecules. Similar intermolecular forces were detected in the synthesized ionic C_{60} fullerene derivative, alkylated *N,N*-dimethylfulleropyrrolidium iodide, besides the salt bridge effect among the ionic moieties (Li et al., 2011). In this zwitterion structure, the aliphatic chains were grafted onto the pyrrolidine ring via a lateral phenyl group.

39.3.5 *π-π INTERACTION IN FULLERENE CONTAINING DYAD*

Self-organization of the C_{60} fullerene−*N,N*-dimethylaminoazobenzene dyad in THF/water binary solvent was studied (Kumar & Patnaik, 2011). A role of intermolecular donor-acceptor, π-π and van der Waals interactions between the electron deficient fullerene core and the *N,N*-dimethylamino-azobenzene electron donor under neutral conditions were revealed. Upon protonation, the electrostatics associated with the charged fullerene−*N,N*-dimethylaminoazobenzene moiety and the dominant intermolecular fullerene-fullerene interactions guided the self-assembly process.

39.3.6 *π-π INTERACTION IN FULLERENE-PORPHYRIN LIKE SUPRAMOLECULAR STRUCTURE*

As porphyrin-like planar aromatic macrocycles with 22 π-electrons, tetrathia [22]-annulene[2,1,2,1] and its derivatives display superior hole-transport properties compared with porphyrins, which can be attributed to their enlarged conjugated systems. In the light of this information, a sulfur-bridged annulene, *meso*-diphenyltetrathia[22]-annulene[2,1,2,1] (donor), was used instead of porphyrin to form co-crystals with C_{60} and C_{70} (acceptors) (Zhang et al., 2013). The co-crystals had a two-dimensional segregated alternating layer structure which displays ambipolar transport properties as well as photoresponsivity. As diphenyltetrathia[22]-annulene[2,1,2,1] was a planar aromatic macrocyclic molecule; π-π attractions, van der Waals forces, and polar electrostatic interactions were the driving forces for the formation of co-crystals with fullerenes. In the two crystals, the confirmation of diphenyltetrathia[22]-annulene[2,1,2,1] deviated from planarity with half of the molecule bent up, and the other half bent down to fit with the curved surface of the fullerenes. Short intermolecular contacts existed between C_{60} molecules, as the closest C–C distance was 3.325–3.388 Å. Meanwhile, shorter intermolecular π-π interactions were also observed between diphenyltetrathia[22]-annulene[2,1,2,1] molecules in the layer of donor molecules, as the shortest intermolecular C–C distance was 3.315 Å, which was less than the 3.40 Å of van der Waals distance between graphite sheets. Such efficient π-π interactions among donor and acceptor molecules provided charge-transport channels for electrons and holes separately.

39.3.7 π-π INTERACTION IN FULLERENE-CARBON NANORING SUPRAMOLECULAR STRUCTURES

The complexes including host-guest π-π interaction formed with two carbon nanoring hosts, [6]cycloparaphenyleneacetylene and its anthracene-containing derivative, and fullerene C_{70} guest, separately, were explored by M06-L and M06-2X Density Functional calculations (Yuan et al., 2014). Besides two previously reported configurations in which C_{70} guest was standing or lying in the cavity of the host (Zhao & Truhlar, 2007), a new kind of configuration in which C_{70} guest was half-lying in the cavity of the host was found. The large host-guest binding energies indicated that both two-host molecules had excellent encapsulation ability for C_{70} guest, and the anthracene-containing derivative even had much better encapsulation ability for C_{70}. Interestingly, the C_{70}-half-lying configurations were more stable than C_{70}-lying configurations whereas the most stable configurations were standing ones.

39.3.8 π-π INTERACTION IN FULLERENE-CYCLIC DECAPEPTIDE SUPRAMOLECULAR STRUCTURE

The dispersion of C_{60} fullerene in water through non-covalent functionalization with a cyclic decapeptide was introduced without the aid of an organic solvent (Bartocci et al., 2015). The presence of two carbobenzyloxy groups on the peptide side chains allowed π-π interactions with C_{60} fullerene leading to the formation of stable supramolecular nanocomposites. These new peptide templates presented two distinct spatially independent functional domains: a) four side chains (all containing amino groups) pointing one face of the cyclic molecule outwards and acting as hydrophilic functionalities; b) other two (both containing carbobenzyloxy groups) pointing to the opposite face and acting as functional ligands for π-π interactions with C_{60} fullerene.

39.4   NANO-CHEMICAL APPLICATIONS

Carbon nanorings are commonly related to π-conjugated hydrocarbon macrocycle structures such as [n]cycloparaphenylene, [n]cycloparaphenyleneacetylene and its derivatives (Yuan et al., 2014). Those carbon nanorings that are used as potential supramolecular elements in their own right also serve as well-defined model compounds for studying concave π-supramolecular interactions. Interestingly, quite stable host-guest complexes can be formed with carbon nanorings and fullerenes via convex-concave π-π interactions. Moreover, the filling of the interior space of the carbon nanorings with nanoscale materials results in novel nano-hybrid with interesting properties and unique functions, which may be very different from the individual components. Supramolecular nanocavities capable of encapsulating large molecules, such as C_{60} fullerene are also possible (Park et al., 2008). In this kind of supramolecular systems, the hosts have a window large enough for a guest to enter but which can also serve as an exit door. π-π interactions are one of the non-covalent interactions which give rise to the flexibility of the host-guest molecule besides other complementary interactions such hydrogen bonding, van der Waals interactions and/or weak metal-ligand interactions (Park et al., 2008).

39.5   MULTI-/TRANS-DISCIPLINARY CONNECTIONS

Because of their combination of unique geometric and physicochemical properties, fullerenes continue to be an exciting research topic in physics, chemistry, biology and materials science. As a good electron acceptor and an n-type semiconducting material, fullerene-based materials have great potential in optoelectronic applications (Li et al., 2011). The supramolecular design principles arising from fullerene-porphyrin studies have potential applications in the preparation of light-harvesting devices, molecular magnets, molecular conductors, porous metal-organic frameworks and medicines (Sun et al., 2002). Study of photoinduced electron transfer in donor-acceptor dyads is a topic of current interest mainly to address the mechanistic details of electron transfer in chemistry and biology, to develop artificial photosynthetic systems for light energy harvesting, and also, to develop molecular optoelectronic devices. Fullerenes as electron acceptors and porphyrins as electron donors have been successfully utilized in the construction of such dyads owing to their rich and well-understood electrochemical, optical and photochemical properties. Fullerenes, because of their unique structure and symmetry, require small reorganization energy in photoinduced electron transfer reactions. As a consequence, fullerenes in donor-acceptor dyads accelerate forward electron transfer (k_{CS}) and slow down backward electron transfer (k_{CR}) resulting in the formation of long-lived charge-separated states, as verified in a number of porphyrin-fullerene dyads (D'Souza et al., 2005).

39.6 OPEN ISSUES

Most of the time, in host-guest supramolecular chemistry, computational studies are performed as predictive studies when experimental quantities cannot be detected. The computation time of fullerene supramolecular complexes exceeds more than usual since they are big (bulky) systems. Therefore, generally, low-level quantum chemical calculations are used to reduce the computation time. However, computational methods and platforms (machines) need to be more sophisticated to get more reliable results.

KEYWORDS

- **fullerenes**
- **host-guest supramolecular complex**
- **Krätschmer**
- **Kroto**
- **π-π interaction**

REFERENCES AND FURTHER READING

Albert, V. V., Chancey, R. T., Oddershede, L. B., Harris, F. E., & Sabin, J. R., (2010). Fragmentation of fullerenes. In: Sattler, K. D., (ed.), *Handbook of Nanophysics: Clusters and Fullerenes* (pp. 1−13). CRC Press: Boca Raton, Florida, USA, Ch. 26.

Anslyn, E. V., & Dougherty, D. A., (2006). *Modern Physical Organic Chemistry* (p. 184, 186). University Science Books: Sausalito, California, USA.

Asanuma, H., Li, H., Nakanishi, T., & Möhwald, H., (2010). Fullerene derivatives that bear aliphatic chains as unusual surfactants: Hierarchical self-organization, diverse morphologies, and functions. *Chemistry−A European Journal, 16*, 9330–9338.

Atkins, P., & De Paula, J., (2002). *Atkins' Physical Chemistry* (7ᵗʰ edn., pp. 697−698). Oxford University Press: New York, USA.

Bartocci, S., Mazzier, D., Moretto, A., & Mba, M., (2015). A peptide topological template for the dispersion of [60] fullerene in water. *Organic & Biomolecular Chemistry, 13*, 348–352.

Briggs, J. B., & Miller, G. P., (2006). [60] Fullerene-acene chemistry: A review. *Comptes. Rendus. Chimie, 9*, 916–927.

Churilov, G. N., Fedorov, A. S., & Novikov, P. V., (2007). Theoretical studies of fullerene formation in plasma. In: Lang, M., (ed.), *Progress in Fullerene Research* (pp. 269–297). NOVA science Pub., NY. USA.

Curl, R. F., (1993). On the formation of the fullerenes. In: Kroto, H. W., & Walton, D. R. M., (eds.), *The Fullerenes: New Horizons for the Chemistry, Physics, and Astrophysics of Carbon* (pp. 19−32). Cambridge University Press: Cambridge, UK.

D'Souza, F., Chitta, R., Gadde, S., Zandler, M. E., McCarty, A. L., Sandanayaka, A. S. D., Araki, Y., & Ito, O., (2005). Effect of axial ligation or π-π-type interactions on photochemical charge stabilization in "two-point" bound supramolecular porphyrin-fullerene conjugates. *Chemistry−A European Journal, 11*, 4416–4428.

Haddon, R. C., Brus, L. E., & Raghavachari, K., (1986). Rehybridization and π-orbital alignment: The key to the existence of spheroidal carbon clusters. *Chemical Physics Letters, 131*, 165–169.

Krätschmer, W., Lamb, L. D., Fostiropoulos, K., & Huffman, D. R., (1990). Solid C_{60}: A new form of carbon. *Nature, 347*, 354–358.

Kroto, H. W., Heath, J. R., O'Brien, S. C., Curl, R. F., & Smalley, R. E., (1985). C_{60}: Buckminsterfullerene. *Nature, 318*, 162–163.

Kumar, K. S., & Patnaik, A., (2011). Tunable one-, two-, and three-dimensional self-assemblies from an acceptor-donor fullerene−*N, N*-dimethylaminoazobenzene dyad: Interfacial geometry and temporal evolution. *Langmuir, 27*, 11017–11025.

Kuramochi, Y., Satake, A., Mitsunari, I., Ogawa, K., Araki, Y., Ito, O., & Kobuke, Y., (2008). Light-harvesting supramolecular porphyrin macrocycle accommodating a fullerene-tripodal ligand. *Chemistry − A European Journal, 14*, 2827–2841.

Kuramochi, Y., Satake, A., Sandanayaka, A. S. D., Araki, Y., Ito, O., & Kobuke, Y., (2011). Fullerene- and pyromellitdiimide-appended tripodal ligands embedded in light-harvesting porphyrin macrorings. *Inorganic Chemistry, 50*, 10249–10258.

Lembo, A., Tagliatesta, P., Cicero, D., Leoni, A., & Salvatori, A., (2009). A glycyl-substituted porphyrin as a starting compound for the synthesis of a π-π-stacked porphyrin-fullerene dyad with a frozen geometry. *Organic & Biomolecular Chemistry, 7*, 1093–1096.

Li, H., Hollamby, M. J., Seki, T., Yagai, S., Möhwald, H., & Nakanishi, T., (2011). Multifunctional, polymorphic, ionic fullerene supramolecular materials: Self-assembly and thermotropic properties. *Langmuir, 27*, 7493–7501.

Limbach, P. A., Schweikhard, L., Cowen, K. A., McDermott, M. T., Marshall, A. G., & Coe, J. V., (1991). Observation of the doubly charged, gas-phase fullerene anions C_{60}^{2-} and C_{70}^{2-}. *Journal of the American Chemical Society, 113*, 6795–6798.

Mukherjee, P., & Bhattacharya, S., (2012). Spectroscopic and theoretical insights on fullerene-octaethylporphyrin self-assembled non-covalent conjugates studied in solution. *Spectrochimica Acta Part A., 90*, 186–192.

Oku, T., (2007). Boron nitride nanocage clusters, nanotubes, and nanohorns. In: Lang, M., (ed.), *Progress in Fullerene Research* (pp. 41–83). Nova Science Publishers: New York, USA.

Park, S. J., Kwon, O. H., Lee, K. S., Yamaguchi, K., Jang, D. J., & Hong, J. I., (2008). Dimeric capsules with a nanoscale cavity for [60] fullerene encapsulation. *Chemistry – A European Journal, 14*, 5353–5359.

Pincak, R., & Pudlak, M., (2007). Electronic structure of spheroidal fullerenes. In: Lang, M., (ed.), *Progress in Fullerene Research* (pp. 235–268). Nova Science Publishers: New York, USA.

Saito, R., Dresselhaus, G., & Dresselhaus, M. S., (1992). Topological defects in large fullerenes. *Chemical Physics Letters, 195*, 537–542.

Scuseria, G. E., (1992). Negative curvature and hyperfullerenes. *Chemical Physics Letters, 195*, 534–536.

Sun, D., Tham, F. S., Reed, C. A., Chaker, L., & Boyd, P. D. W., (2002). Supramolecular fullerene-porphyrin chemistry. Fullerene complexation by metallated "Jaws Porphyrin" hosts. *Journal of the American Chemical Society, 124*, 6604–6612.

Takai, A., Chkounda, M., Eggenspiller, A., Gros, C. P., Lachkar, M., Barbe, J. M., & Fukuzumi, S., (2010). Efficient photoinduced electron transfer in a porphyrin tripod-fullerene supramolecular complex via π-π interactions in nonpolar media. *Journal of the American Chemical Society, 132*, 4477–4489.

Yuan, K., Guo, Y. J., Yang, T., Dang, J. S., Zhao, P., Li, Q. Z., & Zhao, X., (2014). Theoretical insights into the host-guest interactions between [6]cycloparaphenyleneacetylene and its anthracene-containing derivative and fullerene C_{70}. *Journal of Physical Organic Chemistry, 27*, 772–782.

Zhang, J., Tan, J., Ma, Z., Xu, W., Zhao, G., Geng, H., et al., (2013). Fullerene/sulfur-bridged annulene cocrystals: Two-dimensional segregated heterojunctions with ambipolar transport properties and photoresponsivity. *Journal of the American Chemical Society, 135*, 558–561.

Zhao, Y., & Truhlar, D. G., (2007). Size-selective supramolecular chemistry in a hydrocarbon nanoring. *Journal of the American Chemical Society, 129*, 8440–8442.

CHAPTER 40

Proton-Mediated Molecular Clusters

ATTILA BENDE

Molecular and Biomolecular Physics Department, National Institute for Research and Development of Isotopic and Molecular Technologies, Donat Street, No. 67-103, RO-400293 Cluj-Napoca, Romania, Tel: 0040-264-584037, Fax: 0040-264-420042, E-mail: bende@itim-cj.ro

40.1 DEFINITION

With the huge growth of the sophisticated experimental technologies or the computer capacity for molecular modeling, we are able to study in detail the nature of some peculiar physical processes driven by the presence of ionized hydrogen atoms and shed light on the structural complexity of these structures. These small positively charged particles are mainly involved in proton transfer reactions, relaxation processes of molecular excited states, the interaction of gases and aerosol particles in the high atmosphere, as well as cell respiration and energy storage in biological systems.

40.2 HISTORICAL ORIGIN(S)

The key role of the ionized hydrogen atoms (protons) in various chemical and biological processes was already well-known at the beginning of the 19th century when Theodor Grotthuss had proposed its famous theory for water conductivity. Since then it was identified several other chemical and biological phenomena where the basic processes were governed by protons, for example, enzyme-catalyzed reactions, proton pumps across the cell membranes, ATP- or photosynthesis. Besides these dynamic processes, the presence of the proton can initiate special cluster aggregation, intensively studied in the field of atmospheric chemistry or mass spectrometry. Starting in the 1960s, it was proposed different molecular models for the correct structure of water in the liquid phase which have raised the interest for other cluster structures and dynamical processes where protons were involved.

40.3 NANO-SCIENTIFIC DEVELOPMENT(S)

Protons, also known as ionized hydrogen atoms, play an important role in different biological phenomena, like electron transport, ATP synthesis or photosynthesis (Alberts et al., 2008; Reece et al., 2009; Ishikita et al., 2014). In chemistry, for example, they are the main actors in solution or photo-relaxation dynamics processes (Lao-Ngam et al., 2011; Kosower et al., 1986) but can form the starting point of different ionized cluster formations (Di Palma et al., 2013, 2014; Bandyopadhyay et al., 2015). The physical phenomenon starts with a fast electron ionization followed by the dissociative proton transfer reaction (Krasnoholovets et al., 2003). In this way, the freely moving proton can initiate special cluster aggregations by bounding molecular fragments which contain electron acceptor atoms, mainly oxygens or nitrogens. The most representative ionic clusters are the well-known Zundel- and Eigen-clusters of water (Eigen, 1964; Zundel, 1976; Marx et al., 1999; Kirchner, 2007) as well as the Hoogsteen configuration of the DNA base pairs (Hoogsteen, 1963; Segers-Nolten, 1997) (for their geometry configuration, *see* Figure 40.1).

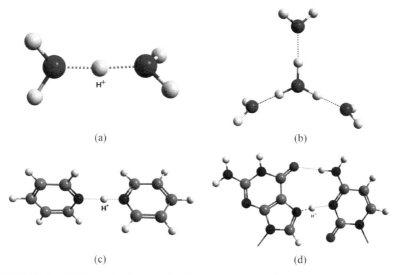

(a)   (b)

(c)   (d)

FIGURE 40.1 Different ionized molecular clusters stabilized by the presence of the positively charged proton. (a) Zundel-ion; (b) Eigen-ion; (c) protonated pyridine dimer; (d) G-C DNA base pair in Hoogsteen configuration.

The proton bridge, basically, is a hydrogen-bond like supramolecular binding builds up by two atoms with high electronegativity where the positively charged proton becomes intercalated between the atoms. In this way, the lone pair orbital of one electronegative atom seizes the positively charged proton making a hydrogen bond donor fragment, while the second electronegative atom behaves as hydrogen bond acceptor.

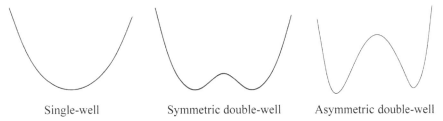

Single-well   Symmetric double-well   Asymmetric double-well

FIGURE 40.2   Different energy profiles of the moving proton between two rigid, electronegative atoms.

Depending on the orbital relaxation of the lone pair electron orbitals different potential energy surface (PES) profiles of a moving proton between two rigid electronegative atoms can form. Their profiles are shown in Figure 40.2. The single-well PES profile is characteristic to the Zundel cation geometry (Figure 40.1a) where the proton stays in the middle of the distance between the two electronegative atoms. Double-well symmetric PES profile can be found in the case of symmetric dimers (Figure 40.1c), while the double-well asymmetric one occurs for a system for systems with electronegative atoms in nonequivalent geometry position (e.g., *see* Figure 40.1d).

40.4   NANO-CHEMICAL APPLICATION(S)

Using the VUV single-photon ionization molecular beam time-of-flight mass spectrometry technique, one can generate and investigates small-ionized molecular clusters which structures are exclusively defined by the presence of the extra proton incorporated in the cluster and ionized through the collision with the photon beam. In this way, several ionized clusters built by the positively charged proton and a certain number of neutral molecules were investigated. Among them we can mention cluster systems $H^+(CH_3OH)_n$ ($n = 1$–12) (Kostko et al., 2008; Bandyopadhyay et al., 2015), $H^+(CH_3OH)_nH_2O$ (Bandyopadhyay et al., 2015), $H^+(CH_3OH)(CO_2)_n$ ($n = 1$–7) (Zhao et al., 2015), $H^+(CH_3CHO)_n$ (Di Palma et al., 2014), $H^+(CH_3COCH_3)_n$ ($n = 1$–7)

(Zhan et al., 2006), $H^+(CH_3COOH)_2$ (Guan et al., 2012), $H^+(Imidazole)_2$ (Lankau et al., 2011), $H^+(CH_3COCH_3)_2$ (Di Palma et al., 2013) investigated using VUV mass spectrometry or *ab initio* quantum chemical methods. The basic mechanism consists of a formation of a core cluster structure formed by two electronegative atoms and the proton with the $X\cdots H^+\cdots X$ (X = O or N) pattern (proton-bridge). This configuration induces an irregular charge distribution of the lone pair orbitals of the electronegative atoms which flank the positively charged proton from left and right sides. Through this are produced secondary binding sites to which other neutral monomers can attach to the cluster core.

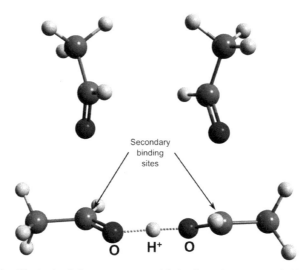

FIGURE 40.3   The ionized cluster structure with the $O\cdots H^+\cdots O$ proton bridge and together with the secondary binding sites.

What's more, these binding sites (hydrogen bond donors or acceptors) become components of a longer hydrogen bond chain displaying the property of the hydrogen bond cooperativity (Jeffrey et al., 1991; Maes et al., 1993) by shortening the individual hydrogen bond distances and strengthen their energies.

40.5   MULTI-/TRANS-DISCIPLINARY CONNECTION(S)

The model is widely used in the field of physics, chemistry, and biology. Through this model several molecular crystal structures, molecular aggregations or structural organization and function of different biomolecular complexes can be easily explained.

40.6 OPEN ISSUES

The most important open question consists in the detailed theoretical description of the proton transfer mechanism from first electronegative atom to the second one when the potential energy surface on which the proton moves is far from the quadratic form and contains an energy barrier between the two minima (Bende et al., 2014). The vibrational oscillation of the proton in a double-well asymmetric potential shows the so-called Duffing-type nonlinearity (Ueda, 1979), which at certain conditions can turn over even to a chaotic behavior (Litak et al., 2006).

KEYWORDS

- **proton bridge**
- **ionized clusters**
- **van der Waals interaction**
- **charge transfer**

REFERENCES AND FURTHER READING

Alberts, B., Johnson, A., Lewis, J., Morgan, D., Raff, M., Roberts, K., & Walter, P., (2008). Chapter 14: Energy Conversion: Mitochondria and Chloroplasts. In: *Molecular Biology of the Cell* (6th ed., pp. 753–812). Publisher: Garland Science, Taylor & Francis Group, LLC, New York, US.

Bandyopadhyay, B., Kostko, O., Fang, Y., & Ahmed, M., (2015). Probing methanol cluster growth by vacuum ultraviolet ionization. *J. Phys. Chem. A., 119*, 4083–4092.

Bende, A., & Muntean, C. M., (2014). The influence of anharmonic and solvent effects on the theoretical vibrational spectra of the guanine-cytosine base pairs in Watson–Crick and hoogsteen configurations. *J. Mol. Model., 20*, 2113.

Di Palma, T. M., & Bende, A., (2013). Vacuum ultraviolet photoionization and *Ab initio* investigations of methyl tert-butyl ether (MTBE) clusters and MTBE-water clusters. *Chem. Phys. Lett., 561–562*, 18–23.

Di Palma, T. M., & Bende, A., (2014). Tautomerism and proton transfer in photoionized acetaldehyde and acetaldehyde-water clusters. *J. Mass Spectrom, 49*(8), 700–708.

Eigen, M., (1964). Proton transfer acid-base catalysis and enzyme hydrolysis. *Angew. Chem. Int. Ed. 3*(1), 1–19.

Guan, J., Hu, Y., Zou, H., Cao, L., Liu, F., Shan, X., & Sheng, L., (2012). Competitive fragmentation pathways of acetic acid dimer explored by synchrotron VUV photoionization mass spectrometry and electronic structure calculations. *J. Chem. Phys., 137*, 124308.

Hoogsteen, K., (1963). The crystal and molecular structure of a hydrogen-bonded complex between 1-methylthymine and 9-methyladenine. *Acta Crystallogr., 16*, 907–916.

Ishikita, H., & Saito, K., (2014). Proton transfer reactions and hydrogen-bond networks in protein environments. *J. R. Soc. Interface, 11*, 20130518.

Jeffrey, G. A., & Saenger, W., (1991). Definitions and concepts. In: Jeffrey, G. A., & Saenger, W., (eds.), *Hydrogen Bonding in Biological Structures* (pp. 15–48). Springer Berlin Heidelberg.

Kirchner, B., (2007). Eigen or Zundel ion: News from calculated and experimental photoelectron spectroscopy. *Chem. Phys. Chem., 8*, 41–43.

Kosower, E. M., & Huppert, D., (1986). Excited state electron and proton transfers. *Ann. Rev. Phys. Chem., 37,* 127–156.

Kostko, O., Belau, L., Wilson, K. R., & Ahmed, M., (2008). Vacuum-ultraviolet (VUV) photoionization of small methanol and methanol-water clusters. *J. Phys. Chem. A., 112*, 9555–9562.

Krasnoholovets, V. V., Tomchuk, P. M., & Lukyanets, S. P., (2003). Proton transfer and coherent phenomena in molecular structures with hydrogen bonds. In: Prigogine, I., & Rice, S. A., (eds.), *Advances in Chemical Physics* (Vol. 125, pp. 351–548). John Wiley & Sons, Inc.

Lankau, T., & Yu, C. H., (2011). A quantum description of the proton movement in an idealized NHN^+ bridge. *Phys. Chem. Chem. Phys., 13*, 12758–12769.

Lao-Ngam, C., Asawakun, P., Wannarat, S., & Sagarik, K., (2011). Proton transfer reactions and dynamics in protonated water clusters. *Phys. Chem. Chem. Phys., 13*, 4562–4575.

Litak, G., & Borowiec, M., (2006). Oscillators with asymmetric single and double well potentials: Transition to chaos revisited. *Acta Mech., 184*, 47–59.

Maes, G., & Smets, J., (1993). Hydrogen bond cooperativity: A quantitative study using matrix-isolation FT-IR spectroscopy. *J. Phys. Chem., 97*(9), 1818–1825.

Marx, D., Tuckermann, M. E., & Parrinello, M., (1999). The nature of the hydrated excess proton in water. *Nature, 397*(6720), 601–604.

Reece, S. Y., & Nocera, D. G., (2009). Proton-coupled electron transfer in biology: Results from synergistic studies in natural and model systems. *Ann. Rev. Biochem., 78*, 673–699.

Segers-Nolten, G. M. J., Sijtsema, N. M., & Otto, C., (1997). Evidence for hoogsteen GC base pairs in the proton-induced transition from right-handed to left-handed poly(dG-dC)·Poly(dG-dC). *Biochem., 36*(43), 13241–13247.

Ueda, Y., (1979). Randomly transitional phenomena in the system governed by Duffing's equation. *J. Stat. Phys., 20*, 181–196.

Zhan, H., Ming-Xing, J., Xue-Song, X., Xi-Hui, C., & Da-Jun, D., (2006). Acetone: Isomerization and aggregation. *Front. Phys. China, 3*, 275–282.

Zhao, Z., Kong, X. T., Lei, X., Zhang, B. B., Zhao, J. J., & Jiang, L., (2015). Early stage solvation of protonated methanol by carbon dioxide. *Chin. J. Chem. Phys., 28*(4), 501–508.

Zundel, G., (1976). Easily polarizable hydrogen bonds – their interactions with the environment – IR continuum and anomalous large conductivity, in the hydrogen bond-recent developments in theory and experiments. In: Schuster, P., Zundel, G., & Sandorfy, C., (eds.), *II Structure and Spectroscopy* (pp. 683–766). North-Holland, Amsterdam.

CHAPTER 41

Pyrimidines

NICOLETA A. DUDAŞ and MIHAI V. PUTZ

Laboratory of Computational and Structural Physical Chemistry for Nanosciences and QSAR, Biology-Chemistry Department, West University of Timişoara, Pestalozzi Str. No. 16, Timişoara 300115, Romania, E-mails: nicole_suceveanu@yahoo.com, mihai.putz@e-uvt.ro, or, mv_putz@yahoo.com

41.1 DEFINITION

Pyrimidines (1,3-diazines or *m*-diazines) are one of three classes of six-membered heterocyclic aromatic compounds containing two nitrogen atoms in their ring, the other two diazines being pyridazines (1,2-diazines) and pyrazines (1,4-diazines). The pyrimidine ring is the basic structural element of nitrogenous bases of DNA and RNA, is one of the most widespread heterocyclic in various biologically important derivatives, as substituted derivatives and ring-fused compounds.

Pyrimidine derivatives are utilized as drugs in the treatment of several types of diseases, agrochemicals, chromophores, complexes, metal sensors, solar cells technology, etc.

Pyrimidines can be obtained by the synthetic chemical route, from biosynthesis (metabolism ↔ catabolism) and also in prebiotic condition.

Because the pyrimidine ring contains two nitrogen atoms it has a weakly basic character, pyrimidine systems are highly electron deficient, and that is why electrophilic aromatic substitution gets more difficult, while nucleophilic aromatic substitution gets easier.

41.2 HISTORICAL ORIGIN(S)

The first pyrimidinic compounds that were discovered are alloxan – 1818, guanine – 1844, barbituric acid – 1864, cytosine – 1894, uracil – 1900, and in 1899 was first synthesized unsubstituted pyrimidine (Figure 41.1) (Brown et al., 1994; Harris & Thimann, 1949; Grimaux, 1879; Fischer, 1897; Kossel & Steudel, 1903; Gabriel & Colman, 1899; Baeyer, 1864). The first simple pyrimidine derivatives that have been extracted from natural sources are vicine and convicine in 1870 from *Vicia sativa* and *Vicia faba L.*, these being the causative agents of favism (Lagoja, 2005; Ritthausen, 1870, 1876, 1881; Brown & Roberts, 1972; Bryant & Hughes, 2011; Mager et al., 1980).

In 1869, DNA was identified by Friedrich Miescher, and the structure and components of DNA and RNA were discovered between 1900–1953 by Phoebus Levene, Erwin Chargaff, James D. Watson and Francis Crick (Watson & Crick, 1953). Therefore, the presence of pyrimidine derivatives was highlighted such as pyrimidine and purine nitrogenous bases in DNA and RNA (cytosine, thymine, uracil, adenine, guanine) (Figure 41.1), which leads us to the conclusion that the pyrimidinic compounds are *vital compounds* of the existence of life on Earth (Dudaş & Putz, 2014). So far, studies of the carbonaceous chondrites such as Allende, Orgueil, Murchison, Lonewolf Nunataks 94102, and Murray have shown the presence of purine- and pyrimidine-based compounds with extraterrestrial origin (like uracil and thymine) in these meteorites (Figure 41.1) (Sandford et al., 2014; Callahan et al., 2011; Martins et al., 2008; Hayes, 1967; Folsome, et al., 1971; Stoks, et al., 1979, 1981; Cronin, et al., 1993; Irvine, 1998). Besides these compounds, other compounds that are part of the essential classes of terrestrial biochemistry have also been identified (e.g., amino acids, sugar derivatives) (Martins et al., 2008). These findings have led researchers to affirm that the organic components of the genetic code were already present in the early solar system and may have played a key role in the origin of life and those pyrimidine derivatives (such as those of uracil) are *prebiotic/primordial compounds* (Martins et al., 2008; Lagoja, 2005; Dudaş & Putz, 2014).

41.3 NANO-SCIENTIFIC DEVELOPMENT(S)

Pyrimidines are very important in dividing and elongating tissues, they are the foundation stone for nucleic acid biosynthesis, also due to the fact that they are an energy source or precursors for the synthesis of primary and secondary products, they take part in various metabolic processes (e.g., pyrimidines are involved in plant carbohydrate metabolism) (Schröder et al., 2005). Nucleotides are key components in a great number of cellular reactions, and they are absolutely essential for normal growth and development of all organisms and plants (Hermansen et al., 2015; Kafer et al., 2004; Schröder et al., 2005; Witz et al., 2012). Nucleotide synthesis uses multiple metabolic pathways across different cell compartments, and the processes are regulated at the transcription level by a set of transcription factors and at the enzyme level by allosteric regulation and feedback inhibition (Lane et al., 2015). For DNA replication and for RNA production (when cells are proliferating) an increased nucleotide synthesis is necessary to support protein synthesis at different stages of the cell cycle (Lane et al., 2015).

Besides pyrimidine and purine nitrogenous bases from DNA and RNA (cytosine, thymine, uracil, adenine, guanine) there are many compounds containing the pyrimidine core which can be found in important classes such as vitamins or α-amino acids. Among them there are: thiamine (vitamin B1), riboflavin (vitamin B2), folic acid (vitamin B9), orotic acid (vitamin B13) and the nonproteogenic α-amino acid willardiine extracted from the seeds of various species of *Acacia sp.* and lathyrine (tingitanine) extracted from *Lathyrus tingitanus* (Figure 41.1) (Dudaş & Putz, 2014; McDowell, 2000; Lanska, 2009; Hoffbrand & Weir, 2001; Lagoja, 2005; Gmelin, 1959; Bell, 1961; Bell & Foster, 1962).

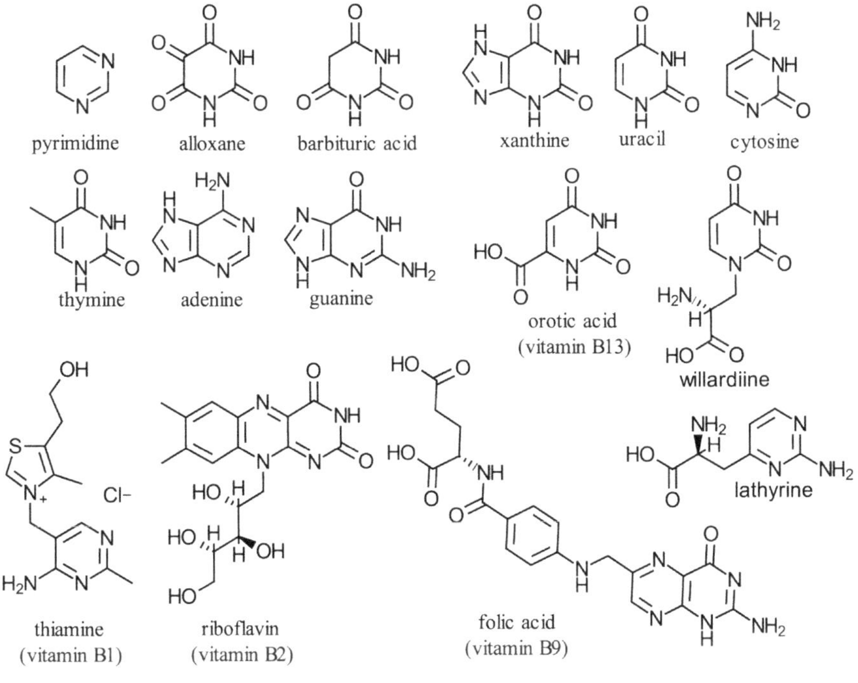

FIGURE 41.1 The first pyrimidine derivatives that were discovered, DNA/RNA bases, examples of vitamins and α-amino acids containing pyrimidine core.

Other pyrimidine derivatives with biological activity extracted from natural sources are: charine isolated from *Momordica charantia* L., used in folk medicine for its hypoglycemic properties, to treat colds, fever, rheumatism, stomach pain, ulcers; bacimethrin isolated from *Bacillus megatherium* and considered to be the easiest antibiotic with pyrimidinic core; blasticidin S, an antibiotic and antifungal agent which is produced by *Streptomyces griseochromogenes*; febrifugine from *Dichroa febrifuga* with antimalarian activity; heteromines F, G, H isolated from *Heterostemma brownii* that are used in the treatment of cancer (Figure 41.2) (Lagoja, 2005; Raman & Lau, 1996; von Angerer, 2007; Dudaş & Putz, 2014).

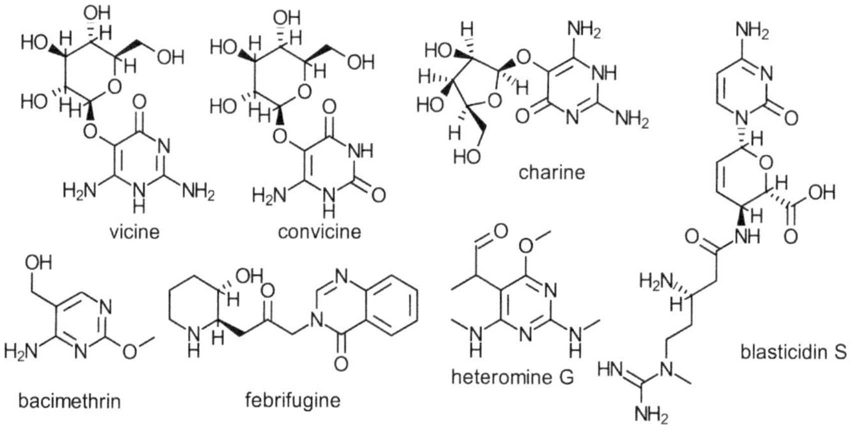

FIGURE 41.2 Pyrimidinic derivatives with biological activity extracted from natural sources.

Many pyrimidinic compounds with biologic activity from marine sources have been isolated such as: from Caribbean sponge *Cryptotethia crypta* spongothymidine (Ara-T) and spongouridine (Ara-U) (also isolated from *Eunicella cavolini* with antiviral activity (Lagoja, 2005; Bhakuni & Rawat, 2005; Utagawa, et al., 1980; Bergmann & Burke, 1955); from starfish *Acanthaster planci* were isolated 2'-deoxyuridine and thymidine derivatives (Kamori et al., 1980); from *Aplysia* sp. was isolated the already known antiviral agent 2,'3'-didehydro-2,'3'-dideoxyuridine (Kondo et al., 1992). Ara-U, Ara-T, 2'-deoxyuridine and thymidine are considered to be the precursors of all nucleosidic drugs (Suckling, 1991), these compounds leading to the development of analogs with biological activity, such as AZT, one of the oldest and most important drugs used in the treatment of infection by HIV (Figure 41.3) (Lagoja, 2005; Gerwick & Moore, 2012; Montaser & Luesch, 2011; Sarfaraj et al., 2012; Bhakuni & Rawat, 2005; Dudaş & Putz, 2014).

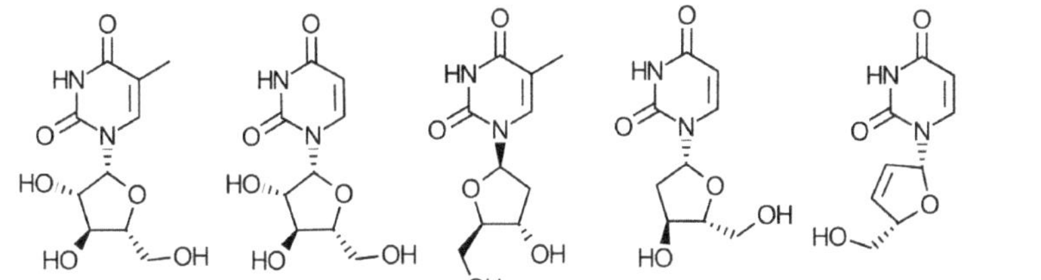

FIGURE 41.3 Pyrimidinic derivatives with biological activity extracted from marine sources.

Pyrimidine derivatives play an important role in the design of anti-HIV drugs. From the compounds approved for the treatment of HIV/AIDS, nearly half of them contain the pyrimidine core in their structure (Dudaş & Putz, 2014; Putz & Dudaş, 2015) (Figure 41.4). Most of them are of the class NRTI (Nucleoside Reverse-Transcriptase Inhibitors), namely: AZT–zidovudine, first drug approved in 1987; didanosine–ddI; zalcitabine–ddC; stavudine–d4T; lamivudine–3TC; abacavir–ABC; emtricitabine–(-)FTC (De Clercq, 2008, 2009a, 2011; Mehellou & De Clercq, 2010; Esposito et al., 2012; Nguyen et al., 2011; Hoffmann & Rockstroh, 2011, 2012); as NtRTI (Nucleotide Reverse-Transcriptase Inhibitors) was aproved in 2001 Teno-fovir (TFV), more specifically it's precursor tenofovir disoproxil fumarate (TDF, PMPA) (De Clercq, 2008, 2009a, 2011; Mehellou & De Clercq, 2010; Esposito et al., 2012; Nguyen et al., 2011); as a second-generation NNRTI (Non-Nucleoside Reverse-Transcriptase Inhibitors) with a much better anti-HIV activity than those of the first generation are diaryl-pyrimidine derivatives (DAPY) etravirine (ETR, TMC125), and rilpivirine (RPV, TMC278), (De Clercq, 2008, 2009a, 2011; Mehellou & De Clercq, 2010; Esposito et al., 2012; Sahlberg & Zhou, 2008; Ravichandran et al., 2008; Usach et al., 2013; Nguyen et al., 2011); in 2000 Lopinavir (ABT-378) an PI (Protease Inhibitors) was placed on the market (De Clercq, 2008, 2009a; Mehellou & De Clercq, 2010; Esposito et al., 2012; Brik & Wong, 2003) and in 2007 the first INI (Integrase Inhibitors), raltegravir (RAL, MK-0518) (De Clercq, 2008, 2009a; Mehellou & De Clercq, 2010; Esposito et al., 2012; Ingale & Bhatia, 2011; Nguyen et al., 2011; Dudaş & Putz, 2014).

The pyrimidine core exists in the structure of many compounds that have been tested for their biological and pharmacological properties, some of them being approved as drugs or pesticides. Among them we can mention (Figure 41.5): veronal (barbital/barbitone) which was the first hypnotic-sleeping drug, synthesized and used for the first time by Fisher and von Mehrin in 1903; thiamylal with sedative, anticonvulsant and hypnotic effects, used as a strong but short acting sedative; minoxidil – an antihypertensive, vasodilator used in the treatment of alopecia; 2-thiouracil and propylthiouracil – for treatment of hyperthyroidism; pyrimethamine with sulfadoxine (a sulphonamide) – for treatment of malaria, toxoplasmosis; bacimethrin, trimethoprim – antibacterial agents, antibiotics; 5-iododeoxyuridine, arabinoside (Ara-A), acyclovir cidofovir – antiviralagents; flucytosine – an antifungal agent; 5-fluorouracil, tegafur, uramustine, cytarabine (Ara-C), gemcitabine – used to treat various types of cancer; pyrimidifen – an acaricide and insecticide; fenarimol–fungicide properties; bromacil, nicosulfuron–herbicides (Fischer & von Mering, 1903; von Angerer, 2007; Lackie, 2010; Daintith, 2008; Jain et al., 2006; Panneer Selvam et al., 2012, 2013; Dudhe et al., 2011; Wilhelmus, 2015; Wood, 2013; Vencill, 2002; Herbicide Classification, 2013; FDA, 2005; Dudaş & Putz, 2014).

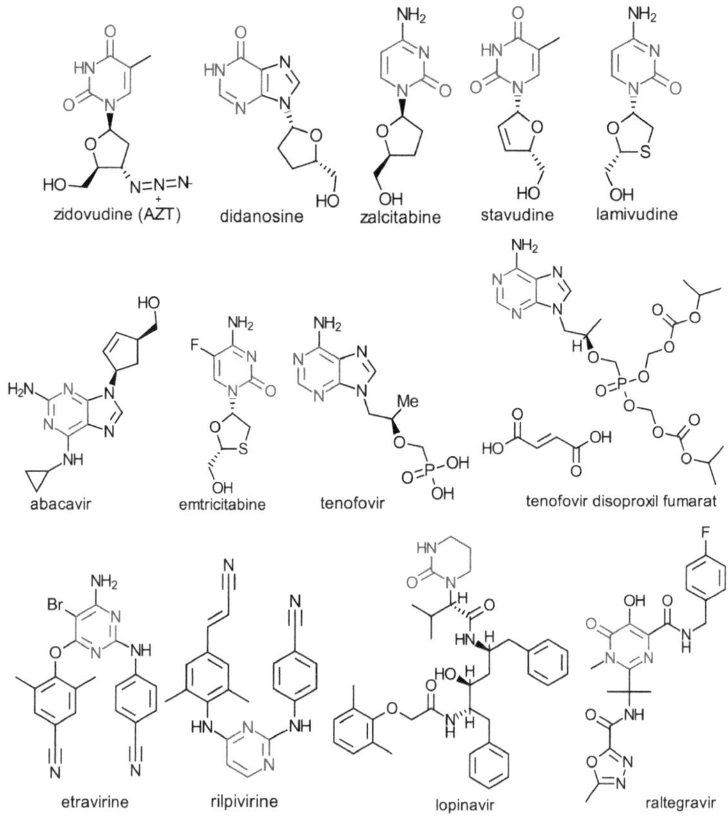

FIGURE 41.4 Pyrimidine derivatives that have been approved for use in therapy of HIV infection/AIDS.

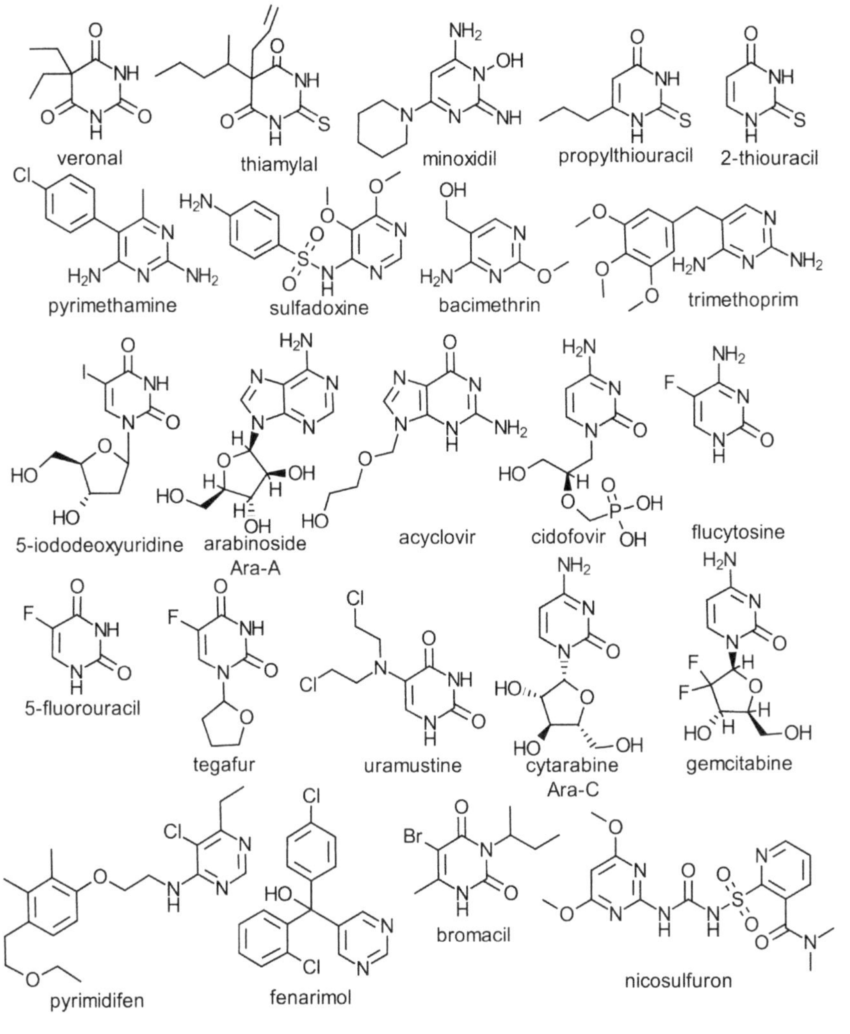

FIGURE 41.5 Pyrimidine derivatives that have been approved for use in therapy of various diseases or as pesticides.

A very large number of various quantitative structure-activity relationship (QSAR) studies were made on different series of pyrimidinic derivatives. These studies reveal the lead compounds which helped the researchers in the development of new drugs. Some examples of SAR studies that have been made on series of pyrimidinic derivatives are: QSAR, quantitative structure-toxicity relationship (QSTR), CoMFA (comparative molecular field analysis) and CoMSIA (comparative molecular field analysis and molecular field similarity indices analysis) 3D-QSAR, SPECTRAL-(DIAGONAL)-SAR, quantitative structure-property relationships (QSPR), quantitative structure-reactivity relationships (QSRRs), quantitative structure-toxicity relationships (QSTRs). In these studies, a large variety of descriptors was used, from 0D- to 4D-descriptors, e.g., empirical descriptors, physical properties, constitutional descriptors, topological indices, geometrical descriptors, quantum descriptors, molecular descriptors, molecular fields (Putz et al., 2015, 2016; Dudaş & Putz, 2014; Putz & Dudaş, 2013a, b; Lauria, 2010; D'Archivio, 2015; Iman & Davood, 2013; Toropova et al., 2015).

Pyrimidine derivatives also find their place in many and diverse applications, in the development of science and nanotechnology. We mention only a few of the applications pyrimidine derivatives: as chemosensor for the detection of Zn^{2+} with high selectivity and sensitivity to Zn^{2+} over other competitive metal ions (Mati et al., 2014); as chromophores that leads to nanoparticles with hydrodynamic radius below 100 nm (Vurth et al., 2012). Pyrimidine can be introduced to serve as a central electron-deficient unit in benzothiadiazole-dicyanovinylene-capped molecules for small-molecule (organic) solar cells (Haijun et al., 2015). Pyrimidine derivatives serve to the development of drug-loaded nanoparticles (Nagarwal et al., 2011), or in the synthesis of nanoscale structures (Hassan et al., 2015), development of hybrid adsorbent materials that can serve as a biochemical sensor (able to absorb the adenine, uracil, and cytosine molecules) (Baei et al., 2014), development of new nanocatalyst as nano-sawdust-OSO_3H (for the preparation of biologically important pyrimidine derivatives) (Sadeghi et al., 2015).

41.4 NANO-CHEMICAL APPLICATION(S)

Because pyrimidine derivatives play a key role in the origin of life on Earth, they are widely spread in nature, have many and diverse applications, their reactivity and all the synthesis pathways: chemical, biosynthesis–metabolism/catabolism, in prebiotic conditions, are very intensely investigated. The chemical structure and reactivity of pyrimidine derivatives have been the subject of many synthetic and theoretical studies, results and progress being periodically summarized in several reviews and books (Brown et al., 1994; Undheim & Benneche, 1996; Hoffmann et al., 1998; Kress, 1989; Kress & Varie, 1990, 1991, 1992; Heinisch & Matuszczak, 1994, 1995; Groziak, 1997, 1998, 1999; Joule & Mills, 2000, 2010; von Angerer, 2007).

41.4.1   REACTIVITY OF PYRIMIDINE DERIVATIVES

Because of the presence of two electronegative nitrogen atoms, (with strong electron-pulling effect) pyrimidine nucleus is π-deficient, the electron densities at 2-, 4- and 6-positions are depleted, and these positions become strongly electron loving, so they are the *electrophilic positions*. The electron density at 5-position is only slightly depleted, is a benzenoid position (Ajani, 2015; Brown, 1994, 2009). At the nitrogen atoms of pyrimidine nucleus, the electron density is greatly enhanced, and the nitrogen atoms constitute the basic and the nucleophilic centers in pyrimidine (Ajani, 2015; Brown, 1994, 2009; Undheim & Benneche, 1996). Due to these characteristics of the pyrimidine core, pyrimidine derivatives are unreactive towards electrophilic aromatic substitution, and reactive towards nucleophilic aromatic substitution. Position 5 of pyrimidine is the most susceptible in the ring to electrophilic attack (Ajani, 2015; Undheim & Benneche, 1996). If they are present on the pyrimidine ring at other positions activating groups (-OH, NH_2), electrophilic substitution (nitration, nitrosation, diazo-coupling) usually occurs, but only at position 5. In a similar way, positions 2-, 4- and 6- of pyrimidine are attacked by nucleophilic reagents. At the ring nitrogen can be carried out alkylation-, acylation reaction, direct N-oxidation takes place with low yields, pyrimidines being susceptible to decomposition (Ajani, 2015; Brown, 1994, 2009; Undheim & Benneche, 1996).

41.4.2   CHEMICAL SYNTHESIS OF PYRIMIDINE DERIVATIVES

There are very diverse and numerous primary synthetic routes that lead to pyrimidines (Brown, 1994, 2009; Ajani et al., 2015; Joule and Mills, 2010). The most widely route used for the synthesis of pyrimidinic derivatives is named *Principal synthesis*, and involves the reaction of one molecule that supply the three-carbon fragment (C4-C5-C6) with another molecule that supply the three-atom fragment (N1-C2-N3) (Brown et al., 1994, 2009; Ajani et al., 2015; Joule & Mills, 2010). These condensations can be made between a β-dialdehydes, β-aldehydoketones, β-diketones, β-aldehydoesters, β-ketoesters, β-diesters (malonic esters), β-aldehydonitriles, β-ketonitriles, β-esternitriles, β-dinitriles (malononitriles), and amidines, guanidines, ureas or thioureas derivatives (Brown et al., 1994, 2009; Ajani et al., 2015; Joule & Mills, 2010). These reactions are usually done under basic (alkaline) condition (e.g., EtONa, EtOH), but in some particular cases can be made in neutral or acidic condition (e.g., HCl, EtOH). (Figure 41.6a, 6b) (Brown et al., 1994, 2009; Ajani et al., 2015; Joule & Mills, 2010).

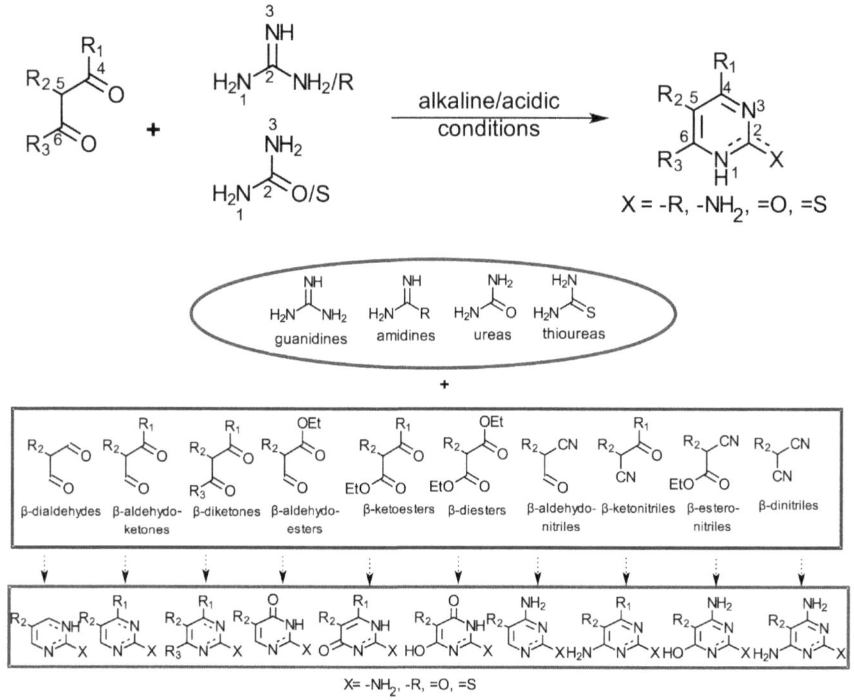

FIGURE 41.6 (a&b) Principal synthesis of pyrimidine derivatives – schematic representation.

Other routes used for the synthesis of pyrimidinic derivatives are the cyclization/synthesis from one or more synthones (Brown et al., 1994, 2009). The cyclization from a single six-atom synthone by completion of the N1-C2 bond can be made using diamides, amino-carbonyl synthones, isocyanato-nitriles, etc., and the one by completion of the N3-C4 bond is the cyclization between an amino and a carbonyl/cyano group or ethylenic bond in the synthone (Figure 41.7a, 7b) (Brown et al., 1994, 2009; Ajani et al., 2015; Undheim & Benneche, 1996). In the synthesis that is using one- and five-atom synthone, the one-atom synthone can provide N1 from the pyrimidinic ring, usually by ammonia or a primary amine and five-atom synthone can be urethanes, isocyanates, isothiocyanates, nitriles, etc. (Brown et al., 1994, 2009). Or the one-atom synthone can provide C2 from pyrimidinic ring, usually by formaldehyde, formamides, formic acid, etc. and five-atom synthone can be β-diamines, β-aminoamides, malonodiamides, malonodiamidines, etc. (Figure 41.7a, 7b) (Brown et al., 1994, 2009; Ajani et al., 2015; Undheim & Benneche, 1996).

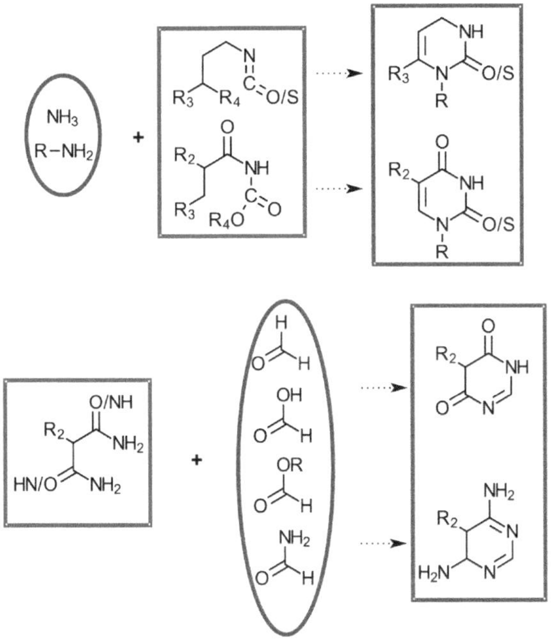

FIGURE 41.7 (a&b) Synthesis of pyrimidine derivatives using one- and five-atom synthone, the one-atom synthone provide N1 – schematic representation.

41.4.3 BIOSYNTHESIS OF PYRIMIDINE DERIVATIVES

Pyrimidine biosynthesis is one of the most ancient of all biochemical pathways; the global metabolism of pyrimidines can be splited into several different phases: de novo synthesis, modification, recycling, salvage, and degradation (Hermansen et al., 2015; Kafer et al., 2004; Schröder et al., 2005; Witz et al., 2012). From an energetic standpoint is beneficial for an organism to recycle nucleotides from RNA degradation (Kafer et al., 2004).

Enzymatically, nucleotide de novo synthesis is highly similar in almost all living organisms, and has six enzymatic steps but the organization into multifunctional enzymes as well as subcellular localization and regulation change with evolution (Hermansen et al., 2015; Kafer et al., 2004; Schröder et al., 2005; Witz et al., 2012; Evans et al., 2004; Löffler et al., 2004, 2013). In plants, the two initial enzymatic reactions of de novo pyrimidine synthesis occur in the plastids, all enzymatic steps are encoded by single genes, have all the necessary enzymes, and lack of function of any of the genes can be deadly (Witz et al., 2012; Schröder et al., 2005; Kafer et al., 2004).

The six-steps pyrimidine de novo biosynthesis is different from purine biosynthesis where the ring is built step by step on ribose-5´-phosphate starting with 5-phosphoribosyl-1-pyrophosphate (PRPP) (Hermansen et al., 2015; Kafer et al., 2004; Löffler & Zameitat, 2004, 2013; Kafer et al., 2004; Rider, 2009). The pyrimidine ring is assembled first and then linked to ribose-phosphate to form a pyrimidine nucleotide (Löffler & Zameitat, 2004, 2013). This de novo biosynthesis is catalyzed by six enzymes: carbamoyl-phosphate synthetase (CPSase), aspartate carbamoyltransferase (ATCase), dihydroorotase (DHOase), dihydroorotate dehydrogenase (DHODHase), orotate phosphoribosyltransferase (OPRTase), and orotidine-5'-phosphate (OMP) decarboxylase/OMP decarboxylase (ODCase) (Figure 41.8) (Hermansen et al., 2015; Kafer et al., 2004; Löffler & Zameitat, 2004, 2013; Rider, 2009; Schröder et al., 2005; Witz et al., 2012). Prokaryotes possess all these enzymes individually encoded and located in chloroplasts (Rider, 2009; Kafer et al., 2004; Löffler & Zameitat, 2004, 2013). In animals, eukaryotes, during evolution, the first three enzymes and glutaminase have been fused together to form a single polypeptide, termed CAD (an acronym of CPSase-ATCase-DHOase) which is located in the cytosol and to small measure in the nucleus (Figure 41.8) (Löffler & Zameitat, 2004, 2013; Grande-García et al., 2014; Kafer et al., 2004; Rider, 2009). Also it was found that in many organisms, the last two reactions are performed by a single polypeptide, OPRTase and ODCase are fused on one bifunctional polypeptide, UMP synthase (uridine 5-monophosphate synthase – UMPS), which is operating in the cytosol (Figure 41.8) (Löffler & Zameitat, 2004, 2013; Grande-García et al., 2014; Rider, 2009; Rider et al., 2009, Schröder et al., 2005; Kafer et al., 2004; Witz et al., 2012).

In the pyrimidine ring, C2 comes from HCO_3^- (CO_2), N3 comes from glutamine, and the remainder of the pyrimidine molecule (C4-C5-C6-N1) comes from a molecule of aspartate. *Step 1* of pyrimidine de novo biosynthesis required two adenosine triphosphate (ATP), and involves formation of carbamoyl phosphate (Figure 41.8) (Löffler & Zameitat, 2004, 2013; Grande-García et al., 2014; Kafer et al., 2004; Rider, L. W. 2009, Schröder et al., 2005; Witz et al., 2012). This step is catalyzed by cytosolic enzyme, carbamoyl phosphate synthetase II (CPS-II), which is different from carbamoylphosphate synthetase I (CPS I) found in mitochondria and used for the urea cycle in mammalian liver, thus CPS II is the first step committed to pyrimidine metabolism in higher animals, PRPP and ATP activate the enzyme and UDP (uridine-5-diphosphate), and UTP (uridine-5-triphosphate) are allosteric inhibitors of its activity (Hermansen et al., 2015; Löffler & Zameitat, 2004, 2013; Grande-García et al., 2014; Rider, 2009; Kafer et al., 2004; Schröder et al., 2005; Witz et al., 2012). *Step 2* leads to the formation of N-carbamoyl aspartate (catalyzed by ATCase), and is the committed step for pyrimidine synthesis in prokaryotes. In *step 3,* the pyrimidine ring is cyclized by DHOase. In *step 4,* from dihydroorotate is formatting the orotate, process catalyzed by DHODHase. *In step 5,* OPRTase catalyze the reaction between orotate and PRPP to form orotidine-5-phosphate (OMP). In *step 6,* catalyzed by OMP decarboxylase, OMP is decarboxylated to form uridine-5′-monophosphate (UMP) (Figure 41.8) (Hermansen et al., 2015; Löffler & Zameitat, 2004, 2013; Grande-García et al., 2014; Rider, 2009; Kafer et al., 2004; Schröder et al., 2005; Witz et al., 2012).

The end product of the pathway, UMP, is further phosphorylated to uridine-5-diphosphate (UDP) and uridine-5-triphosphate (UTP) that can be converted to cytidine triphosphate (CTP) (by CTP Synthetase), and ribonucleotide reductase converts CTP into dCTP (deoxycytidine triphosphate). UMP is also converted into dUMP (deoxyuridine-5-monophosphate) by ribonucleotide reductase and then into dTMP (deoxythymidine-5-monophosphate) by thymidylate synthase. Thus, the canonical pyrimidine bases of DNA and RNA were obtained (Figure 41.8) (Hermansen et al., 2015; Löffler & Zameitat, 2004, 2013; Grande-García et al., 2014; Rider, 2009; Kafer et al., 2004).

Pyrimidine salvage and recycling are energetically expensive, and cells have developed the strategy to continually reuse the preformed nucleotides (Figure 41.9) (Kafer et al., 2004). It is used only when cells are undergoing rapid growth or are required large amounts of new nucleotides (Kafer et al., 2004). For uracil but not for cytosine or thymine, the first type of salvage involves attachment of the base to PRPP with the formation of pyrophosphate (Kafer et al., 2004). The other pathway (for most of pyrimidines) requires the presence of specific kinases to convert the nucleoside to the monophosphate and involves attachment of the base to ribose 1-phosphate (Kafer et al., 2004, Hermansen et al., 2015; Löffler & Zameitat, 2004, 2013; Grande-García et al., 2014; Rider, 2009).

Pyrimidine catabolism in plants has not been well characterized, but it appears to function as a reverse of the de novo biosynthetic pathway (Rider, 2009; Kafer et al., 2004). For pyrimidine catabolism in other organisms three pathways are known, two of these more investigated and spread (Rider, 2009; Rider et al., 2009; Kafer et al., 2004). Pyrimidines are ultimately degraded to CO_2, H_2O, and urea (Rider, 2009; Kafer et al., 2004). The reductive pathway is found in most organisms (Rider, 2009; Kafer et al., 2004). The pyrimidine nucleosides are dephosphorylated, nucleosides are cleaved in ribose 1-phosphate and free pyrimidine bases (Rider, 2009; Kafer et al., 2004). Uracil or thymine undergoes reduction, uracil is broken down

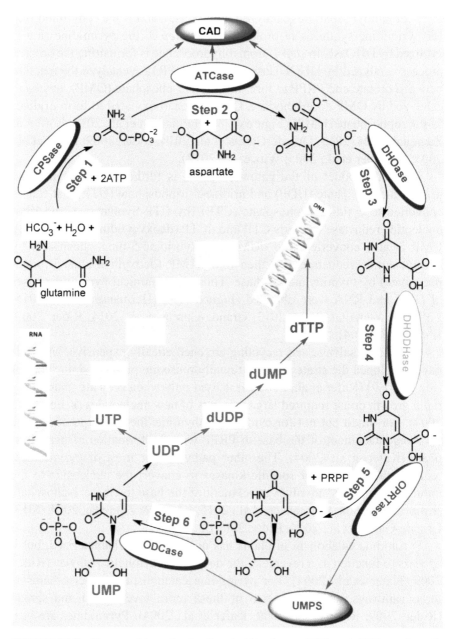

FIGURE 41.8   De novo biosynthesis of pyrimidine bases, DNA and RNA – schematic representation.

to N-carbamoyl-β-alanine which is converted to CO_2, NH_4^+, and β-alanine, while thymine is broken down into CO_2, NH_4^+ and β-aminoisobutyrate, which can be further broken down into intermediates eventually leading into the citric acid cycle (Figure 41.9) (Rider, 2009; Rider et al., 2009; Kafer et al., 2004; Schackerz et al., 2006). The products from the degradation of pyrimidine are excreted into urine or converted into urea (CO_2, NH_4^+, and H_2O) (Rider, 2009; Kafer et al., 2004; Schackerz et al., 2006). In some bacteria was found the oxidative pathway of pyrimidine catabolism: pyrimidines are oxidized to barbituric acid which is then broken down into ureidomalonic acid, and finally into urea and malonic acid (Rider, 2009; Kafer et al., 2004; Loh et al., 2006).

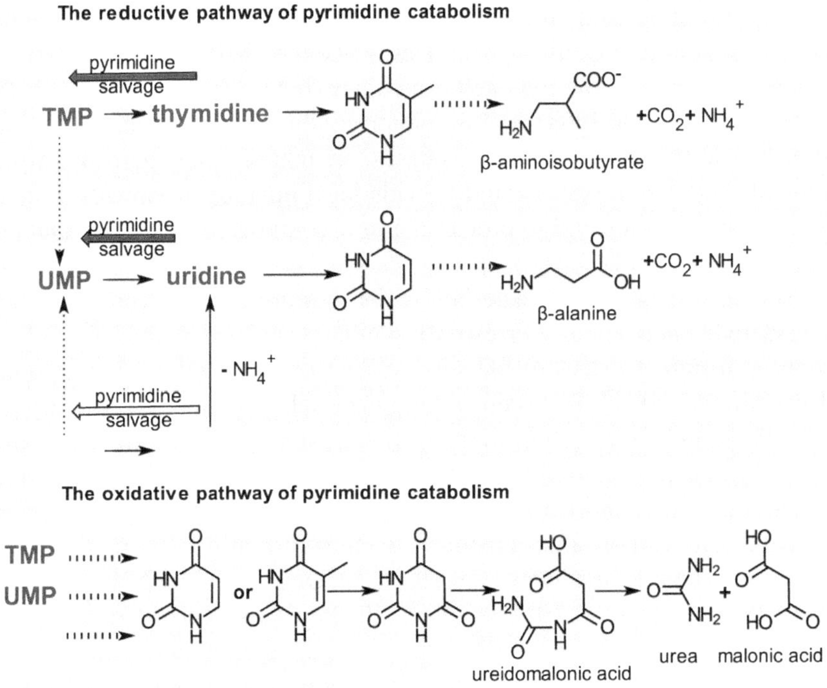

FIGURE 41.9 Pyrimidines catabolism – schematic representation.

Plants produce more than 100,000 secondary metabolites, most of these compounds serve a defensive function to combat bacteria, fungi, insects, or even other plants (Kafer et al., 2004). From secondary metabolic products derived from pyrimidine nucleotides, we can mention lathyrine, willardiine (non-protein amino acids that function in a neuroexcitatory pathway), vicine and convicine (Figures 41.1 and 41.2) (Kafer et al., 2004).

41.4.4 *SYNTHESIS OF PYRIMIDINE DERIVATIVES IN PREBIOTIC EARTH, EXTRATERRESTRIAL CONDITIONS*

Syntheses of nucleobases (nucleosides) in the prebiotic Earth or in extraterrestrial environments have been intensively investigated. The extraterrestrial origin of nucleobases from Murchison has been confirmed (Nuevo, 2012; Botta, 2008) and meteoritic pyrimidines are usually detected in smaller abundances than purines, uracil being found in water extracts of Murchison, Murray, and Orgueil meteorites (Nuevo, 2012). In astrophysical environments pyrimidines and purines should be present, may be condensed on the surface of cold grains, mixed with ices, since they have not been observed in the gas phase, in the interstellar medium (ISM) (Sandford et al., 2014; Nuevo, 2012; Botta, 2008). The formation conditions of nucleobases can be: in drying/wetting cycles, eutectic phases, ice matrices, and aerosols, while minerals and metal oxides could provide local microenvironments (able to concentrate reagents and to protect the new compounds from degradation) (Ruiz-Mirazo, 2014).

Sources of precursor materials under primitive Earth conditions are considered: endogenous (terrestrial) synthesis and exogenous formation, either at different places of the solar system or in the interstellar medium (Ruiz-Mirazo, 2014). The main starting materials in prebiotic chemistry are one-, two-, and three-carbon atom molecules, simple organic molecules that can be synthesized from CO and CO_2, CH_4, NH_3, N_2, H_2O by spark discharge, UV irradiation, or shock waves (Ruiz-Mirazo, 2014; Saladino, 2013a). The plausible one-carbon atom chemical precursors that can be produced by purely geochemical pathways are HCN, HCOH, HCOOH, NH_2CHO, and $HCOONH_4$ (Figure 41.10) (Saladino, 2013a, 2016; Ruiz-Mirazo, 2014), also as starting materials can be cyanate, cyanogen, ammonium cyanide, urea, acetaldehyde, cyanoacetylene, and cyanoacetaldehyde (Ruiz-Mirazo, 2014; Saladino, 2013a).

The start of the prebiotic chemistry of nucleobases is considered to be HCN (Sponer, 2016). In the last decades, many experimental information has been accumulated about this molecule and the conversion from HCN to nucleotides is well characterized (Sponer, 2016). Nucleobases have been synthesized from simple precursor molecules, like HCN, urea or cyanoacetylene (Sponer, 2016). Formamide (NH_2CHO), the hydrolysis product of HCN, an alternative to HCN, or a possible nonvolatile source of HCN is a thermodynamically more stable and less reactive variant, that it could accumulate on the surface of the early Earth (probably as the thermal dissociation product of ammonium-formate) (Figure 41.10) (Saladino, 2012b, 2013a,b). Formamide also have been observed in the comets, in the stellar objects, in the interstellar medium, in carbonaceous chondrites, as a common constituent of star-forming regions that foster planetary systems within the galactic habitable zone (Saladino, 2013a, 2015, 2016).

All four nucleobases and their derivatives have been synthesized starting from formamide in the presence of meteoritic materials or other catalysts, utilizing different energy source (Figure 41.10) (Saladino, 2013a,b, 2015, 2016; Sponer, 2016). It was tested the catalytic effect of several types of meteorites, iron, stony-iron, chondrites, and achondrites, e.g., calcium carbonate, silica, alumina, TiO_2, montmorillonite, olivines, iron sulfides, zirconia, or iron oxides (Figure 41.10) (Saladino, 2013a, b; Ruiz-Mirazo, 2014). Montmorillonite, a phyllosilicate that most likely formed large deposits on the early Earth, this clay mineral is able to catalyze the oligomerization of activated nucleotides as well as the spontaneous conversion of fatty acid micelles into vesicles (Ruiz-Mirazo, 2014).

The high-density-energy events, as well as extraterrestrial chemical processes, have been studied as a possible source of energy in prebiotic synthesis (Sponer, 2016). For the synthesis of nucleobases and their derivatives from formamide different energy sources were tested: thermal energy, [11]B-boron as energy source, ultra-high energy laser experiments were made, or the reaction was initiated by proton irradiation or irradiated with UV light (Ruiz-Mirazo, 2014; Saladino, 2013a, 2015, 2016; Sponer, 2016; Ferus, 2014, 2015).

From this experimental work one or more pyrimidinic derivatives were obtained: uracil, cytosine, is cytosine, thymine, 5,6-dihydrouracil, orotic acid, 4(3H)-pyrimidinone, 2(1H)-pyrimidinone, purine, adenine, guanine, hypoxanthine, 2-aminopurine (Figure 41.10) and the yield of pyrimidine nucleobases increased by increasing the amount of iron in the catalyst (Saladino, 2012, 2013a,b).

Laboratory and theoretical work were done on the ice-phase photochemistry of pyrimidine, and led to the synthesis in extraterrestrial ice of all three pyrimidine-nucleobases (uracil, cytosine, thymine, along with a number of their isomers) (Sandford et al., 2014). Theoretical work pointed out that the presence of condensed phase H_2O in the ice matrices plays a key role in the formation of the nucleobases (this processes requires various combinations of oxidation, amination, and methylation) (Sandford et al., 2014). The experimental work showed the role of water in the formation of relevant prebiotic molecules in a formamide/water environment, a one-pot synthesis, of all the natural nucleobases, of amino acids and carboxylic acids (Rotelli, 2016).

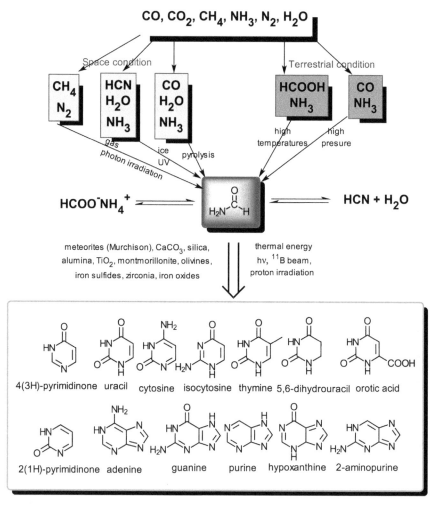

FIGURE 41.10 Schematic representation of the synthesis of pyrimidine derivatives in prebiotic condition.

41.5 OPEN ISSUES

Pyrimidinic compounds may be the answer to the questions: where we came from? When/how will be the end of Earth? That's why the studies for the prebiotic synthesis and degradation of pyrimidine nucleobases should continue and be more intensive.

Pyrimidinic derivatives have been and will be intensely researched, especially for their biological activity and as DNA and RNA nucleobases. For this reason, it is necessary to improve and develop the methods of extraction, identification, and testing of the biological activity of pyrimidinic compounds.

It is also necessary to develop new methods of synthesis for pyrimidine compounds, more efficient, faster, with a yield as high as possible (one-pot synthesis, under microwave/ultrasonic condition, etc.). Considering the fact that in the case of optically active compounds, only one enantiomer shows the desired biological activity, it is absolutely necessary to develop the enantiospecific synthetic methods for pyrimidine derivatives.

To speed up the identification in a short time as possible of new (pyrimidine) derivatives with biological activity is necessary to develop new methods of databases screening to identify leader compounds with a structure as close as possible to "the future drug: that may be used in medicine as a treatment. For the same purpose, QSAR studies are real help for researchers, so it is absolutely necessary to develop the QSAR area, where the techniques and descriptors to be used should be various/targeted, and a double/triple variational vision of the entire ligand-receptor interactions (as (Q)-SAR-docking studies made by Putz & Dudaș, et al. (2015) and Putz et al., 2015) can speed up the identification of the future leaders compounds.

KEYWORDS

- **natural sources**
- **pyrimidines**
- **QSAR**

REFERENCES AND FURTHER READING

Baei, M. T., Taghartapeh, M. R., Lemeski, E. T., & Soltani, A., (2014). A computational study of adenine, uracil, and cytosine adsorption upon ALN and BN nano-cages. *Physica B: Condensed Matter, 444*, 6–13.

Baeyer, A., (1864). Untersuchungen über die Harnsäuregruppe. *Annalen der Chemie und Pharmacie, 131*(3), 291–295.

Barreswil, (1862). Vorkommen des Guanin's van den Schuppen des Weissfisches. *Liebig's Ann., 122,* 128–129.

Bell, E. A., (1961). Isolation of a new amino acid from *Lathyrus tingitanus*. *Biochim. Biophys. Acta, 47,* 602–603.

Bell, E. A., & Foster, R. G., (1962). Structure of Lathyrine. *Nature, 194,* 91–92.

Bergmann, W., & Burke, D. C., (1955). Marine products. XXXIX. The nucleosides of sponges. III. Spongothymidine and spongouridine. *J. Org. Chem., 20,* 1501–1507.

Bhakuni, D. S., & Rawat, D. S., (2005). *Bioactive Marine Natural Products* (pp. 208–234). Springer: New York with Anamaya Publishers: New Delhi, India.

Brik, A., & Wong, C. H., (2003). HIV–1 protease: Mechanism and drug discovery. *Org. Biomol. Chem., 1*(1), 5–14.

Brown, D. J., (2009). *The Chemistry of Heterocyclic Compounds* (1ˢᵗ edn.). The Pyrimidines. John Wiley and Sons, New York, USA.

Brown, D. J., Evans, R. F., Cowden, W. B., & Fenn, M. D., (1994). The pyrimidines. In: *Chemistry of Heterocyclic Compounds* (Vol. 52, pp. 96–106). Taylor EC. Wiley, New York, USA.

Brown, E. G., & Roberts, F. M., (1972). Formation of vicine and convicine by Vicia faba. *Phytochemistry, 11,* 3203–3206.

Bryant, J. A., & Hughes, S. G., (2011). In: Kole, C., (eds.), *Vicia in Wild Crop Relatives: Genomic and Breeding Resources. Legume Crops and Forages* (pp. 273–289). Springer, Berlin, Germany.

Callahan, M. P., Smith, K. E., Cleaves, H. J. II, Ruzicka, J., Stern, J. C., Glavin, D. P., et al., (2011). Carbonaceous meteorites contain a wide range of extraterrestrial nucleobases. *Proc. Natl. Acad. Sci., 108,* 13995–13998.

Cronin, J. R., & Chang, S., (1993). Organic matter in meteorites: Molecular and isotopic analyses of the Murchison meteorite. In: *The Chemistry of Life's Origins* (pp. 209–258). Springer Netherlands.

D'Archivio, A. A., Maggi, M. A., & Ruggieri, F., (2015). Artificial neural network prediction of multilinear gradient retention in reversed-phase HPLC: Comprehensive QSRR-based models combining categorical or structural solute descriptors and gradient profile parameters. *Anal. Bioanal. Chem., 407*(4), 1181–1190. doi: 10.1007/s00216-014-8317-3.

Daintith, J., (2008). *A Dictionary of Chemistry* (6 edn.), Oxford University Press, UK.

De Clercq, E., (2008). Antivirals: Current state of the art. *Future Virology, 3*(4), 393–405.

De Clercq, E., (2009a). Anti-HIV drugs: 25 compounds approved within 25 years after the discovery of HIV. *Int. J. Antimicrob. Ag., 33,* 307–320.

De Clercq, E., (2009b). The history of antiretrovirals: Key discoveries over the past 25 years. *Revi. Med. Virol., 19,* 287–299.

De Clercq, E., (2010). Antiretroviral drugs. *Curr. Opin. Pharmacol., 10*(5), 507–515.

De Clercq, E., (2011). A 40-year journey in search of selective antiviral chemotherapy. *Annu. Rev. Pharmacol. Toxicol., 51,* 1–24.

Dudaş, N. A., & Putz, M. V., (2014). Pyrimidine derivates with biological activity in anti-HIV therapy. The Spectral-Diagonal-SAR approach. *Int. J. Chem. Model., 6,* 95–114.

Dudhe, R., Sharma, P. K., Verma, P., & Chaudhary, A., (2011). Pyrimidine as anticancer agent: A review. *J. Adv. Sci. Res., 2*(3), 10–17.

Evans, D. R., & Guy, H. I., (2004). Mammalian pyrimidine biosynthesis: Fresh insights into an ancient pathway. *J. Biol. Chem., 279*, 33035–33038.

Fan, H., & Zhu, X., (2015). Development of small-molecule materials for high-performance organic solar cells. *Science China Chemistry, 58*(6), 922–936.

FDA (Food and Drug Administration), "Glossary of pesticide chemicals," (US), 2005. URL: http://www.fda.gov/Food/FoodborneIllnessContaminants/Pesticides/ucm113891.htm (accessed on October 2013).

Ferus, M., Michalčíková, R., Shestivská, V., Sponer, J., Sponer, J. E., & Civiš, S., (2014). High-energy chemistry of formamide: A simpler way for nucleobase formation. *J. Phys. Chem. A., 118*(4), 719–736.

Ferus, M., Nesvorný, D., Sponer, J., Kubelík, P., Michalčíková, R., Shestivská, V., Sponer, J. E., & Civiš, S., (2015). High-energy chemistry of formamide: A unified mechanism of nucleobase formation. *Proc. Natl. Acad. Sci. U. S. A., 112*, 657–662.

Finger, S., Boller, F., & Tyler, K. L., (2009). Historical aspects of the major neurological vitamin deficiency disorders: The water-soluble B vitamins. *History of Neurology, 95*, 445–476.

Fischer, E., (1897). Ueber die Constitu Caffeine Caffeins, XHypoxanthinsoxanthins und verwandter. *Basen. Chem. Ber., 30*, 549–573.

Fischer, E., & Von Mering, J., (1903). Ueber eine neue Klasse von Schlafmitteln, *Therap. Gegenw., 44*, 97–101.

Folsome, C. E., Lawless, J., Romiez, M., & Ponnamperuma, C., (1971). Heterocyclic compounds indigenous to the Murchison meteorite. *Nature, 232*, 108–109.

Gabriel, S., & Colman, J., (1899). Ueber das Pyrimidin. Ber. Dtsch. Chem. Ges., 32, 1525–1538.

Gerwick, W. H., & Moore, B. S., (2012). Lessons from the past and charting the future of marine natural products drug discovery and chemical biology. *Chem. Biol., 19*(1), 85–98.

Gmelin, R., (1959). The free amino acids in the seeds of *Acacia willardiana* (Mimosaceae). Isolation of willardiin, a new plant amino acid which is probably L-beta-(3-uracil)-alpha-aminopropionic acid. *Hoppe-Seyler's Z. Physiol. Chem., 316*, 164–169.

Grande-García, A., Lallous, N., Díaz-Tejada, C., & Ramón-Maiques, S., (2014). *Structure, Functional Characterization, and Evolution of the Dihydroorotase Domain of Human CAD Structure, 22*(2), 185–198.

Grimaux, E., (1879). Synthèse des dérivés uriques de la série de l'alloxane. *Bull. Soc. Chim. Paris, [2], 31*, 146–149.

Groziak, M. P., (1997). Diazines and benzo derivatives. *Prog. Heterocycl. Chem., 9*, 249–284.

Groziak, M. P., (1998). Diazines and benzo derivatives. *Prog. Heterocycl. Chem., 10*, 251–293.

Groziak, M. P., (1999). Diazines and benzo derivatives. *Prog. Heterocycl. Chem., 11*, 25–86.

Harris, R. S., & Thimann, K. V., (1949). *Vitamins and Hormones* (Vol. VII, p. 368). Academic Press INC., New York, USA.

Hassan, H. H. A. M., Mansour, E. M. E., Abou, Z. A. M. S., El-Helow, E. R., Elhusseiny, A. F., & Soliman, R., (2015). Synthesis and biological evaluation of new nanosized aromatic polyamides containing amido- and sulfonamidopyrimidines pendant structures. *Chemistry Central Journal, 9*, 44.

Hayes, J. M., (1967). Organic constituents of meteorites–a review. *Geochim. Cosmochim. Acta, 31*, 1395–1440.

Heinisch, G., & Matuszczak, B., (1994). Diazines nd benzo. de. rivatives *Prog. Heterocycl. Chem., 6*, 231–275.

Heinisch, G., & Matuszczak, B., (1995). Diazines and benzo. derivatives. *Prog. Heterocycl. Chem., 7*, 226–257.

Herbicide classification by mode of action, (2003). http://weedscience.org/summary/home.aspx (May 2019).

Hermansen, R. A., Mannakee, B. K., Knecht, W., Liberles, D. A., & Gutenkunst, R. N., (2015). Characterizing selective pressures on the pathway for de novo biosynthesis of pyrimidines in yeast. *BMC Evolutionary Biology, 15*(1), 232.

Hoffbrand, A. V., & Weir, D. G., (2001). The history of folic acid. *Br. J. Haematol., 113*(3), 579–589.

Hoffmann, C., & Rockstroh, J. K., (2011). *HIV 2011, Medizin Fokus Verlag*, Hamburg, Germany.

Hoffmann, C., & Rockstroh, J. K., (2012). *HIV 2012/2013, Medizin Fokus Verlag*, Hamburg, Germany.

Hoffmann, M. G., Nowak, A., & Muller, M., (1998). In: Schaumann, E., (ed.), *Pyrimidines in Methods of Organic Chemistry* (pp. 1–249). George Thiem Verlag, Stuttgart, Germany.

Iman, M., & Davood, A., (2013). QSAR and QSTR study of pyrimidine derivatives to improve their therapeutic index as antileishmanial agents. *Med. Chem. Res., 22*, 5029–5035.

Ingale, K. B., & Bhatia, M. S., (2011). HIV–1 integrase inhibitors: A review of their chemical development. *Antivir. Chem. Chemother., 22*(3), 95–105.

Institute of Medicine (US) Standing Committee on the Scientific Evaluation of Dietary Reference Intakes, (1998). *Dietary Reference Intakes for Thiamin, Riboflavin, Niacin, Vitamin B6, Folate, Vitamin B12, Pantothenic Acid, Biotin, and Choline* (pp. 87–122). National Academies Press (US).

Irvine, W. M., (1998). Extraterrestrial organic matter: A review. *Orig. Life Evol. Biosph., 28*, 365–383.

Jain, K. S., Chitre, T. S., Miniyar, P. B., Kathiravan, M. K., Bendre, V. S., Veer, V. S., Shahane, S. R., & Shishoo, C. J., (2006). Biological and medicinal significance of pyrimidines. *Curr. Sci., 90*(6), 793–803.

Joule, J. A., & Mills, K., (2000). *Heterocyclic Chemistry* (4th ed.). Blackwell Science Ltd.: Cambridge, UK.

Joule, J. A., & Mills, K., (2010). *Heterocyclic Chemistry* (5th edn.). Wiley-Blackwell, Oxford, UK.

Kafer, C., Zhou, L., Santoso, D., Guirgis, A., Weers, B., Park, S., & Thornburg, R., (2004). Regulation of pyrimidine metabolism in plants. *Front. Biosci., 9*, 1611–1625.

Kamori, T., Sanechika, Y., Ito, Y., Matsuo, J., Nohara, T., Kawasaki, T., & Schulten, H. R., (1980). Biologically active glycosides from Asteroidea. I. Structures of a new cerebroside mixture and of two nucleosides from starfish *Acanthaster planci. Leibigs Ann. Chem.*, 653–668.

Kondo, K., Shigemori, H., Ishibashi, M., & Kobayashi, J., (1992). Aplysidine, a new nucleoside from the Okinawan marine sponge *Aplysina* sp. *Tetrahedron, 48*, 7145–7148.

Kossel, A., & Steudel, H. Z., (1903). Weitere Untersuchungen über das Cytosin. *Physiol. Chem., 38*, 49–53.

Kress, T. J., (1989). Diazines and benzo derivatives. *Prog. Heterocycl. Chem., 1*, 243–262.

Kress, T. J., & Varie, D. L., (1990). Diazines and benzo derivatives. *Prog. Heterocycl. Chem., 2*, 185–223.

Kress, T. J., & Varie, D. L., (1991). Diazines and benzo derivatives. *Prog. Heterocycl. Chem., 3*, 205–242.

Kress, T. J., & Varie, D. L., (1992). Diazines and benzo derivatives. *Prog. Heterocycl. Chem., 4*, 186–225.

Lackie, J., (2010). *A Dictionary of Biomedicine*. Oxford University Press, UK.

Lagoja, I. M., (2005). Pyrimaidine as constituent of natural biologically active compounds. *Chem. Biodivers., 2*, 1–50.

Lane, A. N., & Fan, T. W. M., (2015). Regulation of mammalian nucleotide metabolism and biosynthesis. *Nucleic Acids Res., 43*(4), 2466–2485. doi: 10.1093/nar/gkv047.

Lauria, A., Tutone, M., & Almerico, A. M., (2010). Design of new DNA-interactive agents by molecular docking and QSPR approach. *Arkivoc, 11*(1), 13–27.

Löffler, M., & Zameitat, E., (2004). Pyrimidine biosynthesis. In: Lennarz, W. J., & Lan, M. D., (eds.), *Encyclopedia of Biological Chemistry* (pp. 600–605). New York: Elsevier.

Löffler, M., & Zameitat, E., (2013). Pyrimidine biosynthesis and degradation (Catabolism). In: *Encyclopedia of Biological Chemistry* (pp. 712–718). New York: Elsevier.

Loh, D. K., Gyaneshwar, P., Papadimitriou, E. M., Fong, R., Kim, K., Parales, R., Zhou, Z., Inwood, W., & Kustu, S., (2006). A previously undescribed pathway for pyrimidine catabolism. *Proc. Nat. Acad. Sci., 103*, 5114–5119.

Mager, J., Chevion, M., & Glaser, G. F., (1980). In: Lie, Ner, I. E., (eds.), *Toxic Constituents of Plant Foodstuffs* (2nd edn., pp. 265). Academic Press, NUSA, RK, USA.

Martins, Z., Botta, O., Fogel, M. L., Sephton, M. A., Glavin, D. P., Watson, J. S., Dworkin, J. P., Schwartz, A. W., & Ehrenfreund, P., (2008). Extraterrestrial nucleobases in the Murchison meteorite. *Earth Planet. Sci. Lett., 270*, 130–136.

Mati, S. S., Chall, S., Konar, S., Rakshit, S., & Bhattacharya, S. C., (2014). Pyrimidine-based fluorescent zinc sensor: Photo-physical characteristics, quantum chemical inter, relation and application in real samples. *Sensors Actuators B., 201*, 204–212.

McDowell, L. R., (2000). *Vitamins in Animal and Human Nutrition* (2nd edn., pp. 265–346, 479–522, 670). Iowa State University Press, USA.

Mehellou, Y., & De Clercq, E., (2010). Twenty-six years of anti-HIV drug discovery: Where do we stand and where do we go? *J. Med. Chem., 53*(2), 521–538.

Montaser, R., & Luesch, H., (2011). Marine natural products: A new wave of drugs? *Future Med. Chem., 3*(12), 1475–1489.

Nagarwal, R. C., Kumar, R., Dhanawat, M., & Pandit, J. K., (2011). Modified PLA nano in situ gel: A potential ophthalmic drug delivery system. *Colloids Surf B Biointerfaces, 86*(1), 28–34. doi: 10.1016/j.colsurfb.2011.03.023.

Nguyen, B. Y., Isaacs, R., Teppler, H., Leavitt, R. Y., Sklar, P., Iwamoto, M., et al., (2011). Raltegravir: The first HIV–1 integrase strand transfer inhibitor in the HIV armamentarium. *Ann. N. Y. Acad. Sci., 1222*(1), 83–89.

Nuevo, M., Milam, S. N., & Sandford, S. A., (2012). Nucleobases and prebiotic molecules in organic residues produced from the ultraviolet photo-irradiation of pyrimidine in NH$_3$ and H$_2$O+NH$_3$ ices. *Astrobiology, 12*(4), 295–314.

Olayinka, O., Ajani, J. T. Isaac, T. F. O., & Anuoluwa, A. A., (2015). Exploration of the chemistry and biological properties of pyrimidine as a privilege pharmacophore in therapeutics. *Int. J. Biol. Chem., 9*, 148–177.

Panneer, S. T., James, C. R., Phadte, V. D., & Valzita, S. K., (2012). Mini-review Ni review of pyrimidine and fused pyrimidine marketed drugs. *Res. Pharm., 2*(4), 1–9.

Panneer, S. T., Phadte, V. D., & Valzita S. K., (2013). *An Efficient Pyrimidine Scaffolds against Bacterial Strains LAP Lambert*. Academic Publishing AG & Co., Germany.

Putz M. V., Dudaş N. A., & Isvoran A., (2015). Double variational binding-(SMILES) conformational analysis by docking mechanisms for anti-HIV pyrimidine ligands. *Int. J. Mol. Sci., 16*(8), 19553–19601. doi: 10.3390/ijms160819553.

Putz, M. V., & Dudaş, N. A., (2013a). Variational principles for mechanistic quanstructure-activity–activity relationship (QSAR) studies: Application on uracil derivatives' anti-HIV action. *Struct. Chem., 24,* 1873–1893.

Putz, M. V., & Dudaş, N. A., (2013b). Determining chemical reactivity driving biological activity from SMILES transformations: The bonding mechanism of anti-HIV pyrimidines. *Molecules, 18,* 9061–9116.

Putz, M. V., Duda-Seiman, C., Duda-Seiman, D., Putz, A. M., Alexandrescu, I., Mernea, M., & Avram S., (2016). Chemical structure-biological activity models for pharmacophores' 3D-interactions *Int. J. Mol. Sci., 17,* 1087, doi: 10.3390/ijms17071087.

Raman, A., & Lau, C., (1996). Anti-diabetic properties and phytochemistry of *Momordica charantia* L. (Cucurbitaceae). Phytomedi): e, *2*(4), 349–362.

Ravichandran, S., Ravichandran, V., Raman, S., Krishnan, P. N., & Agrawal, R. K., (2008). An overview on HIV–1 reverse transcriptase inhibitors. *Digest J. Nanomat. Biostruct., 3*(4), 171–187.

Rider, L. W., (2009). Flavoenzymes involved in pyrimidine oxidation and reduction, *Pro. Quest LLS, Ann. Arbor., MI,* pp. 1–39.

Rider, L. W., Ottosen, M. B., Gattis, S. G., & Palfey, B. A., (2009). Mechanism of dihydrouridine synthase 2 from yeast and the importance of modifications for efficient tRNA reduction. *The Journal of Biological Chemistry, 284*(16), 10324–10333. http://doi.org/10.1074/jbc.M806137200.

Ritthausen, H., (1870). Kreusler ueber vorkommen von Amygdalin und eine neue dem Asparaginähnliche Substanz in Wickensamen (Vicia sativa). *J. Prakt. Chem., 2*(1), 333–338.

Ritthausen, H., (1876). Ueber Vicin. Bestandtheil der Samen von Vicia sativa. *Ber. Dtsch. Chem. Ges., 9*(1), 301–304.

Ritthausen, H., (1881). Ueber Vicin und eine zweite stickstoffreiche Substanz der Wickensamen, Convicin. *J. Prakt. Chem., 24,* 202–220.

Rotelli, L., Trigo-Rodríguez, J. M., Moyano-Cambero, C. E., Carota, E., Botta, L., Di Mauro, E., & Saladino, R., (2016). The key role of meteorites in the formation of relevant prebiotic molecules in a formamide/water environment. *Scientific Reports, 6,* 38888. doi: 10.1038/srep38888.

Ruiz-Mirazo, K., C. B., & De la Escosura, A., (2014). Prebiotic systems chemistry: New perspectives for the origins of life. *Chem. Rev., 114*(1), 285–366.

Sadeghi, B., & Zarepour, I., (2015). Nano-sawdust–BF3 as a new, cheap, and nanocatalyst catalyst for one-pot synthesis of pyrano[2, 3-d]pyrimidines, *J. Iran. Chem. Soc., 12,* 1801–1808.

Sahlberg, C., & Zhou, X. X., (2008). Development of non-nucleoside reverse transcriptase inhibitors for anti-HIV therapy. *Anti-Infect. Agents Med. Chem., 7,* 101–117.

Saladino, R., Botta, G., Bizzarri, B. M., Di Mauro, E., & Garcia, R. J. M., (2016). A global scale scenario for prebiotic chemistry: Silica-based self-assembled mineral structures and formamide. *Biochemistry, 55*(19), 2806–2811.

Saladino, R., Botta, G., Delfino, M., & Di, Mauro, E., (2013a). Meteorites as catalysts for prebiotic chemistry. *Chem. Eur. J., 19*(50), 16916–16922.

Saladino, R., Botta, G., Pino, S., Costanzo, G., & Di Mauro, E., (2012b). Genetics first or metabolism first? The formamide clue. *Chem. Soc. Rev., 41*(16), 5526–5565.

Saladino, R., Botta, G., Pino, S., Costanzo, G., & Di Mauro, E., (2013b). Materials for the onset. A story of necessity and chance. *Front. Biosci. (Landmark Ed), 18,* 1275–1289.

Saladino, R., Carota, E., Botta, G., Kapralov, M., Timoshenko, G. N., Rozanov, A. Y., et al., (2015). Meteorite-catalyzed syntheses of nucleosides and of other prebiotic compounds from formamide under proton irradiation. *Proc. Natl. Acad. Sci. U.S.A., 112*, E2746–E2755.

Saladino, R., Crestini, C., Pino, S., Costanzo, G., & Di Mauro, E., (2012). Formamide and the origin of life. *Phys Life Rev., 9*(1), 84–104. doi: 10.1016/j. plrev. 2011.12.002.

Sandford, S. A., Bera, P. P., Lee, T. J., Materese, C., & Nuevo, M., (2014). Photosynthesis and photostability of nucleic acids in prebiotic extraterrestrial environments. In: Barbatti, M., Borin, A. C., & Ullrich, S., (eds.), *Photoinduced Phenomena in Nucleic Acids* II (Vol. 356, pp. 123–164). Springer International Publishing, *Top. Curr. Chem.*

Sarfaraj, H. M., Sheeba, F., Saba, A., & Mohd, S. K., (2012). Marine natural products: A lead for anti-cancer. *Indian J. Mar. Sci., 41*, 27–39.

Schackerz, K. D., & Dobritzch, D., (2006). Amidohydrolases of the *R* pyrimidine catabolic pathway purification, characterization, structure, reaction. *Biochem. Biophys. Acta, 67*, 123–130.

Schröder, M., Giermann, N., & Zrenner, R., (2005). Functional analysis of the pyrimidine de novo synthesis pathway in solanaceous species. *Plant Physiol., 138*, 1926–1938.

Sponer, J. E., Szabla, R., Gora, R. W., Saitta, A. M., Pietrucci, F., Saija, F., et al., (2016). Prebiotic synthesis of nucleic acids and their building blocks at the atomic level-merging models and mechanisms from advanced computations and experiments. *Phys. Chem. Chem. Phys., 18*, 20047–20066.

Stoks, P. G., & Schwartz, A. W., (1979). Uracil in carbonaceous meteorites. *Nature, 282*, 709–710.

Stoks, P. G., & Schwartz, A. W., (1981). Nitrogen-heterocyclic compounds in meteorites–significance and mechanisms of formation. *Geochim. Cosmochim. Acta, 45*, 563–569.

Suckling, C. J., (1991). Chemical approaches to the discovery of new drugs. *Sci. Prog., 75*, 323–359.

Toropova, A. P., Toropov, A. A., Veselinović, J. B., Miljković, F. N., & Veselinović, A. M., (2014). QSAR models for HEPT derivates as NNRTI inhibitors based on Monte Carlo method. *Eur. J. Med. Chem., 77*, 298–305.

Undheim, K., & Benneche, T., (1996). Pyrimidines and their benzo derivatives. In: Katritzky, A. R., Rees, C. W., Scriven, E. F. V., & McKillop, A., (eds.), *Comprehensive Heterocyclic Chemistry II* (Vol. 6, pp. 93–231). Pergamon Press, Oxford, UK.

Usach, I., Melis, V., & Peris, J. E., (2013). Non-nucleoside reverse transcriptase inhibitors: A review on pharmacokinetics, pharmacodynamics, safety and tolerability. *J. Int. AIDS Soc., 16*(1), 1–14.

Utagawa, T., Morisawa, H., Miyoshi, T., Yoshinaga, F., Yamazaki, A., & Mitsugi, K., (1980). A novel and simple method for the preparation of adenine arabinoside by bacterial transglycosylation reaction. *FEBS Letters, 109*, 261–263.

Vencill, W. K., (2002). *WSSA Herbicide Handbook* (8th edn.). Weed Science Society of America. Lawrence, KS, USA.

Von Angerer, S., (2007). Product class 12: Pyrimidines. In: Yamamoto, Y., (ed.), *Science of Synthesis* (pp. 379–573). Thieme Publishing Stuttgart, Germany.

Vurth, L., Hadad, C., Achelle, S., García-Martínez, J. C., Rodríguez-López, J., & Stéphan, O., (2012). Pluronic F–68 nanodots incorporating pyrimidine chromophores. *Colloid Polym. Sci., 290*, 1353–1359.

Watson, J. D., & Crick, F., (1953). Molecular structure of nucleic acids. *Nature, 171*, 737–738.

Wilhelmus, K. R., (2015). Antiviral treatment and other therapeutic interventions for herpes simplex virus epithelial keratitis. *The Cochrane Database of Systematic Reviews, 1,* CD002898. doi: 10.1002/14651858.CD002898.pub5.

Witz, S., Jungl, B., Fürst, S., & Möhlmann, T., (2012). De novo pyrimidine nucleotide synthesis mainly occurs outside of plastids, but a previously undiscovered nucleobase importer provides substrates for the essential salvage pathway in Arabidopsis. *The Plant Cell, 24*(4), 1549–1559.

Wood, A., (2013). *Compendium of Pesticide Common Names, (US),* URL: http://www.alanwood. net/pesticides/(May 2019).

CHAPTER 42

Quantum Dot

ARANYA B. BHATTACHERJEE[1] and SUMAN DUDEJA[2]

[1]Department of Physics, Birla Institute of Technology and Science, Pilani, Hyderabad Campus, India

[2]Department of Chemistry, ARSD College, University of Delhi, New Delhi–110021, India

42.1 DEFINITION

A quantum dot (QD) is a semiconductor nanostructure whose diameter is in the range of approximately 2–10 nanometers. In QD motion of conduction band electrons, valence band holes or excitons are confined by a potential barrier to small regions of space where dimensions of confinement are less than their de Broglie wavelength. It is also referred to as an artificial atom. This small confinement results in the density of state being discrete and energy levels quantized. QDs can be produced by colloidal and plasma synthesis.

42.2 HISTORICAL ORIGIN(S)

Quantum dots were discovered in solids (glass crystals) in 1980 by Russian physicist Alexei Ekimov while working at the Vavilov State Optical Institute. In late 1982, American chemist Louis E. Brus then working at Bell Laboratories (and now a professor at Columbia University), discovered the same phenomenon in colloidal solutions (where small particles of one substance are dispersed throughout another; milk is a familiar example). He discovered that the wavelength of light emitted or absorbed by a QD changed over a period of days as the crystal grew, and concluded that the confinement of electrons was giving the particle quantum properties. These two scientists shared the Optical Society of America's 2006 R.W. Wood Prize for their pioneering work.

42.3 NANO-SCIENTIFIC DEVELOPMENT(S)

QDs are used in different material for a different purpose. It can be used as organic dyes, in the computer as qubits, etc. Nowadays, we are using double QDs to study electron transportation in a different potential barrier with Rashba spin interaction and Dresselhaus effect. So far, QDs have attracted because of its interesting optical properties: they are being used for all kinds of applications where precise control of light is important. QDs can also be used instead of pigments and dyes. Embedded in other materials, they absorb incoming light of one color and give out the light of an entirely different color; they're brighter and more controllable than organic dyes (artificial dyes made from synthetic chemicals). QDs are being widely appreciated as a breakthrough technology in the development of more efficient solar cells. In a traditional solar cell, photons of sunlight knock electrons out of a semiconductor into a circuit, making useful electric power, but the efficiency of the process is quite low. QDs produce more electrons (or holes) for each photon that strikes them, potentially offering a boost in the efficiency of perhaps 10% over conventional semiconductors. Charge-coupled devices (CCDs), which are the image-detecting chips, work in a similar way to solar cells, by converting incoming light into patterns of electrical signals; efficient QDs could be used to make smaller and more efficient CCDs for applications where conventional devices are too big and clumsy. In electronic application, it can work as a single electron transistor.

42.4 NANO-CHEMICAL APPLICATION(S)

QDs are used in place of organic dyes in biological research; for instance, they may be used like nanoscopic light bulbs to light up and color specific cells which need to be studied under a microscope. They can also be used as sensors for chemical and biological warfare agents such as anthrax. QDs can also be used as photocatalysts for the chemical conversion of water into hydrogen as a pathway to solar fuel.

42.5 MULTI-/TRANS-DISCIPLINARY CONNECTION(S)

Applications of QDs have been found of great importance in the biological as well as the chemical field. In cancer treatment, QDs are designed, so they accumulate in particular parts of the body and then deliver anti-cancer drugs bound to them. Their big advantage is that they can be targeted at single organs, such as the liver, much more precisely than conventional drugs, so reducing the unpleasant side effects that are characteristic of untargeted, traditional chemotherapy. Another application of QDs has been developed so it can be fitted on top of a fluorescent or LED lamp and convert its light from a bluish color to a warmer, redder, more attractive shade similar to the light produced by old-fashioned incandescent lamps. It can be used in optical computers much the same way transistor in an electronic computer.

OPEN ISSUES

Due to heating effect, the size of the QD cannot be decreased beyond a certain limit.

KEYWORDS

- **nanostructure**
- **quantum dot**
- **semiconductor**

REFERENCES AND FURTHER READING

Algar, W. R., Tavares, A. J., & Krull, U. J., (2010). Beyond labels: A review of the application of quantum dots as integrated components of assays, bioprobes, and biosensors utilizing optical transduction. *Analytica Chimica Acta., 673*, 1.

Zwanenburg, F. A., Dzurak, A. S., Morello, A., Simmons, M. Y., Hollenberg, L. C. L., Klimeck, G., et al., (2013). Silicon quantum electronics. *Rev. Mod. Phys., 85*, 961.

CHAPTER 43

Quantum Electrochemistry

MIRELA I. IORGA[1] and MIHAI V. PUTZ[1,2]

[1]*National Institute for Research and Development in Electrochemistry and Condensed Matter, Department of Applied Electrochemistry, Laboratory of Renewable Energies-Photovoltaics, 300569 Timisoara, Romania, E-mail: mirela_iorga@yahoo.com, mirela.iorga@incemc.ro*

[2]*West University of Timisoara, Biology-Chemistry Department, Laboratory of Structural and Computational Physical Chemistry for Nanosciences and QSAR, 300115 Timisoara, Romania, E-mail: mv_putz@yahoo.com, mihai.putz@e-uvt.ro*

43.1 DEFINITION

Quantum electrochemistry field includes definitions and concepts from quantum mechanics, electrodynamics, and electrochemistry, representing the application of quantum mechanics to chemistry, and in particular to electrochemical processes studies.

43.2 HISTORICAL ORIGIN(S)

The concept of quantum electrochemistry arises from the application of quantum mechanics to chemistry, mainly to the study of electrochemical processes, since electrochemistry deals with interconversion between chemical and electric energy, facilitated by the electron transfer between two species implied in a chemical reaction. The quantum electrochemistry area includes definitions and concepts from various fields of science, such as chemistry, physics, mathematics, electricity, mechanics, and engineering (www.assignmentpoint.com). Since both electrons and protons are quantum particles, their actions could be viewed concerning quantum mechanics. The essential areas in electrochemistry where the quantum feature occurs are interfaces properties, charge transfer kinetics at the interfaces, and proton mobility in solutions (Bockris & Khan, 1993).

The beginnings of quantum electrochemistry could be mentioned in 1931 when the first application of quantum mechanics to electrochemical reactions was made by Gurney in his paper "The Quantum Mechanics of Electrochemistry," published in *Proceedings of Royal Society*. His research work treated the rate of electron transfer through a barrier, to ions in solution; his rate equation could be considered the starting point for all the theories of electron transfer kinetics at interfaces, at quantum and molecular level (Bockris & Khan, 1993). Gurney's research was considered one of the first papers in quantum chemistry discipline (Bockris et al., 1972).

In the 1950s, Taube introduced the concept of different electron transfer mechanisms in the case of inner-sphere or outer-sphere (Marcus, 1993). He was awarded the Nobel Prize "for his work on the mechanisms of electron transfer reactions, especially in metal complexes" (www.nobelprize.org) in 1983.

In the 1960s, under the leadership of Levich was founded the "Moscow School"—a theoretical group of scientists from the Institute of Electrochemistry of the Academy of Science of Moscow. Here were developed mechanical models for proton transfer reactions in chemical systems, theoretical approaches of quantum mechanics of electrode kinetics, etc. (Bockris & Khan 1993). One of the pioneers of quantum electrochemistry was Dogonadze, considered as the first scientist that recognized electron transfer reactions as quantum mechanical transitions between two electronic states (Bard et al., 2012). In their paper "Theory of Non-Radiation Electron Transitions from Ion to Ion in Solutions," published in 1959, Dogonadze and Levich developed the electron transfer reaction theory in homogeneous solutions and at the electrodes. Their research was based on the determination of transition probability of an electron from a donor to an acceptor, using the general expression of time-dependent perturbation theory (Levich & Dogonadze 1959). The advanced quantum mechanical model for proton transfer has established the basis of the quantum mechanical theory of chemical, electrochemical, and biochemical processes in polar liquids (Shukla & Kumar 2008).

Another contributor in the years of 1960s was Gerisher (1960a, 1960b, 1961), whose theoretical participation was to connect the terms of total energy with the single-particle, by his formulation of tunneling electrochemical reactions (Bevan et al., 2016).

Other developments of the quantum effects were done by Sutin in 1962 (Sutin 1962)–treating the quantum effects of the vibrations of the reactants (Marcus 1993).

The Canadian-born scientist, Marcus, started in 1956 with his paper "On the Theory of Oxidation-Reduction Reactions Involving Electron Transfer. I." (Marcus 1956), and further on, in 1956–1965, developed a fundamental theory of electron transfer reactions, concerning the structural changes of the reacting molecules and the molecules of solvent, and calculates the rate of chemical reactions from the energetic transformations of the molecular system (www.nobelprize.org). This theory was named after him and nowadays is known as the Marcus theory. After that, he developed his theory and extended it to quantum effects (Marcus, 1982; Siders & Marcus, 1981), statistical aspects and applications to chemiluminescence and electrode reactions. He is considered the theoretical scientist that "put quantum electrochemistry on the map" (Shukla & Kumar 2008). He was awarded in 1992 the Nobel Prize in Chemistry, "for his contribution to the theory of electron transfer reactions in chemical systems" (www.nobelprize.org).

43.3   NANO-SCIENTIFIC DEVELOPMENT(S)

Quantum electrochemistry represents the application of the quantum mechanical model in chemistry to atoms and molecules, and in particular for the electrochemical reaction studies.

The basic concepts in electrochemistry are derived from the transformation of chemicals in electric energy, facilitated by electron transfer reactions. They are represented by the components of an electrochemical system: an electron-conducting phase (metal or semiconductor), an ion conducting phase (electrolyte), and the interface between those two phases, where the reduction/oxidation processes occur. The most important parameters in the case of an electrochemical system (Mueller et al., 2013) are as follows: the electric potentials (for each phase and the interface), the charge transport rates in the conducting phase, the chemical concentrations, and reaction rates at the interface. Therefore, quantum electrochemistry (Bockris & Khan 1993) is involved in explanations of charge transfer kinetics at the interface, properties of the interface, and mobility of proton in solution, as shown in Figure 43.1.

Quantum mechanics represents the matter and light behavior, especially in the case of atomic objects, such as electrons, protons, neutrons, photons, etc., which could be considered according to Feynman as "particle waves"

(Feynman et al., 1964). A solid representation of the small-scale matter behavior was achieved in 1926, and 1927 by Schrödinger, Heisenberg, and Born. Their expression and theories are the starting points for other methods developed along the time in quantum mechanics, physics, chemistry, and other related disciplines.

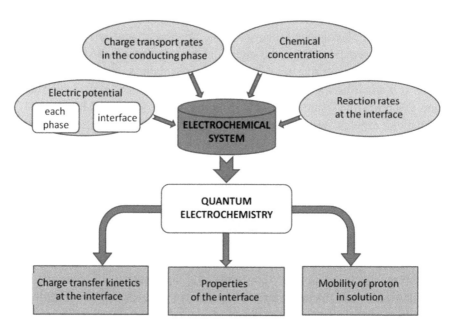

FIGURE 43.1 The general representation of quantum electrochemistry involvement in the study of electrochemical systems.

In the last years, newer and better techniques were developed to observe at the atomic scale and to manipulate the electrode-electrolyte interfaces (Rikvold et al., 2009), which have determined the occurrence of new methods for analysis, synthesis, characterization, and modification of materials. Also, the vast increases in computer industry corroborate with the improved methods and algorithms enabled theoretical chemists or physicians to develop better atomic scale descriptions for surface structures and processes. Advanced mathematical knowledge is necessary, and one of the most important tools is matrix algebra (Szabo & Ostlund, 1996).

In the case of electrochemical processes such computational methods based on quantum mechanical models and statistical mechanics (Mueller et al., 2013), enable to create models and simulations for electrochemical systems starting from the atomistic details of a quantum regime to a description of the electrochemical processes noticed in the case of macro-scopic systems. The atomic level interactions between electrons and nuclei concerning electrochemical processes must be expressed in quantum mechanics terms (Bockris et al., 1972), for example, as space and time-dependent wave function. In this case, the starting point is the hypothesis that a wave function describes every system, and all the system properties are derived from it.

The system dynamics is controlled by the Schrodinger equation (non-relativistic, time-independent) (Mueller et al., 2013; Szabo & Ostlund, 1996):

$$H\Psi(r) = E\Psi(r) \tag{1}$$

If the equation is time-dependent, the expression will be:

$$H\Psi(r,t) = i\hbar\left(\frac{\partial\Psi(r,t)}{\partial t}\right) \tag{2}$$

where: H represents the Hamiltonian operator;

E – the system energy, contains the potential (V) and the kinetic energy (T) terms;

$\hbar$ – Planck's constant;

$\Psi(r,t)$ – space and time-dependent wave-function.

In the case of a molecule, the expression for the Hamiltonian operator is shown in equation (3). The electronic energies (Szabo & Ostlund, 1996) are composed of kinetic energies, interelectronic and internuclear repulsion, and attraction between electrons and nuclei.

$$H = -\sum_{i=1}^{N}\frac{1}{2}\nabla_i^2 - \sum_{A=1}^{M}\frac{1}{2M_A}\nabla_A^2 - \sum_{i=1}^{N}\sum_{A=1}^{M}\frac{Z_A}{r_{iA}} + \sum_{i=1}^{N}\sum_{j>i}^{N}\frac{1}{r_{ij}} + \sum_{A=1}^{M}\sum_{B>A}^{M}\frac{Z_AZ_B}{R_{AB}} \tag{3}$$

where: A, B – represents the nuclei;

i, j – the electrons;

M – the number of nuclei;

N – the number of electrons;

M_A – mass of nucleus A/an electron mass;

Z_A – the atomic number of nucleus A;

∇_i^2, ∇_A^2 – Laplacian operators;

$-\sum_{i=1}^{N}\frac{1}{2}\nabla_i^2$ – the operator for the kinetic energy of electrons;

$$-\sum_{A=1}^{M}\frac{1}{2M_A}\nabla_A^2 \text{ – the operator for the kinetic energy of nuclei;}$$

$$-\sum_{i=1}^{N}\sum_{A=1}^{M}\frac{Z_A}{r_{iA}} \text{ – the Coulomb attraction between electrons and nuclei;}$$

$$\sum_{i=1}^{N}\sum_{j>1}^{N}\frac{1}{r_{ij}} \text{ – the repulsion between electrons;}$$

$$\sum_{A=1}^{M}\sum_{B>A}^{M}\frac{Z_A Z_B}{R_{AB}} \text{ – the repulsion between nuclei.}$$

Because the Schrödinger equation is too complicated for an exact solution, a series of methods for approximate calculation of electronic wave functions or electronic energies occurred.

One of them is the Born-Oppenheimer approximation (Szabo & Ostlund, 1996) assume that in a molecule, the motion of nuclei and electrons could be separated. Since nuclei are much heavier than the electrons and their movement is slower, it may be considered that the electrons are moving in the field of fixed nuclei.

In this case, the approximation which corresponds to Eq. (1) is:

$$\Phi\left(\{r_i\};\{R_A\}\right) = \Phi_{el}\left(\{r_i\};\{R_A\}\right)\cdot\Phi_{nucl}\left(\{r_i\};\{R_A\}\right) \tag{4}$$

An important role in quantum chemistry is played by the Hartree-Fock approximation – the equivalent of the molecular orbital approximation. The equation which defines the best spin orbitals is the Hartree-Fock equation:

$$h(1)\chi_a(1)+\sum_{b\neq a}\left[\int dx_2 |\chi_b(2)|^2 r_{12}^{-1}\right]\chi_a(1)-\sum_{b\neq a}\left[\int dx_2 \chi_x^*(2)\chi_a(2)r_{12}^{-1}\right]\chi_b(1)=\varepsilon_a\chi_a(1) \tag{5}$$

where: $h(1) = -\frac{1}{2}\nabla_1^2 - \sum_A \frac{Z_A}{r_{1A}}$ represents the kinetic and the potential energies of nuclei attraction, in the case of a single electron, considered the first-electron;

χ_a – spin-orbital;
ε_a – orbital energy.

Since the electrons cannot share identical states and must occupy distinct states in a form that their combined wave-function sign is reversed when they are exchanged, according to the Pauli Exclusion Principle: the global electronic wave-function is antisymmetric concerning the interchange of the coordinate (space and spin) of any two electrons (Mueller et al., 2013; Szabo & Ostlund, 1996). In this case, two conditions arise, i.e., all single-electron spatial orbitals must be orthonormal, and the overall wave function is antisymmetric concerning the exchange of any two electrons coordinates, according to relation (6):

$$\Psi\left(x_1,....,x_i,...,x_j,...,x_n\right) = -\Psi\left(x_1,....,x_j,...,x_i,...,x_n\right) \tag{6}$$

The principle mentioned above is often named the antisymmetry principle, and it is an independent postulate of quantum mechanics.

In electrochemistry, the most common processes where the electronic structure calculus is applied are the reactions at solid-liquid interfaces. The basis for a lot of quantum rate calculation in quantum chemistry or physics is Fermi's Golden Rule. According to Bockris & Khan (1993), it could be expressed as a solution of Eq. (2), in the case of weak perturbation, and will have the expression:

$$R_t = \frac{2\pi}{\hbar}\left|\left\langle\Psi_f\left|U_p\right|\Psi_i\right\rangle\right|^2 \cdot \rho\left(E_f\right) \tag{7}$$

where: R_t represents the transition probability/unit time;

$\left\langle\Psi_f\left|U_p\right|\Psi_i\right\rangle$ – the matrix element of transition;

U_p – the perturbation potential from the initial state to the final state;
Ψ_i – the wave function of the initial state;
Ψ_f – the wave function of the final state;
$\rho(E)$ – the weighted densities of final states.

The Fermi-Dirac distribution function (Bonciocat, 1996) gives the probability for an energy level of metal to be occupied with electrons and is expressed as follows:

$$P_E = \frac{1}{1+\exp\dfrac{E-E_f}{KT}} \tag{8}$$

where: P_E represents the probability of electron occupation;
E_f – Fermi energy.

So, there are three possibilities:

- $P_E = 0.5$: the Fermi level is the energy level half occupied with electrons;
- $P_E = 1$: the energy level is entirely occupied with electrons, does not have any gaps, and cannot participate in an electrode reaction;

- $P_E = 0$: the energy level is completely unoccupied with electrons, and cannot participate in an electrode reaction where it is necessary to function as an electron source.

Electron transfer from a metal electrode to the solution takes place according to the general equation:

$$R \leftrightarrow O + e^- \ (Me) \tag{9}$$

According to Marcus theory, the electrochemical reactions could be considered as two types of reactions: outer-sphere reactions and inner-sphere reactions. The outer-sphere consists of free solvent molecules, and the outer-sphere reactions are defined as the reactions where tunneling makes the electron transfer; in this case, the reactant is separated from the substrate by a tunneling barrier. These types of reactions do not change the inner-sphere because any bonds are created or broken. In the case of inner-sphere, this consists of closest molecules of the solvent or ligands in complexes, and in the inner-sphere reactions, the reactant is physically bonded to the solid phase zone by chemisorption. The most studied reactions are outer-sphere reactions (Bevan et al., 2016). The tunnel effect laws explain the passing of an electron from metal to an acceptor species from the solution or vice versa. In these reactions the charge of the reactant and its solvation layer are changing, the electron transfer and the solvent reorganization occurs in an order which follows the Frank-Condon principle (Schmickler & Frank 2007). The transfer of the electron is fast and takes place at a fixed position of the solvent, following a mechanism where first the solvation layer achieves a configuration between initial and final charge states, afterward the electron is transferred, and the system goes to its new equilibrium state. Following Marcus-Hush theory, the rate constant has the following expression, given in Eq. (10) and provides the rate for a homogenous electron-transfer reaction.

$$k = Z \exp\left[-\frac{(\lambda + \Delta G)^2}{4\lambda k_B T} \right] \tag{10}$$

where: ΔG represents the reaction-free energy;

k_B – the Boltzmann's constant.

If applied in electrochemistry, Eq. (10) must be integrated over all distances, and the reaction-free energy must be related to the overpotential, as is shown in Eq. (11).

$$\Delta G = e_0 \eta \tag{11}$$

Thus, the electron transfer rate from metal to reactant will be given by the equation (12):

$$k_{red} = Z \cdot C \int \rho(\varepsilon) f(\varepsilon) \cdot \exp\left[-\frac{(\lambda - \varepsilon + e_0\eta)^2}{4\lambda k_B T} \right] d\varepsilon \tag{12}$$

Schmikler & Frank (2007) proposed a model Hamiltonian for electrochemical electron transfer, in which the solvent reorganization was modeled as a phonon bath or a collection of harmonic oscillators, where is facile to include quantum modes, and must be treated in quantum mechanics terms. The solvent modes are separated in slow vibrational and librational modes, and fast electronic modes. The latter are in electronic equilibrium with the transferred electron, renormalizing the energy of the electronic states of the reactant.

Migliore et al. (2012) studied a molecular conduction junction between a redox molecule bridging metal electrodes, where electron transport follows the Marcus theory and determines that the effect of the solvent reorganization on the conduction properties depends on the particular bridge charging states implied in the conduction process.

43.4 NANO-CHEMICAL APPLICATION(S)

The quantum electrochemistry deals with definitions and concepts from different domains, such as chemistry, physics, electricity, mechanics, and engineering. Also, its applications are extended over a wide range of science.

Bevan et al., (2016) in their paper developed a theoretical connection between the Landauer electron transport and the Gerischer's formulation of tunneling electrochemical reaction. Landauer model (Datta, 2005) is usually used in quantum transport, in which the conductor itself is assumed to be free of interactions, and irreversibility and dissipation occur from the connection with the contacts; it is appropriate to illustrate the coherent transport.

Hossain & Bevan (2016) extended a model for electrochemical charge transfer derived from quantum transport methods, to capture ultrafast linear sweep voltammetry, which was subsequently applied to a model of an electrochemical tunneling system made of a passivating monolayer ended by redox-active groups. They predict the electronic coupling independent voltammetric spectra, from which the redox reorganization energy could be determined.

Rikvold et al. (2009) used computer simulation methods of molecular dynamics (MD) in the case of lithium-battery technology. Also, they used equilibrium and kinetic Monte Carlo (MC) simulations complemented with quantum-mechanical density functional theory (DFT) calculation of interaction energies in the case of major investigations of single-crystal electrodes. The authors developed a method, named the electrochemical first-order reversal curve (EC-FORC), for classifying surface transitions in adsorbate systems, an extension of the conventional cyclic voltammetry.

Albrecht (2012) applied the quantum-mechanical tunneling effect to liquid environments, considering it a sensitive tool in fundamental interfacial processes analyze. He studied chemical reactions at the single-molecule level and established the composition of biopolymers (DNA, RNA, proteins), and finally concluded that this offers the opportunity of a new class of sensor devices.

Gupta et al. (2012) developed a label-free, electronic biomolecular sensing platform to detect and characterize trace amounts of biological toxins, in which for the signal transduction mechanism, the electrostatic coupling of molecule bond vibration to charge transport by tunneling across an insulated electrode-electrolyte interface was used. By experiments and through theoretical modeling was demonstrated that the current at the interface has a component occurring from the rate-limiting transition of a quantum mechanical, electronic relaxation event. In this, the electronic tunneling process between a hydrated proton from the electrolyte and the metal electrode is firmly related to the bond vibrations of molecular species from the electrolyte.

Liska et al. (2013) investigated the reduction of tetranitrocalix[4]arenes by electrochemical methods, followed by correlation of the results of quantum chemical calculations, and described the order of single electron transfers as a consequence of molecular geometry. In general, the calyx[4] arenes are appropriate as building blocks for supramolecular assemblies, and precursors for sensors.

Sandeman et al. (2011) studied the texture and the adsorption features of some activated carbons (ACs), porous poly(vinyl alcohol) gels (PVA), and PVA/AC composites by various methods. They finally modeled the dye-PVA-AC-water interactions by a semi-empirical quantum chemical method that allows a complete describing of these interactions.

The association between electrochemical studies and quantum chemistry was employed by Li et al., (2015) in their paper where they applied cyclic voltammetry and rotation disk electrode analysis to study norbornene-based quinone (NB-quinone). Further, they performed the quantum chemical calculation-based DFT and demonstrated that the calculated potential of the reaction is the same with the one obtained experimentally.

In the latest years novel materials were developed and among them, considerable attention has been granted to quantum dots (QDs), due to their peculiar properties, such as high quantum yield, high stability, and special electrochemical, optical, and luminescence properties.

Huang & Zhu (2013) in their review presented a general description of QDs electrochemical properties and their electrochemical applications. QDs are defined as nanostructured zero-dimensional materials, containing elements in principle from the periodic groups II-VI, III-V, and IV-VI, with semiconductor structures and physical dimensions under the excitation Bohr radius, and with special properties such as piezoelectric, magnetic, electrochemical, optoelectric, photochemical, catalytic, and so on. The presented applications of QDs are from fields such as inorganic or organic substance analysis, immunoanalysis, DNA analysis, aptamer analysis, etc. realized by electrochemical tools.

Other applications of QDs were presented by Liu et al., (2010) resulting from their special properties: small size, high reactivity, and high surface/volume ratio. The main application domains were from electrochemistry (electrochemical luminescence sensors) and biology (immunosensors, DNA sensors, protein sensors, carbohydrate sensors, and pesticide sensors).

In biomedical sensors, Nguyen et al., (2016) reported the improvement of electrochemical signals of etoposide measured by differential pulse voltammetry in the case of a glassy carbon electrode modified with carbon quantum dots (CQDs).

Other applications in biomedical sensors were made by Carrara et al., (2009) by investigations on nanostructured electrodes to detect metabolites, namely glucose, lactate, and cholesterol. They found a general law for the dependence of detection sensitivity and sensor detection limit and demonstrated that electrodes modified with QDs and quantum wires improve the sensor's performance.

Bonu et al., (2016) studied metal nanostructures such as SnO_2 QDs for application in energy as supercapacitors or Li-ion batteries. During electrochemical processes, the smaller nanostructures presented better properties (stability, higher-ion flux, low ion diffusion time, low pulverization, and higher surface).

Daumann et al., (2017) reported a water-free synthesis route to obtain ZnO QDs for implementation as an electron injection layer in light-emitting electrochemical cells.

According to Daniel & Astruc (2004) gold nanoparticles (AuNPs) find a lot of applications, and for example, as a consequence of their ability to correlate the electronic conduction with single-electron tunneling, nanoelectronic digital circuits in connection with self-assembled monolayers could be obtained. In electrochemistry, AuNPs could be used for ultramicroelectrodes, with the possibility to control molecules at a very low resolution. Another domain is for catalytic applications where the large variety of synthetic options using AuNPs components and a better understanding of their role in catalysis will generate new applications.

Tsai et al., (2015) realized a quantum technique for active and high precision temperature control in various conditions. They use fluorescent nanodiamonds with high-density ensembles of nitrogen-vacancy centers conjugated with gold nanorods as a combined nanoheater and nanothermometer for highly localized hyperthermia treatment, with NIR lasers as a heating source, and conducted numerical simulations, to elucidate the nanoscale heating process.

Martin Yerga and Costa Garcia (2017) reported an optimized method for QDs determination in slight concentrations, where electrogenerated copper species are stabilized on the electrode surface by QDs, and further on the amount of QDs is measured by voltammetry technique. The application of this method results in enhancing the detection limit of electrochemical biosensors with QDs as labels.

Martin Yerga et al., (2016) described in their paper a selective electrodeposition process of silver on QDs and subsequently analyzed the nanostructured silver obtained. QDs act as a catalyzer in the electrodeposition process and between electrodeposited silver and QDs a powerful adsorption was noticed. To detect QDs, the application of these processes could generate an ultrasensitive electrochemical method.

Srinivasan et al., (2015) designed and constructed a hybrid electrode structure starting from the hole-conducting polymer poly(3,4-ethylene dioxythiophene) (PEDOT), with electrocatalytic activity and electrochemical stability, which was electrochemically coated with a quantum dot-sensitized semiconductor. The novel electrode containing $Pt/TiO_2/CdS/CdSe/PEDOT$ has an increased photo-stability, higher photocurrent, long-term stability, higher rates for hydrogen generation, and could be applied to other photocatalytic materials and photoelectrochemical reactions.

Nowadays, a new category of low-dimensional materials occurs, such as graphene, graphene quantum dots (GQDs), graphene nanoribbons, etc., which determine the development of higher quality materials with a controllable layer thickness. These materials could be used for applications in biosensing, materials protection, optoelectronics, energy storage, and production, etc.

Bak et al., (2016) presented in a review different synthetic methods for GQDs obtaining in solution, where simple and low-cost processes are implied, with high yields and final resulting products with excellent properties. These are methods applying top-down strategy, starting from the graphene or graphene oxide – such as hydro- and solvothermal, ultrasonication, microwave, and electrochemical methods, from graphite – such as Hummers method and acid-free method, or from carbon nanotubes and carbon fibers. Also, they pointed out various applications of GQDs, such as capacitors, Li-ion batteries, electrodes, solar cells, and due to their high solubility – the possibility to use GQDs in all devices production stages.

Lim et al., (2015) reported the synthesis and characterization of two different types of nanocarbon materials, GQDs and CQDs. They provided an electrochemical comparison between those two categories of materials, and subsequently investigated their potential applications in electrochemical sensing and electrochemistry.

Xu et al., (2014) reviewed in their paper the applications of carbon quantum dots (CQDs) in electrochemiluminescence research studies, starting with the synthesis, the surface states, and finally analyzed the corresponding reaction mechanism of CQDs composed of carbon nanodots and GQDs.

Alwarappan & Pillai (2015) presented in their work the application of graphene, for example, in Li-ion batteries as anode material, and the development of Li-metal anodes to use graphene as a coating material. Due to their excellent photostability, biocompatibility, and capacity to store very small charges are possible to use GQDs in applications such as molecular switches, resonant tunneling diodes, or single electron transistors.

Sharma et al., (2017) reported in their article the synergistic effect of GQDs and silver nanostructures in their nanocomposites, and as a result, the improvement of electrochemical sensors and Surface Enhanced Raman Spectroscopy (SERS) activity.

Wu et al., (2013) reported the development of a new electrode material with improved electrochemical properties, better electrical conductivity, high specific capacitance, and good stability, obtained through in-situ polymerization of pyrrole in the presence of GQDs suspension; polypyrrole/GQDs nanocomposite could be applied for electrochemical supercapacitors.

Tong et al., (2017) reviewed in their article the new trends in the fabrication of graphene-family materials such as GQDs, graphene nanoribbons, and

3D graphene, and their applications as oxygen reduction reaction electro-catalysts in fuel cells and metal-air batteries.

43.5 MULTI-/TRANS-DISCIPLINARY CONNECTION(S)

Quantum electrochemistry implies concepts, models, and approaches from a wide range of science including mathematics, physics, chemistry, mechanics, engineering, energy, biology, genetics, geology, soil sciences, etc.

The rapid development of many areas in nanotechnology has determined the occurrence of new possibilities in the design of functional materials with special properties: optical, photonic, electronic, electrical, thermal, mechanical, chemical, biochemical, and so on. These materials could be realized for specific applications, such as solar cells, rechargeable batteries, supercapacitors, biosensors, and others.

Daniel & Astruc (2004) in their work presented a large variety of gold nanoparticles (AuNPs) possibilities for applications, exploiting their special properties, in various domains. Starting with the optoelectronic properties of AuNPs related to the surface plasmon absorption, relevant to the quantum size effect, the photonics could be combined with biology and medicine, and the result will be new biomolecular applications (drugs and genetic material detection, labeling, and transfer). If the characteristics of AuNPs regarding spectroscopy, fluorescence, luminescence, and electrochemical properties are modified taking into account those of the substrates, such as DNA, or other biological molecules, sensors, and environmental devices could be obtained. A lot of structures, properties, and applications are available in the case of AuNPs, which could be related to studies and applications of other molecular, inorganic or biological nanomaterials in fields like chemistry, physics, biology, medicine, and so on.

43.6 OPEN ISSUES

By quantum electrochemistry arises the possibility to connect electrochem-istry with quantum transport and to expand the concepts in the case of atom-istic approaches. In the future, the conversion between quantum transport and electrochemical concepts into directly measurable electrochemical properties will contribute significantly to the characterization and analysis of nanostructures and nanomaterials.

According to Albrecht (2012), compositional analysis by tunneling future applications could include spatially resolved detection of DNA methylation and other epigenetic markers, analysis of RNA, peptides, proteins, etc.

Alwarappan & Pillai (2015) opine that the nanomaterials offer in perspective the opportunity to study, for example, the electrochemical synthesis of graphene quantum dots with size control and the electrochemical characterization by CV of those materials as a function of the graphene layers number to utilize it as working electrode.

Following Bak et al., (2016), in the case of GQDs in the future, is expected to be developed novel strategies to differentiate GQDs with various sizes to obtain GQDs with uniform properties. The final aim is to reach extremely high yields without the necessity to remove the starting materials and to establish simple purification methods to eliminate residual agents.

ACKNOWLEDGMENT

This contribution is part of the Nucleus-Programme under the project "Deca-Nano-Graphenic Semiconductor: From Photoactive Structure to the Integrated Quantum Information Transport," PN-18-36-02-01/2018, funded by the Romanian National Authority for Scientific Research and Innovation (ANCSI).

KEYWORDS

- **electron-transfer functions**
- **quantum dots**
- **quantum electrochemistry**
- **quantum electrode kinetics**
- **quantum models**
- **quantum system**

REFERENCES AND FURTHER READING

Albrecht, T., (2012). Electrochemical tunneling sensors and their potential applications. *Nature Communications, 3,* 829. doi: 10.1038/ncomms1791.

Alwarappan, S., & Pillai, V. K., (2015). Recent advances in the development of low-dimensional materials for electrochemical applications: CECRI, Karaikudi, *Current Science, 108*, 789–791.

Bak, S., Kim, D., & Lee, H., (2016). Graphene quantum dots and their possible energy applications: A review. *Current Applied Physics, 16*, 1192–1201.

Bard, A. J., Inzelt, G., & Scholz, F., (2012). *Electrochemical Dictionary*. Springer-Verlag Heidelberg.

Bevan, K. H., Hossain, M. S., Iqbal, A., & Wang, Z., (2016). Exploring bridges between quantum transport and electrochemistry. I., *J. Phys. Chem., 120*, 179–187.

Bockris, J. O'M., & Khan, S. U. M., (1993). *Surface Electrochemistry, A Molecular Level Approach*. Plenum Press, New York, U.S.A.

Bockris, J. O'M., Sen, R. K., & Mittal, K. L., (1972). On quantum electrochemical kinetics. *J. Res. Inst. Catalysis, 20*, 153–184. Hokkaido Univ.

Bonciocat, N., (1996). *Electrochimie şi aplicatii* (Engl: Electrochemistry and Applications), Dacia Europa-Nova, Timisoara, Romania.

Bonu, V., Gupta, B., Chandra, S., Das, A., Dhara, S., & Tyagi, A. K., (2016). Electrochemical supercapacitor performance of SnO_2 quantum dots, *Electrochemica Acta, 203*, 230–237.

Carrara, S., Boero, C., & De Micheli, G., (2009). Quantum dots and wires to improve enzymes-based electrochemical bio-sensing. In: Schmid, A., et al., (eds.), *Proceedings of the International Conference Nano-Net 2009* (pp. 189–199). LNICTS 20.

Daniel, M. C., & Astruc, D., (2004). Gold nanoparticles: Assembly, supramolecular chemistry, Quantum-size-related properties, and applications toward biology, catalysis, and nanotechnology. *Chem. Rev., 104*, 293–346.

Datta, S., (2005). *Quantum Transport: Atom to Transistor*. Cambridge University Press, USA, New York.

Daumann, S., Adrzejewski, D., Di Marcantonio, M., Hagemann, U., Wepfer, S., Vollkommer, F., et al., (2017). Water-free synthesis of ZnO quantum dots for application as electron injection layer in light-emitting electrochemical cells, *J. Mater. Chem. C.*, doi: 10.1039/C6TC05571K.

Feynman, R. P., Leighton, R. B., & Sands, M., (1964). *The Feynman Lectures on Physics. III. Quantum Mechanics*. Addison Wesley.

Gerischer, H., (1960a). Über den Ablauf von Redoxreaktionen an Metallen und an Halbleitern. I. Allgemeineszum Elektronenübergangzwischeneinem Festkörper und einem Redoxelektrolyten, *Z. Phys. Chem., 26*, 223–247.

Gerischer, H., (1960b). Über den Ablauf von Redoxreaktionen an Metallen und an Halbleitern. II. Metall-Elektroden, *Z. Phys. Chem., 26*, 325–338.

Gerischer, H., (1961). Über den Ablauf von Redoxreaktionen an Metallen und an Halbleitern. III. Halbleiterelektroden, *Z. Phys. Chem., 27*, 48–79.

Gupta, C., Walker, R. M., Gharpuray, R., Shulaker, M. M., & Zhang, Z., (2012). Electrochemical quantum tunneling for electronic detection and characterization of biological toxins. *Proc. SPIE 8373, Micro- and Nanotechnology Sensors, Systems, and Applications IV*, 837303, http://dx.doi.org/10. 1117/12. 920692.

Hossain, M. S., & Bevan, K. H., (2016). Exploring bridges between quantum transport and electrochemistry. II. A theoretical study of redox-active monolayers. *J. Phys. Chem. C., 120*, 188–194.

Huang, H., & Zhu, J. J., (2013). The electrochemical applications of quantum dots. *Analyst, 138*, 5855–5865.

Levich, V. G., & Dogonadze, R. R., (1959). Theory of non-radiation electron transitions from ion to ion in solutions. *Doklady Akademii Nauk SSSR, 124*, 123–126.

Li, H., Zhao, Y., Pan, J., Li, H., Feng, L., Wang, S., Xie, X., & Huang, H., (2015). Electrochemical study and quantum chemical calculation of norbornene-based quinone in acetonitrile. *Chemistry Letters, 44*, 1443–1445.

Lim, C. S., Hola, K., Ambrosi, A., Zboril, R., & Pumera, M., (2015). Graphene and carbon quantum dots electrochemistry. *Electrochemistry Communications, 52*, 75–79.

Liska, A., Vojtisek, P., Fry, A., & Ludvik, J., (2013). Electrochemical and quantum chemical investigation of tetranitrocalix[4]arenes: Molecules with multiple redox centers. *J. org. Chem., 78*, 10651–10656.

Liu, J., Huang, Q., Wang, X., & Chen, H., (2010). Application of quantum dots in electrochemical biosensors. *Progress in Chemistry, 22*, 2179–2190.

Marcus, R. A., (1956). On the theory of oxidation-reduction reactions involving electron transfer. I., *Journal of Chemical Physics, 24*, 966–978.

Marcus, R. A., (1982). On quantum, classical and semiclassical calculations of electron transfer rates. In: King, T. E., Morrison, M., & Mason, H. S., (eds.), *Oxidases and Related Redox Systems*. Pergamon, New York.

Marcus, R. A., (1993). Electron transfer reactions in chemistry. Theory and experiment. *Reviews of Modern Physics, 65*, 599–610.

Martin, Y. D., & Costa, G. A., (2017). Electrochemical detection of quantum dots by stabilization of electrogenerated copper species, *Electrochemistry Communications, 74*, 53–56.

Martin, Y. D., Costa, R. E., & Costa, G. A., (2016). Electrochemical study and applications of selective electrodeposition of silver on quantum dots. *Anal. Chem., 88*, 3739–3746.

Migliore, A., Schiff, P., & Nitzan, A., (2012). On the relationship between molecular state and single electron pictures in simple electrochemical junctions. *Phys. Chem. Chem. Phys., 14*, 13746–13753.

Mueller, J. E., Fantauzzi, D., & Jacob, T., (2013). Multiscale modeling of electrochemical systems. In: Alkire, R. C., Kibler, L. A., Kolb, D. M., & Lipkowski, J., (eds.), *Electrocatalysis: Theoretical Foundations and Model Experiments.* Wiley-VCH Verlag GmbH & Co. KGaA.

Nguyen, H. V., Richtera, L., Moulick, A., Xhaxhiu, K., Kudr, J., Cernei, N., et al., (2016). Electrochemical sensing of etoposide using carbon quantum dot modified glassy carbon electrode. *Analyst, 141*, 2665–2675.

Quantum Electrochemistry. http://www.assignmentpoint.com/science/chemistry/quantum-electrochemistry.html.

Rikvold, P. A., Hamad, A., Juwonol, T., Robb, D. T., & Novotny, M. A., (2009). *Applications of Computer Simulations and Statistical Mechanics in Surface Electrochemistry* (pp. 1–12). arXiv:0910. 2185v1 [physics. chem-ph].

Rudolph, A., (2014). *Marcus – Facts. Nobelprize.org.* Nobel Media AB. Web. 27 Jan 2017. http://www.nobelprize.org/nobel_prizes/chemistry/laureates/1992/marcus-facts.html.

Sandeman, S. R., Gunko, V. M., Bakalinska, O. M., Howell, C. A., Zheng, Y., Kartel, M. T., et al., (2011). Adsorption of anionic and cationic dyes by activated carbons, PVA hydrogels, and PVA/AC composite. *Journal of Colloid and Interface Science, 358*, 582–592.

Schmickler, W., & Frank, S., (2007). *Quantum Theory of Electrochemical Electron-Transfer Reactions, in Encyclopedia of Electrochemistry*. Wiley-VCH Verlag GmbH & Co. KGaA.

Sharma, S., Prakash, V., & Mehta, S. K., (2017). Graphene/silver nanocomposites-potential electron mediators for proliferation in electrochemical sensing and SERS activity. *TrAC Trends in Analytical Chemistry, 86*, 155–171.

Shukla, A. K., & Kumar, T. P., (2008). Pillars of modern electrochemistry. *The Electrochemical Society Interface*, pp. 31–39.

Siders, P., & Marcus, R. A., (1981). Quantum effects in electron-transfer reactions. *J. Am. Chem. Soc., 103*, 741–747.

Srinivasan, N., Shiga, Y., Atarashi, D., Sakai, E., & Miyauchi, M., (2015). A PEDOT-coated quantum dot as efficient visible light harvester for photocatalytic hydrogen production. *Applied Catalysis B: Environmental, 179*, 113–121.

Sutin, N., (1962). Electron exchange reactions. *Annual Review of Nuclear Science, 12*, 285–328.

Szabo, A., & Ostlund, N., (1996). *Modern Quantum Chemistry: Introduction to Advanced Electronic Structure Theory*. Dover Publications New York.

The Nobel Prize in Chemistry, (1983). *Nobelprize.org.* Nobel Media AB 2014. Web. 30 Jan 2017. http://www.nobelprize.org/nobel_prizes/chemistry/laureates/1983/.

Tong, X., Wei, Q., Zhan, X., Zhang, G., & Sun, S., (2017). The new graphene family materials: Synthesis and applications in oxygen reduction reaction. *Catalysts, 7*, 1–26.

Tsai, P. C., Chen, O. Y., Tzeng, Y. K., Hui, Y. Y., Guo, J. Y., Wu, C. C., Chang, M. S., & Chang, H. C., (2015). Gold/diamond nanohybrids for quantum sensing applications. *EPJ Quantum Technology, 2*, 1–12.

Wu, K., Xu, S. Z., Zhou, X. J., & Wu, H. X., (2013). Graphene quantum dots enhanced electrochemical performance of polypyrrole as supercapacitor electrode. *Journal of Electrochemistry, 19*, 361–370.

Xu, Y., Liu, J., Gao, C., & Wang, E., (2014). Applications of carbon quantum dots in electrochemiluminescence: A mini-review. *Electrochemistry Communications, 48*, 151–154.

CHAPTER 44

Self-Organization

ARANYA B. BHATTACHERJEE[1] and SUMAN DUDEJA[2]

[1]*Department of Physics, Birla Institute of Technology and Science, Pilani, Hyderabad Campus, India*

[2]*Department of Chemistry, ARSD College, University of Delhi, New Delhi–110021, India*

44.1 DEFINITION

Self-organization is a process where global order emerges out of the local interactions/correlations between the constituent parts of an initially disordered system. The process of self-organization can be spontaneous and can be tuned by some external parameter in some systems. It is often induced by random fluctuations that are amplified by positive feedback. The process of self-organization can be classical or quantum in nature, and it involves all the components of the system. The organization is robust against perturbations. Self-organization occurs in a variety of physical, chemical, biological, social, and cognitive systems such as the collection of atoms interacting with the electromagnetic field, chemical oscillations, bacterial colonies, animal swarming and neural networks.

44.2 HISTORICAL ORIGIN(S)

The concept of self-organization goes back to the time of philosopher Descartes. The term "self-organizing" was introduced to contemporary science in 1947 by the psychiatrist and engineer W. Ross Ashby. Physicist Ilya Prigogine won the 1977 Nobel Prize for his work on the thermodynamic concept of self-organization.

44.3   NANO-SCIENTIFIC DEVELOPMENT(S)

Nano-scientific developments of self-organization related to physics include laser, superconductivity, and Bose-Einstein condensation in optical cavities (Dicke super-radiance). In chemistry, self-organization of nanoparticles has bio-medical science. In biology, spontaneous protein folding and pattern formation in bacterial colonies are popular examples.

44.4   NANO-CHEMICAL APPLICATION(S)

Self-organization in chemistry includes molecular self-assembly, oscillating chemical reactions, liquid crystals, colloidal crystals, self-assembled mono-layers, and Langmuir-Blodgett films. A nanoscopic building block that has received special attention is the gold nanoparticle. Gold nanoparticles can be synthesized in all sizes between two and one hundred nanometers. Furthermore, they can easily be tailored with molecules like DNA and proteins, making them ideal as building blocks for self-assembly. Self-organization of gold nanoparticles on surfaces could be tuned by altering the ionic strength of the nanoparticle solution. Spontaneous self-organization enables dielectrophoresis of small nanoparticles and formation of photoconductive microbridges.

44.5   MULTI-/TRANS-DISCIPLINARY CONNECTION(S)

Self-organization behaviors in physical systems include spontaneous magne-tization, laser, reaction-diffusion systems such as Belousov-Zhabotinsky reac-tion, The theory of dissipative structures of Prigogine and Hermann Haken's synergetics were developed to unify the understanding of these phenomena, which include lasers, turbulence, and convective instabilities. Phenomena from mathematics and computer science such as cellular automata, random graphs artificial life exhibit features of self-organization. Self-organization is an impor-tant part to establish networking. Self-organizing behavior of bacterial colonies and in social animals like humans have received much attention recently.

44.6   OPEN ISSUES

Self-organization is controversial in the scientific community, and some criticize it strongly.

KEYWORDS

- **Bose-Einstein condensation**
- **colloidal crystals**
- **Langmuir-Blodgett films**
- **liquid crystals**
- **oscillating chemical reactions**
- **self-assembled monolayers**
- **self-assembly**
- **self-organization**

REFERENCES AND FURTHER READING

Ashby, W. R., (1962). Principals of the self-organizing system, In: Heinz Von, F., & George, W. Z. Jr., (eds.), *Principles of Self-Organization* (pp. 255–278). U.S. Office of Naval Research.

Haken, H., (1983). *Synergetics: An Introduction. Nonequilibrium Phase Transition and Self-Organization in Physics, Chemistry, and Biology*. Third Revised and Enlarged Edition, Springer-Verlag.

Nicolis, G., & Prigogine, I., (1977). *Self-Organization in Nonequilibrium Systems: From Dissipative Structures to Order Through Fluctuations*. Wiley, New York.

Prigogine, I., & Stengers, I., (1984). *Order Out of Chaos: Man's New Dialogue with Nature*. Bantam Books.

CHAPTER 45

Self-Organizing Processes

ISTVÁN LÁSZLÓ

Department of Theoretical Physics, Budapest University of Technology and Economics, H-1521 Budapest, Hungary, E-mail: laszlo@eik.bme.hu

45.1 DEFINITION

In Nature, a system with interacting elements usually has a disordered or chaotic structure in space and time. At special values of the system parameters, it can happen that having the same inter-element interactions some order emerges in space and/or time. We call this self-organizing process. Namely, a system organizes itself in such a way that its order is increasing. We can find self-organizing processes in various domains of the nature as physics, chemistry, biology, robotics, various networks, medicine economy, sociology or politics. These systems can have a few or many components or elements. The study of self-organization can be performed in experimental and in a theoretical way as well, but it must be mentioned that the application of modern computers is essential in the study of such kind of processes. In many cases, long simulation time is needed for obtaining self-organized structures from the locally interacting components of the system. In order to fulfill the laws of thermodynamics, the self-organizing systems are always open systems. Self-organization is often confused with emergence. The appearance of a property or feature not previously seen is the emergence, and the internal dynamical process which maintains the structure without external control is the self-organization (De Wolf & Holvoet, 2005).

45.2 HISTORICAL ORIGIN(S)

The concept of the problem of self-organization goes back to the ancient Greek philosophy to the question of formation organized structures from the chaos in space or in time or in both. The meaning of this term was changing in history, and even in our days, it does not have the same meaning at different authors. Descartes studied the philosophical problem of organized matter formation using the laws of nature (Descartes, 1637). Kant used the first time the term self-organization (Kant, 1790, 1993) when he studied the special property that distinguishes a living system from a collection of inanimate matter. In the 19th century, the invention of the first and second laws of thermodynamics were crucial for understanding self-organization. The first law describes the conservation of energy by supposing transformations between mechanical, thermal, electric, chemical and other versions of energy. The second law of thermodynamics seems to give some constraint to self-organization. It states namely that entropy, the measure of disorder always increases or not changes over time in isolated systems.

Norbert Wiener in 1948 published his book about the "Control and communication in the animal and the machine" (Wiener, 1948) where he defined cybernetics which has many common meaning with self-organization. He discussed in more details the idea of self-organization in the second edition of his book (Wiener, 1961).

The modern concept of self-organization goes back to Ross Ashby (1947, 1962). In his paper, he studied the following topics as: what is an organization, the relation between the whole and parts, what machines mean in general, what good organization means, what are the self-organizing systems? He states that self-organization can have two quite different meaning: (a) A system that starts with its parts separate and whose parts then act so that they change towards forming connections of some type. (b) The other example is a system of electrolytic centers, in which the growth of a filament from one electrode is at first little affected by growth at the other electrodes; then the growths become more and more affected by one another as filaments approach the other electrodes. In general, such a system can be more simply characterized as self-connecting. In this definition of self-organization, there is a change of independence between the parts to some form of connection (Ashby, 1947, 1962). At the end of his paper, Ashby commended as a program for research, the identification of the physical basis of the brain's memory stores.

In 1952, Turing suggested that a system of chemical substances, called morphogens, reacting together and diffusing through a tissue, is adequate to account for the main phenomena of morphogenesis (Turing, 1952). Specifically, he considered a system of two chemicals the inhibitor and the activator. Using mathematical relations, he showed that if the diffusion of the inhibitor was greater than that of the activator, than diffusion driven instability could result. There is a recent explanation of this self-organization process in Maini et al., (1997) as the following. Let be a supply of the inhibitor A and the activator B distributed at local sites with the limitation of its growth concentration. A can diffuse but B has only small diffusivity. The A + B→ 2B reaction will only occur where there has been seeding with a high concentration of B. In this case, Turing had the following two predictions: (a) A structure of high A concentration between the spots of high B concentration develops spontaneously even from an initially almost homogeneous A and B. (b) The final pattern does not necessarily depend on the initial distributions. It depends on the local supply and demand due to diffusion and reaction.

Although there were many experimental and theoretical work on Turing structures, the first experimental realizations were only in the beginning of the 90s (Castets et al., 1990; De Kepper et al., 1991). There is a detailed description of this model in reference (Maini et al., 1997).

In the field of the time-dependent pattern formation, the most known oscillatory reaction is the Belousov-Zhabotinsky reaction. It is a class of chemical reactions occurring in the vibrational mode, in which some parameters of the reaction as color, the concentration of components, temperature, etc. are changed periodically, forming a complex spatial and temporal structure of the reaction medium. In this reaction, bromate ions oxidize malonic acid in a reaction catalyzed by cerium Ce^{3+} and Fe^{3+}. The formed periodic oscillation between the Ce^{3+} and Ce^{4+} concentrations can be visualized if the cerium ions are replaced by the catalysts Fe^{2+} and Fe^{3+}. Belousov wanted to study the Krebs cycle by trying to find its inorganic counterpart. In 1951 he has found that a solution of citric acid in water with acidified bromate as oxidant and yellow ceric ions as a catalyst, turned colorless and returned to yellow periodically for as long as an hour at room temperature while effervescing carbon dioxide. The publication of these results were first rejected by the referees and Belousov could only publish a short abstract in 1959 (Belousov, 1959) about the story of this reaction and on the role of Zhabotinsky (*see* Winfree,1984).

Some other important publications concerning self-organization are about nonequilibrium systems (Nicolis and Prigogine, 1977), on flock of birds and self-driven particles (Reynolds 1987; Vicsek et al., 1995) and on supramolecular chemistry (Lehn, 2002). More detailed historical background of self-organization can be found in references (Keller, 2008, 2009; De Wolf and Holvoet, 2005; Bensaude-Vincent, 2006).

45.3 NANO-SCIENTIFIC DEVELOPMENTS AND NANO-CHEMICAL APPLICATIONS

Self-organization has a vast literature (Glansdorff and Prigogine 1971; Feltz et al., 2006) containing several topics from physics, chemistry, and biology. It can be applied for producing supercapacitor electrodes (Kim et al., 2015) and sub-10 nm nanostructures (Otsuka et al., 2015) or for the dynamics of folded proteins (McCammon et al., 1977). Here we focus on the formation of carbon nanostructures as fullerenes and nanotubes. The most important procedures in which fullerenes can be produced are the laser evaporation (Kroto et al., 1985), and the arc deposition (Krätschmer et al., 1990). In all of the other experimental conditions, it looked that the fullerenes and nanotubes were produced in some kind of self-organizing way. Any traditional chemical synthesis was unsuccessful (Scott et al., 2002; Kabdulov et al., 2013).

In order to find the reasons of the self-organized production of fullerenes, many authors tried to produce the C_{60} fullerene in molecular dynamics calculations, but mostly they could reach a cage-like structure with many defects. They also were using some kind of artificial conditions. Ballone and Milani (1990) kept the carbon atoms on the surface of a sphere for $T > 4000$ K, Chelikowsky (1991) removed and randomly replaced the energetically unfavorable atoms and Wang et al., (1992) confined the carbon atoms into a sphere.

László studied C_{60} fullerene formation in molecular dynamics simulation at helium atmosphere (László, 1998, 1999). He simulated the original laser evaporation experiment (Kroto et al., 1985) by "exploding" four different initial arrangements of carbon atoms (Laszlo, 1998, 1999). The unit cell of size 25.0 Å contained initially 60 carbon atoms and 1372 randomly distributed helium atoms. The carbon-carbon interaction was calculated width the help of the tight-binding potential of Xu et al., (1992), and the carbon-helium and helium-helium interaction were given by the Girifalco potential (Girifalco and Lad 1956; Girifalco, 1992). The first polygons were obtained about 0.7 psec. The emerging of the first cage-like structure was about 5.6 psec, but this structure was not closed cage and had wandering carbon atoms on its surface. The time of closed cage structure is about 19.6 psec. At 30.8 psec they obtained a cage-like structure of 11 pentagons, 19 hexagons, and 1 heptagon.

Irle et al., (2006) developed the "shrinking hot giant" road that leads to the formation of buckminsterfullerene C_{60}, C_{70}, and larger fullerenes.

László and Zsoldos (2012a, 2012b) presented molecular dynamics calculations where the final buckminsterfullerene C_{60}, and C_{70}, fullerenes were coded in the initial structures cut out from the graphene. Thus, the final structure was predetermined by the initial one. They developed this method for the self-organized formation of nanotubes as well (László and Zsoldos 2012a, 2012b, 2014, 2015a, 2015b). Fülep et al. (2015) presented a self-organizing method for the formation of nanotubes from two graphene nanoribbons.

KEYWORDS

- **chemistry**
- **fullerenes**
- **graphene**
- **molecular dynamics**
- **morphology**
- **nanotubes**
- **physics**
- **self-organization**
- **thermodynamics**

REFERENCES AND FURTHER READING

Ashby, W. R., (1947). The principles of the self-organizing dynamic system. *J. Gen. Psychol., 37*, 125–128.

Ashby, W. R., (1962). Principles of the self-organizing dynamic system. In: Von Foerster, H., & Zopf, G. W. Jr., (eds.), *Principles of Self-Organization: Transactions of the University of Illinois Symposium* (pp. 255–278). Pergamon Press: London.

Ballone, P., & Milani, P., (1992). Simulated annealing of carbon clusters. *Phys. Rev., B., 42*, 3201–3204.

Belousov, B. P., (1959). Periodically acting reaction and its mechanism (in Russian). *Sbor. Ref. Po Radio. Med., 147*, 145.

Bensaude-Vincent, B., (2006). *Self-Assembly, Self-Organization: A Philosophical Perspective on Converging Technologies.* Paper prepared for France/Stanford Meeting: Avignon.

Castets, V., Dulos, E., Boissonade, J., & De Kepper, P., (1990). Experimental evidence of a sustained standing Turing-type nonequilibrium chemical pattern. *Phys. Rev. Lett., 64*, 2953–2956.

Chelikowsky, J. R., (1991). Nucleation of C_{60} clusters. *Phys. Rev. Let., 67*, 2970–2973.

De Kepper, P., Castets, V., Dulos, E., & Boissonade, J., (1991). Turing-type chemical patterns in the chlorite-iodide-malonic acid reaction. *Physica. D., 49*, 161–169.

De Wolf, T., & Holvoet, T., (2005). In: Brueckner, et al., (eds.), *Emergence Versus Self-Organization: Different Concepts but Promising When Combined*: *ESOA 2004* (pp. 1–15). LNCS 3464, Springer-Verlag Berlin Heidelberg.

Descartes, R., (1637). *Discours de la Méthode.* Université du québec a Chicoutimi http://classiques.uqac.ca/classiques/Descartes/discours_methode/discours_methode.html (accessed on 19 February 2019).

Feltz, B., Crommelinck, M., & Goujon, P., (2006). *Self-Organization and Emergence in Life Sciences.* Dordrecht, Springer.

Fülep, D., Zsoldos, I., & László, I., (2015). Molecular dynamics simulations for lithographic production of carbon nanotube structures from graphene. In: Mastorakis, N., Pardalos, P. M., & Agrawal, R. P., (eds.), *New Developments in Pure and Applied Mathematics, Proceedings of the International Conference on Pure Mathematics – Applied Mathematics (PM-AM 2015).* Vienna, Austria.

Girifalco, L. A., (1992). Molecular properties of C60 in the gas and solid phase. *J. Phys. Chem., 96*, 858–861.

Girifalco, L. A., & Lad, R. A., (1956). *J. Chem. Phys., 25*, 693–697.

Glansdorff, P., & Prigogine, I., (1971). *Thermodynamic Theory of Structure, Stability, and Fluctuations.* Wiley-Interscience, London.

Irle, S., Zheng, G., Whang, Z., & Morokuma, K., (2006). The C60 formation puzzle "solved": QM/MD simulation reveal the shrinking hot giant road of the dynamic fullerene self-assembly mechanism. *J. Phys. Chem. B., 110*, 14531–14545.

Kabdulov, M., Jansen, M., & Amsharov, K. Y., (2013). Bottom-up C60 fullerene construction from fluorinated C60H21F9 precursor by laser-induced tandem cyclization. *Chem. Eur. J., 19*, 17262–17266.

Kant, I., (1770). *Kritik der Urteilskraft*, Berlin.

Kant, I., (1993). Critique of judgment. In *Great Books* (Vol. 39, p. 558, 66 paragraphs). Chicago: Encyclopedia Britannica.

Keller, E. F., (2008). Organisms, machines, and thunderstorms: A history of self-organization, Part one. *Historical Studies in the Natural Sciences, 38*, 45–75.

Keller, E. F., (2009). Organisms, machines, and thunderstorms: A history of self-organization, Part two. *Historical Studies in the Natural Sciences, 39*, 1–31.

Kim, S. K., Jung, E., Goodman, M. D., Schweizer, K. S., Tatsuda, N., Yano, K., & Braun, P. V., (2015). Self-assembly of monodisperse starburst carbon spheres into hierarchically organized nanostructured supercapacitor electrodes. *ACS Appl. Mater. Interfaces, 7*, 9128–9133.

Krätschmer, W., Lamb, L. D., Fostiropoulos, K., & Huffman, D. R., (1990). Solid C60 a new form of carbon. *Nature, 347*, 354–358.

Kroto, H. W., Heath, J. R., O'Brien, S. C., Curl, R. F., & Smalley, R. E., (1985). C60: Buckminsterfullerene. *Nature, 318*, 162–163.

László, I., (1998). Formation of cage-like C60 clusters in molecular dynamics simulations. *Europhys. Lett., 44*, 741–746.

László, I., (1999). Molecular dynamics study of the C60 molecule. *Journal of Molecular Structure (Theochem), 463*, 181–184.

László, I., & Zsoldos, I., (2012a). Graphene-based molecular dynamics nanolithography of fullerenes, nanotubes and other carbon structures. *Europhys. Lett., 99*, 63001-p5.

László, I., & Zsoldos, I., (2012b). Molecular dynamics simulation of carbon nanostructures: The C60 buckminsterfullerene. *Physica Status Solidi, B., 249*, 2616–2619.

László, I., & Zsoldos, I., (2014). Molecular dynamics simulation of carbon nanostructures: The D5h C70 fullerene. *Physica, E., 56*, 422–426.

László, I., & Zsoldos, I., (2015a). Origami: Self-organizing pole hexagonal carbon structures for the formation of fullerenes, nanotubes and other carbon structures. In: Putz, M. V., & Ori, O., (eds.), *Exotic Properties of Carbon Nanomatter Carbon Materials: Chemistry and Physics.* Springer Science + Business Media Dordrecht.

László, I., & Zsoldos, I., (2015b). Molecular dynamics simulation of carbon nanostructures: The nanotubes. In: Ashrafi, A. R., (ed.), *Distances Symmetry and Topology in Carbon Nanomaterials Carbon Materials: Chemistry and Physics.* Springer Science + Business Media Dordrecht.

Lehn, J. M., (2002). Toward complex matter: Supramolecular chemistry and self-organization. *PNAS, 99*, 4763–4768.

Maini, P. K., Painter, K. J., & Chau, H. N. P., (1997). Spatial pattern formation in chemical and biological systems. *E. J. Chem. Soc. Faraday Trans., 93*, 3601–3610.

McCammon, J. A., Gelin, B. R., & Karplus, M., (1977). Dynamics of folded proteins. *Nature, 267*, 585–590.

Nicolis, G., & Prigogine, I., (1977). *Self-Organization in Nonequilibrium Systems.* Wiley, New York.

Otsuka, I., Zhang, Y., Isono, T., Rochas, C., Kakuchi, T., Satoh, T., & Borsali, R., (2015). Sub–10nm scale nanostructures in self-organized linear di- and triblock copolymers and microarm star copolymers consisting of maltoheptaose and polystyrene. *Macromolecules, 48*, 1509–1517.

Reynolds, C. W., (1987). Flocks, herds, and schools: A distributed behavioral model. *Computer Graphics, 21*, 25–31.

Scott, L. T., Boorum, M. M., McMahon, B. J., Hagen, S., Mack, J., Blank, J., Wegner, H., & De Meijere, A., (2002). A rational chemical synthesis of C60. *Science, 295*, 1500–1503.

Turing, A. M., (1952). MThe chemical basis of morphogenesis. *Phil. Trans. of the Roy. Soc. of London, Series B, Biological Sciences, 237*, 37–72.

Verlet, L., (1967). Computer "experiment" on classical fluids. I. Thermodynamical properties of Lennard-Jones molecules. *Phys. Rev., 159*, 98–103.

Vicsek, T., Czirók, A., Ben-Jacob, E., Cohen, I., & Shochet, O., (1995). Novel type of phase transition in a system of self-driven particles. *Phys. Rev. Lett., 75*, 1226–1229.

Wang, C. Z., Xu, C. H., Chan, C. T., & Ho, K. M., (1992). Disintegration and formation of C60. *J. Phys. Chem., 96*, 3563–3565.

Watanabe, H., Suzuki, M., & Ito, N., (2013). CHuge-scale molecular dynamics simulation of multi-bubble nuclei. *Comput. Phys. Commun., 184*, 2775–2784.

Wiener, N., (1948). *Cybernetics or Control and Communication in the Animal and the Machine.* MIT Press: Cambridge.

Wiener, N., (1961). *Cybernetics or Control and Communication in the Animal and the Machine.* MIT Press, Second edition: Cambridge.

Winfree, A. T., (1984). The prehistory of the Belousov-Zhabotinsky oscillator. *J. of Chem. Edu., 61*, 661–663.

Xu, C. H., Whang, C. Z., Chan, C. T., & Ho, K. M., (1992). A transferable tight-binding potential for carbon tight-binding study of grain boundaries in silicone. *J. Phys.: Condens. Matter, 4*, 6047–6054.

Zheng, G., Irle, S., Elstner, M., & Morokuma, K., (2004). Quantum chemical molecular dynamics model study of fullerene formation from open-ended carbon nanotubes. *J. Phys. Chem. A., 108*, 3182–3194.

CHAPTER 46

Spin Transport Electronics

ARANYA B. BHATTACHERJEE[1] and SUMAN DUDEJA[2]

[1]*Department of Physics, Birla Institute of Technology and Science, Pilani, Hyderabad Campus, India*

[2]*Department of Chemistry, ARSD College, University of Delhi, New Delhi–110021, India*

46.1 DEFINITION

Spin transport electronics or spintronics is the study of the electron spin and associated magnetic field in addition to its charge in quantum transport in solid state devices. All spintronic devices are based on the following three schemes: (a) information is stored in electronic spins as a particular spin orientation (up or down), (b) the spins, being part of mobile electrons, carry the information along a wire when electrons are transported, and (c) the information that is stored in the spins are read at a terminal. Spin orientation of conduction electrons survives for a longer time (nanoseconds), compared to electron momentum (femtoseconds) and this makes spintronic devices attractive for memory storage, magnetic sensors applications, and quantum computing.

46.2 HISTORICAL ORIGIN(S)

The experimental discoveries related to spin-dependent electron transport in solid state physics in the 1970s and 1980s led to the emergence of the concept of spintronics. The first experiments in the 1970s on spintronics were done by Meservey and Tedrow using ferromagnetic/superconducting junctions and by Julliere using magnetic tunnel junctions. In 1985 Johnson and Silsbee observed transport of spin-polarized electron from a ferromagnetic metal to a normal metal, and in 1988, Albert Fert and Peter Gruenberg discovered giant magnetoresistance independently. Pioneers in the theoretical study include Rashba in 1960 and Datta & Das in 1990.

46.3 NANO-SCIENTIFIC DEVELOPMENT(S)

Spin-polarized current can be generated in a metal by passing current through a ferromagnetic material. Giant magnetoresistance is one such example. A GMR device consists of two layers of ferromagnetic material separated by an insulating layer. Metal-based spintronics devices include tunnel magnetoresistance, spin transfer torque, and spin-wave logic device. Motorola developed a first-generation 256 kb magneto-resistive random-access memory (MRAM) based on a single magnetic tunnel junction and a single transistor that has a read/write cycle of under 50 nanoseconds. Spintronics has potential applications in quantum information processing and quantum computation.

46.4 NANO-CHEMICAL APPLICATION(S)

Spintronics applied to organic chemistry has developed into a potentially promising field with rich physics and new applications such as organic spin valves, bipolar spin-valves, and hybrid organic/inorganic light emitting diodes. The proposal of injecting spin-polarized electrons into organic semiconductors and controlling their properties is an attempt to improve the efficiency of organic light-emitting diodes (OLEDs). The spin-valve effect has been observed in a hybrid inorganic-organic device, and the intrinsic resistance variation in an organic semiconductor when a small magnetic field is applied is termed as organic magnetoresistance (OMAR). These observations clearly demonstrate the fact that organic semiconductors can be used for spintronics.

46.5 MULTI-/TRANS-DISCIPLINARY CONNECTION(S)

A solid-state quantum-computing device based on electron spins confined within quantum dots was proposed by Loss and DiVincenzo in 1988. The applications of spin transport can also be extended to information processing devices such as spin transistors, as proposed by Datta & Das in 1990.

46.6   OPEN ISSUES

Coherent control of the ferromagnet-organic interface is one of the major issues of organic spintronics. Conventional micro-fabrication techniques cannot be used because of the extremely fragile nature of organic materials. Non-destructive contacting of organic materials is important in organic spintronics. Not only s reliable contacts but also well-defined and clean organic spacer materials are needed. Defects and impurities in the organic materials also give rise to unreliable observations. Even in the absence of defects or impurities, the combination of different materials can lead to new unknown behavior.

KEYWORDS

- **giant magnetoresistance**
- **spin transport electronics**
- **spintronics**

REFERENCES AND FURTHER READING

Baibich, M. N., Broto, J. M., Fert, A., Nguyen Van Dau, F. N., Petroff, F., Etienne, P., et al., (1988). Giant magnetoresistance of (001) Fe/(001)Cr magnetic superlattices. *Physical Review Letters, 61, 2472.*

Binasch, G., Grünberg, P., Saurenbach, F., & Zinn, W., (1989). Enhanced magnetoresistance in layered magnetic structures with antiferromagnetic interlayer exchange. *Physical Review B., 39, 4828.*

CHAPTER 47

Supramolecular Chemistry

MARGHERITA VENTURI

Department of Chemistry "Giacomo Ciamician," University of Bologna, Italy

47.1 DEFINITION

In a historical perspective, as pointed out by J.-M. Lehn (1992), supramolecular chemistry originated from Paul Ehrlich's receptor idea, Alfred Werner's coordination chemistry, and Emil Fischer's lock-and-key image. It was only in the 1970s, however, that some fundamental concepts such as molecular recognition, pre-organization, self-assembly, etc., were introduced and supramolecular chemistry began to emerge as a discipline. This new, highly interdisciplinary branch of chemistry was consecrated by the award of the Nobel Prize in Chemistry to C. J. Pedersen (1988), D. J. Cram (1988), and J.-M. Lehn (1988) in 1987, and has developed at an astonishingly fast rate during the last three decades (Balzani et al., 2008).

The most authoritative and widely accepted definition of supramolecular chemistry is that given by J.-M Lehn, namely "the chemistry beyond the molecule, bearing on organized entities of higher complexity that result from the association of two or more chemical species held together by intermolecular forces" (Lehn, 1988, 1995).

47.1.1 THE NEED FOR A WIDER DEFINITION

As it is often the case, however, problems arise as soon as a definition is established; for example, the definition of organometallic chemistry (Steed and Atwood, 2000) as "the chemistry of compounds with metal-to-carbon bonds" rules out Wilkinson's compound, $[RhCl(PPh_3)_3]$ (PPh_3 = triphenylphosphine), which is perhaps the most important organometallic catalyst.

A first problem presented by the above-mentioned classical definition of supramolecular chemistry concerns whether or not metal-ligand bonds can be considered intermolecular forces. If yes, complexes like $[Ru(bpy)_3]^{2+}$ (bpy = 2,2'-bipyridine), that are usually considered molecules (Watts 1983; Balzani and Juris 2001), should be defined as supramolecular species; if not, systems like the $[Eu \subset bpy.bpy.bpy]^{3+}$ cryptate that are usually considered supramolecular antenna devices (Lehn 1987) should, in fact, be defined as molecules (Figure 47.1).

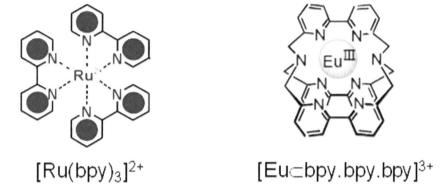

$$[Ru(bpy)_3]^{2+} \qquad\qquad [Eu \subset bpy.bpy.bpy]^{3+}$$

FIGURE 47.1 $[Ru(bpy)_3]^{2+}$ and $[Eu \subset bpy.bpy.bpy]^{3+}$: can these compounds be considered molecular or supramolecular species?

There is, however, a more general problem. Broadly speaking, one can say that with supramolecular chemistry, there has been a change in focus from molecules to molecular assemblies or multicomponent systems. According to the original definition, when the components of a chemical system are linked by covalent bonds, the system should not be considered a supramolecular species, but a molecule. This point is particularly important in dealing with molecular-level devices and machines, that are usually multi-component systems in which the components can be linked by chemical bonds of various nature.

Consider, for example, the three systems (de Rege et al., 1995) shown in Figure 47.2 that play the role of molecular-level charge-separation devices.

In each of them, two components, a Zn^{II}-porphyrin and a Fe^{III}-porphyrin, can be immediately singled out. In **1**, these two components are linked by a hydrogen-bonded bridge, that is, by intermolecular forces, whereas in **2** and **3** they are linked by covalent bonds. According to the above-reported classical definition of supramolecular chemistry, **1** is surely a supramolecular species, whereas **2** and **3** are (large) molecules. In each of the three systems, the two components substantially maintain their intrinsic properties and, upon light excitation, electron transfer occurs from the Zn^{II}-porphyrin unit to the Fe^{III}-porphyrin one. The values of the rate constants for the photoinduced electron transfer ($k_{el} = 8.1\times10^9$, 8.8×10^9, and $4.3\times10^9\,s^{-1}$ for **1**, **2**, and **3**, respectively) show that the electronic interaction between the two components in **1** is comparable to that in **2**, and is even stronger than that in **3**. Clearly, as far as the photoinduced electron transfer is concerned, it would sound strange to say that **1** is a supramolecular species, and **2** and **3** are molecules.

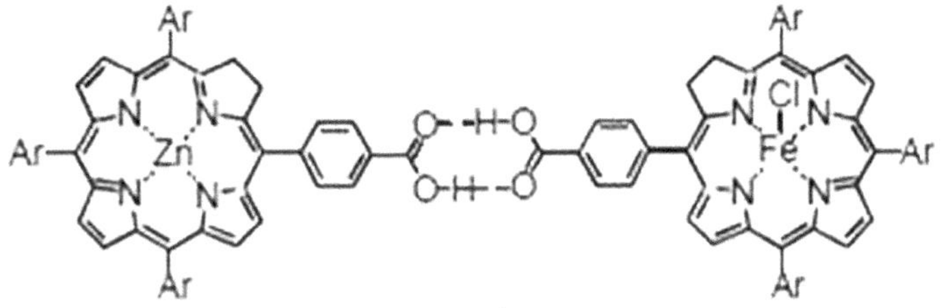

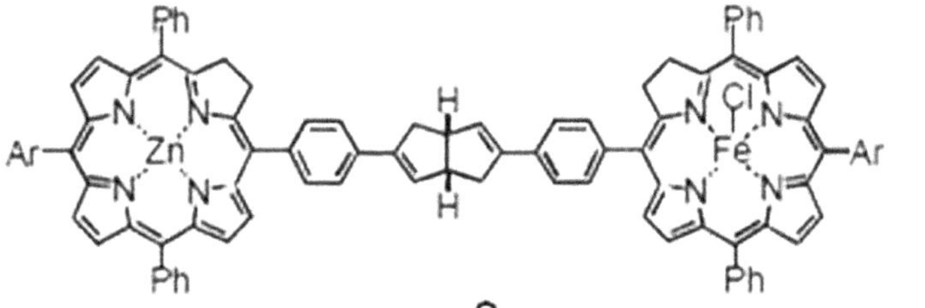

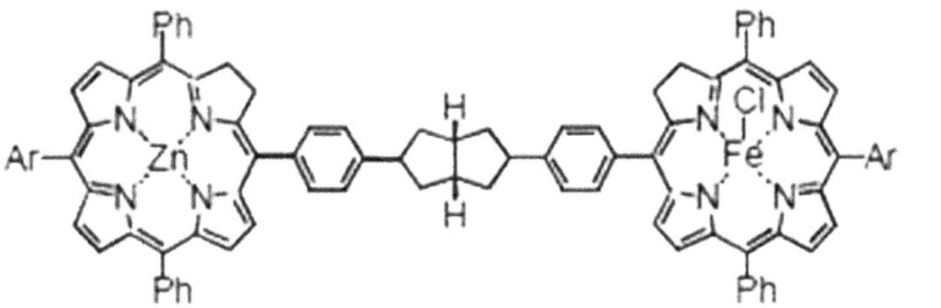

FIGURE 47.2 Three dyads that possess Zn^{II}-porphyrin and Fe^{III}-porphyrin units linked by an H-bonded bridge (**1**, Ar = 3,4,5-trimethoxyphenyl), a partially unsaturated bridge (**2**, Ar = *p*-methoxyphenyl, Ph = phenyl), and a saturated bridge (**3**, Ar = *p*-methoxyphenyl, Ph = phenyl): are they molecular or supramolecular species?

Another example of the difficulty in applying the original definition of supramolecular chemistry is encountered with pseudorotaxanes, rotaxanes, and catenanes (Figure 47.3). A pseudorotaxane, for example, 4^{4+}, as any other type of adduct, can be clearly defined a supramolecular species, whereas a rotaxane, for example, 5^{4+}, and a catenane, e.g., 6^{4+} (Anelli et al., 1992), in spite of the fact that they are more complex species than pseudorotaxanes, should be called molecules according to the classical definition.

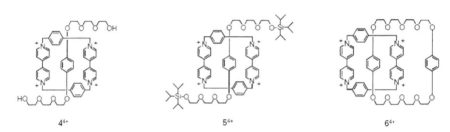

4^{4+} 5^{4+} 6^{4+}

FIGURE 47.3   Pseudorotaxane (4^{4+}), rotaxane (5^{4+}), and catenane (6^{4+}): can they be considered molecular or supramolecular species?

It means that the classical definition of supramolecular chemistry as "the chemistry beyond the molecule" is quite useful, but, from the functional viewpoint, the distinction between what is molecular and what is supramolecular should be better based on other grounds, rather than on the nature of the bonds that link the components.

This aspect can be further clarified by the following example. It has been proposed (Balzani and Scandola, 1991, 1996; Balzani, 1992), and subsequently widely accepted that in the case of chemical systems which are investigated from the viewpoint of the effects caused by external stimulations, the definition of supramolecular species can be based on the degree of intercomponent electronic interactions. This concept is illustrated in Figure 47.4 for the case of a photochemical stimulation. The system A~B, consisting of two units (~ indicates any type of "bond" that keeps the units together), can be defined a supramolecular species if light absorption leads to excited states that are substantially localized on either A or B, or causes an electron transfer from A to B (or vice versa). By contrast, when the excited states are substantially delocalized on the entire system, the species can be better considered as a large molecule. Similarly, oxidation and reduction of a supramolecular species can substantially be described as oxidation and reduction of specific units, whereas oxidation and reduction of a large molecule lead to species in which the hole or the electron are delocalized on the entire system (Balzani and Scandola, 1996).

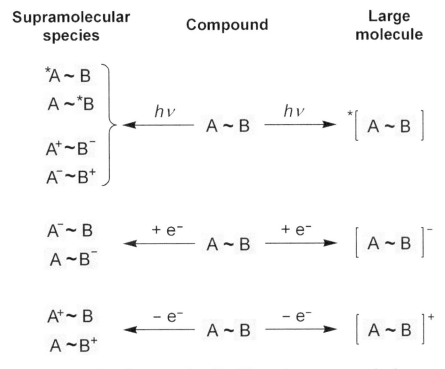

FIGURE 47.4 Schematic representation of the difference between a supramolecular system and a large molecule based on the effects caused by a photon or an electron input.

In more general terms, when the interaction energy between units is small compared to the other relevant energy parameters, a system can be considered a supramolecular species, regardless of the nature of the bonds that link the units. Therefore, species made of covalently-linked (but weakly interacting) components, such as **2** and **3** shown in Figure 47.2, 5^{4+} and 6^{4+} reported in Figure 47.3, as well as polynuclear transition metal complexes (Balzani et al., 1996), like those shown in Figure 47.5, can be taken as belonging to the supramolecular domain when they are stimulated by photons or electrons. It can be noted that the properties of each component of a supramolecular species, that is, of an assembly of weakly interacting molecular components, can be known from the study of the isolated components or of suitable model compounds.

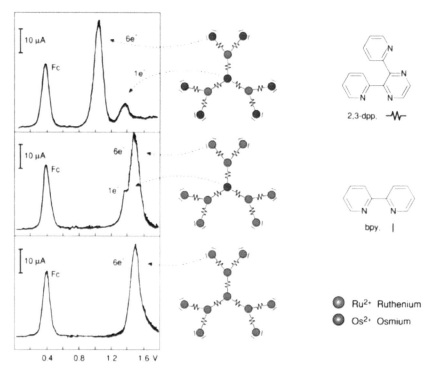

FIGURE 47.5   Redox patterns (differential pulse voltammetric peaks, potentials in V *versus* the SCE) for decanuclear metal complexes showing the selective oxidation of the different metals present in the structure. Peak Fc: oxidation of ferrocene, used as an internal standard.

47.2   ARTIFICIAL SUPRAMOLECULAR SYSTEMS: SELF-ASSEMBLY VERSUS COVALENT SYNTHETIC DESIGN

Self-assembly and self-organization (Lehn 1995; Lindoy and Atkinson, 2000) are dominant processes in the chemistry of living biological systems (Cramer, 1993; Jones, 2004; Goodsell, 2004). For example, light-harvesting antennae of bacterial photosynthesis and protein motors (Figure 47.6) are formed by self-assembling and self-organization of a great number of molecular components (Goodsell, 1996).

It is indeed amazing how nature is capable of mastering weak intermolecular interactions to construct supramolecular devices and machines involved in the processes that lie at the very foundation of the life evolution (Cramer, 1993; Jones, 2004; Goodsell, 2004; Vale and Milligan, 2000; Ritz and Damjanovic, 2002). The first reason that can explain why nature exploits so extensively the molecule self-assembly is that properties emerge not otherwise present, even conceptually, in separated molecules. In fact, because of the interactions established between molecules, supramolecular systems possess superior information contents relative to those of molecules that make up the supramolecular system. Accordingly, supramolecular systems can perform functions not otherwise possible by the component molecules. The second reason relies on the fact that the intermolecular interactions, being weak, can easily occur and broken up thereby conferring to the systems a high flexibility and enabling to correct the initial errors through successive attempts, and explore new structures that could prove very useful from an evolutionary viewpoint.

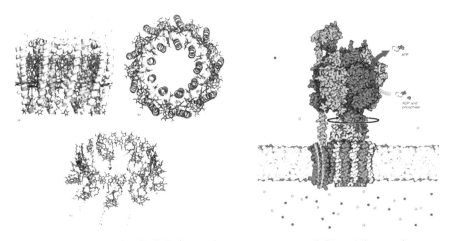

FIGURE 47.6 Example of a light-harvesting antennae system (left), and the protein motor that accomplishes the ATP synthesis (right).

Construction of artificial supramolecular structures by self-organization needs suitably programmed molecular components (Lehn 1995; Balzani et al., 2008) and full control of all the weak intermolecular forces (including solute–solvent interactions) involved in the thermodynamically driven formation of the desired system. This difficult task can be successfully fulfilled by careful chemical design of the molecular components. Several supramolecular structures (e.g., helicates, grids, capsules, Figure 47.7) have indeed been obtained by self-association and self-organization of artificial molecular components.

However, the construction of artificial supramolecular devices and machines by self-assembling and self-organization is a much more difficult task, since the various molecular components have to be programmed not only for their self-assembling into a structurally organized system, but also for their functional integration, as required by the operation that the device or the machine is expected to perform. Furthermore, supramolecular systems based on weak interactions are fragile, since they can be easily disassembled by external perturbations (e.g., change of solvent, change of pH); although this property is very important in nature, as already noticed, in the artificial field it can be exploited only for obtaining particular functions (Balzani et al., 2000, 2002). In most cases, indeed, the device or machine should not undergo disassembly.

Therefore, artificial devices and machines are often constructed by following a design based on covalent interconnecting bonds by using the powerful strategies and techniques of modern synthetic chemistry (Balzani et al., 2008).

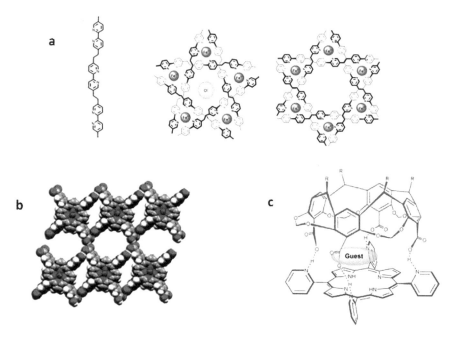

FIGURE 47.7 a) Examples of self-assembled circular helicates; b) a grid formed by Zn-porphyrins capable of undergoing hydrogen bonds; (c) a hydrogen bonded capsule in which small molecules can be hosted.

47.3   SUPRAMOLECULAR CHEMISTRY BETWEEN ART, AND NANOTECHNOLOGY

The supramolecular "bottom-up" approach opens virtually unlimited possibilities concerning design and construction of multicomponent species of nanoscale dimensions. Leonardo da Vinci did not know chemistry; nevertheless, his comment (The literary works of Leonardo da Vinci) *"Where nature finishes producing its species, there man begins with natural things to make with the aid of this nature an infinite number of species,"* is quite appropriate for the outstanding development of artificial supramolecular chemistry (Lehn, 1995).

In several cases, supramolecular species have fascinating shapes (Vögtle, 1992) that recall those of macroscopic objects and structures encountered in everyday life. Two examples of aesthetically appealing supramolecular species are shown in Figure 47.8.

FIGURE 47.8 Two fascinating nanometric-scale supramolecular species and the corresponding macroscopic counterparts. Left: a resorcinarene-calixarene carcerand and the Battistero of Pisa (Italy); right: a norbornylogous-type compound and the medieval bridge in Magliano Vetere (Italy).

As Primo Levi noticed (Levi, 2000), *"In fact it also happens in chemistry as in architecture that 'beautiful' edifices, that is, symmetrical and simple, are also the most sturdy: in short, the same thing happens with molecules as with the cupolas of cathedrals or the arches of bridges."*

Interestingly, while chemists build up supramolecular systems that mimic macroscopic works of art (Soi and Hirsch, 1998) (Figure 47.9), artists (Béla) begin to take supramolecular compounds as models for creating beautiful sculptures (Figure 47.10).

Most importantly, however, the concepts of supramolecular (multicomponent) chemistry can be profitably used to design and construct a great variety of artificial compounds capable of performing useful functions. It is amazing to notice that the bottom-up approach to the construction supramolecular systems was poetically anticipated by Primo Levi in his book *The Monkey's Wrench* (Levi, 1995): *"It is reasonable to proceed a bit at a time, first attaching two pieces, then adding a third, and so on. ... we don't have those tweezers we often dream of at night.... If we had those tweezers (and it's possible that one day, we will), we would have managed to create some lovely things that so far only the Almighty has made, for example, to assemble – perhaps not a frog or a dragonfly – but at least a microbe or the spore of a mold."* Up until now, nobody has succeeded to construct a chemical system as complex as a microbe or the spore of mold. In recent years, however, a number of very simple molecular-level devices and machines have been build up by exploiting the principles of supramolecular chemistry, namely 1) compounds for transfer, transport, and collection of electrons or electronic energy; 2) multistate/multifunctional systems; and 3) compounds capable of performing mechanical movements (machines) (Balzani et al., 2008). This outstanding progress shows that looking at supramolecular chemistry from the viewpoint of functions with references to devices and machines of the macroscopic world, is a very interesting exercise that introduces novel concepts, injects daring ideas, and stimulates creativity in the fields of chemistry and nanotechnology.

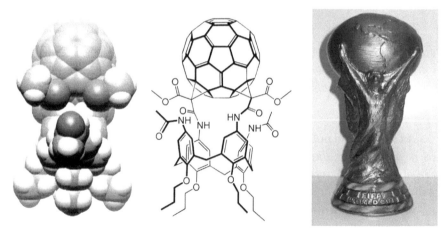

FIGURE 47.9   A supramolecular system consisting of a fullerene covalently linked to a calixarene. Center: a classical chemical representation; left: a PM3-calculated space filling model showing the shape relationship of this supramolecular structure with the football World Cup (right).

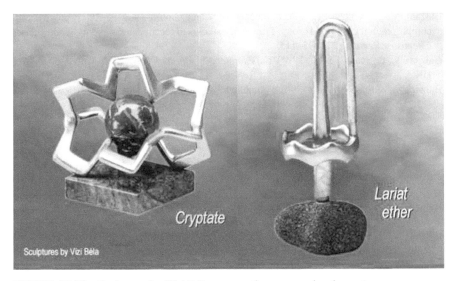

FIGURE 47.10   Sculptures by Vízi Béla representing supramolecular systems.

KEYWORDS

- **Chemistry beyond the molecule**
- **Cram**
- **intermolecular forces**
- **Lehn**
- **multicomponent systems**
- **Pedersen**
- **supramolecular systems vs. large molecules**

REFERENCES AND FURTHER READING

Anelli, P. L., Ashton, P. R., Ballardini, R., Balzani, V., Delgado, M., Gandolfi, M. T., et al., (1992). Molecular Meccano. 1. [2]rotaxanes and a [2]catenane made to order. *Journal of the American Chemical Society, 114,* 193–218.

Balzani, V., (1992). Supramolecular photochemistry. *Tetrahedron, 48,* 10443–10514.

Balzani, V., & Juris, A., (2001). Photochemistry and photophysics of Ru(II)-polypyridine complexes in the Bologna group. From early studies to recent developments. *Coordination Chemistry Reviews, 211*, 97–115.

Balzani, V., & Scandola, F., (1991). *Supramolecular Photochemistry*. Horwood: Chichester, UK.

Balzani, V., & Scandola, F., (1996). Photochemical and photophysical devices. In: Atwood, J. L., Davies, J. E. D., & MacNicol, D. D., & Vögtle, F., (eds.), *Comprehensive Supramolecular Chemistry* (Vol. 10, pp. 687–746). Pergamon Press, Oxford, UK.

Balzani, V., Credi, A., & Venturi, M., (2000). In: Shibasaki, M., Stoddart, J. F., & Vögtle, F., (eds.), *Stimulating Concepts in Chemistry* (pp. 255–266). Wiley-VCH: Weinheim, Germany.

Balzani, V., Credi, A., & Venturi, M., (2002). Controlled disassembling of self-assembling systems: Toward artificial molecular-level devices and machines. *Proc. Natl. Acad. Sci. USA, 99*, 4814–4817.

Balzani, V., Credi, A., & Venturi, M., (2008). *Molecular Devices and Machines – Concepts and Perspectives for the Nanoworld*. Wiley-VCH: Weinheim, Germany.

Balzani, V., Juris, A., Venturi, M., Campagna, S., & Serroni, S., (1996). Luminescent and redox-active polunuclear transition metal complexes. *Chemical Reviews, 96*, 759–833.

Béla, V. *Booklet on Chemistry in Sculptures and Catalog*. Vizi Béla, József A.U.4, Felsoörs H–8227 (Hungary).

Cram, D. J., (1988). The design of molecular hosts, guests, and their complexes (Nobel Lecture). *Angewandte Chemie International Edition, 27*, 1009–1020.

Cramer, F., (1993). *Chaos and Order, the Complex Structure of Living Systems*. VCH: Weinheim, Germany.

de Rege, P. J. F., Williams, S. A., & Therien, M. J., (1995). Direct evaluation of electronic coupling mediated by hydrogen bonds: Implications for biological electron transfer. *Science, 269*, 1409–1413.

Goodsell, D. S., (1996). *Our Molecular Nature: The Body's Motors, Machines, and Messages*. Copernicus: New York, USA.

Goodsell, D. S., (2004). *Bionanotechnology*. Wiley-Liss: Hoboken, USA.

Jones, R. A. L., (2004). *Soft Machines, Nanotechnology and Life*. Oxford University Press: Oxford, UK.

Lehn, J.-M., (1987). Photophysical and photochemical aspects of supramolecular chemistry. In: Balzani, V., (ed.), *Supramolecular Photochemistry* (pp. 29–43). Riedel: Dordrecht, Holland.

Lehn, J.-M., (1988). Supramolecular chemistry – scope and perspectives molecules, supermolecules, and molecular devices (Nobel Lecture). *Angewandte Chemie International Edition, 27*, 89–112.

Lehn, J.-M., (1992). From coordination chemistry to supramolecular chemistry. In: Williams, A. F., Floriani, C., & Merbach, A. E., (eds.), *Perspectives in Coordination Chemistry* (pp. 447–462). VHCA: Basel, Switzerland.

Lehn, J.-M., (1995). *Supramolecular Chemistry: Concepts and Perspectives*. Wiley-VCH: Weinheim, Germany.

Levi, P., (1995). *The Monkey's Wrench* (p. 144). Penguin Books: New York, USA.

Levi, P., (2000). *The Periodic Table* (p. 149). Penguin Books: London, UK.

Lindoy, L. F., & Atkinson, I. M., (2000). *Self-Assembly in Supramolecular Systems*. Royal Society of Chemistry: Cambridge, UK.

Pedersen, C. J., (1988). The discovery of crown ethers (Nobel Lecture). *Angewandte Chemie International Edition, 27*, 1021–1027.

Richter, J. P., & Pedretti, C., (1977). *The Literary Works of Leonardo da Vinci, Compiled and Edited from the Original Manuscripts* (p. 102). By Richter, J. P., commentary by Pedretti, C. Phaidon: Oxford, UK.

Ritz, T., & Damjanovic, A., (2002). The quantum physics of photosynthesis. *ChemPhysChem, 3*, 243–248.

Soi, A., & Hirsch, A., (1998). The molecular world cup: Synthesis of a fullerene-calix[4]arene conjugate containing two malonamide substituents within the upper rim (In a footnote to the illustration reproduced in Figure 10, the authors say that the synthesis of the nanocup was a tribute to the French football team 1998). *New Journal of Chemistry, 22*, 1337–1339.

Steed, J. W., & Atwood, J. L., (2000). *Supramolecular Chemistry*. Wiley, Chichester, UK.

Vale, R. D., & Milligan, R. A., (2000). The way things move: Looking under the hood of molecular motor proteins. *Science, 288*, 88–95.

Vögtle, F., (1992). *Fascinating Molecules in Organic Chemistry*. Wiley: Chichester, UK.

Watts, R. J., (1983). Ruthenium polypyridyls: A case study. *Journal of Chemical Education, 60*, 834–842.

CHAPTER 48

Tetrahedral Dual Coordinate System of Electron Orbital and Quantum Loop Theories in Three and Six Dimensions

JAMES GARNETT SAWYER II[1] and MARLA JEANNE WAGNER[2]

[1]241, Lexington Avenue, Buffalo, NY 14222, USA

[2]407, Norwood Avenue, Buffalo, NY 14222, USA

48.1 DEFINITION

The TDCS represents a deterministic pattern of spinning electrons in the periodic table of the elements (PTE), based on a variation of the electron orbital theory (EO) where electrons spin on four and six axes of symmetry that are in alignment with the (111) triangular planes of crystallography. Internal quantum loop (QL) flow of electromagnetic forces are orthogonal to three axes of symmetry and are directed into the nucleus in alignment with the (100) square planes of crystallography called octahedral ligand bonding.

48.2 HISTORICAL ORIGINS

Crystallography analyzes crystals with three units of measure (h, k, l). When $h = k = l$, a three-dimensional triangular plane is formed, which is called (111) octahedral planes (o) in crystallography. Diamond crystals and Frustrated Geometry (FG) compounds (1, 2) are composed of complex triangular lattice systems. Diamond crystals are in complex octahedral lattice systems. FG pyrochlore compounds are in complex tetrahedral lattice systems. A triangular coordinate system should be applied to crystallography to normalize the Miller index system. Self-assembly of nanocrystals exhibit properties of a dual coordinate system. Crystal formation in solution grows triangular (o) planes which we represent algebraically as (r, s, t, u, v, w).

Crystal formation includes square (*a*) planes (100) based on three axes of symmetry (*x, y, z*). Crystallography needs to be simplified as it currently has 32 classes with 230 space groups.

Ligand octahedral bonding reveals a connection between (*x, y, z*) and (*r, s, t, u, v, w*) coordinate systems. A new paradigm is presented to evaluate the geometric properties of the 92 elements with uniform polyhedral order. The periodic order of shell layers for the elements is geometrically analyzed based on the vertices, faces, and edges for the primary six-dimensional polyhedrons: tetrahedron, octahedron, cuboctahedron, truncated octahedron and truncated rhombic dodecahedron. We contend that there is a fundamental geometric order for the positions of the spinning electrons in quantum space based on the six-dimensional geometry of the cuboctahedron.

The scientific study of minerals began with the study of the geometric properties of crystals. The fundamental exterior geometry of the crystals includes the tetrahedron, octahedron, and cube (Miers, 1902). The regular dodecahedron and icosahedron were not recognized until Dan Shechtman discovered quasicrystals with five-fold symmetry (Dan Shechtman, 2011). The over 800 varieties of crystals contained conflicting identifying properties such as identical chemical compounds with different exterior geometries, and identical polyhedral geometries for different chemical compounds. The (111) triangular planes of the octahedron were recognized as a fundamental crystallographic property of polyhedrons with four axes of symmetry, such as in the cuboctahedron (*see* Figures 48.1 and 48.2). The primary geometry of this crystallographic Miller Indices system (MI) is based on the orthogonal three-dimensional (3D) cube, represented as (*h, k, l*) where (*h = k = l*) coordinates produce triangular-hexagonal planes. The science of crystallography placed known polyhedral crystals inside a cube and applied trigonometry to statically establish (*h, k, l*) ratios to determine the exterior crystal sub-system classification. The triangular planes of crystallography need to be coordinated in triangular space.

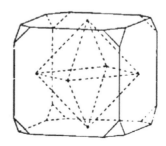

FIGURE 48.1   Triangular (111) parallel planes related to octahedron.

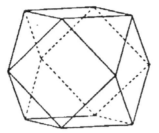

FIGURE 48.2 Triangular (111) parallel planes related to cuboctahedron.

The later development of x-ray technology by William Bragg applied science to the atomic structure of crystals (Bragg, 1915). The early x-ray analysis contained the internal atomic structure of crystals; however, the external sub-system of the MI system classified all triangular polyhedral crystals with the orthogonal cubic system. The internal atomic structures of all crystals were thus geometrically placed inside 3D squares. The Bragg brothers used the cubic KCl and NaCl crystals to model the positions of atoms in atomic space (*see* Figures 48.3–48.5). The known triangular planes of the tetrahedron and octahedron are thus represented as the internal (111) planes of the salt crystals.

All of the complex polyhedrons of crystallography, such as the tetrahedron, octahedron, cuboctahedron, truncated octahedron, rhombic dodecahedron, trapezohedron, pyritohedron, pentagonal dodecahedron, and icosahedron, are modeled on the 3D cube (except the hexagonal MI system).

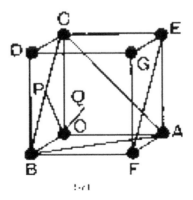

FIGURE 48.3 The internal atomic space of cubic salt crystals as equilateral triangular grids.

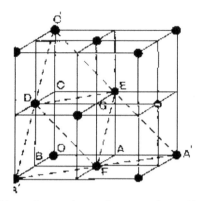

FIGURE 48.4 The positions of atoms in atomic space using cubic KCl crystal.

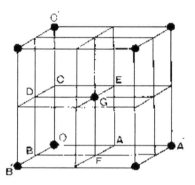

FIGURE 48.5 The positions of atoms in atomic space using cubic NaCl crystal.

The atoms of the PTE include the number of electrons per shell, where Robert Mulliken counted the number of orbitals of electrons per shell as a one-electron orbital wave function, called electron orbital theory (EO) (Mulliken, 1966). This system was applied to the periodic table after Linus Pauling generated an exception for the carbon atom bonding as a sp^3 hybrid Tetrahedral Bonding (TB), which bonds in four directions, technically violating the orthogonal 3D (x, y, z) (p_x, p_y, p_z) orbital geometry (Pauling, 1954). Stephen Hawking contends that all atomic particles are spinning like gyroscopes (Hawking, 2005). We contend that the sp^3 TB is the primary four axes of symmetry atomic bonding theory whereby the four spinning electron orbitals are in tetrahedral symmetry. The electromagnetic forces EO are acting on the four axes of symmetry of the triangular planes of the tetrahedron and/or octahedron in the first and second orbitals of the PTE. The third orbital shell includes transition elements as $(3d_{xy}, 3d_{yz}, 3d_{zx})$ represent

12 EO of Na19 – Zn30 and are acting within the 6D cuboctahedron model on six axes of symmetry. We contend orthogonal octahedral ligand type bonding which is classified as (dz2, and dx2-dy2) is acting without spinning electrons and bonding on the three axes of symmetry at the six square faces of a cuboctahedron. We contend the 12 electron bonding locations for the Fermi model of copper is located at the twelve vertices of the cuboctahedron (Naset et al., 2000). The mechanism for the internal returning force on three axes of symmetry must be based on a combination of a quantum loop gravity (QLG) mechanism (Smolin, 2002) and six- and ten-dimensional string theory (Edward 2002). We contend the electromagnetic forces of spinning electrons are radiating from the nucleus on four and/or six axes of symmetry and looping to the nucleus as theoretical quantum gravity, represented by Lee Smolin on three axes of symmetry (Smolin, 2002). 6D theory presents axes of symmetry for the spinning electrons as asymptotes for the orbital shells of the atomic elements [*see* Figures 48.6 and 48.10 for the six-dimensional coordinate system (6DC) cuboctahedron].

48.3 NANOSCIENTIFIC DEVELOPMENTS

'Six-dimension (6D) theory' presents a new form of algebra to unify triangular planes, octahedral planes of crystallography (111), frustrated geometry, and Kagome and pyrochlore lattice structures. We utilize the algebra of (r, s, t, u, v, w) to geometrically represent the edges and axes of symmetry for the primary six-dimensional polyhedrons: tetrahedron, octahedron, cuboctahedron, truncated octahedron, and truncated rhombic dodecahedron.

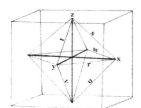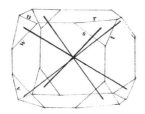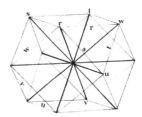

FIGURE 48.6 Three axes of symmetry (x, y, z) through centroids of cube and vertices of octahedron with (r, s, t, u, v, w) edges; four axes of symmetry through centroids of triangle faces of truncated cube with (r, s, t, u, v, w) edges of triangle faces; and six axes of symmetry (r, s, t, u, v, w) through vertices of cuboctahedron with (r, s, t, u, v, w) edges.

There are primary polyhedrons found in crystallography, and nano-crystals, which are geometrically in alignment with triangular and square planes. We see the polyhedrons vertices, faces and edges are in alignment with the three, four and six axes of symmetry of EODCS in the PTE. The dual self-alignment of triangular and square polygons with electron orbitals of Robert Mulliken on multiple axes of symmetry validates the existence of an EODCS.

Polyhedrons with edges, axes of symmetry and/or diagonals congruent to (r, s, t, u, v, w) are as follows:

1. *Platonic Solids:* tetrahedron, octahedron, icosahedron and pentag-onal dodecahedron.
2. *Archimedean Solids:* truncated tetrahedron, cuboctahedron, and truncated octahedron.
3. *Crystallographic Solids:* rhombic dodecahedron, truncated rhombic dodecahedron, double tetrahedron and hybrids such as trapezohedron.

These three groups of polyhedrons all include three, four and six axes of symmetry based on (r, s, t, u, v, w) and (x, y, z) edges. Quasicrystals are composed of icosahedrons and pentagonal dodecahedrons with five-fold symmetry and rotational edges (r, s, t, u, v, w). For the icosahedron, as trian-gular planes rotate on four axes of symmetry, they are no longer congruent to (r, s, t, u, v, w) in five-fold symmetry. These edges rotate into twelve sets of parallel edges (Figure 48.7).

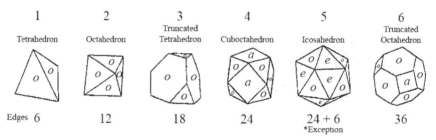

1	2	3	4	5	6
		Truncated			Truncated
Tetrahedron	Octahedron	Tetrahedron	Cuboctahedron	Icosahedron	Octahedron
Edges 6	12	18	24	24 + 6 *Exception	36

FIGURE 48.7 Six-dimensional polyhedrons with edges parallel to (r, s, t, u, v, w) and faces parallel to (o) octahedral planes, (a) cubic planes, and (e) icosahedral planes.

The cuboctahedron has eight equilateral triangular lattice structures parallel to the (o) planes of crystallography and six equilateral square planes parallel to the (a) planes of crystallography. Therefore, the basis of Six Dimension Theory on the cuboctahedron integrates the Cartesian coordinate system with the triangular lattice structures of the six edges of the tetrahedron.

The current classification of crystallography needs development to include triangular analysis of polyhedrons (Table 48.1). We contend that a system in mathematics to represent cubic, octahedral, and cuboctahedral polyhedrons must include three, four and six axes of symmetry. The polyhedrons have axes of symmetry between parallel planes, a plane and a vertex (as in the tetrahedron), and opposite vertices must be recognized in relation to Euler's formula: $V + F = E + 2$ (EF).

Presently, there are 32 classes of crystals based on seven primary crystal systems. Table 48.2 normalizes the crystallography classifications using (111) triangular grids as (r, s, t, u, v, w) planes and (100) square grids as three octahedral axes of symmetry (x, y, z). All of the crystal systems are a synergetic combination of three and six dimensions. This includes the icosahedron in four axes of rotation and the (r, s, t, u, v, w) diagonals of the rhombic dodecahedron.

The pyrite crystal is in the cubic system of crystallography with (111) triangular planes and (100) square planes. Pyritohedron is a polyhedron *and* a specific iron oxide crystal with many forms. This exhibits the rotational geometry and five-fold symmetry of quasicrystals (Figure 48.8).

The isometric system has the most uniformity where three axes of symmetry are equal, and includes forms of all the basic six-dimensional polyhedrons: tetrahedron, octahedron, cuboctahedron, rhombic dodecahedron, and truncated rhombic dodecahedron. The microstructure of the carbon atom is a tetrahedron-octahedron truss, and the macro-structure of the diamond crystal is an octahedron. Galena crystal is an example of the isometric system as it forms cuboctahedrons and double tetrahedrons. Copper crystal is another example, as it forms cuboctahedrons and truncated rhombic dodecahedrons. Nanocrystals are in the isometric system and contain triangular geometry with cuboctahedrons.

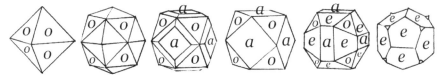

FIGURE 48.8   Pyrite crystal forms related to six-dimensional geometry.

There is a dual square and triangular polyhedral growth system found in platinum nanocrystals. In the study of platinum catalysis, nanocrystals form octahedrons, truncated octahedrons, and cuboctahedrons in a liquid solution. As the nanocrystal polyhedrons settle in solution due to gravity, self-assembly occurs along the symmetry of (100) square (a) planes. Peidong Yang and a team of Berkeley scientists manipulated Pt/Pd to produce a growth pattern

TABLE 48.1 Geometry of Crystallography Polyhedrons

Polyhedron	Faces	Vertices	Edges	Axes of Symmetry
Tetrahedron	4 triangles o	4	6	4: centroid of face to the opposite vertex
Octahedron	8 triangles o	6	12	3: vertex to opposite vertex, 4: centroid of face to centroid of the opposite face
Icosahedron	20 triangles: 8 o +12 e	12	30	3: centroid of edge to the centroid of opposite edge [5 possible sets of (x, y, z)]. 6: vertex to opposite vertex. 10: centroid of face to centroid of the opposite face
Cuboctahedron	8 triangles o 6 squares a	12	24	3: centroid of square face to centroid of the opposite square face; 4: centroid of triangle face to the centroid of opposite triangle face, 6: vertex to opposite vertex
Truncated Octahedron	8 hexagons o 6 square a	24	36	3: centroid of square face to centroid of an opposite square face, 4: centroids of hexagon face to the centroid of opposite hexagon face; 12: vertex to opposite vertex
Cube	6 squares a	8	12	3: centroid of face to centroid of opposite face, 4: vertex to opposite vertex
Pentagonal Dodecahedron	12 Pentagons e	20	30	3: centroid of edge to the centroid of opposite edge, 6: centroid of face to centroid of the opposite face; 10: vertex to opposite vertex
Rhombic Dodecahedron	12 rhombuses	14	30	3: vertex at the intersection of four faces to opposite vertex 4: vertex at the intersection of three faces to opposite vertex 6: centroid of face to centroid of the opposite face

TABLE 48.2 Polyhedral Classification of Crystallography

Name	Face Shape	# of Faces	Forms	Edges
Cube	Square	6	$a(100)$	12
Octahedron	60-degree triangles	8	$o(111)$	12
Rhombic Dodecahedron	Rhombus	12	$(r, s, t, u, v, w)d$	24
Trapezohedron	Isosceles trapezoids	24	$(211)(112)(121)$	48
Icosahedron	$8 + 12$ triangles	20	$8(111) + 12 \ e$	30
Cuboctahedron	8 triangles 6 squares	$8 + 6$	$8(111) + 6(100)$	24
Truncated octahedron	8 hexagons + 6 squares	$8 + 6$	$8(111) + 6(100)$	36

for nanocrystals from cube to cuboctahedron to the octahedron. The scientists determined that there is a dual mechanism involved in atomic formations of polyhedral crystals on the (111) triangular planes (Figure 48.9).

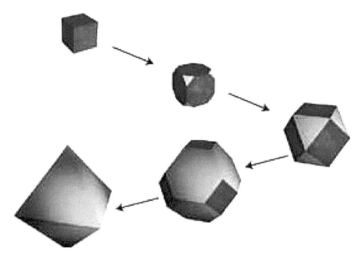

FIGURE 48.9 This figure of platinum nanocrystals is a great example of how three, four, and six axes of symmetry represent nanocrystal structure and growth (Yarris, 2011).

The nanocrystal experiment further presents a 'self-assembly' process where nanoparticles re-assemble into a superlattice system like bricks in a wall where a (*x, y, z*) (100) Three axes of symmetry structural assembly forms (Henzie Joel, Grünwald Michael, et al., 2012). Thus, there exists in nature two separate mechanisms involved in the bonding of atomic particles: Triangular assembly on six axes of symmetry and Square assembly on three axes of symmetry.

Gerhard Ertl's Nobel Prize lecture noted the importance of catalytic platinum as it contains a triangular matrix plane with (111) crystal geometry (Ertl, 2007). Platinum is an active catalyst with a hexagonal structure and morphological properties. In controlled laboratory experiments, scientists manipulate cubic platinum into a cuboctahedron where the (100) square planes diminish, and the (111) triangular planes grow. Nanocrystal diagrams include the same (*a*) square planes and (*o*) triangular planes for polyhedrons as the cubic classification of crystals. There is a geometric relationship between the square (100) planes of the cube and the triangular (111) planes of the octahedron. Nanocrystals present new evidence that the (111) triangular planes are valid growth directions of atomic space.

Groups of twelve electrons are spinning on six axes of symmetry in pairs of two per axis of symmetry, where the opposing electrons appear to spin in opposite directions relative to the center of the nucleus. Therefore, from outside the nucleus, the electrons spin clockwise and counterclockwise. The d_{xy}, d_{xz}, and d_{yz} quantum numbers are transformed into a six-dimensional coordinate system to model the twelve spinning electrons, *see* Figure 48.10.

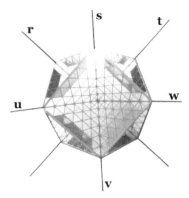

FIGURE 48.10 Twelve electrons spin on six axes of symmetry (r, s, t, u, v, w) through cuboctahedron vertices, two electrons per axis of symmetry.

When quantum numbers d_{xy}, d_{xz}, and d_{yz} are intersected, a cuboctahedron is formed of four triangular hexagon planes; this should be represented with (r, s, t, u, v, w), *see* Figure 48.10. When $d_{x^2-y^2}$ and d_{z^2} are merged, an octahedron is formed; this should be represented as three axes of symmetry (x, y, z) (*see* Figure 48.11).

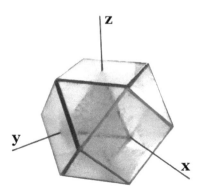

FIGURE 48.11 Ligand bonding occurs at the centroids of cuboctahedron square faces (x, y, z).

Magic Numbers are supporting evidence for the geometrization of quantum polyhedrons when directly applied to the six-dimensional periodic table of the elements (Table 48.2). We contend that six-dimensional geometry applies to the locations of spinning electrons at the vertices and faces of the six-dimensional polyhedrons. The magic numbers for the noble gases are 10, 18, 36, 54, 86 (Pauling, 1954). The six-dimensional magic numbers include:

10 = 2+8 triangular faces octahedron
18 = 10 + 8 triangular faces cuboctahedron
30 = 18+12 vertices cuboctahedron
36 = 18+12 vertices + 6 square faces of cuboctahedron
54 = 30 + 24 vertices of truncated octahedron
86 = 54 + 32 vertices of a truncated rhombic-dodecahedron

The magic number 80 applies to the number of protons for mercury. There are 24 electrons above the nuclear magic number 56 for barium, and a model truncated-octahedron has 24 electrons in the outer shell. Also, the inert properties of mercury would be justified by the 24 shielding spinning electrons in the outer shell.

Three and six-dimensional analysis of the periodic table of the elements reveals a numerical geometric pattern that can be used to deterministically map the rest mass of spinning electrons in polyhedral space for all quantum shell layers in Planck space on three, four and six axes of symmetry (Table 48.3). Px, Py, Pz are re-oriented into the periodic order for transition elements as (x, y, z) axes of symmetry associated with octahedral ligand bonding. EF is applied to this PTE.

We contend that there is a deterministic order for the spinning electrons in the periodic table. The periodic order of the shell layers for the spinning electrons follows three, four and six axes of symmetry as a (TDCS). There are quantum numerical patterns for electrons in each shell layer for all elements. The order of electrons per shell layer is directly connected to the six-dimensional geometry of polyhedrons: tetrahedron, octahedron, cuboctahedron, truncated octahedron and truncated-rhombic dodecahedron. The (TDCS) is applied to the shell layers for the spinning electrons of the periodic table of the elements (Table 48.4). The (TDCS) applies a deterministic polyhedral order for the spinning electrons on three, four and six axes of symmetry. We applied Stephen Hawking's spinning particles to the (TDCS). There is also ligand bonding found in the transition elements which presents evidence of a dual coordinate system of three axes of symmetry and six dimensions.

TABLE 48.3 Electron Locations in Polyhedral Order for the Periodic Table

Elements	Shell Layer	No. of Electrons	No. of Axes	Polyhedral Location
H – He	1	2	3, 4	Tetrahedron
Li – C	2	4	3, 4	4 vertices of tetrahedron
Li – Ne	2	8	3, 4	8 centroids of octahedron faces, in pairs of 2
Na – Si	3	4	3, 4	4 vertices of tetrahedron (shell 2:8 centroids of octahedron faces)
Na – Ar	3	8	3, 4	8 centroids of cuboctahedron triangular faces (shell 2:8 centroids of octahedron faces)
K – Zn	4	12	3, 6	12 vertices of cuboctahedron (shell 3:8 centroids of cuboctahedron triangular faces; shell 2:8 centroids of octahedron faces)
Ga – Kr	4	6	3	6 centroids of cuboctahedron square faces +12 vertices of cuboctahedron (shell 3:8 centroids of cuboctahedron triangular faces; shell 2:8 centroids of octahedron faces)
Rb – Cd	5	12	3, 6	12 centroids of rhombicdodecahedron rhombus faces (shell 4:12 vertices of cuboctahedron + 6 centroids of cuboctahedron square faces;shell 3:8 centroids of cuboctahedron triangular faces;shell 2:8 centroids of octahedron faces)
Ga – Xe	5	24	3, 6	24 vertices of truncated octahedron (shell 4: 12 vertices of cuboctahedron; shell 3:8 centroids of cuboctahedron triangular faces; shell 2:8 centroids of octahedron faces)
Cs – Ba	6	8	3, 6	8 centroids of cuboctahedron or icosahedron triangle faces in rotation (shell 5: 12 vertices of cuboctahedron + 8 centroids of cuboctahedron triangular faces; shell 4: 12 vertices of cuboctahedron; shell 2:8 centroids of octahedron faces) [outer shell could have truncated-rhombicdodecahedron symmetry (Choy, 2000)]
Ce – Rn	7	32	3, 6	32 vertices of truncated-rhombic dodecahedron (shell 6: 8 centroids of icosahedron triangles faces in rotation, shell 5: 12 vertices of cuboctahedron + 8 centroids of cuboctahedron triangular faces; shell 4: 12 vertices of cuboctahedron; shell 3:8 centroids of cuboctahedron triangular faces, shell 2:8 centroids of octahedron faces)

TABLE 48.3 *(Continued)*

Elements	Shell Layer	No. of Electrons	No. of Axes	Polyhedral Location
W – Re	7	12 + 6	3, 6	12 + 6 truncated-rhombic dodecahedron (Choy, 2000)(shell 6: 8 centroids of icosahedron triangles faces in rotation; shell 5:12 vertices of cuboctahedron + 8 centroids of cuboctahedron triangular faces; shell 4: 12 vertices of cuboctahedron; shell 3:8 centroids of cuboctahedron triangular faces, shell 2:8 centroids of octahedron faces)
Fr – U	7	6	3	6 vertices of octahedron + 32 vertices of truncated-rhombic dodecahedron (shell 6: 8 centroids of icosahedron triangles faces in rotation, shell 5: 12 vertices of cuboctahedron + 8 centroids of cuboctahedron triangular faces; shell 4: 12 vertices of cuboctahedron; shell 3:8 centroids of cuboctahedron triangular faces, shell 2:8 centroids of octahedron faces)

TABLE 48.4   Six-Dimensional Periodic Table of the Elements (Sawyer, 2017)

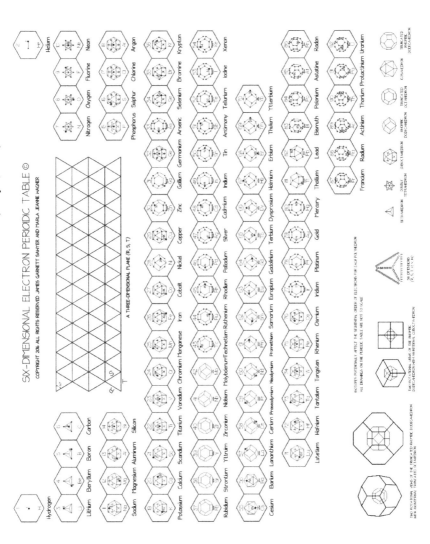

48.4 MULTI-/TRANS-DISCIPLINARY CONNECTIONS

The polyhedral geometry of nanocrystals needs to be applied to the field of Mathematics as multi-dimensional building units such as the octahedron/tetrahedron truss. The application of this polyhedral geometry to architecture and engineering will generate a stronger frame for the structure of buildings. Applying the dynamic geometry of a dual coordinate system to multidimensional electric motors with more axes of symmetry will exponentially increase their power and efficiency. Science will find the answers to the complex dynamic geometries of protein cells which are locked into a dual coordinate mechanism. Automobiles of the future will be able to levitate on quantum gravity electromagnetic fields.

48.5 OPEN ISSUES

The TDSC appears to follow the geometry of clathrate polyhedrons as a fundamental order of nature (Hoffman). The self-assembly process justifies the existence of a (TDCS) based on forces radiation from the nucleus on six dimensions and returning to the nucleus on three dimensions. This explains the dynamic tetrahedral bonding angles of many elements.

Our computer systems need to be upgraded to include triangular pixels to capture nature more accurately. Presently, our entire computer systems are built upon square space. R. Buckminster Fuller once stated in a lecture that computer technology extended the life of our three-dimensional coordinate system (X, Y, Z). The vector equilibrium model of Fuller explains the five-fold icosahedral symmetry of self-assembly in (TDCS).

KEYWORDS

- **electron orbital**
- **self-assembly**
- **three, four, and six axes of symmetry**
- **triangular and square atomic lattice**

REFERENCES

Bragg, W. H., & Bragg, W. L., (1915). *X Rays and Crystal Structure*. London G. Bell and Sons.

Choy, J., Naset, J., Chen, S. H., & Stanton, C., (2000). A database of Fermi surface in virtual reality modeling language (VRML). *Bulletin of the American Physical Society, 45*(1), 36–42. http://www.phys.ufl.edu/fermisurface/ (accessed on 14 February 2019).

Choy, T. S., Naset, J., Hershfield, S., & Stanton, C., (2000). *Periodic Table of the Fermi Surfaces of Elemental Solids*. Physics Department, University of Florida. http://www.phys.ufl.edu/fermisurface/periodic.pdf. 15 March 2000 (accessed on 14 February 2019).

Diep, H. T., (2005). *Frustrated Spin Systems*. World Scientific Pub Co Inc, Singapore.

Ertl, G., (2007). Nobel lecture. *Reactions at Surfaces: From Atoms to Complexity*. Berlin, Germany.

Hawking, S., (2005). *A Brief History of Time*. The Universe in a Nutshell. New York: Bantam Dell Publishing Group.

Hoffman, R., Zeng, T., Nesper, R., Ashcroft, N. W., Strobel, T. A., & Proserpio, D. M., (2015). Li-filled, B-substituted carbon clathrates, JACS. *J. Am. Chem. Soc., 137*, 12639–12652.

Kappraff, J., (1990). *Connections*. New York: McGraw-Hill.

Miers, H. A., (1902). *Mineralogy an Introduction to the Scientific Study of Minerals*. Macmillan and Company, London.

Mulliken, R. S., (1966). *Spectroscopy, Molecular Orbitals, and Chemical Bonding*. Nobel Lecture.

Nelson, S. A. Tulane University, http://www.tulane.edu/~sanelson/eens211/crystal_chemistry.htm.

Pauling, L. *Nobel Prize for Chemistry*. http://www.nobelprize.org/nobel_prizes/chemistry/laureates/1954/. (accessed 14 January 2016).

Sadoc, J. F., & Mosseri, R., (2006). *Geometrical Frustration*. Cambridge University Press, Cambridge.

Sawyer, J., & Wagner, M., (2017). *Periodic Table of the Elements*. Self-Published, Buffalo New York. ISBN: 978-0-692-83598-2.

Shechtman, D., (2011). *The Royal Swedish Academy of Sciences Nobel Prize in Chemistry*.

Smolin, L., (2002). *Three Roads to Quantum Gravity*.

The Quantum Basis of Atomic Orbitals. http://chemistry.umeche.maine.edu/CHY251/Quantum.html. Updated 4 September 2014 (accessed 14 January 2016).

Watkins, T. San Jose State University. Applet-magic.com. *The Magic Numbers for Nuclear Isotope Stability*. http://www.sjsu.edu/faculty/watkins/magicnumber.htm. [accessed on 20 February 2019].

Witten, E., Green, M. B., & Schwarz, J. H., (2002). *Superstring Theory* (Vol. 1, Introduction).

Yarris, L. *On the Road to Plasmonics with Silver Polyhedral Nanocrystals*. http://newscenter.lbl.gov/2011/11/22/silver-polyhedral-nanocrystals/ November 22, 2011 (accessed on 14 January 2016).

CHAPTER 49

Tight Binding Calculation

ISTVÁN LÁSZLÓ

Department of Theoretical Physics, Budapest University of Technology and Economics, H-1521 Budapest, Hungary, E-mail: laszlo@eik.bme.hu

49.1 DEFINITION

A tight-binding method is a simple approach for describing the electronic and geometrical structure of molecules solids or other material compounds. It is a quantum mechanical approximation based on the linear combination of atomic orbitals. There are one orbital one site or more orbital one-site approximations. The Hamiltonian matrix is constructed with the help of the interatomic distances. There is usually a first neighbor or limited distance approximation in the calculation of the matrix elements. The parameterization is based on experimental results or more sophisticated *ab initio* calculations. The total energy contains the quantum mechanical part and an adjusted potential function depending on the interatomic distances. It tends to be used in calculations of very large systems, with more than around a thousand atoms or for a smaller number of atoms if the number of calculated structures is large. The quality of the results depends strongly on the parameterization. In many cases, it grasps only the qualitative properties of the compound, but there are parameterizations giving results approaching the *ab initio* calculations.

49.2 HISTORICAL ORIGIN(S)

The tight binding theory is one version of the linear combination of atomic orbitals (LCAO) method it has its origin in the thesis of Felix Bloch (1928) where he was studying the electronic structure of solids. His method was a one orbital one-site approximation using a linear combination of atomic orbitals with the coefficients of the plane wave $\exp(i\mathbf{k} \cdot \mathbf{R})$ at various positions of R where the atoms were located. This approximation has a great similarity with the Hückel method (Hückel 1931; Coulson et al., 1978). It is interesting to note that Hückel cites Bloch previously mentioned paper. Starting from Bloch paper, the tight binding method developed as a simple method for calculating the electronic structure of solids. The first author who was using more than one orbital on the atomic sites was Wilson (1931) but erroneously supposed that there were no non-diagonal matrix components of energy between Bloch sums formed from the p_x, p_y, and p_z atomic orbitals. Bouckaert, Smoluchowski, and Wigner (Bouckaert et al., 1936) clarified this problem. The first authors who correctly treated the s, p_x, p_y, and p_z orbitals were Jones, Mott, and Skinner (Jones et al., 1934) but later they incorrectly treated the d orbitals (Jones and Mott, 1937). Slater and Koster presented energy integrals for crystals in terms of two-center integrals between s, p_x, p_y, p_z, d_{xy}, d_{yz}, d_{zx}, $d_{x^2-y^2}$ and $d_{3z^2-r^2}$ orbitals (Slater and Koster, 1954). In order to simplify the writing of these integrals in this publication, they introduced the so-called Slater-Koster parameters. A comprehensive application and extension of the tight binding method for solids can be found in the books by Harrison (Harrison 1980, 1989) and Papaconstantopoulos (1986). Chadi wrote the total energy of an electron-ion system as the sum of quantum-mechanical bond energy and a repulsive classical short range energy (Chadi, 1978). Originally the tight-binding method was applied for calculating the electronic structure of solids, but this repulsive potential term made it possible to use it for calculating geometric structures as well. Further justification of this repulsive energy term can be found in Paxton and Sutton publication (Paxton and Sutton 1989). Early versions of tight binding methods were rather successful in describing electronic properties of the atomic arrangements of structures where the parameters were adjusted, but they were not transferable to other structures. The first transferable tight-binding parameters were presented by Goodwin Skinner and Pettifor (Goodwin et al., 1989) and later by Sawada (1990). Adopting the ideas of Goodwin et al., transferable parameters for carbon structures were developed by Xu et al. (1992). Porezag et al. (1995) presented a density-functional-based scheme for determining the necessary parameters of common nonorthogonal tight-binding models.

49.3 NANO-SCIENTIFIC DEVELOPMENT(S)

Here we present the most important formula of the tight binding method. For the case of simplicity, we suppose that the system contains only carbon atoms. Each carbon atom is having one s orbital and three p orbitals the p_x, p_y, and p_z orbitals. The total energy is written in the form of

$$E = \sum_{k=1}^{N} \frac{P_k^2}{2M_k} + E_{bond} + E_{rep}, \tag{1}$$

where

$$E_{bond} = \sum_{i} n_i \varepsilon_i \tag{2}$$

is the tight binding electronic energy where the summation goes over all of the eigenstates of the tight binding Hamiltonian matrix **H**. The value n_i of the occupation number is 2, 1 or 0 depending on the occupation of the eigenfunctions

$$\psi_i(\mathbf{r}) = \sum_{v}^{m} C_{vi} \phi_v (\mathbf{r} \text{-} \mathbf{R}_v). \tag{3}$$

Here the atom-centered basis function $\phi_v((\mathbf{r}\text{-}\mathbf{R}_v))$ is centered around the atom with the position vector $\mathbf{R}_v$ and m equals to the number of basis functions centered on the atoms in the positions $\mathbf{R}_k$ with $1 \leq k \leq N$ where N is the number of atoms and $\mathbf{R}_v$ equals to one of the vectors $\mathbf{R}_k$. $\mathbf{P}_k$ is the momentum of the atom k with the mass M_k. The eigenvalues and the eigenvectors are determined by the equations

$$\sum_{v=1}^{m} C_{vi} \left(H_{\mu v} - \varepsilon_i S_{\mu v} \right) = 0 \tag{4}$$

For each μ and i, where

$$H_{\mu v} = \left\langle \phi_\mu \left| \mathbf{H} \right| \phi_v \right\rangle \tag{5}$$

and

$$S_{\mu v} = \left\langle \phi_\mu \middle| \phi_v \right\rangle \tag{6}$$

is the overlap integral. There are tight binding methods with orthogonalized basis functions as well (Xu et al., 1992). The $H_{\mu v}$ and $S_{\mu v}$ off-diagonal matrix elements are given with the help of the Slater-Koster parameter function $H_{sp\sigma}(R)$, $H_{pp\sigma}(R)$, $H_{ss\sigma}(R)$, $H_{pp\pi}(R)$, $H_{sp\sigma}(R)$, $H_{pp\sigma}(R)$, $H_{ss\sigma}(R)$, $H_{pp\pi}(R)$. Let us suppose that $\phi_v((\mathbf{r}\text{-}\mathbf{R}_v))$ is centered in origin ($\mathbf{R} = (0,00)$) and function $\phi_v((\mathbf{r}\text{-}\mathbf{R}_v))$ is centered on the x-axis ($\mathbf{R} = (R,00)$), in that case,

$$H_{\mu v} = \left\langle \phi_\mu \left| \mathbf{H} \right| \phi_v \right\rangle = H_{sp\sigma}(R) \tag{7}$$

if μ is S orbital and v is $-p_x$ orbital,

$$H_{\mu v} = \langle \phi_\mu | \mathbf{H} | \phi_v \rangle = H_{pp\sigma}(R) \tag{8}$$

if μ is p_x orbital and v is $-p_x$ orbital,

$$H_{\mu v} = \langle \phi_\mu | \mathbf{H} | \phi_v \rangle = H_{pp\pi}(R) \tag{9}$$

if μ and v are s orbitals,

$$H_{\mu v} = \langle \phi_\mu | \mathbf{H} | \phi_v \rangle = H_{pp\pi}(R) \tag{10}$$

if μ and v are p_z orbitals.

Similar is the notation for the overlap Slater-Koster parameter functions. Slater and Koster (Slater and Koster 1954) defined these parameters at given distances, but later version of the tight-binding method distance dependent functions are given.

The Hamiltonian matrix element was given with the help of the Slater-Koster relations as:

$$H_{ss}\left(\mathbf{R_j}\text{-}\mathbf{R_i}\right) = H_{ss\sigma}\left(R_{ij}\right) \tag{11}$$

$$H_{sx}\left(\mathbf{R_j}\text{-}\mathbf{R_i}\right) = \cos\left(\alpha_x\right) H_{sp\sigma}\left(R_{ij}\right) \tag{12}$$

$$H_{xx}\left(\mathbf{R_j}\text{-}\mathbf{R_i}\right) = \cos^2\left(\alpha_x\right) H_{pp\sigma}\left(R_{ij}\right) + \left(1 - \cos^2\left(\alpha_x\right)\right) H_{pp\pi}\left(R_{ij}\right) \tag{13}$$

$$H_{xy}\left(\mathbf{R_j}\text{-}\mathbf{R_i}\right) = \cos\left(\alpha_x\right)\cos\left(\alpha_y\right) H_{pp\sigma}\left(R_{ij}\right) - \cos\left(\alpha_x\right)\cos\left(\alpha_y\right) H_{pp\pi}\left(R_{ij}\right) \tag{14}$$

$$H_{xz}\left(\mathbf{R_j}\text{-}\mathbf{R_i}\right) = \cos\left(\alpha_x\right)\cos\left(\alpha_z\right) H_{pp\sigma}\left(R_{ij}\right) - \cos\left(\alpha_x\right)\cos\left(\alpha_z\right) H_{pp\pi}\left(R_{ij}\right) \tag{15}$$

Here we have supposed that the atomic orbital ϕ_μ was centered at $\mathbf{R_i}$ and the atomic orbital ϕ_v was centered on the atom at $\mathbf{R_j}$. The direction cosines of the vector $\mathbf{R_j}$ - $\mathbf{R_i} \neq 0$ are $\cos(\alpha_x)$, $\cos(\alpha_y)$ and $\cos(\alpha_z)$. R_{ij} is the distance between the two atoms. If the atomic orbitals ϕ_μ and ϕ_v are centered on the same atom, $H_{ss} = E_s$ and $H_{xx} = H_{yy} = H_{zz} = E_p$ for $\mu = v$ are given values, and $H_{\mu v} = 0$ for $\mu \neq v$. The other matrix elements can be generated with the help of symmetry.

Similar relations are valid for the $S_{\mu v}$ overlap matrix elements as well.

The E_{rep} repulsive par potential is usually an interatomic distance function (Porezag et al., 1995).

49.4 NANO-CHEMICAL APPLICATION(S)

The simpler tight-binding methods can give very important qualitative properties of the molecule, and the tight binding methods can give results which are very near to *ab initio* calculations. If the number of atoms reaches the number of several thousand or quantum mechanical approximations are important in molecular dynamics calculations, the *ab initio* methods are unusable because of a long time of calculations, but the tight binding methods give results in calculations of acceptable times. The one orbital one site tight binding method was used to classify the fullerenes (Fowler, Manolopoulos, 1995), and to determine the Dirac points and other important electronic properties of the graphene (Castro Neto et al., 2009). There are many tight-binding molecular dynamics calculations for fullerene and nanotube formations (László, 1998; Zheng et al., 2004; Irle et al., 2006; László and Zsoldos, 2012).

KEYWORDS

- electronic structure
- geometric structure
- optical properties
- tight binding

REFERENCES AND FURTHER READING

Bloch, F., (1928). Über die quantenmechanik der elektronen in kristallgittern. *Z. Physik, 52,* 555–600.

Bouckaert, L. P., Smoluchowski, R., & Wigner, E. P., (1936). Theory of Brillouin zones and symmetry properties of wave functions in crystals. *Phys. Rev., 50,* 58–67.

Castro, N. A. H., Guinea, F., Peres, N. M. R., Novoselov, K. S., & Geim, A. K., (2009). The electronic properties of graphene. *Rev. Mod. Phys., 81,* 108–162.

Chadi, D. J., (1978). Energy minimization approach to the atomic geometry of semiconductor surfaces. *Phys. Rev. Lett., 41,* 1062–1065.

Coulson, C. A., O'Leary, B., & Mallion, R. B., (1978). *Hückel Theory for Organic Chemists.* Academic Press: London New York.

Fowler, P. W., & Manolopoulos, D. E., (1995). *An Atlas of Fullerenes.* Clarendon Press: Oxford.

Goodwin, L., Skinner, A. J., & Pettifor, D. G., (1989). Generating transferable tight-binding parameters: Application to silicon. *Europhys. Lett., 9*, 701–706.

Harrison, W. A., (1980). *Electronic Structure and the Properties of Solids.* W. H. Freeman and Co.: San Francisco.

Harrison, W. A., (1989). *Electronic Structure and the Properties of Solids.* Dover Publications, INC: New York.

Hückel, E., (1931). Quantentheoretische beiträge zum Benzolproblem. I. Die Elektronenkonfiguration des Benzols und verwandter Verbindungen. *Zeitschrift für Physik, 70*, 204–286.

Irle, S., Zheng, G., Whang, Z., & Morokuma, K., (2006). The C_{60} formation puzzle "Solved": QM/MD simulation reveal the shrinking hot giant road of the dynamic fullerene self-assembly mechanism. *J. Phys. Chem. B., 110*, 14531–14545.

Jones, H., & Mott, N. F., (1937). The electronic specific heat and X-ray absorption of metals and some other properties related to electron bands. *Proc. Roy. Soc. (London) A., 162*, 49–62.

Jones, H., Mott, N. F., & Skinner, H. W. B., (1934). A theory of the form of the X-ray emission bands of metals. *Phys. Rev., 45*, 379–384.

László, I., & Zsoldos, I., (2012). Graphene-based molecular dynamics nanolithography of fullerenes, nanotubes and other carbon structures. *Europhys. Lett., 99*, 63001-p5.

László, I., (1998). Formation of cage-like C60 clusters in molecular dynamics simulations. *Europhys. Lett., 44*, 741–746.

Papaconstantopoulos, D. A., (1986). *Handbook of the Band Structure of Elemental Solids.* Plenum Press: New York and London.

Paxton, A. E., & Sutton, A. P., (1989). A tight-binding study of grain boundaries in silicone. *Acta Metall., 37*, 1693–1715.

Porezag, D., Frauenheim, T., Köhler, T., Seifert, G., & Kaschner, R., (1995). Construction of tight-binding-like potentials on the basis of density-functional theory: Application to carbon. *Phys. Rev. B., 51*, 12947–12957.

Sawada, S., (1990). New tight-binding model of silicon for theoretical studies of surfaces. *Vacuum, 41*, 612–614.

Slater, J. C., & Koster, G. F., (1954). Simplified LCAO method for the periodic potential problem. *Phys. Rev. B., 94*, 1498–1524.

Wilson, A. H., (1931). The theory of electronic semiconductors. *Proc. Roy Soc. (London) A., 133*, 458–459.

Xu, C. H., Whang, C. Z., Chan, C. T., & Ho, K. M., (1992). A transferable tight-binding potential for carbon tight-binding study of grain boundaries in silicone. *J. Phys.: Condens. Matter., 4*, 6047–6054.

Zheng, G., Irle, S., Elstner, M., & Morokuma, K., (2004). Quantum chemical molecular dynamics model study of fullerene formation from open-ended carbon nanotubes. *J. Phys. Chem. A., 108*, 3182–3194.

CHAPTER 50

Z-Partition Function

NICOLINA POP

Politehnica University of Timisoara, Department of Fundamental of Physics for Engineers, Vasile Pârvan Blv., 300223 Timisoara, Romania, e-mail: nicolina.pop@upt.ro

50.1 DEFINITION

From a physical point of view, partition function (Z) characterizes the statistical properties of a system in thermodynamic equilibrium. Partition functions are functions of the thermodynamic state variables, such as the temperature and volume. Vibrational contribution to various thermodynamic functions is expressed in terms of a partition function and its first two derivatives (internal energy and respectively, heat capacity), *see* Andrews, Frank C. (1963).

The letter Z stands for the German word Zustandssumme, "sum over states." The meaning and significance of the partition function result from the fact that it can be used to relate macroscopic thermodynamic quantities to the microscopic details of a system through the derivatives of its partition function.

For a canonical ensemble in which the system is allowed to exchange heat with the environment at a fixed temperature, volume, and a number of particles, the *partition function* of the discrete spectrum can be express as:

$$Z = \sum_{n=0}^{n=N} e^{-\beta E_n}, \beta = \frac{1}{k_B T} \tag{1}$$

where n is the index for the microstates of the system; N is the maximal number of state, $n = 0, 1, 2,...N$; E_n is the total energy of the system in the respective microstate; T is thermodynamic temperature; and k_B is the Boltzmann constant.

The *grand canonical partition function* applies to a grand canonical ensemble, in which the system can exchange both heat and particles (N_n

represents total particle number) with the environment, at a fixed temperature, volume, and chemical potential μ:

$$Z = \sum_{n=0}^{n=N} e^{-\beta(E_n - N_n \mu)}. \tag{2}$$

The statistics of a non-interacting many-body quantum gas (Fermi–Dirac
statistics for fermions, Bose-Einstein statistics for bosons) can be derived
using the grand canonical ensemble.

50.2   HISTORICAL ORIGIN(S)

The concept of the partition function comes from a way of counting the
particles of a system when the particles are distributed over the available
energy levels in accord with the Boltzmann distribution as in Eq. (1).

The partition function can also be seen to be the total number of particles
expressed as the sum of the numbers of particles of each energy, with the
population of the ground state being the unit of measure of population (*see*
Addison, 2008).

The population of the ground state is taken to be 1, and the populations
of the other energy states are expressed as fractions of the population of the
ground state. By considering in this way, the partition function can be seen
as a sum of the relative occupancies of states. It is for this reason that the
partition function is also called the sum-overstates.

It has no dimension, but it has a finite range. So, the lowest possible value
for the partition function is 1, the value it could have when the temperature
of the system is 0 K, the temperature at which all particles in the system
are in the ground state, and the fractions in the excited states are zero. The
highest value for the partition function can be very large, but not infinite.
In a real system the number of particles is not infinite, and so the partition
function cannot be infinite. The largest values are attained when the spacing
of the energy levels is very small, and the temperature is very high.

50.3   NANO-SCIENTIFIC DEVELOPMENT(S)

Progress in nanostructured technologies results in the production of
purposely made deformed structures. Planned nanostructures are thin films,
nanocylinders, nanorods, etc. Today three-layer nanofilms are produced
without any problems. In researches of this type, researchers started from

the very complex problems moving to less complex problems which resulted in the slow development of the theory of broken structures. Although it is well known that a similar problem in classical mechanics is solved by Boltzmann's statistics, it took a long time to have Boltzmann's ideas applied to the problem of chaotic structures with broken symmetry. However, Boltzmann's ideas received wide field of application.

So, the probabilistic approach is a unique way of analyzing deformed chaotic systems. Probabilistic nature of statistical analysis approach gives the possibility to obtain the probability of the results of randomly deformed structures and allows the prediction of some fundamental effects.

It should be noted that nanostructures cut from an ideal bulk structure can be analyzed using the stochastic method. It gives the probability to evaluate appropriate behavior of nanostructures.

The partition function Z encodes how the probabilities are partitioned among the different microstates, based on their individual energies.

The probability P_n that the system occupies microstate n is:

$$P_n = \frac{e^{-\beta E_n}}{Z}, \quad \sum_n P_n = 1 \tag{3}$$

and the partition function plays the role of a normalizing constant.

For a chain with N_M equidistant atoms which have different masses m_1, m_2, ... m_m. We assume that the mass m_1 has N_{1M} of atoms, the mass m_2 has N_{2M} of atoms, etc. In this case, the number of atoms N and the mass M of the chain are conserved [*see* Walecka (2000); Nebus (2007)]. This means that the conservation laws are:

$$\sum_{s=1}^{m} N_{sM} = N_M. \tag{4}$$

and

$$\sum_{s=1}^{m} m_s N_{sM} = M. \tag{5}$$

The statistical probability is given by:

$$P = \frac{N!}{\prod\limits_{s=1}^{m} N_s!} = \frac{N^N}{\prod\limits_{s=1}^{m} N_s^{N_s}}. \tag{6}$$

The most probable distribution is obtained by equating the variation function (*see* Sajfert, 2006):

$$\Psi = \ln P - \alpha_M N_M - \gamma M = N_M \ln N_M - \sum_{s=1}^{m}\left(N_{sM} \ln N_{sM} + \alpha_M N_{sM} + \gamma m_s N_{sM}\right). \quad (7)$$

with zero.

The probability that N_s of the atom has a mass m_s is given by:

$$\Omega_s = \frac{e^{-\gamma m_s}}{\sum_{s=1}^{m} e^{-\gamma m_s}} \quad (8)$$

Taking an approximation for the masses:

$$m_s = s m_0 \quad (9)$$

where m_0 is the minimal mass, we obtain the approximate expression for the probability Ω_s:

$$\Omega_s = \frac{N_{sM}}{N_M} = \frac{e^{-\gamma m_0 s}}{\sum_{s=1}^{m} e^{-\gamma m_0 s}} \quad (10)$$

The mean value of the mass, which represents the mass of the equivalent chain is given by:

$$< m_s >= m_{eq} = m_0 \left(\frac{e^{\frac{\gamma m_0}{2}}}{2\sinh\frac{\gamma m_0}{2}} - \frac{m}{e^{m\gamma m_0} - 1} \right); \ \gamma = \frac{N_M}{M} \quad (11)$$

The advancement in information technology, the science of materials, biotechnology, and energy engineering are associated with quantum and molecular dimensions. The obtained probabilities to evaluate appropriate behavior of nanostructures are of the exponential type.

50.4 A METHOD FOR APPROXIMATION OF PARTITION FUNCTION FOR MORSE OSCILLATORS

The partition function of the Morse systems can be satisfactorily approximated using only *elementary functions*.

The Morse potential is one of the most simple and realistic anharmonic potential models, which has proven to be very useful for solving various problems from diverse fields of physics and chemistry (e.g., vibrational motion of diatomic and polyatomic molecules, photodissociation, intermolecular energy transfer, and spectroscopy).

In the case when rotations are neglected, the vibrations of the diatomic molecules are well-described by one-dimensional Morse oscillators governed by the one dimensional Schrödinger equation with Morse potentials [*see* Morse (1929); Sagem (1978); Popov (2001)]:

$$V(y) = D\left[\left(1 - e^{-\alpha(y - y_0)}\right)^2 - 1\right]$$ (12)

which represents wells of depths $-D$ at $y = y_0$, so D is just the dissociation energy of the diatomic molecule. The parameter α is the constant of anharmonicity which can take only positive values.

The energy spectrum has a continuous part in the domain $[0,\infty)$ and a finite number of discrete levels (*see* Cotaescu & Pop, 2009):

$$E_n = -\hbar\omega \frac{(N + \varepsilon - n)^2}{2N + 2\varepsilon + 1}, \quad \omega = \alpha\sqrt{\frac{2D}{m}}, \quad \varepsilon = \frac{2D}{\hbar\omega} - N - \frac{1}{2}$$ (13)

where m is the oscillator mass, the principal quantum number takes the values $n = 0, 1, 2, \ldots, N$. The maximal number of vibrational bound state, N, of the quantum system described by the Morse oscillator is obtainable looking for the maximal condition in which the wave functions remain square integrable. The new parameter satisfies the conditions $0 \le \varepsilon < 1$.

The statistics of an ideal gas of identical diatomic molecules, modeled by Morse oscillators, in thermodynamic equilibrium with a reservoir of temperature T is described by the quantum canonical distribution (Eq. 12). The partition function of the discrete spectrum as is presented in Eq. (1) has the form:

$$Z(x, N) = \sum_{n=0}^{N} x^{\frac{(N + \varepsilon - n)^2}{2N + 2\varepsilon + 1}}, \quad x = e^{\beta\omega}$$ (12)

with variable x is 1 for $T \to \infty$ and tends to infinity when $T \to 0$ K.

For $\varepsilon < 1$, partition function, Z can be written using the Taylor expansion as:

$$Z(x, N) = \sum_{n=0}^{N} x^{\frac{(N - n)^2}{2N + 1}} + \frac{e\beta\hbar\omega}{2} \sum_{n=0}^{N} \left(1 - \frac{(2n + 1)^2}{(2N + 1)^2}\right) x^{\frac{(N - n)^2}{2N + 1}} + O(\varepsilon^2)$$ (13)

If ε is small enough, then we can take $\varepsilon = 0$ remaining only with the first term of Eq. (13). In this case, the partition function can be rewritten as:

$$Z(x,N) = \sum_{n=0}^{N} x^{\frac{(N-n)^2}{2N+1}} = x^{\frac{N^2}{2N+1}} F(x,N) \tag{14}$$

where was introduced the function:

$$F(x,N) = \sum_{n=0}^{N} x^{\frac{-n(2N-n)}{2N+1}} \tag{15}$$

defined on the domain $x \in [1, \infty)$ for any integer N. This function is maximal for $x = 1$ where $F(1, N) = N + 1$ and decreases rapidly to 1 for $x \to \infty$ such that $\lim_{x \to \infty} [x \, \partial_x \, F(x, N)] = 0$.

In what follows were proposed two methods of satisfactory approximating the function $F(x, N)$ by elementary functions in the case of $\varepsilon = 0$ on large domains of temperatures.

50.4.1 APPROXIMATIONS FOR LOW TEMPERATURES

Firstly, were considered functions:

$$\tilde{F}(x,N) = \exp\{\exp[\nu f(x,N)] - 1\} \tag{16}$$

where, $\nu = \ln[\ln(N+1)+1]$ and $f:[1, \infty) \to [0,1]$ must be an elementary function, f is a polynomial of the rank $k + 1$ in $1/x$:

$$f(x,N) = \sum_{i=1}^{i=k} \frac{C_i(N)}{x^i}, \tag{17}$$

whose coefficients represent fit parameters:

$$\sum_{i=1}^{i=k} C_i(N) = 1, \tag{18}$$

The next step is to fix up the integer k and determines the linear independent coefficients, C_1, C_2, C_k, using a convenient fit method. One of the methods is that of the least-squares applied for approaching f to the function $\ln[\ln(F) + 1]$. This method leads to satisfactory approximations even for small values of k.

The functions F and $F\Box$ behave similarly when x goes to one or infinity. So, when $x = 1$, f(1, n) = 1, respectively, when $x \to \infty$, $\lim x \to \infty f(x, n) = 0$.

The partition function (15) may be approximated by the function:

$$\tilde{Z}(x,N) = x^{N^2/2N+1} \tilde{F}(x,N), \tag{19}$$

The functions F and $F\square$ have apparently the same graphics. It results from an evaluation of the errors using the function of relative differences $\delta F = F\square / F - 1$.

In Figure 50.1, the graphic of the function δF is represented in the case of $k = 5$, for $N = 25$ on the domain $x \in [1, 15]$. The accuracy is increasing with **k**, but there are errors arising on the domain of high temperatures, $x \in [1, 2]$.

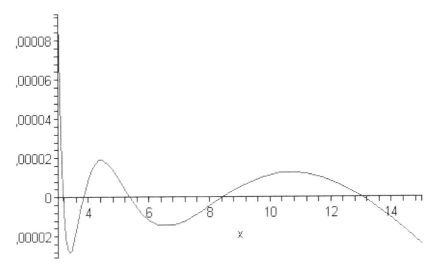

FIGURE 50.1 The function δF on the domain [1, 15] for $k = 5$.

The above method is suitable for approximating the function F only in the domain of low temperatures, $x > 2$. In this domain, relative differences δF remains smaller than $1.5 \ 10^{-3}$ if $k = 5$.

50.4.2 APPROXIMATIONS FOR HIGH TEMPERATURES

For high temperatures it is convenient to introduce the new variable:

$$\tau = \frac{1}{\ln(x)} = \frac{k_B T}{\hbar \omega} \tag{20}$$

and another function:

$$G(\tau, N) = F(e^\tau, N). \tag{21}$$

This function can be approximated by the new function:

$$\tilde{G}(\tau,N) = (N+1)\frac{g(\tau)}{N+g(\tau)} \tag{22}$$

where $g(\tau)$ is a polynomial of the rank l of the form:

$$g(\tau) = 1 + \sum_{i=1}^{i=l} C_i \tau^i \tag{23}$$

depending on the fit constants $C_1, C_2, ...$

The approximate partition function is:

$$\tilde{Z}(\tau,N) = e^{\frac{1}{y}\frac{N^2}{2N+1}}\tilde{G}(\tau,N) \tag{24}$$

The coefficients C_i have to be determined using the method of least-squares for approaching the functions g and $NG/(N+1-G)$.

For a sequence $\tau_{min} = \tau_1, \tau_2, . . ., \tau m = \tau_{max}$ in the domain $[\tau_{min}, \tau_{max}]$, by minimization of the quantity:

$$\Phi(C_1,C_2,...) = \sum_i \left[g(\tau) - \frac{NG(\tau^i,N)}{N+1-G(\tau^i,N)} \right] \tag{25}$$

will obtain the coefficients C_i.

This method of approximation leads to partition functions very close to the real ones as it results from concrete examples. Moreover, the errors can be showed considering the function $\delta G = G\square/G - 1$ which has to measure the degree of accuracy.

The graphic representation from Figure 50.2, for δG for $l = 6$ on the domain [1, 15], suggests that significant errors arise in the domain of low temperatures, for $\tau < 2$. It is enough to consider the case of $l = 6$ for obtaining satisfactory approximations of the partition function at high temperatures, $\tau > 2$. It remains to investigate the sensitive domain $\tau \in [1, 2]$. Our tests indicate that in this domain one can use the same method based on the function $G\square$ since this can be determined with reasonable small errors (less than 5×10^{-3}) taking $l = 6$.

Both methods of approximating the partition function of the Morse oscillators which are complementary to each other, cover the whole domain of temperatures that may be of chemical and physical interest.

The exact Morse oscillator canonical partition function reproduced using Morse coherent states was also used by Toutounji (2014) to derive the auto-correlation function of Morse oscillator.

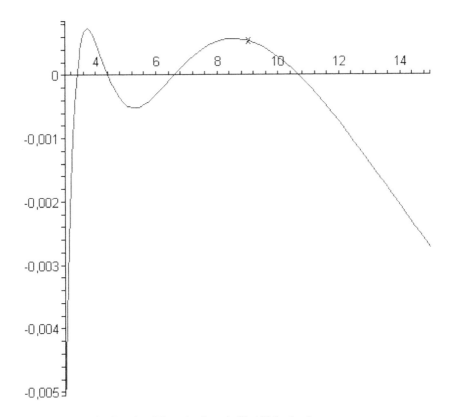

FIGURE 50.2 The function δG on the domain [1, 15] for *l* = 6.

KEYWORDS

- **elementary functions**
- **Morse oscillators**
- **thermodynamic functions**

REFERENCES AND FURTHER READING

Addison, A., (2008). *The Partition Function: If That's What it is Why Don't They Say So!* Cornell College, Mount Vernon, IA 52314.
Andrews, F. C., (1963). *Equilibrium Statistical Mechanics* (p. 35). John Wiley & Sons, Inc.

Cotaescu, I., & Pop, N., (2009). Approximating the partition function of Morse oscillators. *Int. J. Theor. Phys., 48*, 1596–1602.

Lim, C., & Nebus, J., (2007). *Vorticity, Statistical Mechanics, and Monte Carlo Simulation.* Springer, New York.

Morse, (1929). Diatomic molecules according to the wave mechanics, P. M. *Phys. Rev., 70*, 222.

Pop, N., Vlase, G., Vlase, T., & Doca, N., (2009). Study on the mechanism of the selective activation of solids in a homogeneous and isotropic thermal field. *Journal of Thermal Analysis and Calorimetry, 97*, 667–671.

Popov, D., (2001). Considerations concerning the harmonic limit of the Morse oscillator. *Phys. Scr., 63*, 257.

Sage, M. L., (1978). Morse oscillator transition probabilities for molecular bond modes. *Chem. Phys. Lett., 35*, 375.

Sajfert, V., Garić, M., & Tošić, B., (2006). Nanostructure rectangular rod. *J. Comput. Theor. Nanosci., 3*, 431–439.

Toutounji, M., (2014). A new methodology for dealing with time-dependent quantities in anharmonic molecules I: theory. *Theor. Chem. A., 133*, 1461.

Walecka, J. D., (2000). *Fundamentals of Statistical Mechanics: Manuscripts and Notes of Felix Bloch.* Imperial College Press, London, World Scientific, Singapore, London.

Index

T - #0793 - 101024 - C606 - 234/156/27 - PB - 9781774631744 - Gloss Lamination